Signaling and Communication in Plants

Series Editor

František Baluška
Department of Plant Cell Biology, IZMB, University of Bonn, Kirschallee 1,
D-53115 Bonn, Germany

For further volumes:
http://www.springer.com/series/8094

František Baluška
Editor

Long-Distance Systemic Signaling and Communication in Plants

Springer

Editor
František Baluška
University of Bonn
IZMB
Bonn, Nordrhein-Westfalen
Germany

ISSN 1867-9048 ISSN 1867-9056 (electronic)
ISBN 978-3-642-43941-4 ISBN 978-3-642-36470-9 (eBook)
DOI 10.1007/978-3-642-36470-9
Springer Heidelberg New York Dordrecht London

Printed on acid-free paper

Springer is part of Springer Science+Business Media (www.springer.com)

Preface

Higher plants coordinate and integrate their tissues and organs via diverse long-distance signalling and communication circuits. Sophisticated sensory systems sensitively screen both internal and external factors and feed sensory information into chemical and physical systemic long-distance communication cascades. Obviously, our view of plants is changing dramatically. We realize that their long-distance signalling is fast, and signals, both of endogenous and exogenous origin, spread rapidly throughout their bodies. This recent revolution in our understanding of higher plants started more than 40 years ago with the discovery of alarm peptide hormone systemin (Green and Ryan 1972; Ryan and Pearce 2003) and continues with rapid advances further. This volume of the 'Signalling and Communication in Plants' series captures the current dynamic 'state of the art' of this very exciting topic of plant sciences.

In general, there are chemical and physical mechanisms for the long-distance signalling and communication in plants. With respect to chemical communication, the most advanced topics are systemic acquired resistance (SAR), which is an inducible defence syndrome based on salicyl acid signalling (Ross 1966; Sticher et al. 1997; Chaturvedi et al. 2012; Wu et al. 2012), and systemic acquired acclimation (SAA), which is systemic signalling of photo-oxidative stress (Karpinski et al. 1999; Karpinski and Szechynska-Hebda 2010). Both SAR and SAA include several aspects of plant memory and anticipation of future insults via the memorized sensory perceptions, using quantum computing including quantum-redox sensing (Szechynska-Hebda et al. 2010; Karpinski and Szechynska-Hebda 2010). Importantly in this respect, both SAR and SAA are based on ROS and hormonal signalling pathways, but also include very rapid electrical and mechanical long-distance signalling. Another extensively investigated and well-understood topic is the long-distance wound signalling based on the alarm peptide hormone systemin and oxylipin-derived jasmonic acid (Farmer and Ryan 1990; Ryan and Pearce 2003, Sun et al. 2011). The next long-distance system is induced systemic resistance (ISR), which is induced by diverse non-pathogenic agents such as growth-promoting rhizobacteria and other plant beneficial microorganisms (van Wees et al. 2000; Rudrappa et al. 2010; Berendsen et al. 2012; Lee et al. 2012).

The nature of root-to-shoot long-distance communication is still not well understood for the ISR, but besides salicyl acid and jasmonic acid, abscisic acid is also involved (Kumar et al. 2012; Sampath Kumar and Bais 2012). Root-to-shoot long distance is also involved in the initiation and control of the symbiotic Rhizobia bacteria interactions with legume roots via so-called social media pathway (Venkateshwaran et al. 2013). Interestingly, this 'social media' pathway is also supporting long-distance interactions between roots and arbuscular mycorrhizal fungi (Venkateshwaran et al. 2013), which help plants to acquire nutrients, especially phosphate, and solutes. Last but not least, phosphate and iron homeostasis in plants is also safeguarded via long-distance signalling pathways and circuits (Enomoto et al. 2007; Enomoto and Goto 2008; Nagarajan et al. 2011; Smith et al. 2011).

Physical mechanisms of long-distance signalling and communication in plants include both electrical and mechanical/hydraulic mechanisms. In fact, electrical signals were discovered in plants more than 140 years ago (Burdon-Sanderson 1873, 1899; Stahlberg 2006). Although the plant action potentials show the same bioelectric parameters like animal/humans action potentials, they are driven by slightly different ion channels and other molecules (Fromm and Lautner 2007; Hedrich 2012; Baluška and Mancuso 2013). Despite this long tradition in plant electrophysiology, the importance and roles of plant action potentials for plant physiology and plant behaviour are still rudimentary (Brenner et al. 2006). However, it emerges that electric long-distance signalling in plants is more complex than that in animals because it includes also variation potentials, system potentials, and hydraulic signals (Malone 1992; Stahlberg 2006; Stahlberg et al. 2005; Zimmermann et al. 2009). It is also obvious that root apices and phloem represent the most active sites of electric activity in plants (Masi et al. 2006; Fromm and Bauer 1994; Fromm and Lautner 2007; Baluška and Mancuso 2013).

Another important and relatively well-understood topic in plant long-distance signalling and communication is that of mobile RNA molecules that move within the phloem (Lucas et al. 2001; Banerjee et al. 2006, 2009). Besides coding mRNAs, also non-coding regulatory RNAs are moving within plants (Schwab et al. 2009; Molnar et al. 2011), which is related to systemic propagation of the acquired stress-induced epigenetic changes (Molnar et al. 2011). For example, systemic acquired silencing (SAS) is rather a well-understood phenomenon studied in plants for more than a decade (Palauqui et al. 1997). Phloem elements are really unique as they represent supracellular highways for plant long-distance signalling, spanning throughout the whole plant body—integrating it into functional unity, using all kinds of diverse long-distance signalling and communication pathways (Lucas et al. 2001; Van Bel and Hafke 2013).

The final chapter of this volume is devoted to the emerging topic of long-distance signalling and communication in plants: herbivore-induced volatile organic compounds (VOCs) that act as semiochemical signals, playing roles in both the within-plant and plant–plant communication (Baldwin et al. 2006; Girón-Calva et al. 2012; Rodriguez-Saona et al. 2013). One important aspect of this new and important topic is the ability of VOCs to prime defenses in plants by enhancing their resistance and responses to subsequent herbivore attacks (Kobayashi et al. 2006; Ton et al. 2007; Verheggen et al. 2010). Importantly, this long-distance signalling and communication via phytosemiochemicals has great potency for improving crop protection and efficiency of agriculture (Bruce 2010; Jansen et al. 2010; Khan et al. 2010).

Bonn, Germany František Baluška

References

Baldwin IT, Halitschke R, Paschold A, von Dahl CC, Preston CA (2006) Volatile signaling in plant-plant interactions: “talking trees” in the genomics era. Science 311:812–815

Baluška F, Mancuso S (2013) Ion channels in plants: from bioelectricity, via signaling, to behavioral actions. Plant Signal Behav 8:1–4

Banerjee AK, Chatterjee M, Yu Y, Suh SG, Miller WA, Hannapel DJ (2006) Dynamics of a mobile RNA of potato involved in a long-distance signaling pathway. Plant Cell 18:3443–3457

Banerjee AK, Lin T, Hannapel DJ (2009) Untranslated regions of a mobile transcript mediate RNA metabolism. Plant Physiol 151:1831–1843

Berendsen RL, Pieterse CM, Bakker PA (2012) The rhizosphere microbiome and plant health. Trends Plant Sci 17:478–486

Brenner ED, Stahlberg R, Mancuso S, Vivanco J, Baluska F, Van Volkenburgh E (2006) Plant neurobiology: an integrated view of plant signaling. Trends Plant Sci 11:413–419

Bruce TJA (2010) Exploiting plant signals in sustainable agriculture. In: Baluška F, Ninkovic V (eds) Plant communication from an ecological perspective. Springer, Berlin, pp 215–227

Burdon-Sanderson J (1873) Note on the electrical phenomena which accompany irritation of the leaf of *Dionaea muscipula*. Proc R Soc Lond 21:495–496

Burdon-Sanderson J (1899) On the relation of motion in animals and plants to the electrical phenomena, which are associated with it. Proc R Soc Lond 65:37–64

Chaturvedi R, Venables B, Petros RA, Nalam V, Li M, Wang X, Takemoto LJ, Shah J (2012) An abietane diterpenoid is a potent activator of systemic acquired resistance. Plant J 71:161–172

Enomoto Y, Goto F (2008) Long-distance signaling of iron deficiency in plants. Plant Signal Behav 3:396–397

Enomoto Y, Hodoshima H, Shimada H, Shoji K, Yoshihara T, Goto F (2007) Long-distance signals positively regulate the expression of iron uptake genes in tobacco roots. Planta 227:81–89

Farmer EE, Ryan CA (1990) Interplant communication: airborne methyl jasmonate induces synthesis of proteinase inhibitors in plant leaves. Proc Natl Acad Sci USA 87:7713–7716

Fromm J, Bauer T (1994) Action potentials in maize sieve tubes change phloem translocation. J Exp Bot 273:463–469

Fromm J, Lautner S (2007) Electrical signals and their physiological significance in plants. Plant Cell Environ 30:249–257

Girón-Calva PS, Molina-Torres J, Heil M (2012) Volatile dose and exposure time impact perception in neighboring plants. J Chem Ecol 38:226–228

Green TR, Ryan CA (1972) Wound-induced proteinase inhibitor in plant leaves: a possible defense mechanism against insects. Science 175:776–777

Jansen RMC, Wildt JJ, Hofstee JW, Bouwmeester HJ, van Henten EJ (2010) Plant volatiles: useful signals to monitor crop health status in greenhouses. In: Baluška F, Ninkovic V (eds) Plant communication from an ecological perspective. Springer, Berlin, pp 229–247

Karpinski S, Szechynska-Hebda M (2010) Secret life of plants: from memory to intelligence. Plant Signal Behav 5:1391–1394

Karpinski S, Reynolds H, Karpinska B, Wingsle G, Creissen G, Mullineaux P (1999) Systemic signaling and acclimation in response to excess excitation energy in Arabidopsis. Science 284:654–657

Khan ZR, Midega CA, Bruce TJ, Hooper AM, Pickett JA (2010) Exploiting phytochemicals for developing a 'push-pull' crop protection strategy for cereal farmers in Africa. J Exp Bot 61:4185–4196

Kobayashi Y, Yamamura N, Sabelis MW (2006) Evolution of talking plants in a tritrophic context: conditions for uninfested plants to attract predators prior to herbivore attack. J Theor Biol 243:361–374

Kumar AS, Lakshmanan V, Caplan JL, Powell D, Czymmek KJ, Levia DF, Bais HP (2012) Rhizobacteria *Bacillus subtilis* restricts foliar pathogen entry through stomata. Plant J 72:694–706

Lee B, Farag MA, Park HB, Kloepper JW, Lee SH, Ryu CM (2012) Induced resistance by a long-chain bacterial volatile: elicitation of plant systemic defense by a C13 volatile produced by *Paenibacillus polymyxa*. PLoS One 7:e48744

Lucas WJ, Yoo BC, Kragler F (2001) RNA as a long-distance information macromolecule in plants. Nat Rev Mol Cell Biol 2:849–857

Malone M (1992) Kinetics of wound-induced hydraulic signals and variation potentials in wheat seedlings. Planta 187:505–510

Masi E, Ciszak M, Stefano G, Renna L, Azzarello E, Pandolfi C, Mugnai S, Baluška F, Arecchi FT, Mancuso S (2009) Spatiotemporal dynamics of the electrical network activity in the root apex. Proc Natl Acad Sci USA 106:4048–4053

Molnar A, Melnyk C, Baulcombe DC (2011) Silencing signals in plants: a long journey for small RNAs. Genome Biol 12:215

Nagarajan VK, Jain A, Poling MD, Lewis AJ, Raghothama KG, Smith AP (2011) Arabidopsis Pht1;5 mobilizes phosphate between source and sink organs and influences the interaction between phosphate homeostasis and ethylene signaling. Plant Physiol 156:1149–1163

Palauqui JC, Elmayan T, Pollien JM, Vaucheret H (1997) Systemic acquired silencing: transgene-specific post-transcriptional silencing is transmitted by grafting from silenced stocks to non-silenced scions. EMBO J 16:4738–4745

Rodriguez-Saona CR, Mescher MC, De Moraes CM (2013) The role of volatiles in plant-plant interactions. In: Baluška F (ed) Long-distance systemic signaling and communication. Springer, Berlin

Ross AF (1966) Systemic effects of local lesion formation. In: Beemster ABR, Dijkstra J (eds) Viruses of plants. North-Holland, Amsterdam, pp 127–150

Rudrappa T, Biedrzycki ML, Kunjeti SG, Donofrio NM, Czymmek KJ, Paré PW, Bais HP (2010) The rhizobacterial elicitor acetoin induces systemic resistance in *Arabidopsis thaliana*. Commun Integr Biol 3:130–138

Ryan CA, Pearce G (2003) Systemins – a functionally defined family of peptide signals that regulate defensive genes in Solanaceae species. Proc Natl Acad Sci 100:14573–14577

Sampath Kumar A, Bais HP (2012) Wired to the roots: impact of root-beneficial microbe interactions on aboveground plant physiology and protection. Plant Signal Behav 7:1792–1798

Schwab R, Maizel A, Ruiz-Ferrer V, Garcia D, Bayer M, Crespi M, Voinnet O, Martienssen RA (2009) Endogenous TasiRNAs mediate non-cell autonomous effects on gene regulation in *Arabidopsis thaliana*. PLoS ONE 4:e5980

Sibaoka T (1999) Rapid plant movements triggered by action potentials. Bot Mag (Tokyo) 104:73–95

Smith AP, Nagarajan VK, Raghothama KG (2011) Arabidopsis Pht1;5 plays an integral role in phosphate homeostasis. Plant Signal Behav 6:1676–1678

Stahlberg R (2006) Historical overview of plant neurophysiology. Plant Signal Behav 1:6–8

Stahlberg R, Cleland RE, van Volkenburgh E (2005) Decrement and amplification of slow wave potentials during their propagation in *Helianthus annuus* L. shoots. Planta 220:550–558

Sticher L, Mauch-Mani B, Métraux JP (1997) Systemic acquired resistance. Annu Rev Phytopathol 35:235–270

Sun J-Q, Jiang H-L, Li C-Y (2011) Systemin/jasmonate-mediated systemic defense signaling in tomato. Mol Plant 4:607–615

Szechynska-Hebda M, Kruk J, Gorecka M, Karpinska B, Karpinski S (2010) Evidence for light wavelength-specific systemic photoelectrophysiological signalling and cellular light memory of excess light episode in Arabidopsis. Plant Cell 22:2201–2218

Ton J, D'Alessandro M, Jourdie V, Jakab G, Karlen D, Held M, Mauch-Mani B, Turlings TC (2007) Priming by airborne signals boosts direct and indirect resistance in maize. Plant J 49:16–26

Van Bel AJE, Hafke JB (2013) Calcium along the phloem pathway as a universal trigger and regulator of systemic alarms and signals. In: Baluška F (ed) Long-distance systemic signaling and communication. Springer, Berlin

van Wees SCM, de Swart, EAM, van Pelt JA, van Loon LC, Pieterse CMJ (2000) Enhancement of induced disease resistance by simultaneous activation of salicylate- and jasmonate-dependent defense pathways in *Arabidopsis thaliana*. Proc Natl Acad Sci USA 97:8711–8716

Venkateshwaran M, Volkening JD, Sussman MR, Ané JM (2013) Symbiosis and the social network of higher plants. Curr Opin Plant Biol 16(1):118–127

Verheggen FJ, Haubruge E, Mescher MC (2010) Alarm pheromones-chemical signaling in response to danger. Vitam Horm 83:215–239

Wu Y, Zhang D, Chu JY, Boyle P, Wang Y, Brindle ID, De Luca V, Després C (2012) The Arabidopsis NPR1 protein is a receptor for the plant defense hormone salicylic acid. Cell Rep 1:639–647

Zimmermann MR, Maischak H, Mithöfer A, Boland W, Felle HH (2009) System potentials, a novel electrical long-distance apoplastic signal in plants, induced by wounding. Plant Physiol 149:1593–1600

Contents

Long-Distance Signaling in Systemic Acquired Resistance

Jyoti Shah and Ratnesh Chaturvedi

Abstract Systemic acquired resistance (SAR) is an inducible defense mechanism in plants that is activated throughout the foliage in response to a prior localized exposure to a foliar pathogen. The enhanced resistance status resulting from the activation of SAR can be maintained over a couple of generations. Critical to SAR is effective long-distance communication by the pathogen-inoculated organ with rest of the foliage, which requires the lipid transfer protein DIR1. The emerging consensus is that long-distance signaling in SAR involves networking between multiple vascular-translocated signaling molecules. The proposed salicylic acid receptor NPR1 is important for downstream signaling that involves defense priming. Chromatin remodeling is projected as an important mechanism in priming and memory associated with SAR.

Keywords Azelaic acid • Dehydroabietinal • Glycerol-3-phosphate • Methyl salicylate • Pipecolic acid • DIR1

1 Introduction

Plants utilize a combination of preformed and inducible defenses to control diseases (Spoel and Dong 2012). These defenses are manifested in the pathogen-infected organ and can also be activated systemically in tissues located distant to the site of initial infection. Systemic induction of disease resistance was reported as early as the 1930s (Chester 1933). Ross (1966) introduced the term systemic acquired resistance (SAR) to describe the enhanced state of resistance against viral infection in the upper leaves of tobacco (*Nicotiana tabacum*) plants that were previously inoculated on their lower leaves with *Tobacco mosaic virus* (TMV). SAR is now

J. Shah (✉) • R. Chaturvedi
Department of Biological Sciences, University of North Texas, Denton, TX 76203, USA
e-mail: Shah@unt.edu

F. Baluška (ed.), *Long-Distance Systemic Signaling and Communication in Plants*, Signaling and Communication in Plants 19, DOI 10.1007/978-3-642-36470-9_1,

used more generally to refer to systemic enhanced resistance induced by prior local exposure to foliar pathogens (Sticher et al. 1997). SAR confers enhanced resistance against subsequent infection by a broad spectrum of pathogens, an effect that can extend to the roots (Gessler and Kuc 1982; Tahiri-Alaoui et al. 1993). SAR-conferred enhanced disease resistance is associated with the systemic induction of salicylic acid (SA) signaling and requires the *NPR1* (*NON-EXPRESSER OF PR GENES1*) gene (Chaturvedi and Shah 2007; Durrant and Dong 2004; Shah and Zeier 2013), which was recently demonstrated to be one of the receptors for SA (Wu et al. 2012). Light signaling mediated by the red/far-red light-absorbing phytochromes A and B is important for the SAR-associated systemic accumulation of SA and increase in disease resistance (Griebel and Zeier 2008; Zeier et al. 2004). The modulation of SAR strength by light is dependent on the *FMO1* (*FLAVIN-DEPENDENT MONOOXYGENASE1*) gene, which is also required for the SAR-associated systemic accumulation of SA (Mishina and Zeier 2006). In plants exhibiting SAR, defenses are primed to respond faster and stronger in response to challenge inoculation with pathogen (Conrath 2011). Recent studies with the model plant *Arabidopsis thaliana* indicate that once induced, the effect of SAR can be observed over a couple of generations (Luna et al. 2012). Systemic disease resistance in the foliar tissues is also observed in plants with roots colonized by beneficial rhizobacteria, a phenomenon termed induced systemic resistance (ISR) (van Loon 2007). However, SAR and ISR engage different defense mechanisms, and the combined activation of SAR and ISR has an additive effect on disease resistance in foliar tissues (van Wees et al. 2000). Similarly, mycorrhizal associations as well as biocontrol fungi also can promote disease resistance in the foliar tissues (Liu et al. 2007; Shoresh et al. 2010).

The activation of SAR requires long-distance signaling that facilitates communication with the systemic tissues by the organ experiencing the primary infection. The phloem is suggested to provide the conduit for translocation of the "systemic signal" involved in long-distance signaling. Girdling experiments in tobacco and grafting in cucumber (*Cucumis sativus*) suggested that the systemic signal is transported through the phloem (Guedes et al. 1980; Jenns and Kuc 1979; Tuzun and Kuc 1985). In Arabidopsis, the SAR-inducing activity is recovered in vascular sap-enriched petiole exudates (Pex) collected from pathogen-treated leaves (Chaturvedi et al. 2008; Jung et al. 2009; Maldonado et al. 2002). These Pexs are also effective in systemically enhancing disease resistance in other plant species (Chaturvedi et al. 2008). Experiments in *Arabidopsis* indicated that the SAR signal may not be exclusively transported through the phloem, since systemic expression of the *PR1* (*PATHOGENESIS-RELATED1*) gene, which is a molecular marker for SAR, was not limited to the tissues connected by the path of photoassimilate translocation from the primary-infected organ (Kiefer and Slusarenko 2003).

SA levels increase in the phloem sap during SAR (Malamy et al. 1990; Métraux et al. 1990). Hence, for a long time SA was thought to be the systemic signal in SAR (Uknes et al. 1992; Yalpani et al. 1991). However, grafting studies involving tobacco plants expressing the *Pseudomonas putida nahG* gene-encoded salicylate hydroxylase, an enzyme that converts salicylic acid to catechol, confirmed that although required for the manifestation of SAR-conferred enhanced disease

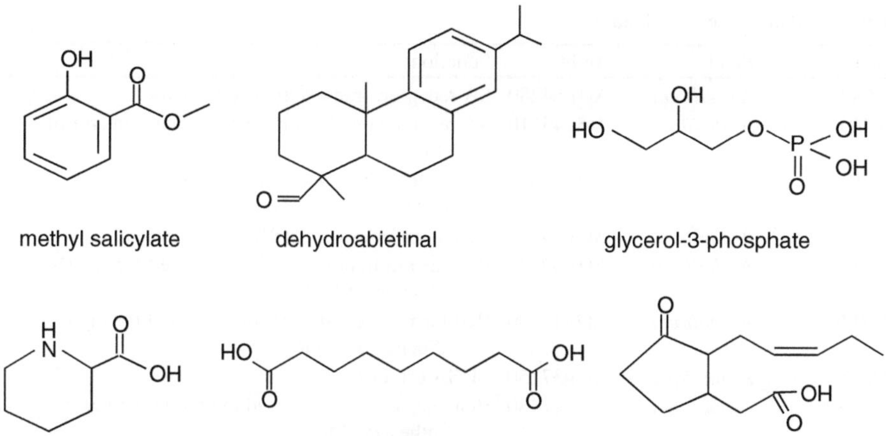

Fig. 1 Structures of metabolites putatively involved in long-distance signaling associated with systemic acquired resistance

resistance, SA per se was not the systemic signal in SAR (Vernooij et al. 1994). Similar conclusions were arrived at with experiments involving tobacco plants rendered SA deficient due to epigenetic suppression of phenylalanine ammonia-lyase expression (Pallas et al. 1996). In recent years, several novel metabolites (Fig. 1) involved in systemic signaling leading to SAR have been described, leading to the suggestion that multiple systemic signals are likely involved in SAR (Depmsey and Klessig 2012; Shah 2009; Shah and Zeier 2013). This chapter will summarize evidence supporting the involvement of these molecules in long-distance communication by the primary-infected organ and the ensuing signaling in the systemic tissues during SAR. Readers are directed to Table 1 for a list of genes that are involved in the synthesis and/or signaling mediated by these SAR signals and biologically (pathogen inoculation) induced SAR.

2 Plant Metabolites Involved in Long-Distance Signaling in SAR

2.1 *Methyl Salicylate*

The role of methyl salicylate (MeSA; Fig. 1) in long-distance signaling in SAR was first reported for tobacco. Increases in MeSA levels were observed in TMV-infected leaves of a TMV-resistant tobacco cultivar (Park et al. 2007). A parallel increase in MeSA was also observed in Pex collected from the TMV-infected leaves and in the systemic leaves. It was noted that the SA-binding protein

Table 1 Plant genes involved in SAR

Gene	Plant	AtG#	Function
NPR1	*A. thaliana*	At1g64280	SA receptor; transcription coactivator
NPR3	*A. thaliana*	At5g45110	SA receptor involved in proteasomal turnover of NPR1
NPR4	*A. thaliana*	At4g19660	SA receptor involved in proteasomal turnover of NPR1
FMO1	*A. thaliana*	At1g19250	Required for systemic SA accumulation
PHYA	*A. thaliana*	At1g09570	Red/far-red light perception; required for light's influence on SAR
PHYB	*A. thaliana*	At2g18790	Red/far-red light perception; required for light's influence on SAR
MES9	*A. thaliana*	At4g37150	MeSA esterase
BSMT1	*A. thaliana*	At3g11480	Benzoic acid/salicylic acid methyl transferase; synthesizes MeSA
ICS1 (*SID2*)	*A. thaliana*	At1g74710	Isochorismate synthase activity involved in SA synthesis
SFD1 (*GLY1*)	*A. thaliana*	At2g40690	Dihydroxyacetonephosphate reductase; synthesizes glycerol-3-phosphate in plastids
DIR1	*A. thaliana*	At5g48485	Lipid-transfer protein
AZI1	*A. thaliana*	At4g12470	Putative lipid-transfer protein
ALD1	*A. thaliana*	At2g13810	Aminotransferase required for pipecolic acid synthesis
ACP4	*A. thaliana*	At4g25050	Acyl-carrier protein required for cuticle development
ACBP3	*A. thaliana*	At4g24230	Acyl-CoA-binding protein required for cuticle development
ACBP4	*A. thaliana*	At3g05420	Acyl-CoA-binding protein required for cuticle development
ACBP6	*A. thaliana*	At1g31812	Acyl-CoA-binding protein required for cuticle development
MPK3	*A. thaliana*	At3g45640	MAP-kinase
MPK6	*A. thaliana*	At2g43790	MAP-kinase
HSFB1	*A. thaliana*	At4g36990	Putative DNA binding protein
NtSABP2	*N. tabaccum*	–	SA-binding protein with MeSA esterase activity
NtSAMT1	*N. tabaccum*	–	SA-methyl transferase; biological synthesis of MeSA
StMES1	*S. tuberosum*	–	Methyl Esterase; release SA from MeSA

SABP2, which is required in the systemic leaves for the activation of SAR, possessed MeSA esterase activity (Forouhar et al. 2005; Kumar et al. 2006; Park et al. 2007), thereby suggesting that MeSA hydrolysis in the systemic leaves may have a role in SAR (Park et al. 2007). Indeed, genetic studies confirmed that MeSA esterase activity of SABP2 was essential for its involvement in SAR; a $Ser_{81} \rightarrow Ala_{81}$ missense mutation that abolished SABP2's MeSA esterase activity was unable to complement the SAR defect of a transgenic line in which expression of the endogenous *SABP2* gene was silenced (Park et al. 2007). Pharmacological experiments provided additional support for the importance of SABP2's MeSA esterase activity in SAR. 2,2,2,2′-tetra-fluoroacetophenone, a competitive inhibitor of SABP2's esterase activity, when applied to wild-type plants blocked the activation of SAR (Park et al. 2009). It was suggested that the conversion of MeSA to SA in the systemic leaves was critical for SAR (Park et al. 2007).

MeSA is synthesized by *S*-adenosyl-L-methionine: salicylic acid carboxyl methyltransferase, which catalyzes the transfer of a methyl group from *S*-adenosyl-L-methionine to SA. RNAi-mediated silencing of the tobacco *SAMT1* (*SA-METHYLTRANSFERASE1*) gene attenuated MeSA accumulation in response to TMV infection and also compromised SAR (Park et al. 2007). Genetic studies indicated that unlike *SABP2*, *SAMT1* function in SAR was required in the primary pathogen-treated organs. In grafting experiments, it was observed that SAR was compromised in the wild-type scion that was grafted on a *NtSAMT1*-silenced root stock, which received the primary TMV inoculation. By contrast, the *NtSAMT1*-silenced scions grafted on a wild-type root stock were SAR competent. It was therefore concluded that MeSA synthesized by SAMT1 in the primary pathogen-inoculated leaves is transported via the phloem to the systemic leaves of tobacco, where it is processed by the esterase activity of SABP2 to yield SA, which is biologically active to trigger SAR (Depmsey and Klessig 2012).

MeSA levels also increased in the pathogen-inoculated and systemic leaves of potato (*Solanum tuberosum*) and *Arabidopsis* plants exhibiting SAR (Manosalva et al. 2010; Vlot et al. 2008). In potato, 2,2,2,2′-tetrafluoroacetophenone application blocked arachidonic acid-induced SAR (Manosalva et al. 2010). Furthermore, RNAi-mediated silencing of the potato *METHYL ESTERASE 1* (*StMES1*) gene, which encodes a SABP2 homolog, compromised arachidonic acid-induced SAR (Manosalva et al. 2010). Similarly, 2,2,2,2′-tetrafluoroacetophenone applied to the systemic leaves inhibited SAR in *Arabidopsis* (Park et al. 2009). SAR was also curtailed in *Arabidopsis* plants in which expression of multiple *AtMES* genes encoding putative MeSA esterases was depressed (Vlot et al. 2008). However, this effect was observed only in 50 % of experiments (Chaturvedi et al. 2012; Vlot et al. 2008), suggesting that the involvement of MeSA in SAR is influenced by other factors. Similarly, while one study demonstrated that *Arabidopsis bsmtl* (*benzoic acid/salicylic acid methyl transferase 1*) mutants, which are deficient in MeSA synthesis, were SAR competent (Attaran et al. 2009), studies by another group demonstrated that SAR was weaker in *bsmtl* mutant plants (Liu et al. 2010, 2011a), providing additional support to the conditional requirement of MeSA in SAR. Liu et al. (2011a) have suggested that light is a likely factor contributing to this variable need of MeSA in SAR. They demonstrated that the time of the day when the plant is inoculated with a SAR-inducing pathogen determines the relative importance of MeSA in SAR. When the inoculations were done closer to the start of the dark period, MeSA was required for SAR. However, when the inoculations were done earlier during the light period, MeSA was less important.

2.2 *Dehydroabietinal*

Chaturvedi et al. (2012) used a biochemical approach to purify the systemic resistance-inducing activity from *Arabidopsis* AvrPex. Their efforts resulted in the identification of dehydorabietinal (DA) (Fig. 1), an abietane diterpenoid, as a potent inducer of SAR (Chaturvedi et al. 2012). Terpenoids include a large group of plant metabolites that have varied functions in plant growth and development, and interaction with other organisms. Picomolar solutions of chemically synthesized DA when applied to a few leaves of *Arabidopsis* induced systemic disease resistance against the bacterial pathogen *Pseudomonas syringae* and the fungal pathogen *Fusarium graminearum* (Chaturvedi et al. 2012). At these low concentrations, DA did not function as an antibiotic, thus suggesting its effect on limiting pathogen growth in plants was indirect.

Locally applied deuterium-labeled DA was systemically transported through *Arabidopsis*. DA application resulted in the local and systemic induction of SA accumulation and expression of the SA-responsive *PR1* gene (Chaturvedi et al. 2012). *FMO1*, which is required for the systemic accumulation of SA during biologically induced SAR (Mishina and Zeier 2006), was also required for the DA-induced systemic increase in SA. However, *FMO1* was not required for SA accumulation in the DA-treated leaves. DA was unable to induce systemic disease resistance in the *fmo1* mutant and in the isochorismate synthase-deficient *ics1 ics2* double mutant, which lacks the ability to synthesize SA via the isochorismate pathway, thus suggesting that DA-induced systemic disease resistance requires SA accumulation. In agreement with the requirement of SA for DA-induced systemic disease resistance, DA was unable to promote resistance in transgenic plants expressing the *nahG* gene. The proposed SA receptor NPR1 (Wu et al. 2012) was required for DA-induced systemic resistance, thus confirming that DA-conferred systemic disease resistance is due to the activation of SAR. Biologically induced SAR was not accompanied by an increase in DA content. Rather, during the biological induction of SAR, DA was redistributed from a biologically inactive (unable to induce SAR) form that elutes in a low-molecular weight range (<30 kDa) to a signaling form (DA*) that is SAR competent and elutes at a higher molecular weight range (>100 kDa). Trypsin treatment abolished the SAR-inducing capabilities of DA* (Chaturvedi et al. 2012), thus suggesting that DA* in AvrPex is associated with one or more proteins that are required for DA-induced systemic disease resistance.

Besides *Arabidopsis*, DA is also present in tobacco and tomato (*Solanum lycopersicum*), and DA application promoted systemic disease resistance in these species, suggesting that DA's role in defense is likely conserved in plants (Chaturvedi et al. 2012). However, whether biological induction of SAR requires DA is currently not known and requires experimentation with plants in which DA accumulation, in particular DA*, is blocked. Although the biosynthesis pathway for DA in angiosperms remains to be elucidated, clues on the synthesis of DA can be drawn from the biosynthesis of abietane family of diterpenoids in conifers, where these metabolites are synthesized by a mechanism that is similar to the biosynthesis of gibberellins (Bohlmann and Keeling 2008; Tholl 2006; Trapp and Croteau 2001).

2.3 A Glycerol-3-Phosphate-Derived Factor

Genetic studies in *Arabidopsis* have implicated the involvement of a glycerol-3-phosphate (G3P)-dependent factor in long-distance SAR signaling. Biologically-induced SAR was compromised in mutant plants lacking *SFD1 (SUPPRESSOR OF FATTY ACID DESATURASE DEFICIENCY1*) activity (Nandi et al. 2004). The systemic increase in SA and *PR1* expression that accompanies SAR was also attenuated in the *sfd1* mutant compared to the wild-type plant (Nandi et al. 2004). AvrPex collected from the *sfd1* mutant were unable to induce systemic disease resistance, when applied to wild-type plants, suggesting that SFD1 activity is required for the synthesis and/or translocation of a long-distance SAR signal (Chaturvedi et al. 2008). By comparison, SAR was restored in *sfd1* mutant plants treated with AvrPex from wild-type plants, thus indicating that the *sfd1* mutant is responsive to the long-distance SAR signal. SFD1 is a dihyroxyacetone phosphate (DHAP) reductase that catalyzes the synthesis of G3P (Fig. 1) from DHAP. G3P is an important precursor for a variety of biomolecules, including membrane and storage lipids. The *sfd1* mutant contained lower level of 34:6-monogalactosyldiacylglycerol, a major galactolipid in *Arabidopsis* that is synthesized in the plastids. This decrease in 34:6-monogalactosyldiacylglycerol was accompanied by a compensatory increase in 36:6-monogalactosyldiacylglycerol. Missense mutations that abolish SFD1's DHAP reductase activity were unable to complement the 34:6-monogalactosyldiacylglycerol deficiency and the SAR defect of the *sfd1* mutant, indicating that SFD1's involvement in galactolipid synthesis and SAR requires its DHAP reductase activity (Lorenc-Kukula et al. 2012). SFD1 contains a leader sequence at its N-terminus that is required for targeting SFD1 to the plastids. Although SFD1 lacking this leader sequence retains DHAP reductase activity, the N-terminus-deleted SFD1 was unable to complement the SAR and 34:6-monogalactosyldiacylglycerol deficiency of *sfd1*, suggesting that SFD1's DHAP reductase activity is required in the plastids for SAR and galactolipid synthesis (Lorenc-Kukula et al. 2012).

More recently SAR was also shown to be compromised in the *gly1* mutant, which is allelic with *sfd1* (Chanda et al. 2011). However, unlike the *sfd1* mutants, which are in the accession Nössen, systemic increase in SA and *PR1* expression that accompanies SAR was not attenuated in the *gly1* mutant, which is in the accession Columbia. Chanda et al. (2011) demonstrated that SAR was accompanied by an increase in G3P content in leaves treated with a SAR-inducing pathogen. G3P levels were also elevated in AvrPex and in the distal leaves of these plants. Local application of G3P with AvrPex or Avr pathogen restored SAR in the *gly1* mutant (Chanda et al. 2011). These pharmacological studies along with the genetic studies with plants expressing DHAP reductase-deficient SFD1 (Lorenc-Kukula et al. 2012), confirm an important role for SFD1-derived G3P in SAR. Chanda et al. (2011) further noted that ^{14}C-labeled G3P infiltrated into *Arabidopsis* leaves could not be recovered in the distal leaves as [^{14}C]G3P. Thus, G3P per se is likely not systemically translocated and the systemic increase in G3P observed during SAR is likely due to the de novo synthesis of G3P in the distal leaves. Further work is needed to identity the G3P-dependent factor associated with long-distance signaling in SAR.

G3P when applied by itself to wild-type *Arabidopsis* was not sufficient to induce systemic resistance. However, when co-applied with Pex from either $MgCl_2$-treated or Avr pathogen-inoculated plants, G3P was capable of enhancing systemic disease resistance, thus suggesting that a factor present in Pex is required for G3P-promoted systemic disease resistance (Chanda et al. 2011). Local application of G3P resulted in the enhanced expression of the *MES9* gene in the systemic tissues of *Arabidopsis* (Chanda et al. 2011). *MES9* encodes a homolog of the tobacco SABP2. By comparison, expression of the *BSMT1* gene was downregulated, thus predicting increased conversion of MeSA to SA in the systemic leaves of plants that were locally treated with G3P. However, comparable to plants that received a local control (mock) treatment, no increase in SA or SAG was observed in the systemic leaves of *Arabidopsis* that were treated on other leaves with G3P (Chanda et al. 2011). Thus the significance of the altered expression of *MES9* and *BSMT1* to G3P-induced SAR is unclear.

2.4 *Azelaic Acid*

As mentioned above, in tissues exhibiting SAR, defenses are primed to respond faster and stronger in response to pathogen infection. However, how these defenses are primed is poorly understood. Jung and coworkers (2009) suggested that azelaic acid (Fig. 1), a nine-carbon dicarboxylic acid, is involved in priming of systemic defenses. GC-MS scans for small molecules (70–550 Da) revealed elevated levels of azelaic acid in AvrPex compared to Pex collected from mock-treated *Arabidopsis* leaves (Jung et al. 2009). Locally applied deuterium-labeled azelaic acid could be recovered in PeX and the distal leaves, indicating that azelaic acid is systemically transported. When applied at concentrations greater than 10 μM, azelaic acid systemically enhanced disease resistance. Azelaic acid-induced systemic resistance in *Arabidopsis* required genes involved in SA synthesis and signaling, *DIR1* (*DEFECTIVE IN INDUCED RESISTANCE1*), *FMO1*, and *ALD1* (*AGD2-LIKE DEFENSE RESPONSE PROTEIN1*), which encodes an aminotransferase that is involved in pipecolic acid synthesis (see Sect. 2.5). However, unlike MeSA and DA, azelaic acid applied to *Arabidopsis* foliage did not increase SA content and *PR1* expression. Instead, azelaic acid-treated plants were primed for the enhanced accumulation of SA and *PR1* expression when challenged with a pathogen. Although azelaic acid treatment did not have a major impact on the plant transcriptome, one of the genes that was transiently expressed at elevated levels in azelaic acid-treated plants was *AZI1* (*AZELAIC ACID-INDUCED 1*) (Jung et al. 2009), which encodes a protein with homology to lipid-transfer proteins. *AZI1* expression was also induced in leaves treated with AvrPex (Jung et al. 2009). *AZI1* is required for priming associated with azelaic acid- and biologically induced SAR.

A likely mechanism for the synthesis of azelaic acid involves the sequential action of 9-lipoxygenase (9-LOX) and hydroperoxide lyases on fatty acids to yield 9-oxononanoic acid that is subsequently oxidized to yield azelaic acid. To determine if *LOX1* and *LOX5*, the two 9-LOX-encoding genes in *Arabidopsis*, are involved in pathogen infection associated accumulation of azelaic acid, Zoeller et al. (2012) compared azelaic acid levels in the *Arabidopsis lox1 lox5* double mutant plant after inoculation with pathogen. However, azelaic acid levels were found to increase to comparable levels in the pathogen-inoculated leaves of wild-type and the *lox1 lox5* plant, suggesting that *LOX1* and *LOX5* do not contribute to azelaic acid synthesis in pathogen-inoculated *Arabidopsis*. Instead, it was suggested that azelaic acid is synthesized in plastids by a free radical-based galactolipid fragmentation mechanism (Zoeller et al. 2012). Zoeller and coworkers (2012) further suggested that azelaic acid is a general marker for lipid peroxidation.

2.5 Pipecolic Acid

In addition to azelaic acid, the lysine catabolite pipecolic acid (Pip) (Fig. 1) also has been implicated in priming and amplification of plant defenses that contribute to SAR-conferred enhanced disease resistance. In addition, Pip is also required for local defenses against virulent and avirulent pathogen (Návarová et al. 2012). The levels of Pip increase in the pathogen-inoculated and the systemic pathogen-free leaves. Pip application promotes local and systemic disease resistance in *Arabidopsis*. Pip accumulation in *Arabidopsis* infected with pathogen requires the *ALD1*-encoded aminotransferase, which is also required for SAR (Jing et al. 2011; Song et al. 2004a, b). Pip application restored disease resistance in the *ald1* mutant. *ALD1* expression is induced in pathogen-infected and systemic leaves. Since lysine can be utilized as a substrate by ALD1 in vitro (Návarová et al. 2012), the aminotransferase activity of ALD1 likely is directly involved in Pip synthesis in vivo.

Pip also accumulates at elevated levels in Pex collected from pathogen-inoculated leaves; thus Pip could be systemically transported. The low level of Pip that accumulates in the systemic uninfected leaves of plants exhibiting SAR likely promotes its own synthesis when challenged with pathogen by inducing *ALD1* expression. Since *ALD1* is also involved in a SA amplification loop (Song et al. 2004b), Pip therefore might contribute to signal amplification by priming SA accumulation in response to challenge with pathogen. Indeed, pre-treatment with Pip resulted in a faster increase in SA content in response to subsequent pathogen inoculation. *FMO1*, which is required for systemic accumulation of SA, is also required for the systemic induction of *ALD1* expression during SAR and for Pip-induced systemic disease resistance, leading to a model in which Pip acting through *FMO1*, promotes *ALD1* expression and thus its own synthesis in the distal leaves, thereby priming the rapid increase in SA content upon pathogen infection.

2.6 *Jasmonates*

Jasmonic acid (JA) (Fig. 1) and its derivatives, collectively known as jasmonates, are involved in systemic signaling associated with wounding in tomato (Lee and Howe 2003) and have also been suggested to be involved in the manifestation of SAR (Truman et al. 2007). The *OPR3* (*12-OXOPHYTODIENOATE REDUCTASE 3*) gene, which is involved in JA synthesis, and the *JIN1* and *JAI4* genes, both of which are associated with JA signaling, were required for the activation of SAR in *Arabidopsis* (Truman et al. 2007). Furthermore, JA rapidly accumulated in the AvrPex. This accumulation of JA in AvrPex was paralleled by the systemic induction of JA-responsive genes and preceded the expression of the SA-responsive genes. MeJA application induces expression of the *Arabidopsis BSMT1* gene (Koo et al. 2007), thus suggesting that jasmonates could potentially promote MeSA synthesis in the primary pathogen-inoculated leaves and thus contribute to long-distance signaling in SAR. However, results from other studies have questioned the involvement of JA as a systemic signal in SAR. Unlike Truman et al. (2007), Attaran et al. (2009) reported that the *opr3* and *jin1* mutants were SAR competent. The ability to induce SAR was also retained in the JA-insensitive *coi1* (*coronatine insensitive1*) and the *jar1* (*jasmonate resistant1*) mutants (Attaran et al. 2009; Cui et al. 2005; Mishina and Zeier 2007). Furthermore, when AvrPex was fractionated by molecular-sieve chromatography, JA did not copurify in fractions that contained the SAR-inducing activity (Chaturvedi et al. 2008). It is plausible that other environmental factors might influence the involvement of JA in SAR, thus explaining the differences between Truman et al. (2007) and the other studies.

3 Lipid-Transfer Proteins in Long-Distance Signaling

3.1 *DIR1*

The *Arabidopsis DIR1* was one of the first genes to be identified that is critical for long-distance signaling in SAR (Maldonado et al. 2002). The *Arabidopsis dir1* mutant was incapable of developing SAR in response to primary inoculation with an Avr pathogen. AvrPex collected from the *dir1* mutant were unable to systemically enhance *PR1* expression and disease resistance when applied to wild-type plants (Chaturvedi et al. 2008; Maldonado et al. 2002). However, the *dir1* mutant was responsive to the SAR signal present in AvrPex collected from wild-type plants, thus suggesting that *DIR1* is required for the accumulation and/or long-distance translocation of a SAR signal(s) (Chaturvedi et al. 2008; Maldonado et al. 2002). Basal resistance against pathogen was not impacted in the *dir1* mutant (Maldonado et al. 2002), thus suggesting that DIR1 is specifically required for SAR. More recently, DIR1 homologs were also demonstrated to have an important function in SAR in tobacco, as well (Liu et al. 2011b).

Liu et al. (2011b) demonstrated that the *Arabidopsis BSMT1* gene, which encodes a MeSA-synthesizing methyltransferase, was expressed at elevated levels in the Avr pathogen-inoculated and the systemic leaves of the *dir1* mutant compared to the wild-type plant. Furthermore, MeSA content was higher and free SA and its glucoside, SAG, levels were lower in these *dir1* tissues, thus suggesting that in the systemic leaves of wild-type plants, DIR1 depresses the conversion of SA to MeSA, and thereby promotes SA accumulation. Similarly, in tobacco silenced for expression of the *DIR1* gene, a correlation was observed between the SAR-deficient phenotype and elevated *SAMT1* expression level in the pathogen-treated and distal tissues (Liu et al. 2011b).

Overexpression of DIR1 in *Arabidopsis* did not lead to the constitutive activation of SAR-like responses (Maldonado et al. 2002). This suggests that additional factors are required for systemic signaling in SAR. Unlike AvrPex from the *dir1* and *sfd1* mutants, which when applied individually to wild-type plants were unable to induce SAR, when mixed together, *dir1* plus *sfd1* AvrPexs were effective in systemically inducing *PR1* expression and disease resistance (Chaturvedi et al. 2008). These results implicate a combined requirement of DIR1 and a G3P-dependent factor in SAR. In support of this view, Chanda et al. (2011) observed that DIR1 when co-applied with G3P was capable of inducing SAR in *Arabidopsis*. It was suggested that G3P promotes the systemic translocation of DIR1 (Chanda et al. 2011). In light of the observations that G3P promotes the expression of *MES9* and depresses the expression of *BSMT1* in *Arabidopsis* (Chanda et al. 2011), and *DIR1* promotes systemic SA accumulation (Liu et al. 2011b), it would be important to know whether co-application of DIR1 + G3P impacts systemic accumulation of SA and *PR1* expression. DIR1 is also required for DA- and azelaic acid-induced SAR (Chaturvedi et al. 2012; Jung et al. 2009). Likewise, *SFD1*, and hence presumably a G3P-dependent factor, is also required for the full potential of DA-induced SAR.

The crystal structure of DIR1 indicates that it shares some similarities to LTP2 family of lipid-transfer proteins (Lascombe et al. 2008). Recombinant DIR1 can bind lipids. DIR1 contains two SH3 domains, which in other proteins facilitate protein–protein interaction. Lascombe et al. (2008) suggested that DIR1 likely interacts with other proteins, as well. Indeed, compared to the relatively small size of DIR1 (<10 kDa), the SAR-inducing activity in AvrPex, which is trypsin sensitive, elutes in a range that is larger than 100 kDa (Chaturvedi et al. 2012). DIR1 is present in this SAR activating fraction derived from AvrPex (R. Chaturvedi and J. Shah, unpublished), thus supporting the opinion that DIR1 associates with other proteins.

3.2 *AZI1*

As mentioned above, expression of *AZI1*, which encodes a putative lipid-transfer protein, is induced in AvrPex- and azelaic acid-treated leaves (Jung et al. 2009).

Biologically activated, and AvrPex- and azelaic acid-induced systemic disease resistance was compromised in the *azi1* mutant (Chaturvedi et al. 2012; Jung et al. 2009). The SAR-associated priming of SA accumulation and *PR1* expression were weaker in the *azi1* mutants than the wild-type plant (Jung et al. 2009). Azelaic acid and AvrPex when applied to the *azi1* mutant were capable of inducing disease resistance in the treated leaves, thus indicating that the *azi1* mutant is sensitive to azelaic acid and the SAR-inducing signal present in AvrPex. By comparison, AvrPex collected from the *azi1* mutant were unable to enhance disease resistance in the foliage of wild-type plants, thus suggesting that *AZI1* is likely involved in the production and/or the translocation of a long-distance signal involved in defense priming. Whether the local accumulation and/or systemic translocation of azelaic acid or any of the other signaling metabolites described above is impacted in the *azi1* mutant, remains to be determined. Although not essential for systemic disease resistance induced by DA applied at concentrations above 10 pM, *AZI1* was required for systemic disease resistance induced by lower concentrations of DA, thus suggesting that *AZI1*- and azelaic acid-mediated priming promotes DA's effectiveness in inducing SAR. It would be of particular interest to determine if AZI1 is part of the high-molecular weight complex that contains DA* and DIR1.

4 Perception of the SAR Signals and Ensuing Signaling

4.1 Perception of the SAR Signals

How some of the systemic signals are perceived in the systemic leaves is not known. In case of MeSA, binding to MeSA esterase might be a mechanism by which MeSA is perceived during SAR. As discussed below, an intact cuticle has been suggested to be important for perception of the SAR signal.

The cuticle, which is composed of waxes and cutin monomers, forms a hydrophobic barrier on the surface of most foliar tissues. The cuticle also provides a physical barrier to pathogens. It has also been suggested to serve as a source for signals that promote resistance against the necrotrophic fungus *Botrytis cinerea* (Chassot and Métraux 2005; Chassot et al. 2007). In other cases, damaged cuticle has been associated with increased susceptibility to pathogens (Xia et al. 2012). An intact cuticle has also been demonstrated to be required for SAR. SAR was compromised when the cuticle was mechanically damaged in *Arabidopsis* (Xia et al. 2009). Furthermore, SAR was compromised in the cuticle-defective *acp4* (*acyl carrier protein 4*) mutant. The *acp4* mutant was impaired in its ability to respond to the SAR signal present in AvrPex from wild-type plants (Xia et al. 2009, 2010). In comparison, AvrPex from *acp4* was capable of inducing SAR when applied to wild-type plants, thus indicating that the *acp4* mutant is capable of producing the systemically translocated SAR signal, but is deficient in the perception and/or response to this signal. Mutations in some genes encoding acyl

CoA-binding proteins (ACBPs) that are required for proper cuticle development also resulted in attenuated SAR (Xia et al. 2012). However, unlike the *acp4* mutant, these *acbp* mutants were responsive to the SAR signal. Instead, AvrPex from these *acbp* mutants lacked the ability to induce SAR when applied to wild-type leaves. Xia et al. (2012) have suggested that these *acbp* mutants are defective in the generation of the long-distance SAR signal. Thus, cuticular components could be involved in both, signal generation and perception.

AZI1 expression is also elevated in transgenic *Arabidopsis* with cuticular defects resulting from expression of a fungal cutinase (Chassot et al. 2007). Whether *AZI1* expression is similarly altered in the *acp4* and/or the *acpb* mutants remains to be determined. However, the *acbp* mutants were not affected in pathogen-induced accumulation of azelaic acid (Xia et al. 2012), thus indicating that their SAR-deficiency is not due to defects in azelaic acid accumulation. It remains to be determined if the different classes of cuticular mutants have defects in the accumulation and/or response to one or more of the long-distance signaling molecules reviewed here.

4.2 Downstream Signaling in SAR

4.2.1 Priming

Priming involves mechanisms that make the primed cells more sensitive to perceive and/or respond to a stress, than non-primed cells (Conrath 2011). A primed state is also one of the characteristics of SAR. During SAR, SA accumulation and SA signaling are primed to respond more strongly when the tissue is challenged by a pathogen. Azelaic acid and Pip have been implicated in priming increases in SA content in response to challenge inoculation with pathogen (Jung et al. 2009; Návarová et al. 2012). The *FMO1* gene, which is required for azelaic acid- and Pip-induced SAR, has been suggested to participate in a feedback loop involving Pip and the *ALD1* gene to promote SA accumulation during SAR (Návarová et al. 2012).

Recent studies indicate that priming in SAR is associated with alterations in MAPK pathway activity and also epigenetic alterations of transcription regulatory genes. *MPK3* and *MPK6* transcripts and the corresponding proteins accumulated at elevated levels in the systemic tissues of *Arabidopsis* in which SAR was induced by inoculating the lower leaves with an Avr pathogen (Beckers et al. 2009). When infiltrated with water or pathogen, the levels of MPK3 and MPK6 proteins increased further in the systemic leaves of plants exhibiting SAR, than control plants in which SAR was not induced. This increase in MPK3 and MPK4 correlated with the higher expression of *PR1* and systemic disease resistance. The primed expression of *PR1* was not observed in the *mpk3* mutants, thus confirming the involvement of *MPK3* in priming associated with SAR. By contrast to MPK3, MPK6 had a weaker contribution to priming in SAR. The SA receptor NPR1 was required for this priming. MPK3 and MPK4 transcript and protein also accumulate at elevated levels when low levels of the SA analogue benzo(1,2,3)thiadiazole-7-carbothioic acid *S*-methyl ester (BTH) were applied to *Arabidopsis*. The requirement of NPR1 and the ability of BTH to prime MPK3 and MPK6 are suggestive of the involvement of SA signaling in this priming of MPKs. Whether azelaic acid and Pip involvement in priming involves MPK3 and MPK6 is not known.

Heat shock factors (HSFs) are DNA binding proteins that are involved in regulating expression of the heat shock proteins. In *Arabidopsis*, the *HSFB1* gene (also referred as *TBF1* and *HSF4*) is required for SA-induced disease resistance (Pajerowska-Mukhtar et al. 2012) and for priming defense gene expression during SAR (Pick et al. 2012). Expression of *HSFB1* was upregulated in the systemic leaves of plants that were inoculated on their lower leaves with Avr strains of *P. syringae*. *hsfb1* mutant plants were defective in the BTH-induced priming of the defense genes *PAL1* (*PHENYLALANINE AMMONIA-LYASE1*) and *WRKY29*. Furthermore, local inoculation with an Avr pathogen was unable to enhance systemic disease resistance in the *hsfb1* mutant plants. Since *PR1* expression was induced in the systemic leaves, but disease resistance was not, and BTH was unable to prime gene expression in the *hsfb1*plants, Pick et al. (2012) suggested that the *hsfb1* mutant is not defective in systemic long-distance signaling, but rather in the priming of defenses associated with SAR.

Chromatin remodeling involving histone modifications has also been implicated in priming and memory associated with SAR. Jaskiewicz et al. (2011) observed that *Arabidopsis* plants that were inoculated on their lower leaves with a pathogen exhibited enhanced level of histone modifications on the *WRKY6*, *WRKY29*, and *WRKY53* genes. Despite these chromatin modifications, expression of these three genes was not substantially altered in the systemic pathogen-free leaves. However, when these systemic leaves were stressed by infiltrating water, expression of these *WRKY* genes increased substantially over that in the non-primed plants. Similarly, BTH-promoted priming and stress-induced expression of these genes were accompanied by increased histone modifications. The *NPR1* gene was required for the increase in histone modifications and for the stress-induced expression of these genes in response to pre-treatment with BTH, thus confirming a role for *NPR1* in BTH-promoted chromatin modification and in priming the stress-induced expression of these *WRKY* genes. Readers are directed to two excellent reviews (Conrath 2011; van den Burg and Takken 2009) on the involvement of chromatin modification in basal and induced expression of SA-responsive genes and in defense priming. Determining whether there is a relationship between histone modifications and the MAPK cascade, and azelaic acid- and Pip-promoted priming in SAR will be of particular interest.

4.2.2 The SA Receptors

The NPR1 protein exists in the nucleus and the cytosol. In the cytosol, NPR1 is suggested to exist as an oligomer. Its conversion to the monomeric form promotes

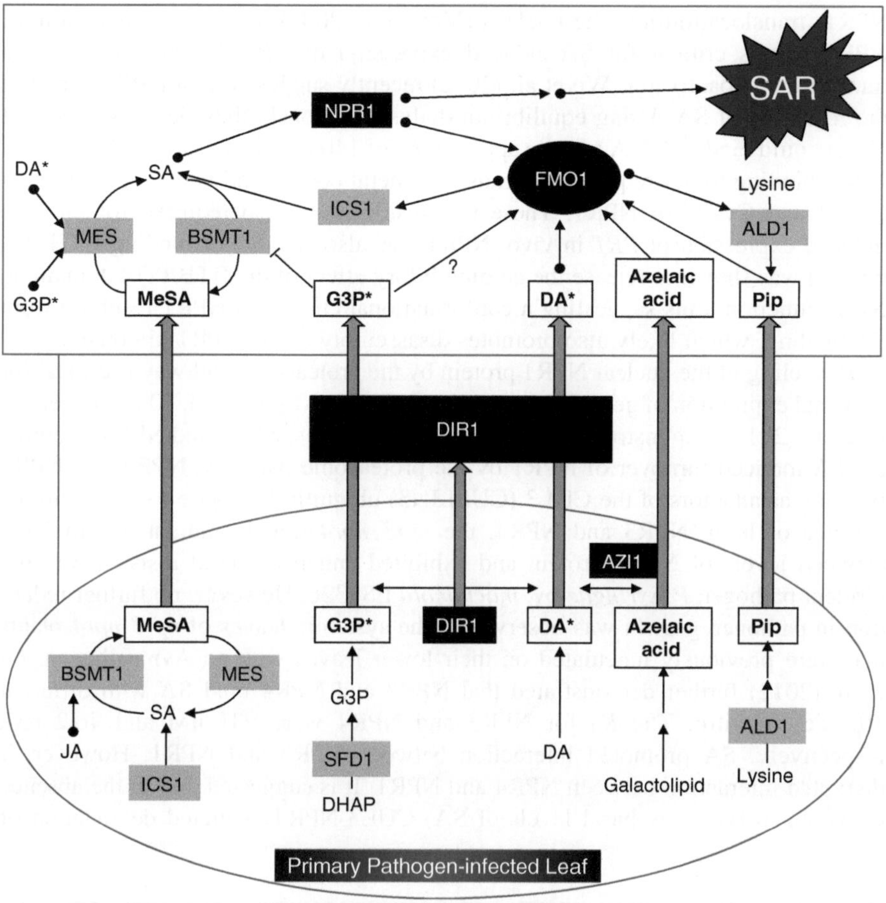

Fig. 2 Networking between SAR signaling molecules. *Events in the primary pathogen-infected leaf:* Increased activity of ICS1, resulting from increased expression of the corresponding gene, contributes to increase in SA content. SA is converted to MeSA by BSMT1. In tobacco, the high level of SA was simultaneously shown to inhibit the MeSA esterase (MES) activity of SABP2, thus ensuring increase in MeSA level. JA is known to promote expression of *BSMT1*. Simultaneously, glycerol-3-phosphate levels increase. SFD1 (GLY1) catalyzes the synthesis of glycerol-3-phosphate from dihydroxyacetone phosphate (DHAP). Azelaic acid and pipecolic (Pip) levels also increase. Azelaic acid has been suggested to be synthesized on galactolipids by a nonenzymatic method, while Pip synthesis from lysine requires the ALD1 aminotransferase. Expression of the *ALD1* gene is induced in response to pathogen inoculation. Absolute levels of DA do not change. However, DA is mobilized from a non-signaling low-molecular weight to a high-molecular weight signaling DA (DA*) complex in response to pathogen inoculation. Trypsin treatment destroys the high-molecular weight DA* complex, suggesting the presence of proteins in this complex. DIR1 is one of the proteins in this high-molecular weight complex. The azelaic acid-inducible *AZI1* gene is required for accumulation and/or transport of the SAR signal. *AZI1* is required for azelaic acid-induced SAR and also promotes DA*-induced SAR. However, its involvement in SAR induced by the other factors is not known. *Events in the distal (systemic) leaf:* Systemic transport of MeSA, a G3P-derived factor (G3P*), DA*, azelaic acid, Pip, and DIR1 from the pathogen-inoculated leaf to the distal leaves occurs via the vasculature, most probably the

NPR1s translocation into the nucleus (Mou et al. 2003). It is the nuclear form of NPR1 that is critical for SA-induced expression of *PR1*, by functioning as a transcription coactivator. Wu et al. (2012) recently suggested that NPR1 is one of the receptors of SA. Using equilibrium dialysis approach, they demonstrated that NPR1 could bind [^{14}C]SA with an apparent K_d of 140 nM. They further showed that NPR1 binding to SA requires the transition metal copper and is mediated through Cys^{521} and Cys^{529} of NPR1. These two residues are also required for the SA-induced expression of *PR1* in vivo. NPR1 was also capable of binding BTH. SA binding was shown to release the autoinhibitory effect of the BTB/POZ domain on NPR1 function, thus suggesting a conformational change in NPR1 resulting from SA binding, which likely also promotes disassembly of the NPR1 oligomers.

Recycling of the nuclear NPR1 protein by the proteasome pathway is critical for maximal expression of genes that are targets of NPR1 (Spoel et al. 2009). Recently, Fu et al. (2012) demonstrated that the NPR1 paralogues, NPR3 and NPR4, promote the SA-induced turnover of NPR1 by the proteasome pathway. NPR3 and NPR4 function as adaptors of the CUL3 (CULLIN3) ubiquitin E3 ligase to NPR1. In the absence of both, NPR3 and NPR4, the *npr3 npr4* double mutant accumulated elevated levels of NPR1 protein and exhibited enhanced basal resistance to the virulent pathogen *P. syringae* pv. *maculicola* ES4326. However, no further reduction in pathogen growth was observed in the systemic leaves of *npr3 npr4* plants that were previously inoculated on their lower leaves with an Avr pathogen. Fu et al. (2012) further demonstrated that NPR3 and NPR4 bind SA with different affinities in vitro. The K_d for NPR3 and NPR4 were 981 nM and 46.2 nM, respectively. SA promoted interaction between NPR3 and NPR1. However, it disrupted interaction between NPR4 and NPR1. It is suggested that in the absence of pathogen (i.e., low basal levels of SA) CUL3-NPR4-mediated degradation of

---

Fig. 2 (continued) phloem. G3P* and DIR1 have been suggested to facilitate long-distance transport of each other. DA* and G3P* promote accumulation of MES transcript (and likely the corresponding protein). Simultaneously, G3P* and DIR1 downregulate the expression of *BSMT1*, thus ensuring that the equilibrium is in favor of conversion of MeSA to SA. An amplification loop involving SA, the SA receptor NPR1, FMO1, and ICS1 promotes SA accumulation. NPR1 activation by SA leads to the expression of defense genes that contribute to SAR. *FMO1* is required for the induction of *ICS1* expression and accumulation of SA in the pathogen-free distal leaves. DA*, azelaic acid, and Pip signals converge at *FMO1*, which is required for activation of SAR by these signal molecules. It is likely that *FMO1* is also required for G3P*- and MeSA-induced SAR. However, this needs to be tested. *ALD1* is a point of convergence of the azelaic acid and Pip pathways. Pip acting through an amplification loop involving *FMO1* promotes *ALD1* expression and thus its own synthesis. DIR1 is essential for SAR induced by MeSA, G3P*, DA*, and azelaic acid. Whether it is required for Pip-induced SAR is not known. DA is shown to interact synergistically with azelaic acid and the *SFD1*-dependent mechanism. *White* and *gray boxes* represent the signaling molecules and biosynthetic enzymes, respectively. Signaling/transport proteins are represented by *black boxes/circles*. *Gray-filled arrows* represent long-distance transport. *Black arrows* ending in *black circles* indicate positive regulation (induction), while *black lines* ending with a *bar* indicate negative regulation. *Bidirectional arrows* indicate known synergistic interactions

NPR1 prevents spurious activation of defenses. In pathogen-inoculated plants, SA levels increase in the pathogen-bearing and systemic pathogen-free organs. The high levels of SA near the infection site result in the CUL3-NPR3-mediated turnover of NPR1, thus allowing cell death to be turned on. In the systemic tissues that have comparatively lower levels of SA, the turnover of NPR1 by CUL3-NPR3 is suggested to facilitate binding of newly synthesized NPR1 to the promoters of NPR1-regulated genes, thus promoting reinitiation of transcription at these promoters.

5 Concluding Remarks

The effects of SAR can be transmitted for a couple of generations (Luna et al. 2012). In addition, the manifestation of SAR confers a fitness advantage when plants are cultivated under disease pressure (Luna et al. 2012; Traw et al. 2007). However, SAR is an energy-driven process that requires diversion of resources from growth (Heidel et al. 2004; Pajerowska-Mukhtar et al. 2012). In addition, pathogen-derived effectors target genes and mechanisms that contribute to defense. The ability to recruit multiple signals empowers plants with better control over the activation of SAR. Several metabolites putatively involved in long-distance signaling have been described above. Figure 2 summarizes our current understanding of the potential interactions between these signal molecules during SAR. Progress also has begun to be made on understanding the mechanism of priming and the putative involvement of chromatin remodeling in SAR. The next 5 years will be important for unraveling the networking between these SAR signaling molecules and their liaison with priming and chromatin remodeling in SAR.

Acknowledgments The authors thank Jürgen Zeier for sharing unpublished results. This work was supported by grants from the National Science Foundation (IOS-1121570 and MCB-0920600) and the U.S. Department of Agriculture as a cooperative project with the U.S. Wheat & Barley Scab Initiative (Agreement No. 59-0790-8-060).

References

Attaran E, Zeier TE, Griebel T, Zeier J (2009) Methyl salicylate production and jasmonate signaling are not essential for systemic acquired resistance in *Arabidopsis*. Plant Cell 21:954–971

Beckers GJ, Jaskiewicz M, Liu Y, Underwood WR, He SY, Zhang S, Conrath U (2009) Mitogen-activated protein kinases 3 and 6 are required for full priming of stress responses in *Arabidopsis thaliana*. Plant Cell 21:944–953

Bohlmann J, Keeling CI (2008) Terpenoid biomaterials. Plant J 54:656–669

Chanda B, Xia Y, Mandal MK, Yu K, Sekine K-T, Gao Q-M, Selote D, Hu Y, Stromberg A, Navarre D, Kachroo A, Kachroo P (2011) Glycerol-3-phosphate is a critical mobile inducer of systemic immunity in plants. Nat Genet 43:421–427

Chassot C, Métraux J-P (2005) The cuticle as source of signals for plant defense. Plant Biosyst 139:28–31
Chassot C, Nawrath C, Métraux J-P (2007) Cuticular defects lead to full immunity to a major plant pathogen. Plant J 49:972–980
Chaturvedi R, Shah J (2007) Salicylic acid in plant disease resistance. In: Hayat S, Ahmad A (eds) Salicylic acid – a plant hormone. Springer, Dordrecht, pp 335–370
Chaturvedi R, Krothapalli K, Makandar R, Nandi A, Sparks A, Roth MR, Welti R, Shah J (2008) Plastid ω-3 desaturase-dependent accumulation of a systemic acquired resistance inducing activity in petiole exudates of *Arabidopsis thaliana* is independent of jasmonic acid. Plant J 54:106–117
Chaturvedi R, Venables B, Petros RA, Nalam V, Li M, Wang X, Takemoto LJ, Shah J (2012) An abietane diterpenoid is a potent activator of systemic acquired resistance. Plant J 71:161–172
Chester KS (1933) The problem of acquired physiological immunity in plants. Q Rev Biol 8:275–324
Conrath U (2011) Molecular aspects of defence priming. Trends Plant Sci 16:524–531
Cui J, Bahrami AK, Pringle EG, Hernandez-Guzman G, Bender CL, Pierce NE, Ausubel FM (2005) *Pseudomonas syringae* manipulates systemic plant defenses against pathogens and herbivores. Proc Natl Acad Sci USA 102:1791–1796
Depmsey DA, Klessig DF (2012) SOS – too many signals for systemic acquired resistance? Trends Plant Sci 17:538–545
Durrant WE, Dong X (2004) Systemic acquired resistance. Annu Rev Phytopathol 42:185–209
Forouhar F, Yang Y, Kumar D, Chen Y, Fridman E, Park SW, Chiang Y, Acton TB, Montelione GT, Pichersky E, Klessig DF, Tong L (2005) Structural and biochemical studies identify tobacco SABP2 as a methyl salicylate esterase and implicate it in plant innate immunity. Proc Natl Acad Sci USA 102:1773–1778
Fu ZQ, Yan S, Saleh A, Wang W, Ruble J, Oka N, Mohan R, Spoel SH, Tada Y, Zheng N, Dong X (2012) NPR3 and NPR4 are receptors for the immune signal salicylic acid in plants. Nature 486:228–232
Gessler C, Kuc J (1982) Induction of resistance to *Fusarium* wilt in cucumber by root and foliar pathogens. Phytopathology 72:1439–1441
Griebel T, Zeier J (2008) Light regulation and daytime dependency of inducible plant defenses in Arabidopsis: phytochrome signaling controls systemic acquired resistance rather than local defense. Plant Physiol 147:790–801
Guedes MEM, Richmond S, Kuc J (1980) Induced systemic resistance to anthracnose in cucumber as influenced by the location of the inducer inoculation with *Colletotrichum lagenarium* and the onset of flowering and fruiting. Physiol Plant Pathol 17:229–233
Heidel AJ, Clarke JD, Antonovics J, Dong X (2004) Fitness costs of mutations affecting the systemic acquired resistance pathway in *Arabidopsis thaliana*. Genetics 168:2197–2206
Jaskiewicz M, Conrath U, Peterhänsel C (2011) Chromatin modification acts as a memory for systemic acquired resistance in the plant stress response. EMBO Rep 12:50–55
pJenns A, Kuc J (1979) Graft transmission of systemic resistance of cucumber to anthracnose induced by *Colletotrichum lagenarium* and tobacco necrosis virus. Phytopathology 69:753–756
Jing B, Xu S, Xu M, Li Y, Li S, Ding J, Zhang Y (2011) Brush and spray: a high-throughput systemic acquired resistance assay suitable for large-scale genetic screening. Plant Physiol 157:973–980
Jung HW, Tschaplinski TJ, Wang L, Glazebrook J, Greenberg JT (2009) Priming in systemic plant immunity. Science 324:89–91
Kiefer IW, Slusarenko AJ (2003) The pattern of systemic acquired resistance induction within the Arabidopsis rosette in relation to the pattern of translocation. Plant Physiol 132:840–847
Koo YJ, Kim MA, Kim EH, Song JT, Jung C, Moon J-K, Kim J-H, Seo HS, Song SI, Kim JK, Lee JS, Cheong JL, Choi YD (2007) Overexpression of salicylic acid carboxyl methyltransferase reduces salicylic acid-mediated pathogen resistance in *Arabidopsis thaliana*. Plant Mol Biol 64:1–15

Kumar D, Gustafsson C, Klessig DF (2006) Validation of RNAi silencing specificity using synthetic genes: salicylic acid-binding protein 2 is required for innate immunity in plants. Plant J 45:863–868

Lascombe M-B, Bakan B, Buhot N, Marion D, Blein J-P, Larue V, Lamb C, Prange T (2008) The structure of "defective in induced resistance" protein of *Arabidopsis thaliana*, DIR1, reveals a new type of lipid transfer protein. Protein Sci 17:1522–1530

Lee GI, Howe GA (2003) The tomato mutant *spr1* is defective in systemin perception and the production of a systemic wound signal for defense gene expression. Plant J 33:567–576

Liu J, Maldonado-Mendoza I, Lopez-Meyer M, Cheung F, Town CD, Harrison MJ (2007) Arbuscular mycorrhizal symbiosis is accompanied by local and systemic alterations in gene expression and an increase in disease resistance in the shoots. Plant J 50:529–544

Liu P-P, Yang Y, Pichersky E, Klessig DF (2010) Altering expression of *Benzoic acid/salicylic acid carboxyl methyltransferase 1* compromises systemic acquired resistance and PAMP-triggered immunity in *Arabidopsis*. Mol Plant-Microbe Interact 23:82–90

Liu PP, von Dahl CC, Klessig DF (2011a). The extent to which methyl salicylate is required for signaling systemic acquired resistance is dependent on exposure to light after infection. Plant Physiol 157:2216–2226. doi:10.3389/fpls.2013.00030

Liu PP, von Dahl CC, Park SW, Klessig DF (2011b) Interconnection between methyl salicylate and lipid-based long distance signaling during the development of systemic acquired resistance in Arabidopsis and tobacco. Plant Physiol 144:1762–1768

Lorenc-Kukula K, Chaturvedi R, Roth M, Welti R, Shah J (2012) Biochemical and molecular-genetic characterization of SFD1's involvement in lipid metabolism and defense signaling. Front Plant Sci 3:26. doi:10.3389/fpls.2012.00026

Luna E, Bruce TJA, Roberts MR, Flors V, Ton J (2012) Next generation systemic acquired resistance. Plant Physiol 158:844–853

Malamy J, Carr JP, Klessig DF, Raskin I (1990) Salicylic acid: a likely endogenous signal in the resistance response of tobacco to viral infection. Science 250:1002–1004

Maldonado AM, Doerner P, Dixon RA, Lamb CJ, Cameron RK (2002) A putative lipid transfer protein involved in systemic acquired resistance signalling in Arabidopsis. Nature 419:399–403

Manosalva PM, Park SW, Forouhar F, Tong L, Fry WE, Klessig DF (2010) *Methyl Esterase 1* (*StMES1*) is required for systemic acquired resistance in potato. Mol Plant-Microbe Interact 23:1151–1163

Métraux JP, Signer H, Ryals J, Ward E, Wyss-Benz M, Gaudin J, Raschdorf K, Schmid E, Blum W, Inverard B (1990) Increase in salicylic acid at the onset of systemic acquired resistance in cucumber. Science 250:1004–1006

Mishina TE, Zeier J (2006) The Arabidopsis favin-dependent monooxygenase FMO1 is an essential component of biologically induced systemic acquired resistance. Plant Physiol 141:1666–1675

Mishina TE, Zeier J (2007) Pathogen-associated molecular pattern recognition rather than development of tissue necrosis contributes to bacterial induction of systemic acquired resistance in Arabidopsis. Plant J 50:500–513

Mou Z, Fan W, Dong X (2003) Inducers of plant systemic acquired resistance regulate NPR1 function through redox changes. Cell 113:935–944

Nandi A, Welti R, Shah J (2004) The *Arabidopsis thaliana* dihydroxyacetone phosphate reductase gene *SUPPRESSOR OF FATTY ACID DESATURASE DEFICIENCY1* is required for glycerolipid metabolism and for the activation of systemic acquired resistance. Plant Cell 16:465–477

Návarová H, Bernsdorff F, Döring A-C, Zeier J (2012) Pipecolic acid, an endogenous mediator of defense amplification and priming, is a critical regulator of inducible plant immunity. Plant Cell 24:5123–5141.

Pajerowska-Mukhtar KM, Wang W, Tada Y, Oka N, Tucker CL, Fonesca JP, Dong X (2012) The HSF-like transcription factor TBF1 is a major molecular switch for plant growth-to-defense transition. Curr Biol 22:103–112

Pallas JA, Paiva NL, Lamb C, Dixon RA (1996) Tobacco plants epigenetically suppressed in phenylalanine ammonia-lyase expression do not develop systemic acquired resistance in response to infection by tobacco mosaic virus. Plant J 10:281–293

Park S-W, Kaimoyo E, Kumar D, Mosher S, Klessig DF (2007) Methyl salicylate is a critical mobile signal for plant systemic acquired resistance. Science 318:113–116

Park S-W, Liu P-P, Forouhar F, Vlot AC, Tong L, Tietjen K, Klessig DF (2009) Use of a synthetic salicylic acid analog to investigate the roles of methyl salicylate and its esterases in plant disease resistance. J Biol Chem 284:7307–7317

Pick T, Jaskiewicz M, Peterhänsel C, Conrath U (2012) Heat Shock Factor HsfB1 primes gene transcription and systemic acquired resistance in Arabidopsis. Plant Physiol 159:52–55

Ross AF (1966) Systemic effects of local lesion formation. In: Beemster ABR, Dijkstra J (eds) Viruses of plants. North-Holland, Amsterdam, pp 127–150

Shah J (2009) Plants under attack: systemic signals in defence. Curr Opin Plant Biol 12:459–464

Shah J, Zeier J (2013) Long-distance communication and signal amplification in systemic acquired resistance. Frontiers Plant Sci. 4:30

Shoresh M, Harman GE, Mastouri F (2010) Induced systemic resistance and plant responses to fungal biocontrol agents. Annu Rev Phytopathol 48:21–43

Song J, Lu H, Greenberg J (2004a) Divergent roles in *Arabidopsis thaliana* development and defense of two homologous genes, *ABERRANT GROWTH AND DEATH2* and *AGD2-LIKE DEFENSE RESPONSE PROTEIN1*, encoding novel aminotransferases. Plant Cell 16:353–366

Song J, Lu H, McDowell JM, Greenberg J (2004b) A key role for *ALD1* in activation of local and systemic defenses in Arabidopsis. Plant J 40:200–212

Spoel SH, Dong X (2012) How do plants achieve immunity? Defence without specialized immune cells. Nat Rev Immunol 12:89–100

Spoel SH, Mou Z, Tada Y, Spivey NW, Genschik P, Dong X (2009) Proteasome-mediated turnover of the transcription coactivator NPR1 plays dual role in regulating plant immunity. Cell 317:860–872

Sticher L, Mauch-Mani B, Métraux JP (1997) Systemic acquired resistance. Annu Rev Phtopathol 35:235–270

Tahiri-Alaoui A, Dumas-Gaudot E, Gianinazzi S, Antoniw JF (1993) Expression of the PR-b1 gene in roots of two *Nicotiana* species and their amphidiploid hybrid infected with virulent and avirulent races of *Chalara elegans*. Plant Pathol 42:728–736

Tholl D (2006) Terpene synthases and the regulation, diversity and biological roles of terpene metabolism. Curr Opin Plant Biol 9:297–304

Trapp S, Croteau R (2001) Defensive resin biosynthesis in conifers. Annu Rev Plant Physiol Plant Mol Biol 52:689–724

Traw MB, Kniskern JM, Bergelson J (2007) SAR increases fitness of Arabidopsis thaliana in the presence of natural bacterial pathogens. Evolution 61:2444–2449

Truman W, Bennett MH, Kubigsteltig I, Turnbull C, Grant M (2007) Arabidopsis systemic immunity uses conserved defense signaling pathways and is mediated by jasmonates. Proc Natl Acad Sci USA 104:1075–1080

Tuzun S, Kuc J (1985) Movement of a factor in tobacco infected with *Peranospora tabacina* Adam which systemically protects against blue mold. Physiol Plant Pathol 26:321–330

Uknes S, Mauch-Mani B, Moyer M, Potter S, Williams S, Dincher S, Chandler D, Slusarenko A, Ward E, Ryals J (1992) Acquired resistance in Arabidopsis. Plant Cell 4:645–656

van den Burg HA, Takken FLW (2009) Does chromatin remodeling mark systemic acquired resistance? Trends Plant Sci 14:286–294

van Loon LC (2007) Plant responses to plant growth-promoting rhizobacteria. Eur J Plant Pathol 119:243–354 van Wees SCM, de Swart EAM, van Pelt JA, van Loon LC, Pieterse CMJ (2000) Enhancement of induced disease resistance by simultaneous activation of salicylate- and jasmonate-dependent defense pathways in *Arabidopsis thaliana*. Proc Natl Acad Sci USA 97:8711–8716

Vernooij B, Friedrich L, Morse A, Reist R, Kolditz-Jawhar R, Ward E, Uknes S, Kessmann H, Ryals J (1994) Salicylic acid is not the translocated signal responsible for inducing systemic acquire. Plant Cell 6:959–965

Vlot AC, Liu P-P, Cameron RK, Park S-W, Yang Y, Kumar D, Zhou F, Padukkavidana T, Gustafsson C, Pichersky E, Klessig DF (2008) Identification of likely orthologs of tobacco salicylic acid-binding protein 2 and their role in systemic acquired resistance in *Arabidopsis thaliana*. Plant J 56:445–456

Wu Y, Zhang D, Chu JY, Boyle P, Wang Y, Brindle ID, De Luca V, Després C (2012) The Arabidopsis NPR1 protein is a receptor for the plant defense hormone salicylic acid. Cell Rep 1:639–647

Xia Y, Gao Q-M, Yu K, Navarre D, Hildebrand D, Kachroo A, Kachroo P (2009) An intact cuticle in distal tissues is essential for the induction of systemic acquired resistance in plants. Cell Host Microbe 5:151–165

Xia Y, Yu K, Navarre D, Seebold K, Kachroo A, Kachroo P (2010) The *glabra1* mutation affects cuticle formation and plant responses to microbes. Plant Physiol 154:833–846

Xia Y, Yu K, Gao Q-M, Wilson EV, Navarre D, Kachroo P, Kachroo A (2012) Acyl CoA binding proteins are required for cuticle formation and plant responses to microbes. Front Plant Sci 3:224. doi:10.3389/fpls.2012.00224

Yalpani N, Silverman P, Wilson TMA, Kleier DA, Raskin I (1991) Salicylic acid is a systemic signal and an inducer of pathogenesis-related proteins in virus-infected tobacco. Plant Cell 3:809–818

Zeier J, Pink B, Müeller MJ, Berger S (2004) Light conditions influence specific defence responses in incompatible plant-pathogen interactions uncoupling systemic resistance from salicylic acid and PR-1 accumulation. Planta 219:673–683

Zoeller M, Stingl N, Krischke M, Fekete A, Waller F, Berger S, Mueller MJ (2012) Lipid profiling of the Arabidopsis hypersensitive response reveals specific lipid peroxidation and fragmentation processes: biogenesis of pimelic and azelaic acid. Plant Physiol 160:365–378

Salicylic Acid-Induced Local and Long-Distance Signaling Models in Plants

Tomonori Kawano and François Bouteau

Abstract Salicylic acid (SA) is one of the key hormonal factors determining the fate of plants exposed to stressful conditions, which is naturally found in plants and shown to be involved in the plant defense-related actions against infection by various pathogens. Recently, intracellular SA receptors were finally identified after a long survey of SA-binding proteins. In this chapter, the modes of both the short- and long-distance signaling events involving the actions of SA, a defense-related key signaling molecule, are compared by covering both the biochemical and electrophysiological views. Here, two distinct models for local SA perception and signaling mechanisms involved in the extracellular and intracellular paths (referred to as models i and ii), and the three different models for long-distance signaling mediated by SA are reviewed (referred to as models iii–v). The local SA signaling events can be attributed to (i) the extracellular SA perception model in which reactions between SA and apoplastic proteins result in acute oxidative burst

T. Kawano (✉)
Faculty and Graduate School of Environmental Engineering, The University of Kitakyushu, Kitakyushu, Japan

University of Florence LINV Kitakyushu Research Center (LINV@Kitakyushu), Kitakyushu, Japan

Université Paris Diderot, Sorbonne Paris Cité, Paris 7 Interdisciplinary Energy Research Institute, 75103 Paris, France

International Plant Neurobiology Laboratory, University of Florence, Sesto Fiorentino, Italy
e-mail: kawanotom@kitakyu-u.ac.jp

F. Bouteau
University of Florence LINV Kitakyushu Research Center (LINV@Kitakyushu), Kitakyushu, Japan

Institut de Biologie des Plantes, Bât 630 91405, Orsay cedex, France

International Plant Neurobiology Laboratory, University of Florence, Sesto Fiorentino, Italy

Institut des Energies de Demain (FRE3597), Université Paris Diderot, Sorbonne Paris Cité, 75103 Paris, France

F. Baluška (ed.), *Long-Distance Systemic Signaling and Communication in Plants*, Signaling and Communication in Plants 19, DOI 10.1007/978-3-642-36470-9_2,

followed by internalization of the derived signals via activation of calcium channels, and/or (ii) intracellular SA perception mechanism by which the action and life cycle of NPR1 protein are determined depending on the concentration of SA in both the infected cells and neighboring cells. On the other hand, the long-distance SA action could be attributed to three different modes, namely, (iii) local increase in SA followed by phloem transport of SA, (iv) systemic propagation of SA-derived mobile signals with both electrical and chemical natures without direct movement of SA, and (v) synergistic propagation of both SA and derived signals through the tissues and phloem. We view here that the long-distance SA signaling events (models iii–v) inevitably involve the mechanisms described in the local signaling models (models i and ii) as the key pieces of the puzzle.

Keywords Long-distance signaling • Phloem-mobile signal • Salicylic acid • Signal transduction

1 Introduction

Recent studies have elucidated that plant responses to different stresses are highly complex and involve the changes at physiological, cellular, and transcriptome levels (Atkinson and Urwin 2012). For finely geared controls of plant behaviors leading to homeostasis, plants are also equipped with stress-responsive signaling mechanisms as such mediated by hormonal regulations. Salicylic acid (SA) is one of the key phytohormones involved in both the abiotic (Kunihiro et al. 2011; Liu et al. 2012; Drzewiecka et al. 2012) and biotic (Vlot et al. 2008, 2009; Dempsey et al. 2011) stress adaptation.

For regulation of the growth and development within entire plant bodies, the long-distance signal translocation machineries are inevitable components because some of the local information or locally sensed input data such as wounding, viral infection, changes in nutritional condition, or water potential sensed by root hairs or stomata must be signaled to entire plant bodies to adapt to environments (Furuichi et al. 2007). Thus, locally targeted stimuli are rapidly converted to intracellular signals and then the re-generated extracellular signals by the stimulated cells are transferred to the parts or organs distant from the site of stimulus perception. One of the paths for long-distance signaling in plants is the highly systematized vascular bundle system which basically plays central roles in the absorption and translocation of water, minerals, and other nutrients to systemically support and maintain the growth of tissues and cells (Furuichi and Kawano 2006).

In this chapter, the modes of both the short- and long-distance signaling events involving the actions of SA, a defense-related key signaling molecule, are compared by covering both the biochemical and electrophysiological views. In the below sections, two distinct models for local SA perception and signaling mechanisms involved in the extracellular and intracellular paths (models i and ii), and the three different models for long-distance signaling mediated by SA are reviewed (models iii–v).

The local SA signaling events can be attributed to (i) the extracellular SA action model and/or (ii) intracellular SA perception model. On the other hand, the long-distance SA action could be attributed to three different modes, namely, (iii) local increase in SA followed by transportation of SA, (iv) systemic propagation of SA-derived mobile signals with both electrical and chemical natures without direct movement of SA, and (v) synergistic propagation of both SA and derived signals through the tissues and phloem. This includes the alternately repeated secondary signal propagation and production and/or release of SA finally contributing to the systemic spread of SA-derived signals.

Although five different models (i–v) are reviewed here, some models can be considered as a part of one model. For instance, we view here that the long-distance SA signaling events inevitably involve the mechanisms or modes of SA actions described in the local signaling models as the key pieces of the puzzle.

2 Early Defense Signaling Models and Salicyl Acid

2.1 SA and Systemic Acquired Resistance

In the environments, living plants must respond to and combat a variety of stressful stimuli with biotic and abiotic natures, which often threaten the life of plants. Biotic factors menacing the plants include animal and insect herbivores (Barbehenn et al. 2010) and a wide range of pathogenic microbes such as viruses, bacteria, and fungi (Kerchev et al. 2012; Kangasjärvi et al. 2012).

SA is one of the key hormonal factors determining the fate of plants exposed to stressful conditions, which is naturally found in plants and shown to be involved in the plant defense-related actions against infection by various pathogens (Vlot et al. 2009). The name of SA and related compounds originally came from the *Salix helix* (willow), since they were discovered as the major components in the extracts from the tree barks of willow and also from poplar, which had been used as natural anti-inflammatory drugs over centuries until the eighteenth century (Rainsford 1984; Weissman 1991).

The first study focusing on the role of salicylates as disease resistance-inducing chemicals (White 1979; Antoniw and White 1980) revealed that treatment of the leaves of tobacco (*Nicotiana tabacum* L.) with aspirin (acetylsalicylic acid) drastically enhances the resistance to subsequent infection by tobacco mosaic virus (TMV). Later studies demonstrated that aspirin and SA induce systemic acquired resistance (SAR) represented by systemic accumulation of pathogenesis-related (PR) proteins in plants through activation of corresponding cellular signaling mechanisms (Malamy et al. 1990; Métraux et al. 1990; Kessmann and Ryals 1993; Fu et al. 2012).

2.2 *Oxidative Signaling Events Leading to Abiotic and Biotic Stress Adaptation*

Reactive oxygen species (ROS) are in fact inevitably produced as by-products from normal metabolic reactions including mitochondrial respiration, photosynthetic processes, and fatty acid metabolism (Møller 2001; Baker et al. 2006; Noctor et al. 2007). A common property of all ROS types is that they can cause oxidative damage to cellular components such as proteins, DNA, and membranes (Møller et al. 2007). The specificity of the biological response of living plant cells to ROS depends on the chemical identity of ROS, intensity of the signal, sites of production, and developmental stages (Del Río et al. 2002). Therefore, apart from their harmful action, generation of ROS could be potentially beneficial to living organisms depending on the conditions (Apel and Hirt 2004).

Induced production of ROS such as superoxide anion radicals ($O_2^{\bullet -}$), hydrogen peroxide (H_2O_2), and hydroxyl radicals ($HO^{\bullet}$) at the cell surface or apoplastic space well known as the "oxidative burst" is one of the earliest events detectable during the incompatible interactions between plants and pathogens (Yoshioka et al. 2008). Three decades ago, Doke, a plant pathologist at Nagoya University (Nagoya, Japan), first reported on the involvement of ROS in the plant–pathogen interaction, after observing that infection by *Phytophthora infestans* in potato tubers causes the generation of $O_2^{\bullet -}$ at the host cells' plasma membrane (PM), only in the incompatible interactions (Doke 1983a). *P. infestans* is a typical pseudofungal species which is now classified within the class of Oomycetes (Subclass, Peronosporomycetidae; Order, Pythiales) that causes the serious potato disease known as late blight or potato blight.

A series of Doke's works (Doke 1983a, b, 1985) demonstrated for the first time that ROS generation occurs in plants upon attacks by a pathogenic microorganism and that the members of ROS possibly function as the chemical signals required for induction of hypersensitive response (HR) as typified by host cell death, now often referred to as plant apoptosis (Coll et al. 2011; De Pinto et al. 2012). Doke also demonstrated that the treatment of potato tuber protoplasts with the cell wall preparation from *P. infestans* effectively induces the oxidative burst, suggesting that chemical components derived from pathogenic microorganisms (elicitors) trigger the burst of ROS production in order to stimulate the plant defense mechanisms (Doke 1983b).

Nowadays, a number of active teams working on plant ROS biology are distributed worldwide and their studies concern numerous aspects of the plant physiology throughout the plants' life cycle (Yoshioka et al. 2008). ROS production is actually recognized as a common denominator to not only biotic stresses such as plant–microbe interactions and plant–herbivore interactions, but also abiotic environmental stressful conditions such as exposure to soil metals, high salinity, drought, high intensity light, and low or high temperature stresses that cause major crop losses worldwide (Kawano et al. 2001, 2003; Yamamoto et al. 2003; Mittler et al. 2011; Swanson et al. 2011; Yokawa et al. 2011).

Exposures to both biotic and environmental abiotic stresses reportedly increase the intra- and intercellular levels of H_2O_2 by modulating the finely elaborated ROS-detoxification and regeneration networks, composed of ROS-producing enzymes, antioxidant enzymes, and biosynthetic pathways for low molecular antioxidants, all responsible for maintaining the homeostasis of ROS levels under tight control (Bolwell et al. 2002; Del Río et al. 2002; Kawano 2003; Kotchoni and Gachomo 2006; Yoshioka et al. 2008). This allows ROS to serve as signaling molecules in regulation of plant metabolism and cellular signal transduction pathways activated in response to environmental stresses (Gechev et al. 2006; Mittler et al. 2011).

Accumulated pieces of evidence suggested that hormonal signaling pathways leading to development of SAR are regulated under controlled ROS production as observed for SA, abscisic acid, jasmonic acid (JA), and ethylene (Gaupels et al. 2011). Such ROS-mediated hormonal regulations might play some crucial roles in the cross talk between biotic and abiotic stress signaling (Kawano 2003; Ströher and Dietz 2006; Mori and Schroeder 2004). Although many components of oxidative signaling networks have recently been identified, the mechanisms for orchestrated control of the diversified ROS production mechanisms at different cellular sites through fine-tuning of ROS feedback control to meet the physiological requirements such as plant growth, development, stress adaptation, and programmed cell death (PCD) are now actively studied (Coll et al. 2011).

2.3 *ROS-Triggered Calcium Signaling Events*

As discussed in our previous review (Kawano and Furuichi 2007), early studies have indicated that SA is an oxidative signal inducer which is essentially involved in the development of SAR against various pathogens with various natures. In the early 1990s, it was proposed that SA signal transduction leading to SAR is mediated by ROS derived from H_2O_2, since SA binds and inhibits H_2O_2-detoxifying enzymes, catalase (Chen et al. 1993a; Durner and Klessig 1996) and ascorbate peroxidase (Durner and Klessig 1995). While the proposed enzyme inhibition models suggested the involvement of passive mechanisms supporting the increases in ROS, an active mechanism involving extracellular peroxidase and NADPH oxidase that directly generates ROS in the presence of SA was reported in the late 1990s as discussed below.

In addition to ROS, the changes in cytosolic free calcium ion concentration ($[Ca^{2+}]_c$) are another key factor of SA-mediated signaling, and certain number of reports indicated that an increase in $[Ca^{2+}]_c$ is essential for the action of SA during plant defense, since $[Ca^{2+}]_c$ plays roles as a secondary messenger for certain processes in plant defense mechanisms (Knight et al. 1991; Sanders et al. 1999). First data suggesting the involvement of Ca^{2+} during the action of SA was obtained after removal of Ca^{2+}. Inhibition of SA-dependent induction and accumulation of chitinase by depletion or chelation of free Ca^{2+} was observed in tobacco cells and leaves (Raz and Fluhr 1992), and carrot cell suspension culture (Schneider-Müller et al. 1994).

Direct evidence for the actions of SA leading to rapid generation of ROS (especially $O_2^{\bullet -}$) and increase in $[Ca^{2+}]_c$ was obtained through experiments using $O_2^{\bullet -}$-specific chemiluminescent probe-treated and aequorin-expressing tobacco BY-2 cells (Kawano et al. 1998). Treatment of tobacco BY-2 cells with sub-mM order of SA resulted in rapid and transient generation of $O_2^{\bullet -}$, and in turn, $O_2^{\bullet -}$ triggered the influx of Ca^{2+} into the cells by stimulating the opening of ROS-responsive calcium channels. Further works have revealed that the SA-induced extracellular oxidative burst (generation of $O_2^{\bullet -}$) is catalyzed by apoplastic free and cell wall-bound peroxidases (Kawano et al. 1998; Kawano and Muto 2000). Peroxidase activity is often enhanced upon challenges by microbes and insects and induced enzyme activity contributes to production of semiquinone free radicals and quinones through oxidation of phenolics (Barbehenn et al. 2010). Thus, SA could be one of such active peroxidase substrates leading to the production of radical species, chiefly SA radicals, which are further converted to stable compounds while yielding ROS members as by-products (Kawano et al. 1998; Kawano and Muto 2000; Gozzo 2003). Interestingly, induced peroxidase activity in plants further contributes to protection of plants from insects by damaging the plant feeding caterpillars partly via post-ingestive phenol-derived radical production mechanisms in the midguts of larvae (Barbehenn et al. 2010).

2.4 SA Receptors Identified

Klessig and his colleagues have conducted a series of pioneering works on SA-binding proteins (SABPs), putative SA receptors. The first SABP isolated from tobacco was shown to be catalase (Chen et al. 1993a, b; Conrath et al. 1995). As discussed above, an idea came out from these works suggesting that the increases in H_2O_2 and/or other ROS derived from H_2O_2 may be the key events acting downstream of SA, since the binding of SA to purified catalase resulted in the inhibition of spectroscopically monitored decay of H_2O_2. Similarly, Kawano et al. (1998) have observed the SA-dependent inhibition of catalase in suspension cultured tobacco cells in vivo by monitoring the H_2O_2-dependent evolution of oxygen. However, contribution of SA binding to catalase in SA biology still remains uncertain. Similar model for SA binding to ascorbate peroxidase has been proposed, but this could be excluded from the candidate for SA-signaling mediators. Concerning the involvement of ascorbate peroxidase, both positive (Durner and Klessig 1995) and negative (Miyake et al. 1996; Kvaratskhelia et al. 1997; Tenhaken and Rübel 1997; Kawano et al. 1998, 2004a) views have been presented.

From tobacco, SABP2 was isolated as a putative SA receptor (Du and Klessig 1997) which is highly required for TMV-induced SAR development (Kumar and Klessig 2003), and finally its role as a SA-stimulated lipase (Kumar and Klessig 2003) or a methyl salicylate (SAMe) esterase that demethylates SAMe to produce free SA has been determined (Forouhar et al. 2005). SABP3 isolated from tobacco was identified as a chloroplast-targeted carbonic anhydrase that shows antioxidative activity when expressed in yeast (Slaymaker et al. 2002). Since possible role of chloroplast as the site of SA biosynthesis is highlighted through the study of *Arabidopsis thaliana* sid2 mutant lacking the chloroplast-localized ICS enzyme (Wildermuth et al. 2001), we can expect that a SABP specifically localized in chloroplasts may play some key roles. As Shah (2003) has predicted in his review, chloroplasts and plastids might be the source of signals affecting the response to pathogens, by analogy to the mitochondrial roles in mammals. As expected, recent study has provided the molecular link between chloroplasts and the cytoplasmic-nuclear immune system. Nomura et al. (2012) have shown that (a) pathogen-associated molecular pattern (PAMP) signals are quickly relayed to chloroplasts and evoke specific Ca^{2+} signatures in the stroma, (b) a chloroplast-localized calcium-sensing receptor designated as CAS is involved in stromal Ca^{2+} transients and responsible for both the basal resistance and R gene-mediated hypersensitive cell death induced by PAMP, (c) CAS acts upstream of SA accumulation, and (d) CAS is involved in PAMP-induced expression of defense genes and suppression of chloroplast gene expression possibly via singlet oxygen ($^1O_2$)-mediated path.

Apart from SABPs, NONEXPRESSOR OF PATHOGENESIS-RELATED PROTEINS1 (NPR1) was identified as a key molecule involved in SA signaling. Importantly, SA actually regulates the conversion of NPR1 from an oligomeric form to a monomeric form that migrates from cytosol into nucleus (Mou et al. 2003) and phosphorylation of NPR1 for facilitating NPR1's recruitment to a Cullin3 (CUL3) E3 ligase, subsequently leading to proteasome-mediated degradation of NPR1 protein (Spoel et al. 2003). Recently, Klessig's group (Moreau et al. 2012) has stated that "despite identification of the aforementioned SABPs, SA's signaling mechanisms remain unclear" and continued that "considering SA's many roles in plants, these SABPs may constitute only a small portion of SA's targets; moreover, the SA receptor remained to be found" and introduced the model for NPR1-dependent defense responses involving novel SABPs (NPR3 and 4) recently proposed by Xinnian Dong's team. Thus, SA receptors acting in the intracellular space were finally identified after a long journey.

3 Apoplastic and Intracellular SA Perception Models

3.1 Intracellular SA Perception

Recent demonstration by Xinnian Dong and her colleagues (Fu et al. 2012) represents a major step forward in our understanding of SA signaling mechanisms during plant defense responses. Although NPR1 has been long known as a key player in most of the SA-mediated defense signal transduction, it does not appear to be an SA receptor since SA does not directly bind to NPR1. The recent demonstration by Dong's team can be summarized that the NPR1 paralogues, namely, NPR3 and NPR4, are adaptor proteins for the CUL3 E3 ligase that specifically target NPR1 for degradation in an SA concentration-dependent manner (Fu et al. 2012). Since the npr3/4 single and double mutants reportedly show elevated levels of NPR1, it is viewed that NPR3 and NPR4 directly interact with and determine the fate of NPR1. The newly proposed model views NPR3 and NPR4 as SA receptors that differ in their affinity to SA. The newly proposed roles for SA are to disrupt the NPR1–NPR4 interaction, leaving NPR1 free from the NPR4-mediated degradation in the presence of low concentrations of SA (nanomolar range), and to promote the NPR1–NPR3 interaction, allowing NPR1 degradation when SA is excess (over micromolar range). Thus the signaling action of NPR1 is allowed only in the moderate range of SA concentration. This model explains the relationship between the homeostasis of NPR1 level and SA action at different concentrations. The above mechanism corresponds to the model (i) for intracellular SA perception and induced signaling.

3.2 Apoplastic SA Perception and Signaling Events

As discussed in the earlier sections, the first report connecting ROS and Ca^{2+} in plant SA signaling has reported the direct measurements of the SA-induced $O_2^{\bullet-}$ and the SA-induced increase in $[Ca^{2+}]_c$ in aequorin-expressing tobacco BY-2 cells (Kawano et al. 1998). Addition of SA to tobacco BY-2 cells reportedly resulted in rapid generation of H_2O_2 (Kawano and Muto 2000) and $O_2^{\bullet-}$ (Kawano et al. 1998), and a transient increase in $[Ca^{2+}]_c$ (Kawano et al. 1998), in this order. In the model, ROS actively triggers the influx of Ca^{2+} into the cells, and this early oxidative burst induced by SA was shown to be an extracellular event involving the action of extracellular free and cell wall-bound peroxidases (Kawano 2003).

Action of SA mediated by both the cell wall peroxidase-dependent ROS production and Ca^{2+} influx was also observed in *Vicia faba* epidermis (Mori et al. 2001) and the cell suspension-cultured *A. thaliana* (Kadono et al. 2010). Both SA-induced $O_2^{\bullet-}$ and chemically generated $O_2^{\bullet-}$ were shown to induce the closure of stomata, which is known as a Ca^{2+}-dependently regulated event studied in *Commelina communis* L. (Lee 1998), *V. faba* L. (Manthe et al. 1992; Mori et al. 2001), and *A. thaliana* (Khokon et al. 2011). The SA-induced stomatal closure in *V. faba* and Arabidopsis was reportedly inhibited by ROS scavengers and inhibitors of peroxidase such as salicylhydroxamic acid, suggesting the involvement of peroxidise-mediated redox signaling during the action of SA leading to stomatal closure (Mori et al. 2001; Khokon et al. 2011).

According to Chen et al. (2002), induction of a PR-protein (protein N) is differently regulated by the two distinct signaling mechanisms corresponding to high- and low-dose of exogenously supplied SA. The process which is dependent on the higher concentration of SA (ca. 200 μM) reportedly relays the SA signal to induce protein N through ROS production, Ca^{2+} signaling, and protein phosphorylation, while lower dose (20 μM) of SA induces protein N via alternative mode of signaling independent from ROS, Ca^{2+}, and protein phosphorylation events.

The above local SA perception mechanism involving apoplastic oxidative burst and calcium signaling corresponds to the model (ii) for extracellular SA perception and induced signaling which may form the earliest signaling events upon treatment of plant cells with SA. In addition to extracellularly localized peroxidases, PM-localized NADPH oxidases known as respiratory burst oxidase homologues (RBOHs) are likely to be involved in the apoplastic SA-dependent ROS production model (Yoshioka et al. 2001).

3.3 Cross talk Between the Extracellular and Intracellular SA Perception Mechanisms via SA Transport

In earlier publications, we made an overview on the possible control of local and systemic SA levels through movements of SA "in" and "out" of the cells, tissues, and organs (Kawano et al. 2004a; Kawano and Furuichi 2007). As SA is produced inside the cells, the first step in SA movement is secretion of SA by the cells. Secretion of SA mediates the cell–cell interaction among neighboring cells, through the extracellular SA perception mechanism and also via intracellular SA perception mechanism after internalization of SA. Also in case of experimental SA treatment, exogenously applied SA (extracellular SA) must be internalized in order to elicit the intracellular SA perception mechanism. This SA internalization process and SA excretion process are likely under the control of the extracellular SA perception mechanism, thus potentially impacting the homeostasis for intracellular SA concentration.

According to the works by a Taiwanese group, SA smoothly moves in and out of the plant cells and these processes are finely regulated by ROS and Ca^{2+} (Chen and Kuc 1999; Chen et al. 2001). They showed that ^{14}C-labeled free SA added to tobacco cells in suspension can be rapidly absorbed by the cells (within 5 min), and with time the majority (over 90 %) of the radioactivity can be released back to the extracellular medium as free SA (by 5 h). In this model, de novo induction of SA excretion process reportedly takes place when the cells are exposed to a relatively high dose of SA (ca. 200 μM). Interestingly, this process requires the production of ROS and subsequent cascades of Ca^{2+} signaling and protein phosphorylation (Chen et al. 2001), confirming the roles for ROS-triggered calcium signaling events following the extracellular SA perception as described by Kawano and his colleagues (Kawano et al. 1998; Kawano and Muto 2000; Mori et al. 2001), in the systemic spread of SA. A possible mechanism for excretion of SA at cellular level was proposed (Chen and Kuc 1999; Chen et al. 2001). This model partially explains the mode of long-distance SA spread through cell-to-cell SA translocation reported by Ohashi et al. (2004). In addition to the mechanism of SA transport

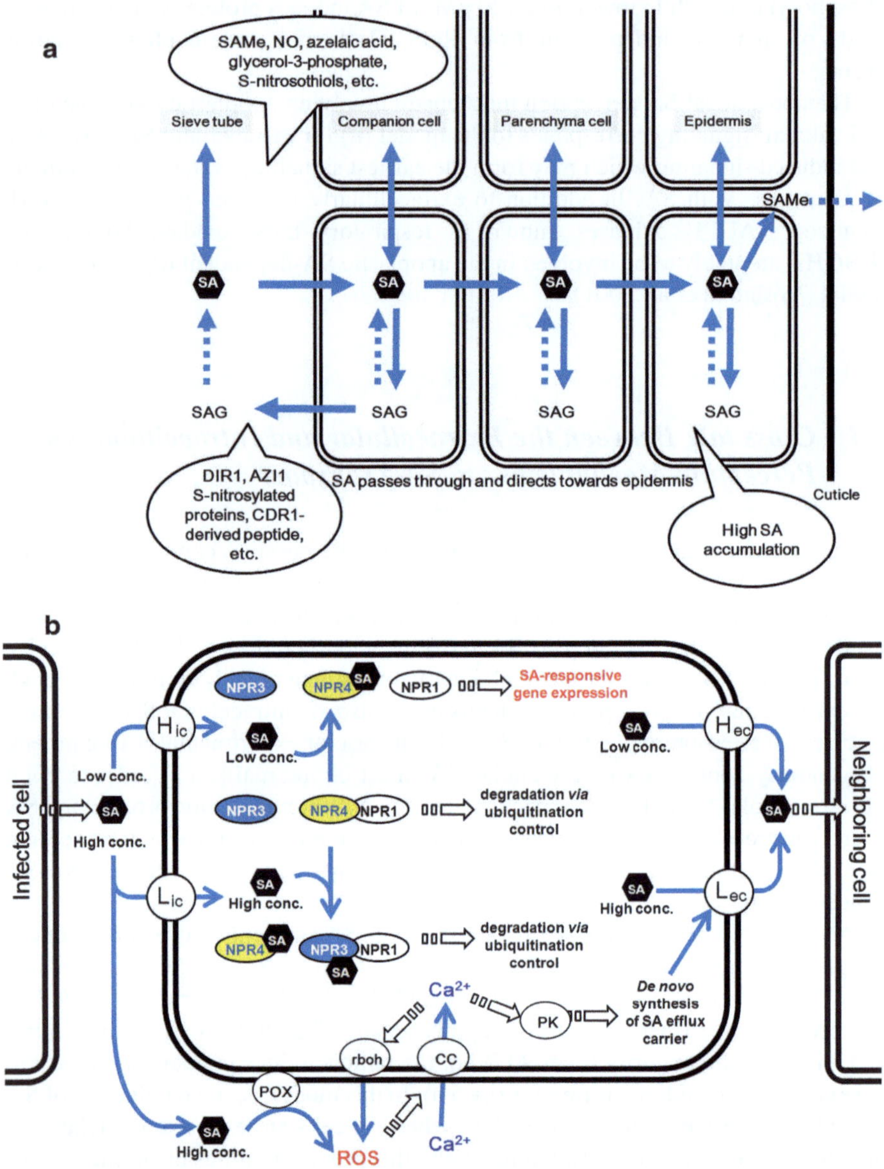

Fig. 1 Models for SA translocation at cellular and tissue levels. (**a**) Models for SA translocation in plants. Vertical movement of SA involving vascular transport and cell-to-cell transport is faster than horizontal movement of SA. In horizontal SA movement, cell-to-cell exchange of SA with preferential direction toward epidermis takes place. The cuticle layer is hardly permeable to SA under the physiological pH. *Arrows* represent the massive and rapid movement or interconversion of SA (to SAG and SAMe). *Broken arrows* indicate slower and minor paths of movement or interconversion. In addition to SA, SAG, and SAMe, putative mobile signals (both small molecules and proteins) contributing to development of SAR are listed. (**b**) SA concentration-dependent SA perception and SA

responsive to high dose of exogenous SA, there would be an alternative SA transport process responsive to low dose of exogenously applied SA (below 20 μM; Chen and Kuc 1999; Chen et al. 2001). The low-dose SA-responsive SA excretion process requires no de novo protein synthesis and is constitutively active independent of ROS, Ca^{2+}, and protein kinase. Thus, two distinct SA efflux carrier (s) constitutively present and newly produced in response to exogenous SA contribute to both the translocation and homeostasis of SA level in the cells (Kawano et al. 2004a). Figure 1 summarizes the modes of cell–cell and trans-phloem transports of SA and related chemical signals SA-dependently regulated in the pathogen-challenged plant tissue.

In *Arabidopsis* cells, a pH-dependent SA transport system has been reported (Clarke et al. 2005). Accordingly, SA is rapidly taken up by Arabidopsis cells, and SA uptake coincides with the alkalization of media and acidification of cytosol, and SA uptake was shown to be inhibited by the ionophore nigericin, suggesting that SA import is driven by a proton gradient. Importantly, against the ongoing $[H^+]$-dependent SA import, SA was shown to be exported back into the media as free SA after initial uptake of SA (Clarke et al. 2005). As mentioned earlier, the SA transport model through the cells reported in tobacco cells was also confirmed in Arabidopsis cells.

Since the above SA transport system is dependent on extracellular pH, homeostasis of extracellular pH could be the target of pathogen challenges combating against the systemic spread of defense signals chiefly SA, especially during incompatible plant–pathogen interactions. From this view point, Clarke et al. (2005) have investigated how SA transport may be modulated during incompatible defense responses by employing the bacterial harpin Pss as a model elicitor. As expected, harpin induced a rapid and sustained alkalization of the cell suspension media, reaching the critical pH (pH 5.9–6.1), and importantly, under this condition, SA import was shown to be inhibited for ca. 1 h.

The pH-dependent SA transport system was also found in mammalian cells such as human trophoblast cells, human choriocarcinoma cell lines, and hamster cheek pouch mucosa cells. The presence of carrier-mediated SA absorption mechanism has been elucidated by tracing the fate of ^{14}C-SA (Utoguchi et al. 1999; Emoto et al. 2002) and expression of SA-efflux transporter mRNA (Ikeda et al. 2012). In these animal models, SA influx carriers pH-dependently support the uptake of SA by cells

Fig. 1 (continued) translocation. Incorporation of low-dose and high-dose SA determines the action and fate of NPR1 through binding to NPR4 and/or NPR3. SA excretion mechanism responsive to high-dose SA reportedly requires the generation of ROS and subsequent Ca^{2+} and protein phosphorylation signaling cascades. The SA excretion mechanism responsible for low-dose SA is constitutively active independent of ROS, Ca^{2+}, and protein phosphorylation, requiring no de novo protein synthesis. L_{ic} and H_{ic}, low-affinity and high-affinity SA-influx carriers, respectively. L_{ec} and H_{ec}, low-affinity and high-affinity SA-efflux carriers, respectively. *CC* calcium channel, *PK* protein kinase, *POX* peroxidase. For details, see main text

and tissues both in vivo and in vitro, requiring low extracellular pH and higher intracellular pH, thus sensitive to protonophores and NaN_3. In contrast, a pH-independent SA uptake across the basolateral membrane of Malpighian tubules which is mediated by a non-electrogenic, α-cyano-4-hydroxycinnamic acid-sensitive, Na^+: salicylate co-transport system has been elucidated by tracing the fate of orally applied ^{14}C-SA in *Drosophila melanogaster* (Ruiz-Sanchez and O'Donnell 2006).

4 Models for Long-Distance Signaling

Generally, the long-distance signaling events in plants are possibly manifested through transport of active signaling molecules, the relays and exchanges of signals (or migration of secondary signals faster than the spread of signal-triggering molecules) without direct movement of the signal-triggering molecules, or combination of secondary signaling events followed by systemic production of the signal-triggering molecules.

In addition to two distinct local SA perception models (i) and (ii), we wish to propose that the long-distance SA action could be attributed to three different modes, namely, (iii) local increase in SA followed by transport of SA, (iv) systemic propagation of SA-derived signals with both electrical and chemical natures without direct movement of SA, and (v) alternately repeated secondary signal propagation and biosynthesis of SA and/or conversion of inert SA intermediates to free SA finally contributing to the systemic spread of SA-derived signals.

4.1 Long-Distance SA Transport

The induction of SAR following a localized infection must be mediated by some kinds of long-distance communication mediators moving through the phloem since earlier demonstrations showed that blocking of phloem transmission by stem girdling prevents the induction of SAR in leaves distal to the block (Delaney 2004). Later observations in tobacco and cucumber suggested that SA shows upward migration from the site of infection to the upper noninfected leaves through phloem (Métraux et al. 1990; Rasmussen et al. 1991; Yalpani et al. 1991).

In TMV-resistant Xanthi-nc tobacco, SA levels increase systemically after the single leaf inoculation with TMV. By tracing the $^{18}O_2$-labeled SA produced in TMV-infected leaves of Xanthi-nc tobacco, systemic spread of SA was observed (Shulaev et al. 1995). Similarly, Mölders et al. (1996) have shown that radioactivity of ^{14}C-labeled benzoic acid applied together with tobacco necrosis virus (TNV) to cotyledons of cucumber seedlings can be transported through phloem to upper leaves only after conversion to ^{14}C-SA. This study has concluded that SA rather than its precursor is translocated from the site of virus inoculation to the upper young leaves, finally leading to the development of SAR.

Ohashi's team has examined the possible mechanisms for SA translocation and interconversion between SA and SA derivatives such as salicylic acid β-glucoside (SAG) and SAMe at cellular level leading to systemic spread of SA in tobacco plants, and concluded that vertical movement of SA involving vascular transport and cell-to-cell transport is faster than horizontal movement of SA (Seo et al. 1995; Ohashi et al. 2004). In horizontal SA movement, cell-to-cell exchange of SA with preferential direction from phloem toward epidermis takes place. As the cuticle layer is hardly permeable to SA under the physiological pH, massive movement of SA in horizontal direction ends there unless converted to SAMe.

Niederl et al. (1998) has evaluated the role of cuticle as the barrier at the plant surface for preventing the passive penetration of SA. Their work has shown that ^{14}C-SA hardly penetrates through the specific path on the cuticle, which is utilized for water penetration, thus this path allows no or negligible level of SA transport at normal physiological range of pH (between 3 and 6). This was also confirmed by Ohashi et al. (2004). Notably, the suggested form of the SA derivative which is active in cuticular penetration was shown to be SAMe, suggesting that methylation of SA is one of the key steps allowing pH-independent diffusion of SA-related molecules across the outer physical barriers of plants (Niederl et al. 1998). Emission of this volatile derivative of SA may partially contribute to the long-distance SA signaling events.

Using ^{14}C-labeled SA, without inoculation with pathogen, Ohashi et al. (2004) observed that translocation of SA is unexpectedly rapid when artificially applied onto the cut end of petiole from young and adult tobacco plants. When the spread of ^{14}C signal was monitored after feeding of ^{14}C-SA to the petiole end of the adult plants with expanded leaves, the signal reached the six neighboring upper leaves and three adjacent lower leaves within 10 min, and accumulated throughout the plant body in further 50 min in each replicate. Data also suggested that the majority of SA migrate as free form rather than glucosylated form SAG, especially in the early phase of SA translocation (within 10 min after SA addition). Thus SAG is a storage form requiring conversion to free SA in order to be translocated and utilized at the site of SA action. As mentioned above, capacity for rapid translocation of free SA inside the plants is high enough to allow the systemic spread of SA within short period. The above phenomena may fulfill the model (iii) for long-distance SA transport following the local increase in SA.

Controversially, an experiment opposing to the view that SA is a vascular-mobile signal in induction of virus-induced SAR was also obtained (Vernooij et al. 1994). Grafting experiments with wild-type and transgenic tobacco plants showed that *NahG* root-stocks (lacking SA accumulation due to expression of a bacterial SA-degrading enzyme) inoculated with TMV were fully capable of delivering a signal allowing the non-transgenic parts to resist the secondary TMV infection, suggesting the presence of additional vascular-mobile SAR-inducing signal(s) which is not SA. Although the above experiment was elegantly designed, the results must be dealt with caution since *NahG* transgene merely contributes to the removal of SA without affecting the SA biosynthesis, and therefore, a small amount of residual SA can be found in *NahG* plants (Delaney 2004).

4.2 Long-Distance Communications by SA-Derived and SA-Induced Secondary Signals

Systemic signals are perceived in distant plant tissues and initiate systemic stress resistance through priming or induction of defense responses (Thompson et al. 2012). Recently, the knowledge on such long-distance signaling has been documented through the studies on SAR, systemic acquired acclimation (SAA), and systemic wound response (SWR). According to the above studies, phloem is the likely path for systemic transmission or movement of signals associated with SAR, SAA, and SWR. Similarly to SAR induction following the challenges by pathogens, abiotic environmental stresses can be the triggers for SA-centered signaling cascade finally leading to SAA in the challenged plants.

There are common views that both plasmodesmata and phloem are the likely paths for the spread of ROS during oxidative burst (Miller et al. 2009; Suzuki et al. 2011). However, the evidence for traveling of SA-induced ROS through phloem contributing to the long-distance signaling is not strong enough to conclude their roles, despite the key involvement of ROS in local SA action leading to SAR and SAA (Alvarez et al. 1998; Thompson et al. 2012).

In search for SA derivatives possibly acting as SA-derived signals, Pastora et al. (2012) have employed the precursor ion scan methods using an electrospray ionization tandem mass spectrometric technique (a multi-residue method with, liquid chromatography quadrupole time-of-flight mass spectrometry (LC-Q-TOF) and liquid chromatography multiple-reaction monitoring (LC-MRM)) and identified SAG and salicylic acid glucosyl ester (SGE), a new member of SA intermediates, as SA derivatives present in *Arabidopsis* cells challenged by virulent *Pseudomonas syringae* pv tomato DC3000.

To date, the candidates for the chemical signals systemically transported through phloem leading to induction of SAR, thus detectable in sieve elements or sieve-tube exudates, include small molecules such as SA (Yalpani et al. 1991; Shulaev et al. 1995; Mölders et al. 1996; Rocher et al. 2006, 2009), SAMe (Park et al. 2007; Rocher et al. 2009), azelaic acid and azelaic acid insensitive 1 (AZI1) protein (Jung et al. 2009), nitric oxide (NO; Song and Goodman 2001), and S-nitrosothiols (Rustérucci et al. 2007). Tran et al. (2012) have recently reported that oxalate, the simplest dicarboxylic acid, behaves as one of the signaling molecules in ozone (O_3)-exposed *Arabidopsis* cells; it is likely that the mode of oxalate action is similar to that of the action of azelaic acid. Among simple dicarboxylic acids sharing structural similarity as $(CH_2)n(CO_2H)_2$ with differed n (0–8), only oxalate ($n = 0$) and azelaic acid ($n = 7$) showed $O_2^{\bullet-}$-generating activity in cell suspension cultures of tobacco and rice (Kawano and Bouteau, unpublished results). Therefore, in addition to azelaic acid, we are focusing the possible role for oxalate as a mobile signal under oxidative stress in plants.

Macromolecules such as S-nitrosylated proteins (Rustérucci et al. 2007; Leitner et al. 2009), constitutive disease resistance 1 (CDR1)-derived peptide (Xia et al. 2004), and defective in induced resistance 1 (DIR1) protein accompanying the glycerolipid-derived compounds such as glycerol-3-phosphate are also listed as candidates for phloem mobile signals during SAR development (Maldonado et al. 2002; Nandi et al. 2004; Mitton et al. 2009; Chaturvedi et al. 2008, 2012; Chanda et al. 2011). These molecules are all related to the signaling actions or metabolism of SA.

Apart from the SA-dependent mechanism, another plant defense-related phytohormone often antagonizing to SA induces SAR, and JA was also found to be translocated through phloem (Li et al. 2002; Hause et al. 2003; Truman et al. 2007; Chaturvedi et al. 2008; Glauser et al. 2008; Koo et al. 2009; Robert-Seilaniantz et al. 2011). In addition to induction of SAR, JA also induces SWR in plants. Similarly, a polypeptide systemin also induces SWR by acting as a phloem-mobile signaling molecule (McGurl et al. 1992).

4.3 Possible Long-Distance Communications via RNA-Binding Proteins

As RNA-binding proteins (RBP) can control gene expression at both transcriptional and posttranscriptional levels, involvement of RBP in plants' response to pathogen infection with rapid reprogramming of gene expression would be a likely mechanism. Recently, a putative model for SA-induced immunity in Arabidopsis mediated by RBP has been proposed (Qi et al. 2010). Reportedly, *A. thaliana* RNA-binding protein-defense related 1 (AtRBP-DR1) is involved in resistance to *P. syringae* pv. *tomato* DC3000. Notably, susceptibility and resistance to the above-mentioned pathogen were enhanced in *AtRBP-DR1* loss-of-function mutants and *AtRBP-DR1*-overexpressed plants, respectively (Qi et al. 2010). AtRBP-DR1 could be a positive regulator of SA-mediated immunity, possibly acting on SA signaling-related genes at a posttranscriptional level since the free SA level was maintained at low in the *Atrbp-dr1* mutant and high in the *AtRBP-DR1*overexpression line, and *AtRBP-DR1* overexpression lines showed spontaneous cell death in mature leaves accompanied by higher mRNA levels of SA-inducible *SID2* and *PR1*. Furthermore, a putative RBP from Arabidopsis, glycine-rich RNA-binding protein7 (AtGRP7) has been shown to confer plant defense against *P. syringae* DC3000 and other diverse pathogens such as *Pectobacterium carotovorum* SCC1 and TMV (Lee et al. 2012).

Although, to date, the involvement of RBP transport in long-distance SA signaling is obscure, it is tempting to expect that long-distance RBP transport takes place in the systemic plant defense mechanism, by analogy to the models reported for the macromolecular transport through phloem playing pivotal roles in viral protein propagation and endogenous plant-transport-mediated growth and development (Ruiz-Medrano et al. 2012). Delivery of RNA to distant tissues might reflect a mechanism used by plants to regulate developmental and defense processes (Jorgensen et al. 1998; Lucas et al. 2001). Data presented by Aoki et al. (2005) introducing phloem RBP from pumpkin into rice plants indicated that, in addition to passive shoot-ward bulk flow transport, a destination-selective process is involved in long-distance root-ward movement of proteins through protein–protein interaction in the phloem sap. Recent study also suggested a regulatory mode of the phloem transport of RBP that involves the protein phosphorylation events in order to form a stable phloem-mobile complex (Li et al. 2011).

4.4 Long-Distance Electric Signaling

In animal systems, chiefly in species with developed brains, studies have revealed that the environmental stresses are recognized and transmitted via neurons. At the sensory cells of neurons, external signals directly or indirectly activate a number of ion channels, thus allowing the flux of ions across PM. These events effectively result in rapid changes in membrane potential, and these electrical responses, termed action potentials, are utilized to convey the information to the brain.

Although plants possess no neurons morphologically comparable to that of the animal system, the generation of action potentials in plant materials has been well documented (Meimoun et al. 2009; etc.). Dating back to the late sixteenth century, the most pioneering researcher in the field of electrophysiology, Luigi Galvani (1737–1798) provided the first evidence for plant electric signaling (Galvani 1791). Alexander von Humboldt (1769–1859), a German natural scientist, concluded that both animals and plants possess the bioelectrical feature, and suggested that the excitability of plant cells could be involved in long-distance signal translocation in plants (Botting 1973). Today, we can name two distinct types of electrical signals in plants, viz. the rapid action potentials and slow variation potentials, to be initiated and propagated through the activation of mechanosensitive and voltage-dependent channels which permeate cations or anions (Furuichi and Kawano 2006).

As demonstrated by the group at the University of East Anglia, the most successful model for systemic electrical signaling in plants is a wound response, in which the electrical activity spreads from the site of wounding (e.g., chewing by insect at cotyledons of tomato seedlings) to the whole plants, finally leading to systemic expression of a series of defense-related proteins, proteinase inhibitors, that inhibit the digestion in insects (Wildon et al. 1989, 1992; Rhodes et al. 1996). According to the above works, the expression of proteinase inhibitors in the intact tissues distal from the directly wounded tissue was not inhibited when chemical translocation was inhibited by chilling of the petiole of wounded leaf, suggesting that electrical potentials, caused by a transient activation of ion channels, are one of the main components for wound-inducible long-distance signaling.

As mentioned earlier, the electrical response in a single cell can be considered as the triggering event for long-distance propagation of electric signals; we would like to overview if SA can cause such electrical response. The following section describes the cases of SA involvement possibly stimulating the electrophysiological signaling path in abiotic stress-challenged plant cells.

5 Electrophysiological Actions of SA: Cases of Anion Channel and Calcium Channel Activations by Abiotic and Biotic Stresses

5.1 Plant Responses to Air Pollution

Ozone (O_3) produced by a complex series of photochemical reactions from primary precursors emitted as nitrogen oxides and volatile organic compounds is a major secondary air pollutant often reaching high concentrations in urban areas under strong daylight, and studies are now suggesting that a steep increase in global background concentrations of O_3 is in progress and thus the impact of atmospheric O_3 on plants including valuable crops might be severer in the future world (Ashmore 2005). Several studies have shown that exposure to O_3 elicits the production of ROS as key mediators of stress response in growing plants. The most widely accepted model describing the nature of O_3 toxicity/tolerance is the oxidative stress model in which generation of ROS and release of oxidation products are involved in the generation and propagation of toxic compounds throughout the plants (Fiscus et al. 2005).

Chronic expositions to low concentrations of O_3 reportedly show a negative impact on crop yields by reducing photosynthesis and growth, and moreover, inducing senescence in premature leaf of sensitive plants (Pell et al. 1997). On the other hand, acute and transient exposures to O_3 induce the development of O_3 lesions on the leaves (Kangasjärvi et al. 1994). The lesions induced by O_3 highly resemble PCD that takes place in HR in plant–pathogen interactions (Kangasjärvi et al. 2005; Overmyer et al. 2005; Pasqualini et al. 2003). Such localized cell death is a common feature of O_3 phytotoxicity and is generally thought to be initiated by strong oxidizing action of O_3 itself as well as by O_3-derived ROS intermediates (Schraudner et al. 1998).

By focusing on the induction of PCD-like cell death, Kadono et al. (2006) have examined the development of O_3-induced cell death in two suspension-cultured cell lines of tobacco derived from Bel-W3 (hypersensitive to O_3) and Bel-B (highly tolerant to O_3) and observed the difference in sensitivity to O_3 as observed in their original plants. As high production of $^1O_2$ and H_2O_2 in the O_3-sensitive Bel-W3 cells, but not in the O_3-tolerant Bel-B cells, was observed upon exposure to O_3

(Kadono et al. 2006) and ROS scavengers and chelators of Fenton reagents effectively rescued the cells from the PCD induction by O_3, involvements of $^1O_2$, H_2O_2, $HO^•$, and redox-active metals such as Fe^{2+} in O_3-induced acute damages to the cells have been suggested.

5.2 *SA Is One of the Hubs Mediating the Responses to Abiotic Stresses*

Upon exposure to O_3, a rapid increase in $[Ca^{2+}]_c$ occurs in the plant cells (Clayton et al. 1999; Evans et al. 2005; Kadono et al. 2006; Tran et al. 2012). As the O_3-dependent increases in $[Ca^{2+}]_c$ is sensitive to treatment with Ca^{2+} chelators, ion channel blockers, and ROS scavengers, the ROS-dependently induced flux Ca^{2+} from the apoplast into the cells might play a role as a signaling path initiating the oxidative cell death (Overmyer et al. 2005; Kadono et al. 2006). Interestingly, it has been pointed that there are the similarities between the defense activation mechanism and O_3 response in plants (Sandermann et al. 1998; Kadono et al. 2010). In fact, in O_3-sensitive line of tobacco (Bel-W3), SA is known as a potential biomarker of PCD development induced in response to ground-level ozone under ambient conditions (Drzewiecka et al. 2012). The sequence of the events starting from ROS generation and induction of Ca^{2+} signaling finally leading to both rapid and long-lasting cellular responses such as PCD-like localized cell death, stomatal closure, and SAR-related gene expression largely resembles the SA-induced phenomena (Kawano et al. 1998, 2004a; Mori et al. 2001). At the molecular level, involvement of *NPR1* in O_3-induced cell death in *A. thaliana* has been demonstrated (see Fig. 2b; Kadono et al. 2010).

Abiotic stresses involving SA-dependent signaling events (*NahG*-sensitive and *NPR1*-dependent path) include metal toxicities as such observed in aluminum-treated cells of *A. thaliana* (Kunihiro et al. 2011). As mentioned earlie, molecular evidences on the involvement of SA in the plant abiotic responses have been recently provided and accumulated; thus our understanding of the SA signaling pathways and mechanisms by which SA performs its role as the mediator of stress responses has been largely advanced.

5.3 *Activation of Ion Channels by SA*

As the possible involvement of SA in responses to O_3 was suggested, Kadono et al. (2010) have compared the electrophysiological impact of O_3 and SA in the cells of *A. thaliana*. Both O_3 and SA induced a rapid but slight hyperpolarization of the cells followed by a larger depolarization event within a few minutes. The temporal changes in the PM potential and the changes in anion channel activity were

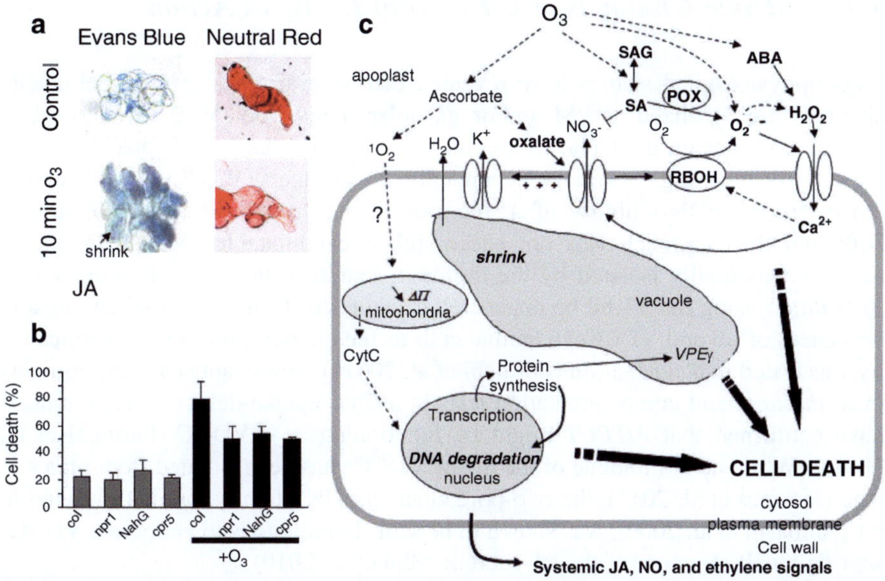

Fig. 2 Involvement of SA in the O_3-induced cell death in suspension cultured cells of *Arabidopsis thaliana*. (**a**) Microscopic images of O_3-induced cell death in wild-type cells (col) visualized after staining with Evans blue and Neutral red. (**b**) Effect of SA-related mutations on the O_3-induced cell death. (**c**) Hypothetical scheme for the action of SA interacting with ion channels and ROS-generating enzymes upon exposure to O_3, finally leading to PCD in *A. thaliana* cells. *CytC* cytochrome C, *VPEγ* vacuolar processing enzyme γ; Data and illustration are adapted from Kadono et al. (2010)

examined after treatments of the cells with O_3 and SA. As the delayed depolarization is correlated with an increase in anion channel activity, SA could not be responsible for the early depolarization induced by O_3, but SA can fuel the generation of H_2O_2 and Ca^{2+} influx involved in O_3-induced cell death. The SA-induced generation of H_2O_2 via stimulation of peroxidase and/or NADPH-oxidase is known to lead to Ca^{2+} influx (Kawano and Muto 2000; Kawano et al. 2004a) which could explain the delayed increase in anion currents observed in response to exogenous application of free SA. Upon stimulation, SA could be released from apoplastic SAG pool through the action of SAGase (Kawano et al. 2004b; Umemura et al. 2009).

As mentioned, SA was shown to be converted to electrophysiological signals correlated with biochemical changes leading to PCD development in Arabidopsis cells. As for future experiments, it is tempting to testify if these locally identified changes could be propagated as the chain of electric signal in the tissue as observed for SWR (Wildon et al. 1989, 1992; Rhodes et al. 1996).

5.4 Calcium Channels as a Target of Early SA Action

Electrophysiological studies have revealed that several types of Ca^{2+}-permeable channels are localized at PM and/or vacuolar membrane (VM) in many plant species (Jammes et al. 2011). Muto and his colleagues have attempted to isolate the cDNAs encoding for voltage-dependent Ca^{2+} channels (VDCCs) by expressing clones from a cDNA library of *A. thaliana* in the yeast *cch1* and *mid1* strains. Although this approach was not successful, a candidate for VDCC in plants, *AtTPC1* was finally isolated by the thorough search of the genomic sequence of *A. thaliana* using the 30–60 bp degenerated sequences from the partial amino acid sequences of several VDCCs in animal cells as the queries, and its Ca^{2+} permeability was tested in a *cch1* strain (Furuichi et al. 2001a). Sense–antisense experiments in *A. thaliana* and complementation tests in a Ca^{2+} uptake-defective yeast mutant have confirmed that *AtTPC1* might be functioning as a VDCC (Furuichi et al. 2001a). While no homologue of the major VDCCs has been isolated from plants to date (Jammes et al. 2011), the two-pore channel (TPC) family, originally found in rat (Ishibashi et al. 2000), was shown to be semi-homologous to vertebrate VDCCs sharing a half structure of the α1-subunit (Zhu et al. 2010).

Recent demonstration revealed that human TPCs mediate the nicotinic acid adenine dinucleotide phosphate (NAADP)-induced Ca^{2+} release from the acidic organelles in HEK293 cells (Calcraft et al. 2009). Prior to elucidation of the function of mammalian TPCs, plant biologists have shown that Arabidopsis's *AtTPC1* behaves as a slow-activating vacuolar cation channel (Peiter et al. 2005) which is involved in the sucrose-induced elevation of $[Ca^{2+}]_c$ (Furuichi et al. 2001b), extracellular Ca^{2+}-induced stomatal movement (Peiter et al. 2005; Islam et al. 2010), and abscisic acid-mediated prevention of seed germination (Islam et al. 2010).

Notably, *TPC1* family is the most likely group of VDCCs involved in ROS responses, assuming the well-conserved negatively charged residues within voltage-sensor (S4 of Shaker-units) are responsive to ROS-dependent voltage changes as previously demonstrated and reviewed (Kawano et al. 2004c; Kawano and Furuichi 2007). Orthologs belonging to the plant *TPC1* family were isolated from tobacco (Kadota et al. 2004) and rice (Hashimoto et al. 2005; Kurusu et al. 2004), and demonstrated to exist in some other plant species such as corn, broad bean, pea, spinach, and turnip (White et al. 2002; Kawano and Furuichi 2007). In *A. thaliana* and rice, such genes exist as single copy genes in the genome and are expressed in the entire plants, suggesting their systemic roles (Furuichi et al. 2001a, b; Kurusu et al. 2004). In tobacco BY-2 cultured cell line, there are two copies of genes with high similarity (97.1 % identity) with slightly different molecular masses (Kadota et al. 2004), apparently detectable with the specific antibody against *AtTPC1* (Kawano and Furuichi 2007).

Concerning the distortion of Ca^{2+} homeostasis by toxic ions in plants, it has been shown that action of Al^{3+} as a channel blocker specific for ROS-responsive Ca^{2+} influx (Kawano et al. 2003) is mediated by inhibition of *TPC1* channels (Kawano et al. 2004c). Since the use of Al enables the dissection of *TPC1*-mediated ROS-responsive Ca^{2+} influx from the Ca^{2+} influx stimulated by other stimuli such as the mechano-sensitive nonselective cation channel-mediated osmotic stress-responsive Ca^{2+} influx, the involvement of plant *TPC1* channels in SA-induced and ROS-mediated Ca^{2+} influx in plant cells was tested (Lin et al. 2005). The inhibitory effect of Al^{3+} as a putative and selective blocker of *TPC1* channels supported the view that the *TPC1* type channels is involved in the SA-induced Ca^{2+} influx in tobacco BY-2 cells.

The roles of plant *TPC1* channels in overall defense response against the pathogens were further studied by several groups. Kadota et al. (2004) showed that *TPC1*s from tobacco (*NtTPC1*s) possess several physiological roles such as regulation of hypersensitive cell death and defense-related gene expression triggered by cryptogein, an elicitor from an oomycete, through elevation of $[Ca^{2+}]_c$ in tobacco BY-2 cells. In addition, *TPC1s* from rice and wheat appear to function in responses to abiotic stresses (Kurusu et al. 2004; Wang et al. 2005).

5.5 Localization of TPC1 Channels in Plant Cells

Molecular and electrophysiological studies have shown that Arabidopsis *TPC1* is mainly localized at the VM and functions as a slow-activating vacuolar cation channel (Peiter et al. 2005; Ranf et al. 2008; Dadacz-Narloch et al. 2011; Hedrich and Marten 2011). In contrast, TPCs in monocots including *OsTPC1* have been suggested to be localized at the PM and independently confirmed by several groups (Wang et al. 2005; Kurusu et al. 2005; Hamada et al. 2012; Hashimoto et al. 2005). Similarly, Kawano et al. (2004c) have reported the partial but significant localization of active *AtTPC1* fused with GFP at the plasma membrane when expressed in tobacco BY-2 cells. Thus, the presence of minor TPC1 population in PM should not be ignored as there would be physiological roles in both forms of TPC1 proteins localized in VM and PM.

6 Conclusions

In this chapter, the modes of the local and long-distance signaling events involving the actions of SA are compared by covering both the biochemical and electrophysiological views. Especially, we aimed at outlining the modes of SA action by allocating the newly elucidated model for the action of two SA receptors, NPR3 and 4, involved in regulation of NPR1's action and life cycle, in addition to the previously known SABPs and SA-reacting enzymes.

Here, two distinct models for local SA perception and signaling mechanisms, namely, the extracellular and intracellular paths, are listed. One of local SA perception model responsive to extracellular SA involves the ROS-generating reaction catalyzed by apoplastic (cell wall-bound and free) peroxidases and PM-localized RBOHs followed by calcium signaling. The other local SA perception mechanism is responsive for intracellular SA, by which the action and life cycle of NPR1 protein are regulated depending on the concentration of SA in the cells. The model with NPR1 paralogues clearly explains how NPR1 escapes the ubiquitination-controlled turnover active in the absence and in the presence of excess level of SA in the cells. The intracellular and extracellular SA signaling mechanisms are likely linked through the import and export of SA by the cells. Since the low-affinity SA efflux carrier required for excretion of highly accumulated SA is reportedly induced by extracellular SA-induced oxidative burst and calcium signaling mechanism, the involvement of NPR4 could be minimized by enhanced SA efflux.

We view here that the long-distance SA signaling events inevitably involve the components in the local SA perception and signaling mechanisms. Here, three different models of SA mediated for long-distance signaling are described. The long-distance SA action could be attributed to the phloem-mediated and cell–cell transports of SA following the local increase in SA in the pathogen infected cells, SA-derived mobile signals with electrical and chemical natures systemically propagated without direct movement of SA, and synergistic propagation of both SA and derived signals throughout the tissues and phloem.

As known, long-distance signaling models also predicted the involvement of electrophysiological events at the site of the initial signal perception followed by systemic long-distance relaying of the electrically detectable signals; the last part of this article was used for reviewing the actions of SA on activation of ion channels such as anion channels and calcium channels. Electrophysiological models are also examples of our views that the long-distance SA signaling events inevitably involve the components or events in the local SA perception and signaling events.

Acknowledgment This work was supported by a grant from Regional Innovation Strategy Support Program 2012, MEXT, Japan.

References

Alvarez ME, Pennell RI, Meijer P-J, Ishikawa A, Dixon RA, Lamb C (1998) Reactive oxygen intermediates mediate a systemic signal network in the establishment of plant immunity. Cell 92:773–784

Antoniw JF, White RF (1980) The effects of aspirin and polyacrylic acid on soluble leaf proteins and resistance to virus infection in five cultivars of tobacco. Phytopathol Z 98:331–341

Aoki K, Suzui N, Fujimaki S, Dohmae N, Yonekura-Sakakibara K, Fujiwara T, Hayashi H, Yamaya T, Sakakibara H (2005) Destination-selective long-distance movement of phloem proteins. Plant Cell 17:1801–1814

Apel K, Hirt H (2004) Reactive oxygen species: metabolism, oxidative stress, and signal transduction. Annu Rev Plant Biol 55:373–399

Ashmore MR (2005) Assessing the future global impacts of ozone on vegetation. Plant Cell Environ 28:949–964

Atkinson NJ, Urwin PE (2012) The interaction of plant biotic and abiotic stresses: from genes to the field. J Exp Bot 63:3523–3544

Baker A, Graham IA, Holdsworth M, Smith SM, Theodouloue FL (2006) Chewing the fat: β-oxidation in signalling and development. Trends Plant Sci 11:124–132

Barbehenn R, Dukatz C, Holt C, Reese A, Martiskainen O, Salminen J-P, Yip L, Tran L, Constabel CP (2010) Feeding on poplar leaves by caterpillars potentiates foliar peroxidase action in their guts and increases plant resistance. Oecologia 164:993–1004

Bolwell GP, Bindschedler LV, Blee KA, Butt VS, Davies DR, Gardner SL, Gerrish C, Minibayeva F (2002) The apoplastic oxidative burst in response to biotic stress in plants: a three-component system. J Exp Bot 53:1367–1376

Botting D (1973) Humboldt and the cosmos. Sphere Books, London

Calcraft PJ, Ruas M, Pan Z, Cheng X, Arredouan A, Hao X, Tang J, Rietdorf K, Teboul L, Chuang K-T, Lin P, Xiao R, Wang C, Zhu Y, Lin Y, Wyatt CN, Parrington J, Ma J, Evans AM, Galione A, Zhu MX (2009) NAADP mobilizes calcium from acidic organelles through two-pore channels. Nature 459:596–600

Chanda B, Xia Y, Mandal MK, Yu K, Sekine K-T, Gao Q-M, Selote D, Hu Y, Stromberg A, Navarre D, Kachroo A, Kachroo P (2011) Glycerol-3-phosphate is a critical mobile inducer of systemic immunity in plants. Nat Genet 43:421–427

Chaturvedi R, Krothapalli K, Makandar R, Nandi A, Sparks AA, Roth MR, Welti R, Shah J (2008) Plastid ω3-fatty acid desaturase-dependent accumulation of a systemic acquired resistance inducing activity in petiole exudates of *Arabidopsis thaliana* is independent of jasmonic acid. Plant J 54:106–117

Chaturvedi R, Venables B, Petros RA, Nalam V, Li M, Wang X, Takemoto LJ, Shah J (2012) An abietane diterpenoid is a potent activator of systemic acquired resistance. Plant J 71:161–172

Chen H-J, Kuc J (1999) Ca^{2+}-dependent excretion of salicylic acid in tobacco cell suspension culture. Bot Bull Acad Sin 40:267–273

Chen Z, Silva H, Klessig DF (1993a) Active oxygen species in the induction of plant systemic acquired resistance induced by salicylic acid. Science 262:1883–1886

Chen Z, Ricigliano JR, Klessig DF (1993b) Purification and characterization of soluble salicylic acid binding protein from tobacco. Proc Natl Acad Sci USA 90:9533–9537

Chen H-J, Hou W-C, Kuc J, Lin Y-H (2001) Ca^{2+}-dependent and Ca^{2+}-independent excretion modes of salicylic acid in tobacco cell suspension culture. J Exp Bot 52:1219–1226

Chen H-J, Hou W-C, Kuc J, Lin Y-H (2002) Salicylic acid mediates alternative signal transduction pathways for pathogenesis-related acidic β-1,3-glucanase (protein N) induction in tobacco cell suspension culture. J Plant Physiol 159:331–337

Clarke A, Mur LAJ, Darby RM, Kenton P (2005) Harpin modulates the accumulation of salicylic acid by *Arabidopsis cells via* apoplastic alkalization. J Exp Bot 56:3129–3136

Clayton H, Knight MR, Knight H, McAinsh MR, Hetherington AM (1999) Dissection of the ozone-induced calcium signature. Plant J 17:575–579

Coll NS, Epple P, Dangl JL (2011) Programmed cell death in the plant immune system. Cell Death Differ 18:1247–1256

Conrath U, Chen Z, Ricigliano JR, Klessig DF (1995) Two inducers of plant defense response, 2,6-dichloroisonicotinic acid and salicylic acid, inhibit catalase activities in tobacco. Proc Natl Acad Sci USA 92:7143–7147

Dadacz-Narloch B, Beyhl D, Larisch C, López-Sanjurjo EJ, Reski R, Kuchitsu K, Müller TD, Becker D, Schönknecht G, Hedrich R (2011) A novel calcium binding site in the slow vacuolar cation channel TPC1 senses luminal calcium levels. Plant Cell 23:2696–2707

De Pinto MC, Locato V, de Gara L (2012) Redox regulation in plant programmed cell death. Plant Cell Environ 35:234–244

Del Río LA, Corpas FJ, Sandalio LM, Palma JM, Gómez M, Barroso JB (2002) Reactive oxygen species, antioxidant systems and nitric oxide in peroxisomes. J Exp Bot 53:1255–1272

Delaney TP (2004) Salicylic acid. In: Davies PJ (ed) Plant hormones. Biosynthesis, signal transduction, action. Kluwer Academic, Dordrecht, pp 635–653

Dempsey DA, Vlot AC, Wildermuth MC, Klessig DF (2011) Salicylic acid biosynthesis and metabolism. Arabidopsis Book 9:e0156
Doke N (1983a) Involvement of superoxide anion generation in the hypersensitive response of potato tuber tissues to infection with an incompatible race of *Phytophthora infestans* and to the hyphal wall components. Physiol Plant Pathol 23:345–357
Doke N (1983b) Generation of superoxide anion by potato tuber protoplasts during the hypersensitive response to hyphal wall components of *Phytophthora infestans* and specific inhibition of the reaction by suppressors of hypersensitivity. Physiol Plant Pathol 23:359–367
Doke N (1985) NADPH-dependent O_2^- generation in membrane fractions isolated from wounded potato tubers inoculated with *Phytophthora infestans*. Physiol Plant Pathol 27:311–322
Drzewiecka K, Borowiak K, Bandurska H, Golinski P (2012) Salicylic acid – a potential biomarker of tobacco Bel-W3 cell death developed as a response to ground level ozone under ambient conditions. Acta Biol Hung 63:231–249
Du H, Klessig DF (1997) Identification of a soluble, high-affinity salicylic acid-binding protein in tobacco. Plant Physiol 113:1319–1327
Durner J, Klessig DF (1995) Inhibition of ascorbate peroxidase by salicylic acid and 2,6-dichloroisonicotinic acid, two inducers of plant defense responses. Proc Natl Acad Sci USA 92:11312–11316
Durner J, Klessig DF (1996) Salicylic acid is a modulator of tobacco and mammalian catalases. J Biol Chem 272:28492–28501
Emoto A, Ushigome F, Koyabu N, Kajiya H, Okabe K, Satoh S, Tsukimori K, Nakano H, Ohtani H, Sawada Y (2002) H^+-linked transport of salicylic acid, an NSAID, in the human trophoplast cell line BeWo. Am J Physiol 282:C1064–C1075
Evans NH, McAinsh MR, Hetherington AM, Knight MR (2005) ROS perception in *Arabidopsis thaliana*: the ozone-induced calcium response. Plant J 41:615–626
Fiscus EL, Booker FL, Burkey KO (2005) Crop responses to ozone: uptake, modes of action, carbon assimilation and partitioning. Plant Cell Environ 28:997–1011
Forouhar F, Yang Y, Kumar D, Chen Y, Fridman E, Park SW, Chiang Y, Acton TB, Montelione GT, Pichersky E, Klessig DF, Tong L (2005) Structural and biochemical studies identify tobacco SABP2 as a methyl salicylate esterase and implicate it in plant innate immunity. Proc Natl Acad Sci USA 102:1773–1778
Fu ZQ, Yan Y, Saleh A, Wang W, Ruble J, Oka N, Mohan R, Spoel SH, Tada Y, Zheng N, Dong X (2012) NPR3 and NPR4 are receptors for the immune signal salicylic acid in plants. Nature 486:228–232
Furuichi T, Kawano T (2006) Biochemistry and cell biology of calcium channels and signaling involved in plant growth and environmental responses. In: Teixeira da Silva JA (ed) Floriculture, ornamental and plant biotechnology, vol III. Global Science Books, London, pp 26–36
Furuichi T, Cunningham KW, Muto S (2001a) A putative two pore channel AtTPC1 mediates Ca^{2+} flux in Arabidopsis leaf cells. Plant Cell Physiol 42:900–905
Furuichi T, Mori IC, Takahashi K, Muto S (2001b) Sugar-induced increase in cytosolic Ca^{2+} in *Arabidopsis thaliana* whole plants. Plant Cell Physiol 42:1149–1155
Furuichi T, Kawano T, Tatsumi H, Sokabe M (2007) Roles of ion channels in environmental responses of plants. In: Martinac B (ed) Sensing with ion channels. Springer series in biophysics, vol 11. Springer, Berlin, pp 47–62
Galvani L (1791) De viribus electricitatis in motu musculari commentarius. Bon Sci Art Inst Acad Comm 7:363–418
Gaupels F, Kuruthukulangarakoola GT, Durner J (2011) Upstream and downstream signals of nitric oxide in pathogen defence. Curr Opin Plant Biol 14:707–714
Gechev TS, Van Breusegem F, Stone JM, Denev I, Laloi C (2006) Reactive oxygen species as signals that modulate plant stress responses and programmed cell death. Bioessays 28:1091–1101

Glauser G, Grata E, Dubugnon L, Rudaz S, Farmer EE, Wolfender JL (2008) Spatial and temporal dynamics of jasmonate synthesis and accumulation in Arabidopsis in response to wounding. J Biol Chem 283:16400–16407

Gozzo F (2003) Systemic acquired resistance in crop protection: from nature to a chemical approach. J Agric Food Chem 51:4487–4503

Hamada H, Kurusu T, Okuma E, Nokajima H, Kiyoduka M, Koyano T, Sugiyama Y, Okada K, Koga J, Saji H, Miyao A, Hirochika H, Yamane H, Murata Y, Kuchitsu K (2012) Regulation of a proteinaceous elicitor-induced Ca^{2+} influx and production of phytoalexins by a putative voltage-gated cation channel, OsTPC1, in cultured rice cells. J Biol Chem 287:9931–9939

Hashimoto K, Saito M, Iida H, Matsuoka H (2005) Evidence for the plasma membrane localization of a putative voltage-dependent Ca^{2+} channel, OsTPC1, in rice. Plant Biotechnol 22:235–923

Hause B, Hause G, Kutter C, Miersch O, Wasternack C (2003) Enzymes of jasmonate biosynthesis occur in tomato sieve elements. Plant Cell Physiol 44:643–648

Hedrich R, Marten I (2011) TPC1-SV channels gain shape. Mol Plant 4:428–441

Ikeda K, Yamasaki K, Homemoto M, Yamaue S, Ogawa M, Nakao E, Fukunaga Y, Nakanishi T, Utoguchi N, Myotoku M, Hirotani Y (2012) Efflux transporter mRNA expression profiles in differentiating JEG-3 human choriocarcinoma cells as a placental transport model. Pharmazie 67:86–90

Ishibashi K, Suzuki M, Imai M (2000) Molecular cloning of a novel form (two-repeat) protein related to voltage-gated sodium and calcium channels. Biochem Biophys Res Commun 270:370–376

Islam MM, Munemasa S, Hossain MA, Nakamura Y, Mori IC, Murata Y (2010) Roles of AtTPC1, vacuolar two pore channel 1, in Arabidopsis stomatal closure. Plant Cell Physiol 51:302–311

Jammes F, Hu HC, Villiers F, Bouten R, Kwak JM (2011) Calcium-permeable channels in plant cells. FEBS J 278:4262–4276

Jorgensen RA, Atkinson RG, Foster RL, Lucas WJ (1998) An RNA-based information superhighway in plants. Science 279:1486–1487

Jung HW, Tschaplinski TJ, Wang L, Glazebrook J, Greenberg JT (2009) Priming in systemic plant immunity. Science 324:89–91

Kadono T, Yamaguchi Y, Furuichi T, Hirono M, Garrec JP, Kawano T (2006) Ozone-induced cell death mediated with oxidative and calcium signaling pathways in tobacco Bel-W3 and Bel-B cell suspension cultures. Plant Signal Behav 1:312–322

Kadono T, Tran D, Errakhi R, Hiramatsu T, Meimoun P, Briand J, Iwaya-Inoue M, Kawano T, Bouteau F (2010) Increased anion channel activity is an unavoidable event in ozone-induced programmed cell death. PLoS One 5:e13373

Kadota Y, Furuichi T, Ogasawara Y, Goh T, Higashi K, Muto S, Kuchitsu K (2004) Identification of putative voltage dependent Ca^{2+} permeable channels involved in cryptogein-induced Ca^{2+} transients and defense responses in tobacco BY-2 Cells. Biochem Biophys Res Commun 317:823–830

Kangasjärvi J, Talvinen J, Utriainen M, Karjalainen R (1994) Plant defence systems induced by ozone. Plant Cell Environ 17:783–794

Kangasjärvi J, Jaspars P, Kollist H (2005) Signaling and cell death in ozone-exposed plants. Plant Cell Environ 28:1–16

Kangasjärvi S, Neukermans J, Li S, Aro E-M, Noctor G (2012) Photosynthesis, photorespiration, and light signalling in defence responses. J Exp Bot 63:1619–1636

Kawano T (2003) Roles of the reactive oxygen species-generating peroxidase reactions in plant defense and growth induction. Plant Cell Rep 21:829–837

Kawano T, Furuichi T (2007) Salicylic acid as a defense-related plant hormone: roles of oxidative and calcium signaling paths in salicylic acid biology. In: Hayat S, Ahmad A (eds) Salicylic acid – a plant hormone. Springer, Dordrecht, pp 277–321

Kawano T, Muto S (2000) Mechanism of peroxidase actions for salicylic acid-induced generation of active oxygen species and an increase in cytosolic calcium in tobacco suspension culture. J Exp Bot 51:685–693

Kawano T, Sahashi N, Takahashi K, Uozumi N, Muto S (1998) Salicylic acid induces extracellular generation of superoxide followed by an increase in cytosolic calcium ion in tobacco suspension culture: the earliest events in salicylic acid signal transduction. Plant Cell Physiol 39:721–730

Kawano T, Kawano N, Muto S, Lapeyrie F (2001) Cation-induced superoxide generation in tobacco cell suspension culture is dependent on ion valence. Plant Cell Environ 24:1235–1241

Kawano T, Kadono T, Furuichi T, Muto S, Lapeyrie F (2003) Aluminum-induced distortion in calcium signaling involving oxidative bursts and channel regulations in tobacco BY-2 cells. Biochem Biophys Res Commun 308:35–42

Kawano T, Furuichi T, Muto S (2004a) Controlled free salicylic acid levels and corresponding signaling mechanisms in plants. Plant Biotechnol 21:319–335

Kawano T, Kadono T, Fumoto K, Lapeyrie F, Kuse M, Isobe M, Furuichi T, Muto S (2004b) Aluminum as a specific inhibitor of plant *TPC1* Ca^{2+} channels. Biochem Biophys Res Commun 324:40–45

Kawano T, Tanaka S, Kadono T, Muto S (2004b) Salicylic acid glucoside acts as a slow inducer of oxidative burst in tobacco suspension culture. Z Naturforsch 59c:684–692

Kerchev PI, Fenton B, Foyer CH, Hancock RD (2012) Plant responses to insect herbivory: interactions between photosynthesis, reactive oxygen species and hormonal signalling pathways. Plant Cell Environ 35:441–453

Kessmann H, Ryals J (1993) Requirement of salicylic acid for the induction of systemic acquired resistance. Science 261:754–756

Khokon MAR, Okuma E, Hossain MA, Munemasa S, Uraji M, Nakamura Y, Mori IC, Murata Y (2011) Involvement of extracellular oxidative burst in salicylic acid-induced stomatal closure in Arabidopsis. Plant Cell Environ 34:434–443

Knight MR, Campbell AK, Smith SM, Trewavas AJ (1991) Transgenic plant aequorin reports the effects of touch and cold-shock and elicitors on cytoplasmic calcium. Nature 352:524–526

Koo AJK, Gao X, Jones AD, Howe GA (2009) A rapid wound signal activates systemic synthesis of bioactive jasmonates in Arabidopsis. Plant J 59:974–986

Kotchoni SO, Gachomo EW (2006) The reactive oxygen species network pathways: an essential prerequisite for perception of pathogen attack and the acquired disease resistance in plants. J Biosci 31:389–404

Kumar D, Klessig DF (2003) High-affinity salicylic acid-binding protein 2 is required for plant innate immunity and has salicylic acid-stimulated lipase activity. Proc Natl Acad Sci USA 100:16101–16106

Kunihiro S, Hiramatsu T, Kawano T (2011) Involvement of salicylic acid signal transduction in aluminum-responsive oxidative burst in Arabidopsis thaliana cell suspension culture. Plant Signal Behav 6:611–616

Kurusu T, Sakurai Y, Miyao A, Hirochika H, Kuchitsu K (2004) Identification of a putative voltage-gated Ca^{2+}-permeable channel (*OsTPC1*) involved in Ca^{2+} influx and regulation of growth and development in rice. Plant Cell Physiol 45:693–702

Kurusu T, Yagala T, Miyao A, Hirochika H, Kuchitsu K (2005) Identification of a putative voltage-gated Ca^{2+} channel as a key regulator of elicitor-induced hypersensitive cell death and mitogen-activated protein kinase activation in rice. Plant J 42:798–809

Kvaratskhelia M, George SJ, Thorneley RNF (1997) Salicylic acid is a reducing substrate and not an effective inhibitor of ascorbate peroxidase. J Biol Chem 272:20998–21001

Lee J-S (1998) The mechanism of stomatal closing by salicylic acid in *Commelina communis* L. J Plant Biol 41:97–102

Lee HJ, Kim JS, Yoo SJ, Kang YY, Han SH, Yang K-Y, Kim YC, Gardener BM, Kang H (2012) Different roles of glycine-rich RNA-binding protein7 in plant defense against *Pectobacterium carotovorum*, *Botrytis cinerea*, and tobacco mosaic viruses. Plant Physiol Biochem 60:46–52

Leitner M, Vandelle E, Gaupels F, Bellin D, Delledonne M (2009) NO signals in the haze: nitric oxide signalling in plant defence. Curr Opin Plant Biol 12:451–458

Li L, Li C, Lee GI, Howe GA (2002) Distinct roles for jasmonate synthesis and action in the systemic wound response of tomato. Proc Natl Acad Sci USA 99:6416–6421

Li P, Ham B-K, Lucas WJ (2011) CmRBP50 protein phosphorylation is essential for assembly of a stable phloem-mobile high-affinity ribonucleoprotein complex. J Biol Chem 286:23142–23149

Lin C, Yu Y, Kadono T, Iwata M, Umemura K, Furuichi T, Kuse M, Isobe M, Yamamoto Y, Mastumoto H, Yoshizuka K, Kawano T (2005) Action of aluminum, novel TPC1-type channel inhibitor, against salicylate-induced and cold shock-induced calcium influx in tobacco BY-2 cells. Biochem Biophys Res Commun 332:823–830

Liu N, You J, Shi W, Liu W, Yang Z (2012) Salicylic acid involved in the process of aluminum induced citrate exudation in *Glycine max* L. Plant Soil 352:85–97

Lucas WJ, Yoo BC, Kragler F (2001) RNA as a long-distance information macromolecule in plants. Nat Rev Mol Cell Biol 2:849–857

Malamy J, Carr JP, Klessig DF, Raskin I (1990) Salicylic acid: a likely endogenous signal in the resistance response of tobacco to viral infection. Science 250:1002–1004

Maldonado AM, Doerner P, Dixon RA, Lamb CJ, Cameron RK (2002) A putative lipid transfer protein involved in systemic resistance signalling in Arabidopsis. Nature 419:399–403

Manthe B, Schulz M, Schnable H (1992) Effects of salicylic acid on growth and stomatal movement on *Vicia faba* L.: evidence for salicylic acid metabolism. J Chem Ecol 18:1525–1539

McGurl B, Pearce G, Orozco-Cardenas M, Ryan CA (1992) Structure, expression, and antisense inhibition of the systemin precursor gene. Science 255:1570–1573

Meimoun P, Vidal G, Bohrer AS, Lehner A, Tran D, Briand J, Bouteau F, Rona JP (2009) Intracellular Ca^{2+} stores could participate to abscisic acid-induced depolarization and stomatal closure in *Arabidopsis thaliana*. Plant Signal Behav 4:830–835

Métraux J-P, Signer H, Ryals J, Ward E, Wyss-Benz M, Gaudin J, Raschdorf K, Schmid E, Blum W, Inverardi B (1990) Increase in salicylic acid at the onset of systemic acquired resistance in cucumber. Science 250:1004–1006

Miller G, Schlauch K, Tam R, Cortes D, Torres MA, Shulaev V, Dangl JL, Mittler R (2009) The plant NADPH oxidase RBOHD mediates rapid systemic signaling in response to diverse stimuli. Sci Signal 2:ra45

Mittler R, Vanderauwera S, Suzuki N, Miller G, Tognetti VB, Vandepoele K, Gollery M, Shulaev V, Van Breusegem F (2011) ROS signaling: the new wave? Trends Plant Sci 16:300–309

Mitton FM, Pinedo ML, de la Canal L (2009) Phloem sap of tomato plants contains a DIR1 putative ortholog. J Plant Physiol 166:543–547

Miyake C, Sano S, Asada K (1996) A new assay of ascorbate peroxidase using the coupled system with monodehydroascorbate radical reductase. In: Obinger C, Burner U, Ebermann R, Penel C, Greppin (eds) Plant peroxidases: biochemistry and physiology. University of Geneva, Vienna, pp 386–389

Mölders W, Buchala A, Métraux J-P (1996) Transport of salicylic acid in tobacco necrosis virus-infected cucumber plants. Plant Physiol 112:787–792

Møller IM (2001) Plant mitochondria and oxidative stress: electron transport, NADPH turnover, and metabolism of reactive oxygen species. Annu Rev Plant Physiol Plant Mol Biol 52:561–591

Møller IM, Jensen PE, Hansson A (2007) Oxidative modifications to cellular components in plants. Annu Rev Plant Biol 58:459–481

Moreau M, Tian M, Klessig DF (2012) Salicylic acid binds NPR3 and NPR4 to regulate NPR1-dependent defense responses. Cell Res 2012:1–3

Mori IC, Schroeder JI (2004) Reactive oxygen species activation of plant Ca^{2+} channels. A signaling mechanism in polar growth, hormone transduction, stress signaling, and hypothetically mechanotransduction. Plant Physiol 135:702–708

Mori IC, Pinontoan R, Kawano T, Muto S (2001) Involvement of superoxide generation in salicylic acid-induced stomatal closure in *Vicia faba*. Plant Cell Physiol 42:1383–1388

Mou Z, Fan WH, Dong X (2003) Inducers of plant systemic acquired resistance regulate NPR1 function through redox changes. Cell 113:935–944

Nandi A, Welti R, Shah J (2004) The *Arabidopsis thaliana* dihydroxyacetone phosphate reductase gene *SUPPRESSOR OF FATTY ACID DESATURASE DEFICIENCY1* is required for glycerolipid metabolism and for the activation of systemic acquired resistance. Plant Cell 16:465–477

Niederl S, Kirsch T, Riederer M, Schreiber L (1998) Co-permeability of ^{3}H-labeled water and ^{14}C-labeled organic acids across isolated plant cuticles. Plant Physiol 116:117–123

Noctor G, De Paepe R, Foyer CH (2007) Mitochondrial redox biology and homeostasis in plants. Trends Plant Sci 12:125–134

Nomura H, Komori T, Uemura S, Kanda Y, Shimotani K, Nakai K, Furuichi T, Takebayashi K, Sugimoto T, Sano S, Suwastika IN, Fukusaki E, Yoshioka H, Nakahira Y, Shiina T (2012) Chloroplast-mediated activation of plant immune signalling in Arabidopsis. Nat Commun 3:926

Ohashi Y, Murakami T, Mitsuhara I, Seo S (2004) Rapid down and upward translocation of salicylic acid in tobacco plants. Plant Biotechnol 21:95–101

Overmyer K, Brosche M, Pellinen R, Kuittinen T, Tuominen H, Ahlfors R, Keinaenen M, Saarma M, Scheel D, Kangasjaervi J (2005) Ozone-induced programmed cell death in the Arabidopsis radical-induced cell death 1 mutant. Plant Physiol 137:1092–1104

Park S-W, Kaimoyo E, Kumar D, Mosher S, Klessig DF (2007) Methyl salicylate is a critical mobile signal for plant systemic acquired resistance. Science 318:113–116

Pasqualini S, Piccioni C, Reale L, Ederli L, Torre GD, Ferranti F (2003) Ozone-induced cell death in tobacco cultivar Bel W3 plants. The role of programmed cell death in lesion formation. Plant Physiol 133:1122–1134

Pastora V, Vicentb C, Cerezoa M, Mauch-Manic B, Deand J, Flors V (2012) Detection, characterization and quantification of salicylic acid conjugates in plant extracts by ESI tandem mass spectrometric techniques. Plant Physiol Biochem 53:19–26

Peiter E, Maathuis FJ, Mills LN, Knight H, Pelloux J, Hetherington AM, Sanders D (2005) The vacuolar Ca^{2+}-activated channel TPC1 regulates germination and stomatal movement. Nature 434:404–408

Pell EJ, Schlagnhaufer CD, Arteca RN (1997) Ozone-induced oxidative stress: mechanisms of action and reaction. Physiol Plant 100:264–273

Qi Y, Tsuda K, Joe A, Sato M, Nguyen LV, Glazebrook J, Alfano IR, Cohen JD, Katagiri F (2010) A putative RNA-binding protein positively regulates salicylic acid-mediated immunity in Arabidopsis. Mol Plant-Microbe Interact 23:1573–1583

Rainsford DK (1984) Aspirin and salicylates. Butterworth, London

Ranf S, Wünnenberg P, Lee J, Becker D, Dunkel M, Hedrich R, Scheel D, Dietrich P (2008) Loss of the vacuolar cation channel, AtTPC1, does not impair Ca^{2+} signals induced by abiotic and biotic stresses. Plant J 53:287–299

Rasmussen JB, Hammerschmidt R, Zook MN (1991) Systemic induction of salicylic acid accumulation in cucumber after inoculation with *Pseudomonas syringae* pv. syringae. Plant Physiol 97:1342–1347

Raz V, Fluhr R (1992) Calcium requirement for ethylene-dependent responses. Plant Cell 4:1123–1130

Rhodes JD, Thain JF, Wildon DC (1996) The pathway for systemic electrical signal conduction in the wounded tomato plant. Planta 200:50–57

Robert-Seilaniantz A, Grant M, Jones JDG (2011) Hormone crosstalk in plant disease and defense: more than just jasmonate-salicylate antagonism. Annu Rev Phytopathol 49:317–343

Rocher F, Chollet J-F, Jousse C, Bonnemain J-L (2006) Salicylic acid, an ambimobile molecule exhibiting a high ability to accumulate in the phloem. Plant Physiol 141:1684–1693

Rocher F, Chollet J-F, Legros S, Jousse C, Lemoine R, Faucher M, Bush DR, Bonnemain J-L (2009) Salicylic acid transport in *Ricinus communis* involves a pH-dependent carrier system in addition to diffusion. Plant Physiol 150:2081–2091

Ruiz-Medrano R, Kragler F, Wolf S (2012) Signaling and phloem-mobile transcripts, In: Kragler F, Hülskamp M (eds) Short and long distance signaling. Advances in plant biology, vol 3. Springer, Berlin, pp 151–177

Ruiz-Sanchez E, O'Donnell MJ (2006) Characterization of salicylate uptake across the basolateral membrane of the Malpighian tubules of *Drosophila melanogaster*. J Insect Physiol 52:920–928

Rustérucci C, Espunya MC, Díaz M, Chabannes M, Martínez MC (2007) S-nitrosoglutathione reductase affords protection against pathogens in *Arabidopsis*, both locally and systemically. Plant Physiol 143:1282–1292

Sandermann H, Ernst D, Heller W, Langebertles C (1998) Ozone: an abiotic elicitor of plant defense reaction. Trend Plant Sci 3:47–50

Sanders D, Brownlee C, Harper JF (1999) Communicating with calcium. Plant Cell 11:691–706

Schneider-Müller S, Kurosaki F, Nishi A (1994) Role of salicylic acid and intracellular Ca^{2+} in the induction of chitinase activity in carrot suspension culture. Physiol Mol Plant Pathol 45:101–109

Schraudner M, Moeder W, Wiese C, Van Camp W, Inze D, Langebartels C, Sandermann H (1998) Ozone-induced oxidative burst in the ozone biomonitor plant, tobacco Bel W3. Plant J 16:235–245

Seo S, Ichizawa K, Ohashi Y (1995) Induction of salicylic acid-β-glucosidase in tobacco leaves by exogenous salicylic acid. Plant Cell Physiol 36:447–453

Shah J (2003) The salicylic acid loop in plant defense. Curr Opin Plant Biol 6:365–371

Shulaev V, Leon J, Raskin I (1995) Is salicylic acid a translocated signal of systemic acquired resistance in tobacco? Plant Cell 7:1691–1701

Slaymaker DH, Navarre DA, Clark D, Del Pozo O, Martin GB, Klessig DF (2002) The tobacco salicylic acid-binding protein 3 (SABP3) is the chloroplast carbonic anhydrase, which exhibits antioxidant activity and plays a role in the hypersensitive defense response. Proc Natl Acad Sci USA 99:11640–11645

Song F, Goodman RM (2001) Activity of nitric oxide is dependent on, but is partially required for function of, salicylic acid in the signaling pathway in tobacco systemic acquired resistance. Mol Plant-Microbe Interact 14:1458–1462

Spoel SH, Koornneef A, Claessens SM, Korzelius JP, Van Pelt JA, Mueller MJ, Buchala AJ, Metraux JP, Brown R, Kazan K, Van Loon LC, Dong X, Pieterse CM (2003) NPR1 modulates cross-talk between salicylate- and jasmonate-dependent defense pathways through a novel function in the cytosol. Plant Cell 15:760–770

Ströher E, Dietz KJ (2006) Concepts and approaches towards understanding the cellular redox proteome. Plant Biol 8:407–418

Suzuki N, Miller G, Morales J, Shulaev V, Torres A, Mittler R (2011) Respiratory burst oxidases: the engines of ROS signalling. Curr Opin Plant Biol 14:691–699

Swanson SJ, Choi W-G, Chanoca A, Gilroy S (2011) *In vivo* imaging of Ca^{2+}, pH, and reactive oxygen species using fluorescent probes in plants. Annu Rev Plant Biol 62:273–297

Tenhaken R, Rübel C (1997) Salicylic acid is needed in hypersensitive cell death in soybean but does not act as a catalase inhibitor. Plant Physiol 115:291–298

Thompson GA, van Bel AJE, Gaupels F, Vlot AC (2012) Plant defense and long-distance signaling in the phloem. In: Thompson GA, van Bel AJE (eds) Phloem. Molecular cell biology, systemic communication, biotic interactions. Wiley-Blackwell, Oxford, pp 227–248

Tran D, Kadono T, Molas ML, Errakhi R, Briand J, Biligui B, Kawano T, Bouteau F (2012) A role for oxalic acid generation in ozone-induced signalization in Arabidopsis cells. Plant Cell Environ. doi:10.1111/j.1365-3040.2012.02596.x

Truman W, Bennett MH, Kubigsteltig I, Turnbull C, Grant M (2007) Arabidopsis systemic immunity uses conserved defense signaling pathways and is mediated by jasmonates. Proc Natl Acad Sci USA 104:1075–1080

Umemura K, Satoh J, Iwata M, Uiozumi N, Koga J, Kawano T, Koshiba T, Anzai H, Mitomi M (2009) Contribution of salicylic acid glucosyltransferase, OsSGT1, to chemically induced disease resistance in rice plants. Plant J 57:463–472

Utoguchi N, Watanabe Y, Takase Y, Suzuki T, Matsumoto M (1999) Carrier-mediated absorption of salicylic acid from hamster cheek pouch mucosa. J Pharmacol Sci 88:142–146

Vernooij B, Friedrich L, Morse A, Reist R, Kolditz-Jawhar R, Ward E, Uknes S, Kessmann H, Ryals J (1994) Salicylic acid is not the translocated signal responsible for inducing systemic acquired resistance but is required in signal transduction. Plant Cell 6:959–965

Vlot AC, Klessig DF, Park S-W (2008) Systemic acquired resistance: the elusive signal(s). Curr Opin Plant Biol 11:436–442

Vlot AC, Dempsey DM, Klessig DF (2009) Salicylic acid, a multifaceted hormone to combat disease. Annu Rev Phytopathol 47:177–206

Wang YJ, Yu JN, Chen T, Zhang ZG, Hao YJ, Zhang JS, Chen SY (2005) Functional analysis of a putative Ca^{2+} channel gene TaTPC1 from wheat. J Exp Bot 56:3051–3060

Weissman G (1991) Aspirin. Sci Am 264:84–90

White RF (1979) Acetylsalicylic acid induces resistance to tobacco mosaic virus in tobacco. Virology 99:410–412

White PJ, Bowen HC, Demidchik V, Nichols C, Davies JM (2002) Genes for calcium-permeable channels in the plasma membrane of plant root cells. Biochim Biophys Acta 1564:299–309

Wildermuth MC, Dewdney J, Wu G, Ausubel FM (2001) Isochorismate synthase is required to synthesize salicylic acid for plant defense. Nature 414:562–565

Wildon DC, Doherty HM, Eagles G, Bowles DJ, Thain JF (1989) Systemic responses arising from localized heat stimuli in tomato plants. Ann Bot 64:691–695

Wildon DC, Thain JF, Minchin PEH, Gubb IR, Reilly AJ, Skipper YD, Doherty HM, O'Donnell PJ, Bowles DJ (1992) Electrical signalling and systemic proteinase inhibitor induction in the wounded plant. Nature 360:62–65

Xia Y, Suzuki H, Borevitz J, Blount J, Guo Z, Patel K, Dixon RA, Lamb C (2004) An extracellular aspartic protease functions in Arabidopsis disease resistance signaling. EMBO J 23:980–988

Yalpani N, Siverman P, Wilson TMA, Kleier DA, Raskin I (1991) Salicylic acid is a systemic signal and an inducer of pathogenesis-related proteins in virus-infected tobacco. Plant Cell 3:809–818

Yamamoto Y, Kobayashi Y, Devi SR, Rikiishi S, Matsumoto H (2003) Oxidative stress triggered by aluminum in plant roots. Plant Soil 255:239–243

Yokawa K, Kagenishi T, Kawano T, Mancuso S, Baluška F (2011) Illumination of Arabidopsis roots induces immediate burst of ROS production. Plant Signal Behav 6:1457–1461

Yoshioka H, Sugie K, Park HJ, Maeda H, Tsuda N, Kawakita K, Doke N (2001) Induction of plant $gp91^{phox}$ homolog by fungal cell wall, arachidonic acid, and salicylic acid in potato. Mol Plant-Microbe Interact 14:725–736

Yoshioka H, Bouteau F, Kawano T (2008) Discovery of oxidative burst in the field of plant immunity: looking back at the early pioneering works and towards the future development. Plant Signal Behav 3:153–155

Zhu MX, Ma J, Parrington J, Galione A, Evans AM (2010) TPCs: endolysosomal channels for Ca^{2+} mobilization from acidic organelles triggered by NAADP. FEBS Lett 584:1966–1974

Long-Distance Signaling via Mobile RNAs

David J. Hannapel

Abstract Beyond its role in moving sugars from source leaves to sinks, the phloem is an important conduit for transporting information molecules that function as signals responding to environmental and internal cues related to numerous aspects of physiology. One of the most prominent groups of these signaling molecules is full-length mRNAs. Thousands of full-length transcripts have been identified in sieve elements of the phloem, but only a few have been confirmed to be mobile. This chapter focuses on six RNAs that move long distance through the plant and have a documented role in regulating development. These include *StBEL5* and *POTH1* of potato, *CmGAI* of pumpkin, *PFP-LeT6* from tomato, and *Aux/IAA* and *FLOWERING LOCUS T* from *Arabidopsis*. Their impact in controlling development and the mechanisms that facilitate their movement are discussed.

Keywords Non-cell-autonomous • Phloem • Polypyrimidine tract-binding proteins • Potato • Signal • Vascular biology

1 Introduction

The phloem of plants is a complex, dynamic system for transporting numerous molecules throughout the plant. Phloem plays a pivotal role in plant development not only by transporting sugars but also as a conduit for moving signal RNAs and proteins. From extensive, groundbreaking research, we now know there are, at least, three types of mobile RNAs in plants (1) pathogenic viral and viroid RNAs, (2) small RNAs including siRNAs and microRNAs, and (3) full-length cellular mRNAs (Kehr and Buhtz 2008). The model of viral RNA movement was critical in gaining an understanding of the mechanism by which RNA/protein complexes

D.J. Hannapel (✉)
Plant Biology Major, Iowa State University, Ames, IA 50011-1100, USA
e-mail: djh@iastate.edu

F. Baluška (ed.), *Long-Distance Systemic Signaling and Communication in Plants*, Signaling and Communication in Plants 19, DOI 10.1007/978-3-642-36470-9_3,

transversed cell boundaries through plasmodesmata connections (Lucas 2006; Lucas et al. 2009). Essentially viral RNAs are spread from cell to cell by mimicking the plant's endogenous transport system. Eloquent studies on viroid movement have revealed distinct RNA domains that facilitate directional transport of RNA throughout the plant (Qi et al. 2004; Zhong et al. 2007). Recent studies on small RNAs have uncovered numerous examples of transport and movement through the phloem system involved with systemic gene silencing, root-to-shoot communication, and response to stress (Martin et al. 2009; Bai et al. 2011; Hyun et al. 2011; Kasai et al. 2011; Melnyk et al. 2011). During the establishment of parasitic connections to host plants, numerous full-length mRNAs traffic from the host to the parasite (Roney et al. 2007; LeBlanc et al. 2012). Host RNAs from pumpkin and tomato confirmed to move into dodder in this system include *CmNAC*, *CmWRKYP*, *CmSUT*, *LeGAI*, and *LeBeclin 1* (Roney et al. 2007).

Either by specific phloem cell isolation or analysis of phloem sap, the transcriptome of phloem includes thousands of full-length mRNAs with putative functions ranging from metabolism to gene expression. Full-length, phloem-mobile mRNAs function to integrate environmental cues for plant development via this long-distance signaling pathway (Haywood et al. 2005; Banerjee et al. 2006). Despite the huge number of full-length transcripts that have been identified in the phloem, only a few have been confirmed to be mobile through the phloem translocation stream by grafting analyses. These mobile RNAs have been identified in diverse species such as melon, pumpkin, potato, tomato, *Arabidopsis*, apple, and broccoli (Tables 1 and 2) with known annotated functions in hormone response, transcriptional control, and organ development.

Examples of mobile RNAs associated with an observable phenotype are even more rare (Table 2). These include *StBEL5* (Banerjee et al. 2006) and *POTH1* (Mahajan et al. 2012) of potato, *CmGAI* of pumpkin (Haywood et al. 2005), *PFP-LeT6* from tomato (Kim et al. 2001), and *Aux/IAA* (Notaguchi et al. 2012) and *FLOWERING LOCUS T* (Li et al. 2011; Lu et al. 2012) from *Arabidopsis*. *GA INSENSITIVE* (*GAI*) is notable in that long-distance movement of its mRNA has been established in several plant species including cucumber, tomato, pumpkin (Haywood et al. 2005; Ham et al. 2009), apple (Xu et al. 2010), and *Arabidopsis* (Huang and Yu 2009). As mentioned previously, during infection of tomato by the parasitic plant dodder (*Cuscuta pentagona*), *GAI* mRNA was also observed to move from host to dodder through the parasite's vascular connection (David-Schwartz et al. 2008). Based on their functional significance and our understanding of the mechanisms that facilitate their movement, these six mobile RNAs will be the focus of this chapter.

2 Long-Distance Movement of Two Knotted1-Type mRNAs, *PFP-LeT6* and *POTH1*

Knotted1-like transcription factors are members of the *t*hree *a*mino acid *l*oop *e*xtension (TALE) superfamily that are ubiquitous among plants and are involved in numerous aspects of development (Hake et al. 2004). The maize KNOTTED1

Table 1 Phloem-mobile mRNAs that move across heterografts

RNA	Annotation	Possible function	Reference
MpSLR/IAA14	Auxin response factor	Transcriptional factor	Kanehira et al. (2010)
CmSCL14P	Scarecrow like	Transcription factor	Ham et al. (2009)
CmSTM	Shoot meristemless	Meristem regulator	Ham et al. (2009)
CmERF	Ethylene response factor	Ethylene signaling	Ham et al. (2009)
CmNAC	NAM, ATAF1/2, and CUC2	Meristem development	Ruiz-Medrano et al. (1999)
CmPP16-1	Phloem protein 16	RNA-binding protein	Ruiz-Medrano et al. (1999)
CmMyb	Myb-like transcription factor	Transcriptional activator	Ham et al. (2009)
BoFVE	Mammalian retinoblastoma-associated protein	Floral regulator	Yang and Yu (2010)
BoAGL24	Agamous like	Floral regulator	Yang and Yu (2010)
CmeAux/IAA	Auxin response factor	Auxin signaling	Omid et al. (2007)
CmeSAUR	Small auxin-up RNA	Auxin response factor	Omid et al. (2007)

Cm, *Cucurbita maxima*; *Cme*, *Cucumis melo*; *Mp*, *Malus prunifolia*; *Bo*, *Brassica oleracea*

Table 2 Long-distance, full-length, mobile RNAs associated with a phenotype

RNA	Annotation	Function	Reference
CmGAI	GA Insensitive	Leaf morphology ⇑	Haywood et al. (2005)
StBEL5	Potato BEL1-like family	Tuber growth ⇓	Banerjee et al. (2006)
POTH1	Potato Knotted1-type	Vegetative growth ⇓	Mahajan et al. (2012)
PFP-LeT6	Tomato Knotted1-type function	Leaf morphology ⇑	Kim et al. (2001)
FT	Arabidopsis Flowering locus T	Flowering ⇑	Li et al. (2011)
Aux/IAA	Auxin regulators	Root growth ⇓	Notaguchi et al. (2012)

The arrows indicate direction of movement of the RNA. An arrow up indicates movement through the graft union from the stock into the scion, an arrow down, the opposite. *Cm*, *Cucurbita maxima*; *St*, *Solanum tuberosum*; *PFP* pyrophosphate-dependent fructose 6-phosphate phosphotransferase

(KN1) homeodomain protein was one of the first plant proteins shown to traffic cell to cell through leaf mesophyll cells (Lucas et al. 1995), onion epidermal cells, and the shoot apical meristem (Kim et al. 2002). Class I KNOX proteins can also promote transport of the *KN1* mRNA through plasmodesmata (Lucas et al. 1995; Kim et al. 2005). Using a novel trafficking assay in *Arabidopsis* trichomes, Kim et al. (2005) showed that the KNOX homeodomain functioned as the minimal sequence required for trafficking the *KN1* mRNA.

Several observations now suggest that Knotted1-type mRNAs are also transported long distance through the sieve element system. Transcripts of the *SHOOT MERISTEMLESS* ortholog of pumpkin were identified as a component in a phloem-mobile ribonucleoprotein complex (Ham et al. 2009). This *CmSTM* RNA contains cytosine/uracil motifs in its sequence that facilitate binding to the RNA-binding protein, CmRBP50 (Ham et al. 2009). Using heterografting experiments, long-distance transport of a *Knotted1* mRNA of tomato was demonstrated (Kim et al. 2001). The mobile transcript was a chimeric mRNA composed of sequence from a pyrophosphate-dependent phosphofructokinase and *LeT6*, a class I KNOTTED1-like transcription factor of tomato. Overexpression of this fusion sequence resulted in the mouse-ear (Me) leaf phenotype in tomato. When this mutant was used as the stock, the *PFP-LeT6* fusion RNA was transported up from the Me stocks into the wild-type heterografted scions and produced the mouse-ear phenotype in leaves of the wild-type scion. This mobile transcript was detected in phloem sieve tubes and associated companion cells and accumulated in shoot apices and leaf primordial of the wild-type scion (Kim et al. 2001).

Similar to the *LeT6* fusion, the RNA of a Knotted1-type transcription factor of potato, *POTH1*, was also confirmed to be phloem mobile (Mahajan et al. 2012). POTH1 interacts with BEL1-like proteins to facilitate binding to specific target genes to modulate hormone levels, mediate leaf architecture, and enhance tuber formation (Rosin et al. 2003; Chen et al. 2004). Using in situ hybridization and laser capture microdissection, its RNA was previously identified in phloem cells (Rosin et al. 2003; Yu et al. 2007) and its promoter activity was detected in leaf veins and in phloem cells of both petioles and stems (Mahajan et al. 2012). Because the overexpression phenotype of POTH1 in potato produces plants that are dwarf and slow growing with reduced leaf size (Rosin et al. 2003) and are difficult to use in heterografts, overexpression lines of tobacco were engineered instead using a constitutive promoter. In soil-grown heterografts of transgenic scions and wild-type stocks, transgenic *POTH1* mRNA was detected in stem, roots, and newly emerging leaves arising from axillary buds of the stock (Fig. 1a; Mahajan et al. 2012). These actively growing leaves from the wild-type stock containing the mobile *POTH1* RNA exhibited a reduction in leaf size relative to leaves from wild-type grafts (Fig. 1b). RNA/protein-binding assays demonstrated that the untranslated regions (UTRs) of *POTH1* bind to two RNA-binding proteins, a polypyrimidine tract-binding (PTB) protein and an alba-domain type (Mahajan et al. 2012). The PTB ortholog of pumpkin, designated CmRBP50, has been identified as the core protein in a phloem-mobile RNA/protein complex (Ham et al. 2009), and the potato PTB protein that binds to *POTH1* also interacts with the 3′ UTR of *StBEL5* (Mahajan et al. 2012). An alba-domain protein similar to the potato ortholog that binds the *POTH1* UTR has been detected in phloem sap of pumpkin (Lin et al. 2009). Alba-domain proteins function to regulate RNA metabolism by binding to a conserved sequence element present in the 3′ UTR of its target RNAs (Aravind et al. 2003; Mani et al. 2011). This glycerol-responsive element of 25 nucleotides is present in a secondary loop structure, and this motif is conserved in both the 5′ and 3′ UTRs of *POTH1* (Mahajan et al. 2012). These compelling results indicate that *POTH1* functions as a mobile signal involved in controlling vegetative development and that this movement is likely mediated by specific RNA-binding proteins.

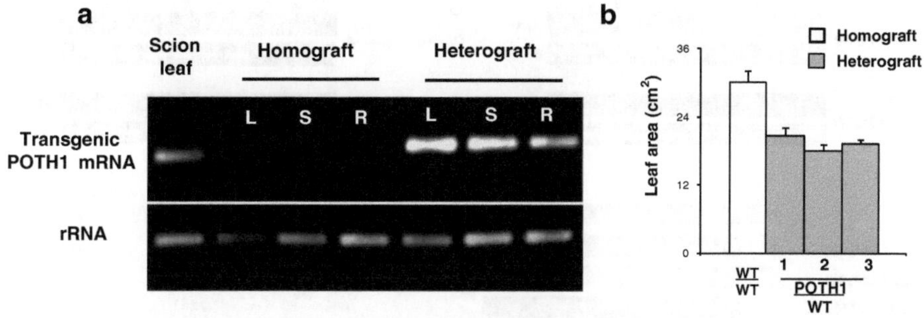

Fig. 1 RT-PCR detection of transgenic *POTH1* RNA moving across a graft union. (**a**) Grafts were generated with 35S:POTH1-His tobacco lines as scion and wild-type tobacco lines as stocks. WT/WT tobacco homografts were used as a graft control. Leaf (L), stem (S), and root (R) samples were harvested 30 days post-grafting from wild-type stock material. Samples collected from stocks for the RT-PCR analyses were stem sections 1.0 cm above the soil line, new leaves arising from axillary shoots, and root tips approximately 2.0 cm in length. RNA was extracted and a nonplant sequence tag fused to the transgenic construct was used to detect transgenic-specific transcripts. PCR reactions were performed twice with nested gene-specific primers. Scion leaf RNA was used as a positive control with wild-type gene-specific primers. 18 s rRNA was used as a PCR control. Axillary shoot branches grew from the wild-type stocks of both 35S:POTH1/WT and WT/WT grafts, and leaves from these branches were scored for size (**b**). Leaf area data from *panel* (**b**) are the means of three replicates ± SE. Modified from Figure 5 of Mahajan et al. (2009) with permission from Anjan Banerjee

3 Movement of *GAI* mRNAs

Phloem mobility of *GA INSENSITIVE* (*GAI*) has been verified in cucumber, tomato, pumpkin (Haywood et al. 2005; Ham et al. 2009), apple (Xu et al. 2010), and *Arabidopsis* (Huang and Yu 2009) and was the first mobile RNA associated with a phenotype (Haywood et al. 2005) and with a mobile RNA/protein complex (Ham et al. 2009). *GAI* encodes a protein that belongs to the GRAS family of transcriptional regulators and functions as a negative regulator of gibberellic acid (GA) responses (Peng et al. 1997; Pysh et al. 1999; Richards et al. 2001). The *Arabidopsis* gibberellic acid-insensitive (gai) mutant displays a dark-green dwarf phenotype (Koornneef et al. 1985), and overexpression can affect leaf architecture. In studies on both pumpkin and *Arabidopsis GAI*, selective movement from stock phloem to the shoot apical meristem of the scion was demonstrated independent of tissue sink strength (Haywood et al. 2005). Using heterografts of tomato, *GAI* RNA movement was shown to mediate phenotypic changes in tomato leaf morphology. Using both constitutive and companion cell-specific promoters and *GAI* and control RNA sequence, only *GAI* transcripts were able to cross the graft union and move into the scion tissue (Haywood et al. 2005). These experiments showed that control over phloem transport of *GAI* is likely regulated by conserved sequence motifs present in the mobile RNA itself. By using a series of deletion mutants and movement assays with grafts of *Arabidopsis*, Huang and Yu (2009) showed that the UTRs of *GAI*

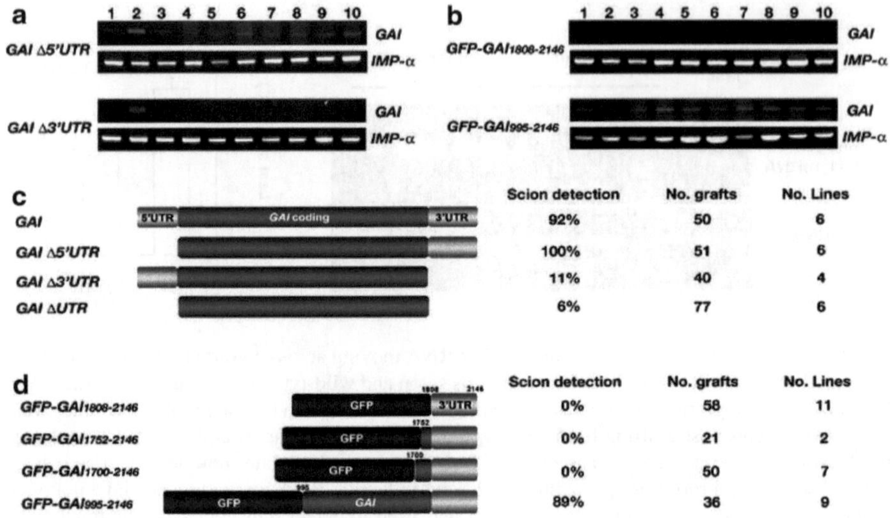

Fig. 2 Deletion analysis to identify motifs required for movement of *GAI* RNA. Reverse transcription-PCR analyses using RNA extracted from wild-type scions grafted onto (**a**) *PCaMV35S-GAI Δ5′ UTR* or *PCaMV35S–GAI Δ3′ UTR*, or (**b**) *PCaMV35S–GFP-GAI*$_{1808–2146}$ or *PCaMV35S-GFP-GAI*$_{995\text{-}2146}$ transformant stocks. The PCR was conducted with 35 cycles and the primers *GAI–RT–For* and *NOS term Rev* (**a**), or *GFP-For* and *GFP-Rev* (**b**). The gene *IMPORTIN-α* (*IMP-α*) was used as a loading control. (**c**) Statistical data from *Arabidopsis* grafting experiments performed with wild-type scions and transformant stocks carrying different UTR-truncated *GAI*s. For each construct, at least four independent transformants were used as stocks. The PCR was conducted with 35 cycles and the non-detection samples were further confirmed with 40 cycles. (**d**) *Arabidopsis* grafting experiments performed with transformant stocks carrying different *GFP–GAI* transgenes. The RNA of *GAI*$_{995–2146}$ is sufficient to target *GFP* for long-distance movement. With permission from Figure 3 of Huang and Yu (2009)

were essential for movement of *GAI* coding sequence and that the 3′ UTR was most critical (Fig. 2a–c). The 3′ UTR alone, however, could not move a green fluorescent protein (GFP) control tag without *GAI* coding sequence in the construct (Fig. 2d). From these results, it would appear that whereas specific sequences are critical, the overall secondary structure of the RNA may also influence its capacity to bind a protein partner and to facilitate long-distance movement. Through a linker-scanning analysis, specific regions of the *GAI* RNA were mapped for facilitating movement (Huang and Yu 2009). Three regions were identified: two within the coding sequence, designated A and B, and one in the 3′ UTR, designated motif C. In summary, these results suggest that both specific UTR sequence and folding of the *GAI* coding sequence regulate long-distance trafficking of *GAI* RNA.

As clearly shown in numerous studies on mobile RNA of animals, protein partners are involved in mediating movement and translation through an interaction with the target RNA (St. Johnston et al. 1991; Ferrandon et al. 1994; Gu et al. 2004; Lewis et al. 2004; King et al. 2005). Recently, an RNA/protein complex that transports *GAI* RNA of pumpkin has been identified that contains six mRNAs and up to sixteen proteins (Ham et al. 2009). This ribonucleoprotein complex is phloem mobile moving across pumpkin and cucumber heterograft unions, but the precise role of the proteins in this complex in transport and RNA metabolism is still not entirely clear. The core protein of this complex was RBP50, a PTB protein that recognizes groups of cytosine/uracil motifs. RNA gel shift assays confirmed binding of RBP50 with three regions of the *CmGAI* transcript. Two were located in the UTRs and the third was present in the coding sequence. All three contained clusters of cytosine/uracil motifs. Mutational analysis verified that the cytosine/uracil clusters mediated this RNA/protein interaction (Ham et al. 2009). PTB proteins have been implicated in numerous aspects of RNA metabolism and have been shown to bind select mobile RNAs of plants (Ham et al. 2009; Mahajan et al. 2012). They usually contain four RNA-recognition motifs (RRMs) and interact with sequence in UTRs containing groups of four cytosine/uracil motifs at least 4 nt in length (Oberstrass et al. 2005; Auweter and Allain 2008). The RNA sequences that interact with PTB proteins are typically uracil rich, with interspersed cytosines (Maynard and Hall 2010). The lengths of these cytosine/uracil motif sequences can vary enormously, from tetramers to hundreds of nucleotides, but each of the four RRMs recognizes one cytosine/uracil motif (Auweter and Allain 2008). Because of the importance of PTB proteins and UTR sequence in mediating RNA/protein complexes, the UTRs of three mobile RNAs, *CmGAI*, *AtGAI*, and *StBEL5*, were carefully examined for cytosine/uracil motifs (Table 3). PTB interaction has been previously demonstrated for both *CmGAI* and *StBEL5* (Ham et al. 2009; Mahajan et al. 2012). For purposes of this chapter, cytosine/uracil tetramers (and longer) were scored on the UTR sequences of these three RNAs.

As expected, *CmGAI* contained eight CU clusters in its 5′ UTR and nine in its 3′ UTR both located within regions of approximately 100 nt in length (underlined sequence for *CmGAIP*, Table 3). This 100-nt sequence in the 3′ UTR corresponds to the *GAIP*(3) probe confirmed as a positive interactor to RBP50 in Ham et al. (2009). *AtGAI* contained five CU clusters in its 5′ UTR within a 40-nt region and seven in its 3′ UTR located within a region of 102 nt in length (underlined sequence for *AtGAI*, Table 3). Using RNA-mobility assays discussed previously (Huang and Yu 2009), this 3′ UTR sequence that partially overlaps motif C (Fig. 4, Huang and Yu 2009) was established as critical for regulating movement of full-length *AtGAI*. *StBEL5* sequence contained only one CU motif in its 5′ UTR and a pair of five CU-motif clusters in its 3′ UTR located within two separate regions of approximately 130 nt in length (Table 3, *StBEL5*-b, underlined sequence). The first of these 130-nt regions, located in the 5′ end of the UTR and designated *T1*, was experimentally confirmed to bind to a potato PTB protein (Mahajan et al. 2012). Mobility assays for *StBEL5* indicated that the 3′ UTR of its transcript was most influential in transporting the RNA into stolon tips (Banerjee et al. 2009). Similarly, the mobile RNA and transcription partner of StBEL5, *POTH1* (Mahajan et al. 2012; Chen et al. 2004), contains four CU motifs in a 126-nt region of its 3′ UTR and interacted with the potato PTB protein, StPTB6 (Mahajan et al. 2012). Overall, these results strongly suggest that RBP50-like RNA-binding proteins play a critical role in mediating long-distance movement of select phloem mRNAs in a range of plant species.

Table 3 Cytosine/uracil (CU) motifs present in the 5′ (a) and 3′ (b) UTRs of three mobile RNAs, *CmGAIP* (Ham et al. 2009), *AtGAI* (Huang and Yu 2009), and *StBEL5* (Banerjee et al. 2006)

a ***CmGAIP***

1) caaucauucagaagaugcggguucacaccauuuuccuuugauucucucuccuucuuccc
ccauuuuucuacccaaucuccaccgucuaaacccaacccuuuuuuccucccuccgccacc
gccgccguuucu acacacg (138)

b

1878) gtcaauuccacucaguggaccgagucaggucaagaaugaaacaggacugacugucagcuug
acagagugacccaccacucgcaaatguggccatcacuuuugcctgtcauccugacccuucgg
cuucccgcccuucgucuuaguuggugguaucauaugaagggcucuucugtaguucuuc
ugagugguaaauuucggaagaagaaaaggauuuauguauuuauguuauugaauuaagu
agtaaauaaauguggaaaugaauuaauuagauuggtgggaataaaagaaaggagaagga
agaacaagagggguguuuugga (2202)
!

a ***AtGAI***

1) uaauaaucauuuuuuuucuuauaaccuuccucucuauuuuuacaauuuauuuuguuau
uagaagugguaguggagugaaaaaacaaauccuaagcaguccuaaccgauccccgaag
cuaagauucuucaccuucccaaauaaagcaaaaccuagatccgacattgaaggaaaaacc
uuuuagauccaucucugaaaaaaaccaacc (208)

b

1808) gaugguggcucaaugaauugaucuguugaaccgguuaugaugauagauuuccgaccga
agccaaacuaaauccuacuguuuuucccuuugucacuuguuaagaucuuaucuuuca
uuauauuagguaauugaaaauuuuaaucucgcuuuggagaguuuuuuuuuuuugca
ugugacauuggagguaaauuggauaggcagaaauagaaguauguguuaccaaguaug
ugcaauugguugaaauaaaaucaucuugagugucaccaucuauaaaauucauuguaau
gacuaaugagccugauuaaacugucucuuaugauaaugugcugauucucaug (2148)

a ***StBEL5***

1) caugcagagauaaaaauauagaucagucugacaagaaggcaacuucucaaagcuuaga
gagcuaccacccgaagauagacaguuaguuacauguacuguuauagauaaaaggagaa
auccgaagaagaaagaauuuuuuuugcagau (147)

b

2215) auaccgaaagucucguauugauagcugaaaagauaaaaggaaguuagggauuacucuu
auauugugugaggccuucuggcccaagucggaggacccaauuugauacaaccuaucaua
ggagaaaagaaguggagacuaaauuaaaguaacaaaauuuuaaagcacacuuucuagu
auauauacuucuuuuuuuuauaguauagaaaagaagagauuuugugcuuuaguguau
agauagagucuacuuaguauuagguuauacuucuagcuuccuugagaagauugauacaa
cuaguaguauuuuuuuucuuuuggguuggcuuggaguacuauuuuaaguuauuggaaa
cuagcuauaguaaauguuguaaaguugugauauuguuccucucaauuugcauauaauu
ugaaauauuuuguaccuacuagcuagucucuaaauuauguuuccauugcuuguaauugc
aauuuuauuugaauuuugugcuaucauuauuagauuagcaaaaaaaaaaaa (2717)

(continued)

Table 3 (continued)

CU motifs of four nt or more are designated in a bold red font. Cassettes of four or more motifs discussed in the text are underlined. Numbering for the 5′ UTR ends at the start codon. Numbering for the 3′ UTRs begins immediately after the stop codon for each. GenBank accession numbers are AY326306 for *CmGAIP*, Y15193 for *AtGAI*, and AF406697 for *StBEL5*. These CU motifs are putative targets for the four RNA-recognition motifs present in polypyrimidine tract-binding proteins that bind numerous RNAs to regulate their metabolism (Auweter and Allain 2008)

4 *FT* mRNA Moves to the Shoot Apex

The possibility that RNA of FLOWERING LOCUS T (FT) is moving long distance from leaf to shoot apex may be the most intriguing example of the six mobile RNAs presented here. In the current model, FT represents the floral signal that is transcribed and translated in leaves exposed to photoperiodic conditions inductive for flowering (Turck et al. 2008). The protein then moves upward through the vascular system into the shoot apical meristem. In the shoot apex, the FT protein interacts with the transcription factor FD to active the floral pathway. Initial studies suggested that *FT* mRNA moved to the shoot apex in *Arabidopsis* to induce flowering (Huang et al. 2005), but a subsequent retraction disavowed these early results (Böhlenius et al. 2007). A series of reports followed documenting the movement of FT protein correlated with floral induction in several plant species (Corbesier et al. 2007; Lin et al. 2007; Tamaki et al. 2007). Using heterografts with pumpkin as stock, pumpkin FT protein, but not its mRNA, was detected in the phloem sap of the short-day flowering cucurbit scion, *Cucurbita moschata* (Lin et al. 2007). Recent results suggest, however, that in addition to FT protein, *FT* RNA may also be moving to shoot apices to contribute to systemic floral signaling in some species (Li et al. 2011; Lu et al. 2012). Grafting experiments with *Arabidopsis* (Lu et al. 2012) demonstrated that *FT* RNA moves long distance from the stock to the scion apex. The *cis*-acting element that facilitated this movement was mapped to the first 102 nt of the *FT* mRNA coding sequence (Li et al. 2009).

Further evidence for the non-cell-autonomous mobility of *FT* RNA was generated by using a model transport system that utilizes two distinct movement-defective plant viruses, *potato virus X* (PVX, Fig. 3a) and *turnip crinkle virus*, and an agroinfiltration mobility assay (Li et al. 2009). This study demonstrated that (1) nontranslatable *FT* mRNA, independent of the FT protein, moves throughout *Nicotiana benthamiana* and *Arabidopsis* plants; (2) sequences from the *FT* RNA fused to GFP and viral RNAs made them more mobile in the plant; and (3) viral ectopic expression of *FT* in leaves and subsequent movement of its RNA induced flowering in *Nicotiana tabacum* cv. Maryland Mammoth under noninductive conditions (Fig. 3b, c). This induced flowering was correlated with the accumulation of functional *FT* transcripts in newly emerging leaves from the shoot apex (Fig. 3d). *Arabidopsis FT* mRNA, independent of the FT protein, could readily move into the shoot apical meristem (Li et al. 2011). Both WT and nontranslatable *FT* mRNA can mediate delivery of *potato virus X* sequence into the SAM, whereas

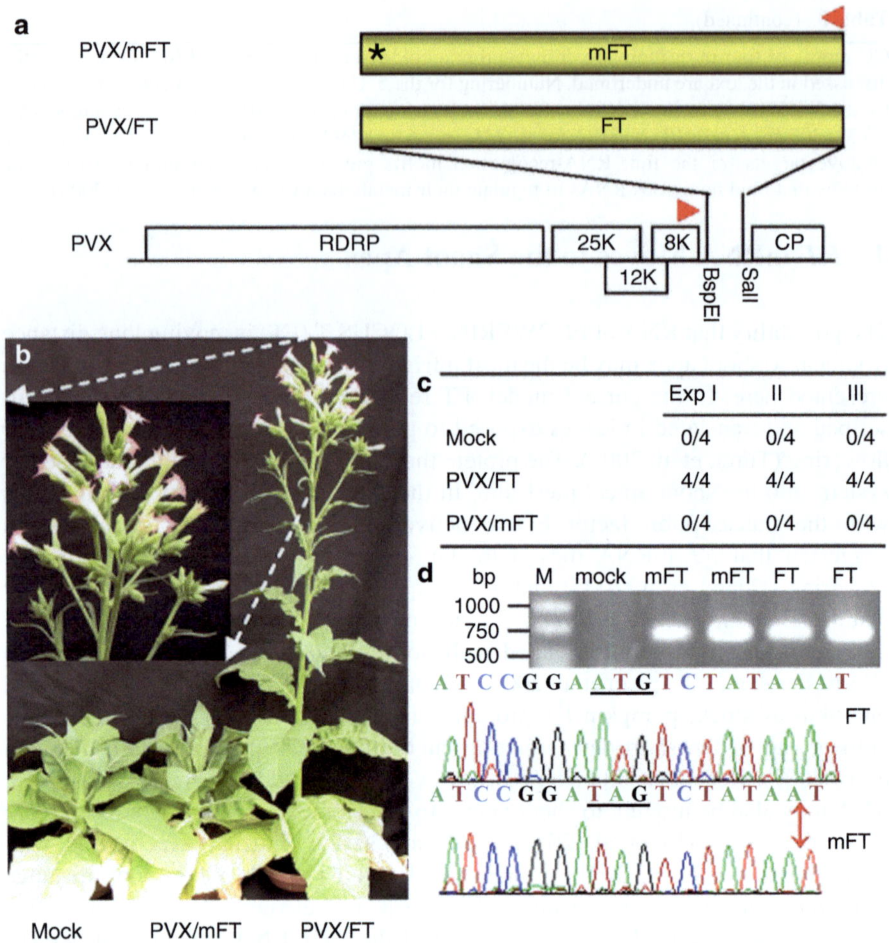

Fig. 3 Ectopic expression of *FT* induces flowering. (**a**) The translatable and mutated (*asterisk*) nontranslatable *Arabidopsis* FT coding sequences were cloned into wild-type PVX vector to produce PVX/FT and PVX/mFT, respectively. (**b** and **c**) Floral induction caused by viral expression of FT protein. Young SD *N. tabacum* Maryland Mammoth plants were mock inoculated or infected with PVX/FT or PVX/mFT and grown under a noninducing long-day photoperiod. Twelve plants infected by PVX/FT in three separate experiments started bolting at ~20 dpi, flowered at ~35 dpi, and were photographed at 42 dpi (**b** and *inset image*). Tobacco mock inoculated or infected with PVX/mFT did not flower (**b** and **c**). (**d**) Detection of viral transient *FT* RNA. Viral transient *FT* RNA was detected by RT-PCR using primers PP82 (▸) and PP356 (◂) in systemic young leaves from two separate plants infected with PVX/mFT (mFT) or PVX/FT (FT) but not in a mock-infected plant (mock). The position and the sizes of 1-kb DNA ladder (*lane* M) are indicated. Direct sequencing of RT-PCR products (648 bp) verified the presence of virally expressed wild-type and mutant *FT* RNA in flowering and nonflowering plants, respectively. The native *FT* ATG (*underlined*) in PVX/FT and its TAG replacement (*underlined*) together with a nucleotide deletion (*double-arrow*) in PVX/mFT are indicated. By permission from Li et al., J Virol. 2009 April; 83(8):3540-3548

PVX fused to GFP sequence could not (Li et al. 2011). The authors concluded that *Arabidopsis FT* mRNA could pass through the surveillance system that excludes viral RNAs from the SAM (Foster et al. 2002; Schwach et al. 2005) and that this selective entry occurs independently of suppression of RNA silencing. These experiments demonstrate that along with FT protein, *FT* RNA also moves long distance from source leaves through phloem cells to the shoot apex to regulate the photoperiodic floral signal.

5 *Aux/IAA* RNA Movement

Recent transcriptome analysis has shown that select *IAA* transcripts, designated *F-308* and *F-571*, were present in phloem sap of melon (Omid et al. 2007). Exogenous application of indole-3-acetic acid on hypocotyls resulted in a significant increase in levels of the two phloem sap *Aux/IAA* RNAs, indicating that both genes are auxin responsive. Indole-3-acetic acid (IAA) proteins are important negative regulators of auxin signaling and function in a diverse range of developmental processes. Using a melon:pumpkin heterografting system, Notaguchi et al. (2012) confirmed movement of the *F-308* transcripts across the graft union but not *F-571*. *Arabidopsis* orthologs of *F-308* did not move across a heterograft union, but two other *IAA* RNAs, *IAA18* and *IAA28*, transversed the union. These RNAs are transcribed in vascular cells of mature leaves and then transported to the roots where they function to inhibit lateral root development.

Further grafting experiments confirmed that *IAA18* and *IAA28* transcripts, but not their proteins, were delivered long distance into the root tip. In this study, the authors made use of a heterografting system using dominant mutants of Arabidopsis (designated *diaa18*) for scion grafted to wild-type or knock-out *iaa* mutant stocks to test for mobility of *IAA* transcripts and their effect on root growth. This work was unique in that phloem-mobile transcripts were delivered to a region of the plant in which their transcription was not detected. The *IAA18* system differs from previous reports on phloem-mobile transcripts in that examples discussed elsewhere in this chapter focus on genes that are expressed in both source and target organs (Kim et al. 2001; Haywood et al. 2005; Banerjee et al. 2006). Through this long-distance signaling process, Notaguchi et al. (2012) provide the experimental basis for a new concept in which auxin activity can be regulated by both auxin distribution and phloem-mobile *Aux/IAA* mRNAs.

6 *StBEL5* Acts as a Mobile Signal in Potato

StBEL5, a BEL1-like transcription factor of the TALE superfamily, regulates development in potato by binding to upstream sequences of numerous target genes. Working in tandem with the Knotted1-type transcription factor, POTH1,

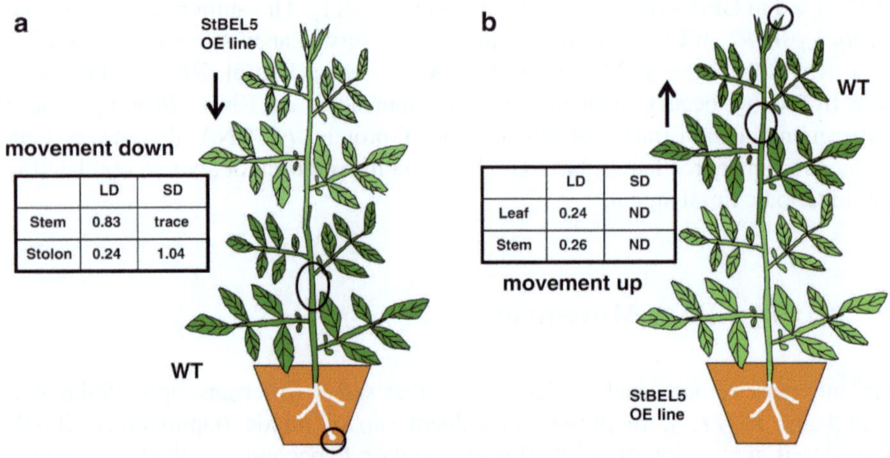

Fig. 4 Full-length *StBEL5* RNA movement in a downward (**a**) or upward (**b**) direction in response to photoperiod in the photoperiod-responsive potato subspecies, *Solanum tuberosum* ssp. *andigena*. *StBEL5* RNA was quantified using gene-specific primers that amplified only transgenic RNA in real-time qRT-PCR. *BEL5* RNA was calculated in pg of product in the wild-type stock (**a**) or scion (**b**) per 100 pg of cDNA from a transgenic source leaf. The product was quantified using DNA markers of known concentration as a reference. Wild-type sample sources are indicated by *circles*. The transgenic lines expressed full-length *StBEL5* RNA driven by the 35S CaMV promoter. Grafts were wrapped in plastic and allowed to heal for 2 weeks at 25 °C under long-day conditions (16 h light, 8 h dark) and were then transferred to short days (8 h light, 16 h dark) or maintained under long days for 2 more weeks. *ND* not detected

StBEL5 mediates vegetative development by regulating hormone levels (Chen et al. 2004). The BEL5/POTH1 complex binds specifically to a double TTGAC core motif. In both in vitro and soil-grown plants, overexpression of StBEL5 has been consistently correlated with enhanced tuber formation (Chen et al. 2003; Banerjee et al. 2009). Using a leaf-specific promoter in transgenic plants, tuberization was again enhanced in concordance with accumulation of its mRNA in stolon tips, the site of tuber initiation. Both in situ hybridization and isolation of phloem RNA using laser capture microdissection confirmed the presence of *StBEL5* transcripts in phloem cells (Banerjee et al. 2006). Heterografting experiments utilizing StBEL5 transgenic lines as scions confirmed the downward movement of its full-length RNA across the graft union to localize in stolon tips (Banerjee et al. 2006). Together these results implicate *StBEL5* RNA as a mobile signal that induces tuber formation.

Promoter analysis showed that *BEL5* transcription originated in phloem cells of leaf veins and petioles and was induced by low levels of light, but was not active in stems despite a substantial level of transcript accumulation (Banerjee et al. 2006). Whereas photoperiod had no effect on transcription, three separate experiments, one using heterografts (Fig. 4) and two utilizing whole transgenic plants (Fig. 5), confirmed that movement of *StBEL5* transcripts was enhanced by a short-day photoperiod. In day length-responsive potato species, a short-day photoperiod is

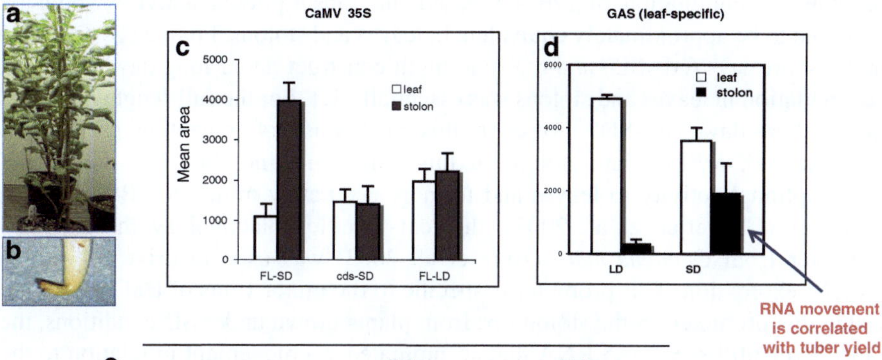

e

line	Tuber number	Tuber yield
GAS::GUS	1.2 ± 0.4	26 ± 2.6
WT 7540	1.3 ± 0.8	22 ± 2.4
GAS::StBEL5	7.5 ± 1.0	45 ± 6.6

Fig. 5 The effect of untranslated regions (UTRs) of the *StBEL5* RNA and photoperiod on RNA movement to stolon tips. Movement of transcripts with (FL) and without (cds) the UTRs were assayed. Whole transgenic plants (**a**) were grown under long days (16 h light, 8 h dark) for 3 weeks and then were either transferred to short days (8 h light, 16 h dark) or maintained under long days for 14 days. RNA was then extracted from 0.5-cm stolon tips (**b**) and new leaves from three separate plants for each construct (**c**, **d**). One-step RT-PCR was performed using 20 (**c**) or 500 (**d**) ng of total RNA and a nonplant sequence tag fused to all transgenic RNAs and a gene-specific primer. The PCR reactions were standardized and optimized to yield product in the linear range, normalized using rRNA, and quantified by using ImageJ software (Abramoff et al. 2004). Harvested plants were scored for tuber numbers after 10 days and tuber yields after 28 days under short day conditions (**e**). Standard errors of the means of three biological replicates are shown for *panels* (**c**, **d** and **e**). Constructs were driven by either the CaMV 35S promoter (**c**) or the leaf-specific GAS promoter (**d** and **e**; Ayre et al. 2003). *Open bars*, leaf; *closed bars*, stolon. In potato, the GAS promoter is active only in the minor veins of leaves (reprinted from Banerjee et al. 2006; Copyright American Society of Plant Biologists, http://www.plantcell.org)

inductive for tuber formation, whereas long days are suppressive. In the heterografting experiment using transgenic BEL5 overexpression lines as both scion and stock, full-length *BEL5* transgenic RNA moved in both an upward and downward direction but with differences depending on day length (Fig. 4). Based on quantitative RT-PCR, some upward movement of transgenic *BEL5* RNA into wild-type stems and new leaves was detected under long days but not short days (Fig. 4b). Movement downward into wild-type stolon tips was detected under both photoperiods but was enhanced by more than fourfold under short days (Fig. 4a). Using two different promoters in transgenic plants, movement of *StBEL5* into stolon tips was also monitored in whole plants (Fig. 5a, b). In the first experiment, movement of both the full-length *BEL5* transcript and a truncated version containing only the coding sequence (cds) without UTRs driven by the constitutive CaMV 35S promoter was evaluated under long and short days (Fig. 5c). Using this promoter, accumulation of *StBEL5* based on transcriptional activity would be expected to be approximately equivalent in leaves and stolons. For the cds construct under short days (cds-SD) and the full-length construct under long days (FL-LD), accumulation in leaves and stolons was essentially 1:1. For the full-length construct under short days (FL-SD), however, this ratio was 1:4, reflecting preferential movement of *StBEL5* into stolons of this transgenic line (Fig. 5c). To restrict transcriptional activity to leaves and to more accurately mimic the BEL5 native promoter (Chatterjee et al. 2007), the leaf-specific galactinol synthase (GAS) promoter from *Cucumis melo* (Ayre et al. 2003) was used to drive full-length *StBEL5* expression. This promoter is specific to the minor veins of leaf mesophyll. Using this promoter, in the stolon tips from plants grown under SD conditions, the ratio of quantified *StBEL5* RNA that accumulated via movement in relation to the source leaf was sevenfold greater in SD compared to LD plants (Fig. 5d). This increased RNA mobility and accumulation in stolons was correlated with earliness (more tubers after 10 days of SD inductive conditions) and enhanced tuber yields (Fig. 5e). In summary, three important conclusions can be drawn from these results (1) the UTRs of *StBEL5* affect movement of its RNA to stolon tips, (2) short-day photoperiodic conditions enhance movement, and (3) accumulation of *StBEL5* in stolons as a result of its long-distance movement is again associated with yield. Using both heterografts and mobility assays in whole plants, the movement of *StBEL5* RNA into roots was also recently confirmed (Lin et al. 2013). This movement was correlated with increased growth, changes in morphology, and the accumulation of specific transcripts for genes associated with hormone metabolism.

7 Conclusions

Both transcriptomic and proteomic analyses have revealed the breadth of presence of mobile signals moving long distances through the plant's vascular system. Through phloem cell and sap analyses, we now know there are thousands of potential mobile RNAs transported through the phloem with putative functions in stress response, metabolism, and development. Our knowledge of the function of these non-cell-autonomous RNAs is meager at best. Yet still we may pose the question, why move so many RNAs? The transport and localization of full-length mRNA in plants serves as a dynamic cellular tool for rapidly responding to environmental cues and tightly regulating developmental processes. From animal systems, we clearly understand that movement of transcripts within a cell is an efficient, well-regulated process that involves transport and translation suppression mediated by RNA/protein, RNA/RNA (Ferrandon et al. 1997), and protein/protein interactions (Elvira et al. 2006). Repression of translation ensures that the mobile mRNAs are functional only at the target site (King et al. 2005).

Transporting RNA is cost efficient since translating an mRNA on site reduces the energy cost of transporting a protein (Du et al. 2007). In addition, localizing RNA provides spatial control over protein synthesis that ensures the protein is not translated at any other location than the functional site. Transport of the mRNA and local, targeted translation of the protein minimizes the possibility of misexpression that could potentially lead to aberrant growth patterns. Mobilizing RNA and regulating translation can channel the localization of a specific regulatory protein like a transcription factor to control site-specific growth or overcome a threshold gradient. Movement of *StBEL5* through the stem into the stolon tips where it is functional to activate the tuberization pathway follows this rationale. As another example, the polarity of the developing *Drosophila* embryo is determined by the transport and localization of the select RNAs: *oskar*, *bicoid*, *gurken*, and *nanos*. This chaperone-mediated localization leads to morphogen gradients establishing the pattern of embryogenesis (Ephrussi et al. 1991; Mahowald 2001).

A more expansive characterization of the RNA/protein complexes that move through the phloem is a critical challenge inherent in understanding this system. RNA-binding proteins control stability and translation and most certainly regulate direction and final destination of mobile RNAs. Linear and structural motifs in RNA sequence mediate these protein interactions. The role of PTB proteins in mediating RNA metabolism and movement provides a valuable model for exploring the process. Three of the mobile RNAs discussed here contained clusters of polypyrimidine motifs and interacted with PTB proteins in RNA/protein-binding assays. These results suggest that PTB proteins play a general role in regulating the movement and translatability of non-cell-autonomous RNAs in plants. Trafficking and delivery of mRNAs provide a unique mechanism for selective control of targeted gene expression in plants. Much more work is still needed, however, to better understand the processes of recognition, delivery, and release inherent in phloem transport of mobile RNAs.

Acknowledgements Thanks to Anjan Banerjee, Yiguo Hong, and Tien-Shin Yu for allowing use of their data for this book chapter. Thanks also to Anjan Banerjee, Mithu Chatterjee, Ameya Mahajan, and Xin Meng for their dedicated contributions to this work.

References

Abramoff MD, Magelhaes PJ, Ram SJ (2004) Image processing with ImageJ. Biophoton Int 11:36–42

Aravind L, Iyer LM, Anantharaman V (2003) The two faces of Alba: the evolutionary connection between proteins participating in chromatin structure and RNA metabolism. Genome Biol 4: R64

Auweter SD, Allain FHT (2008) Structure-function relationships of the polypyrimidine tract binding protein. Cell Mol Life Sci 65:516–527

Ayre BG, Blair JE, Turgeon R (2003) Functional and phylogenetic analyses of a conserved regulatory program in the phloem of minor veins. Plant Physiol 133:1229–1239

Bai S, Kasai A, Yamada K, Li T, Harada T (2011) A mobile signal transported over a long distance induces systemic transcriptional gene silencing in a grafted partner. J Exp Bot 62:4561–4570

Banerjee AK, Chatterjee M, Yu Y, Suh SG, Miller WA, Hannapel DJ (2006) Dynamics of a mobile RNA of potato involved in a long-distance signaling pathway. Plant Cell 18:3443–3457

Banerjee AK, Lin T, Hannapel DJ (2009) Untranslated regions of a mobile transcript mediate RNA metabolism. Plant Physiol 151:1831–1843

Böhlenius H, Eriksson S, Parcy F, Nilsson O (2007) Retraction. Science 316:367

Chatterjee M, Banerjee AK, Hannapel DJ (2007) A BELL1-like gene of potato is light activated and wound inducible. Plant Physiol 145:1435–1443

Chen H, Rosin FM, Prat S, Hannapel DJ (2003) Interacting transcription factors from the TALE superclass regulate tuber formation. Plant Physiol 132:1391–1404

Chen H, Banerjee AK, Hannapel DJ (2004) The tandem complex of BEL and KNOX partners is required for transcriptional repression of ga20ox1. Plant J 38:276–284

Corbesier L, Vincent C, Jang S, Fornara F, Fan Q, Searle I, Giakountis A, Farrona S, Gissot L, Turnbull C, Coupland G (2007) FT protein movement contributes to long-distance signaling in floral induction of *Arabidopsis*. Science 316:1030–1033

David-Schwartz R, Runo S, Townsley B, Machuka J, Sinha N (2008) Long-distance transport of mRNA via parenchyma cells and phloem across the host-parasite junction in *Cuscuta*. New Phytol 179:1133–1141

Du TG, Schmid M, Jansen RP (2007) Why cells move messages: the biological functions of mRNA localization. Semin Cell Dev Biol 18:171–177

Elvira G, Massie B, DesGroseillers L (2006) The zinc-finger protein ZFR is critical for Staufen 2 isoform specific nucleocytoplasmic shuttling in neurons. J Neurochem 96:105–117

Ephrussi A, Dickinson LK, Lehmann R (1991) *oskar* organizes the germplasm and directs localization of the posterior determinant *nanos*. Cell 66:37–50

Ferrandon D, Elphick L, Nusslein-Volhard C, St. Johnston D (1994) Staufen protein associates with the 3′ UTR of bicoid mRNA to form particles that move in a microtubule-dependent manner. Cell 79:1221–1223

Ferrandon D, Koch I, Westhof E, Nusslein-Volhard C (1997) RNA-RNA interaction is required for the formation of specific bicoid mRNA 3′ UTR-STAUFEN ribonucleoprotein particles. EMBO J 16:1751–1758

Foster TM, Lough TJ, Emerson SJ, Lee RH, Bowman JL, Forster RLS, Lucas WJ (2002) A surveillance system regulates selective entry of RNA into the shoot apex. Plant Cell 14:1497–1508

Gu W, Deng Y, Zenklusen D, Singer RH (2004) A new yeast PUF family protein, Puf6p, represses ASH1 mRNA translation and is required for its localization. Genes Dev 18:1452–1465

Hake S, Smith HM, Holtan H, Magnani E, Mele G, Ramirez J (2004) The role of knox genes in plant development. Annu Rev Cell Dev Biol 20:125–151

Ham BK, Brandom JL, Xoconostle-Cazares B, Ringgold V, Lough TL, Lucas WJ (2009) A polypyrimidine tract binding protein, pumpkin RBP50, forms the basis of a phloem-mobile ribonucleoprotein complex. Plant Cell 21:197–215

Haywood V, Yu TS, Huang NC, Lucas WJ (2005) Phloem long-distance trafficking of GIBBERELLIC ACID-INSENSITIVE RNA regulates leaf development. Plant J 42:49–68

Huang NC, Yu TS (2009) The sequences of *Arabidopsis* GA-INSENSITIVE RNA constitute the motifs that are necessary and sufficient for RNA long-distance trafficking. Plant J 59:921–929

Huang T, Böhlenius H, Eriksson S, Parcy F, Nilsson O (2005) The mRNA of the *Arabidopsis* gene FT moves from leaf to shoot apex and induces flowering. Science 309:1694–1696

Hyun TK, Uddin MN, Rim Y, Kim JY (2011) Cell-to-cell trafficking of RNA and RNA silencing through plasmodesmata. Protoplasma 248:101–116

Kanehira A, Yamada K, Iwaya T, Tsuwamoto R, Kasai A, Nakazono M, Harada T (2010) Apple phloem cells contain some mRNAs transported over long distances. Tree Genet Genomes 5:635–642

Kasai A, Bai S, Li T, Harada T (2011) Graft-transmitted siRNA signal from the root induces visual manifestation of endogenous post-transcriptional gene silencing in the scion. PLoS One 6: e16895

Kehr J, Buhtz A (2008) Long distance transport and movement of RNA through the phloem. J Exp Bot 59:85–92

Kim M, Canio W, Kessler S, Sinha N (2001) Developmental changes due to long-distance movement of a homeobox fusion transcript in tomato. Science 293:287–289

Kim JY, Yuan Z, Cilia M, Khalfan-Jagani Z, Jackson D (2002) Intercellular trafficking of a KNOTTED1 green fluorescent protein fusion in the leaf and shoot meristem of *Arabidopsis*. Proc Natl Acad Sci USA 99:4103–4108

Kim JY, Rim Y, Wang J, Jackson D (2005) A novel cell-to-cell trafficking assay indicates that the KNOX homeodomain is necessary and sufficient for intercellular protein and mRNA trafficking. Genes Dev 19:788–793

King ML, Messitt TJ, Mowry KL (2005) Putting RNAs in the right place at the right time: RNA localization in the frog oocyte. Biol Cell 97:19–33

Koornneef M, Elgersma A, Hanhart CJ, van Loenen-Martinet EP, van Rign L, Zeevaart JAD (1985) A gibberellin insensitive mutant of *Arabidopsis thaliana*. Physiol Plant 65:33–39

LeBlanc M, Kim G, Westwood JH (2012) RNA trafficking in parasitic plant systems. Front Plant Sci 3:203

Lewis RA, Kress TL, Cote CA, Gautreau D, Rokop ME, Mowry KL (2004) Conserved and clustered RNA recognition sequences are a critical feature of signals directing RNA localization in *Xenopus* oocytes. Mech Dev 121:101–109

Li C, Zhang K, Zeng X, Jackson S, Zhou Y, Hong Y (2009) A *cis* element within flowering locus T mRNA determines its mobility and facilitates trafficking of heterologous viral RNA. J Virol 83:3540–3548

Li C, Gu M, Shi N, Zhang H, Yang X, Osman T, Liu Y, Wang H, Vatish M, Jackson S, Hong Y (2011) Mobile FT mRNA contributes to the systemic florigen signalling in floral induction. Sci Rep 1:73

Lin MK, Belanger H, Lee YJ, Varkonyi-Gasic E, Taoka K, Miura E, Xoconostle-Cázares B, Gendler K, Jorgensen RA, Phinney B, Lough TJ, Lucas WJ (2007) FLOWERING LOCUS T protein may act as the long-distance florigenic signal in the cucurbits. Plant Cell 19:1488–1506

Lin MK, Lee YJ, Lough TJ, Phinney BS, Lucas WJ (2009) Analysis of the pumpkin phloem proteome provides insights into angiosperm sieve tube function. Mol Cell Proteomics 8:343–356

Lin T, Sharma P, Gonzalez DH, Viola IL, Hannapel DJ (2013) The impact of the long-distance transport of a BEL1-like mRNA on development. Plant Physiol 161:760–772

Lu KJ, Huang NC, Liu YS, Lu CA, Yu TS (2012) Long-distance movement of Arabidopsis FLOWERING LOCUS T RNA participates in systemic floral regulation. RNA Biol 9:653–662

Lucas WJ (2006) Plant viral movement proteins: agents for cell-to-cell trafficking of viral genomes. Virology 344:169–184

Lucas WJ, Bouché-Pillon S, Jackson DP, Nguyen L, Baker L, Ding B, Hake S (1995) Selective trafficking of KNOTTED1 homeodomain protein and its mRNA through plasmodesmata. Science 270:1980–1983

Lucas WJ, Ham BK, Kim JY (2009) Plasmodesmata – bridging the gap between neighboring plant cells. Trends Cell Biol 19:495–503

Mahajan A, Bhogale S, Kang IH, Hannapel DJ, Banerjee AK (2012) The mRNA of a Knotted1-like transcription factor of potato is phloem mobile. Plant Mol Biol 79:595–608

Mahowald AP (2001) Assembly of the Drosophila germplasm. Int Rev Cytol 203:187–213

Mani J, Güttinger A, Schimanski B, Heller M, Acosta-Serrano A, Pescher P, Späth G, Roditi I (2011) Alba-domain proteins of Trypanosoma brucei are cytoplasmic RNA-binding proteins that interact with the translation machinery. PLoS One 6:e22463

Martin A, Adam H, Diaz-Mendoza M, Zurczak M, Gonzalez-Schain ND, Suarez-Lopez P (2009) Graft-transmissible induction of potato tuberization by the microRNA miR172. Development 136:2873–2881

Maynard CM, Hall KB (2010) Interactions between PTB RRMs induce slow motions and increase RNA binding affinity. J Mol Biol 397:260–277

Melnyk CW, Molnar A, Bassett A, Baulcombe DC (2011) Mobile 24 nt small RNAs direct transcriptional gene silencing in the root meristems of *Arabidopsis thaliana*. Curr Biol 21:1678–1683

Notaguchi M, Wolf S, Lucas WJ (2012) Phloem-mobile *Aux/IAA* transcripts target to the root tip and modify root architecture. J Integr Plant Biol 54:760–772

Oberstrass FC, Auweter SD, Erat M, Hargous Y, Henning A, Wenter P, Reymond L, Amir-Ahmady B, Pitsch S, Black DL, Allain FHT (2005) Structure of PTB bound to RNA: specific binding and implications for splicing regulation. Science 309:2054–2057

Omid A, Keilin T, Glass A, Leshkowitz D, Wolf S (2007) Characterization of phloem-sap transcription profile in melon plants. J Exp Bot 58:3645–3656

Peng J, Carol P, Richards DE, King KE, Cowling RJ, Murphy GP, Harberd NP (1997) The Arabidopsis GAI gene defines a signaling pathway that negatively regulates gibberellin responses. Genes Dev 11:3194–3205

Pysh LD, Wysocka-Diller JW, Camilleri C, Bouchez D, Benfey PN (1999) The GRAS gene family in Arabidopsis: sequence characterization and basic expression analysis of the SCARECROW-LIKE genes. Plant J 18:111–119

Qi Y, Pelissier T, Itaya A, Hunt E, Wassenegger M, Ding B (2004) Direct role of a viroid RNA motif in mediating directional RNA trafficking across a specific cellular boundary. Plant Cell 16:1741–1752

Richards DE, King KE, Aitali T, Harberd NP (2001) How gibberellin regulates plant growth and development: a molecular genetic analysis of gibberellin signaling. Annu Rev Plant Physiol Plant Mol Biol 52:67–88

Roney JK, Khatibi PA, Westwood JH (2007) Cross-species translocation of mRNA from host plants into the parasitic plant dodder. Plant Physiol 143:1037–1043

Rosin FM, Hart JK, Horner HT, Davies PJ, Hannapel DJ (2003) Overexpression of a knotted-like homeobox gene of potato alters vegetative development by decreasing gibberellin accumulation. Plant Physiol 132:106–117

Ruiz-Medrano R, Xoconostle-Cazares B, Lucas WJ (1999) Phloem long-distance transport of CmNACP mRNA: implications for supracellular regulation in plants. Development 126:4405–4419

Schwach F, Vaistij FE, Jones L, Baulcombe DC (2005) An RNA-dependent RNA polymerase prevents meristem invasion by potato virus X and is required for the activity but not the production of a systemic silencing signal. Plant Physiol 138:1842–1852

St. Johnston D, Beuchle D, Nusslein-Volhard C (1991) Staufen, a gene required to localize maternal RNAs in the Drosophila egg. Cell 66:51–63

Tamaki S, Matsuo S, Wong HL, Yokoi S, Shimamoto K (2007) Hd3a protein is a mobile flowering signal in rice. Science 316:1033–1036

Turck F, Fornara F, Coupland G (2008) Regulation and identity of florigen: FLOWERING LOCUS T moves center stage. Annu Rev Plant Biol 59:573–594

Xu H, Zhang W, Li M, Harada T, Han Z, Li T (2010) Gibberellic acid insensitive mRNA transport in both directions between stock and scion in Malus. Tree Genet Genomes 6:1013–1019

Yang HW, Yu TS (2010) Arabidopsis floral regulators *FVE* and *AGL24* are phloem-mobile RNAs. Bot Stud 51:17–26

Yu YY, Lashbrook CC, Hannapel DJ (2007) Tissue integrity and RNA quality of laser microdissected phloem of potato. Planta 226:797–803

Zhong X, Tao X, Stombaugh J, Leontis N, Ding B (2007) Tertiary structure and function of an RNA motif required for plant vascular entry to initiate systemic trafficking. EMBO J 26:3836–3846

Systemic Epigenetic Signaling in Plants

Andriy Bilichak and Igor Kovalchuk

Abstract Plants have developed a plethora of signaling pathways which allow them to respond quickly to the alterations in the environmental conditions in order to reduce their negative impact. Emerging evidence indicates that vascular plants can memorize changes in the transcriptome profile after stress exposure and in some cases propagate it into the next generation. This phenomenon is termed "transgenerational inheritance." Curiously, short-term and transgenerational plasticity of plant phenotypes does not involve changes in the DNA sequence, but instead involve reversible changes in chromatin structure that determine DNA accessibility for transcriptional factors. Chromatin structure reshaping depends on epigenetic factors, such as DNA methylation, histone posttranslational modifications/replacements, and small RNA (smRNA) metabolism, which form a flexible self-reinforcing loop of gene regulation. In the following chapter, we will provide some examples of gene activity regulation through alterations in the epigenetic profile in response to environmental stimuli. Additionally, we will discuss a systemic propagation of the acquired stress-induced epigenetic changes into the progeny and the possible contribution of epigenetic components to the process of plant adaptation and acclimation.

Keywords Transgenerational inheritance • Systemic epigenetic signalling • Small non-coding RNAs • Chromatin structure • Epigenetic reprogramming

1 Introduction

Organisms are in constant interaction with environmental cues which can either benefit or jeopardize their homeostasis, depending on the intensity of abiotic and biotic factors encountered. If perceived environmental factors deviate substantially

A. Bilichak • I. Kovalchuk (✉)
Department of Biological Sciences, University of Lethbridge, Lethbridge, AB, Canada
e-mail: a.bilichak@uleth.ca; igor.kovalchuk@uleth.ca

F. Baluška (ed.), *Long-Distance Systemic Signaling and Communication in Plants*, Signaling and Communication in Plants 19, DOI 10.1007/978-3-642-36470-9_4,

from the organism's optimal range, the organism initiates specific or nonspecific stress responses which in animals can simply lead to the avoidance of or escape from unfavorable conditions, while in plants, they can result in metabolic and physiologic alterations. A rapid adaptation to adverse abiotic or biotic cues is the main prerequisite of survival in plants. A memory of a robust transcriptional response to the imposed stress can be propagated into the next generation in order to minimize the amount of energy used to acquire the same parental phenotype by the progeny. At the same time, the memory of stress exposure has to be reversible by its nature to bring the flexibility into the process of adaptation and acclimation. The only pathway that meets all the above-mentioned requirements known thus far involves epigenetic factors that regulate gene activity. Known as "soft inheritance" compared to hard or Mendelian inheritance, an epigenetic memory is a rapid and plastic system which may enable both short-term tuning of gene expression immediately after stress and inheritance into the following generation (Youngson and Whitelaw 2008). In contrast to genetics, epigenetics at the molecular level involves the modulation of gene expression through the structural adaptation of chromosomal regions that is not accompanied by any changes in DNA sequence (Bird 2007; Kovalchuk and Kovalchuk 2012).

Nowadays, plant epigenetic stands on the three main whales of gene expression regulation: DNA methylation, histone posttranslational modifications/replacement, and small RNA metabolism. All three components interact with each other forming a self-reinforcing loop of gene expression regulation. Chromatin structure is shaped by the activity of enzymes which either methylate cytosine residues in DNA and bring chemical modifications to histone tails or use the energy released from ATP hydrolysis to disrupt histone-DNA interactions (de la Serna et al. 2006). These fluctuations in chromatin properties are observed throughout the life of organisms and undergo precise and coordinated changes at the defined stages of development. Moreover, in recent years, it has become apparent that dynamic alterations in chromatin structure and the metabolism of smRNAs also contribute to the modulation of gene expression, which is important for stress response. Emerging evidence indicates that a stress response mediated by epigenetics carries a systemic component, with smRNAs being possible messengers (see Antoniades and Watts 2013).

Global changes in the epigenetic landscape of exposed tissues after stress treatment are well documented; but unfortunately, studies are scarce that indicate the acquired epigenetic alterations in systemic untreated tissues in plants. In this chapter, we will provide a brief overview of documented alterations in DNA methylation, chromatin modifications, and smRNA metabolism in response to stress and provide some examples of the systemic propagation of stress memory into the following unstressed generation.

2 Systemic Epigenetic Signaling

The coining of the term "systemic epigenetic signaling" (SES) can be attributed to Wingard (1928) who described acquired resistance against viral invasion in noninfected tissues of tobacco plants. In his experiments, localized infection of lower leaves of tobacco plants with the *Tobacco ringspot virus* triggered strong symptoms of the infection, while the upper uninfected leaves did not exhibit any signs of pathogen attack and became resistant to the following infection with the same virus (Wingard 1928). Later, confirmation of this phenomenon was provided by David Baulcombe's and Sandrine Balzergue's groups who demonstrated the propagation of posttranscriptional transgene silencing to bystander tissues in transgenic tobacco plants triggered by the localized ectopic expression of homologous sequences (Voinnet et al. 1998; Palauqui and Balzergue 1999; Palauqui et al. 1997). Almost simultaneously, both groups obtained intriguing data on the generation of a specific nucleic acid signal that can be produced at the place of homologous sequence delivery, can be conveyed to systemic untreated tissue, and can trigger epigenetic variations of gene expression. The initiation of systemic acquired silencing (SAS) (Palauqui and Balzergue 1999; Palauqui et al. 1997) was achieved regardless of a method for delivery of homologous DNA, transgene orientation, and the availability of the promoter in DNA constructs. Nevertheless, DNA length and its concentration used for bombardment had a significant impact on the efficiency of both localized acquired silencing and SAS (Palauqui and Balzergue 1999). Both these groups suggested a possible involvement of the systemic spreading of posttranscriptional gene silencing as a defense system against viral invasion. This speculation was further supported by a number of studies providing an evidence of an ancient immune system against pathogen attacks, transposons, and transgenes (Yaegashi et al. 2008; Vance and Vaucheret 2001).

Hence, we can define systemic epigenetic signaling as a specific response of an organism to stress that triggers the appearance of a wide spectrum of measurable alterations in bystander tissues during plant ontogenesis, thus resulting in a signal that is faithfully transmitted to gametes and eventually is detected in the next generation as an altered epigenetic landscape.

Possible nucleotide messengers that can trigger SES involve small RNA sequences generated at the place of a localized viral invasion, transgene delivery, pathogen attack, or stress exposure. Epigenetic suppression of gene expression can act through two distinct but not mutually exclusive mechanisms: posttranscriptional and transcriptional inhibition of gene activity called posttranscriptional gene silencing (PTGS) and transcriptional gene silencing (TGS). Apart from playing a defense role in plant ontogenesis, PTGS and TGS pathways were found to be the key components in guiding chromatin remodeling, development, and propagation (Molnar et al. 2011).

Previous research has provided an explicit proof that short-range (less than 15 cells) (Dunoyer et al. 2005; Hamilton et al. 2002; Himber et al. 2003) and long-range epigenetic signaling (between tissues and organs) (Palauqui et al. 1997; Voinnet and Baulcombe 1997) are the two distinct mechanisms with possibly different messengers. Whereas in the former case the accumulation of short (21–22 nt) smRNAs correlates with mRNA decay but not with systemic signaling or RNA-directed DNA methylation (RdDM), in the latter case, the accretion of long (24–26 nt) smRNAs is associated with silencing in systemic tissues and methylation of homologous DNA (Hamilton et al. 2002).

Environmental factors are perceived and transmitted by a wide range of plant signal receptors which by turning on specific transcription factors trigger modulations of the activities of genes encoding effector proteins that enable plant adaptation (Mirouze and Paszkowski 2011). Emerging evidence indicates that epigenetic components are at the forefront of plant stress response, albeit studies of the perception of environmental cues by the epigenetic machinery are still in its infancy.

3 Stress Perception and Epigenetics

Despite numerous complex feedback controls that are known to be activated in response to stress, we are still missing a link between stress perception and the epigenetic machinery. Only recent research sheds new light on this phenomenon, bringing epigenetic factors into a complex stress-sensing mechanism (Kumar and Wigge 2010). A direct connection between the perception ambient temperature fluctuations and modifications of the epigenetic landscape comes from the research of Kumar and Wigge (2010). Searching for genes that orchestrate the plant transcriptome in response to heat stress, the authors identified the nuclear *ACTIN-RELATEDPROTEIN 6* (*ARP6*) mutants with more than 5,000 genes that were constitutively mis-regulated regardless of the ambient temperature. A product of the *ARP6* gene is a component of the SWR1 chromatin remodeling complex which substitutes histone H2A for the alternative histone variant H2A.Z in euchromatin nucleosomes (Deal et al. 2007; Kobor et al. 2004). An increase in the ambient temperature was positively correlated with the eviction of H2A.Z from the nucleosomes at the 5′ region of heat-responsive genes. Intriguingly, it has recently been shown that *ARP6* manages the expression of phosphate starvation genes through the deposition of histone variant H2A.Z at transcription-start sites (TSS) of the respective genes (Smith et al. 2010). Moreover, H2A.Z histone replacement has been found to be vital to control immunity of Arabidopsis plants against pathogen attacks (March-Diaz et al. 2008; van den Burg and Takken 2009). Nevertheless, the result of the enrichment of histone variant H2A.Z at TSS is not unidirectional since it has both the negative and positive effect on the expression of encoded genes, thus suggesting the involvement of additional activators or repressors of the gene activity (Kumar and Wigge 2010; Guillemette et al. 2005; March-Diaz et al. 2008). The modulation of the activity at TSS enriched with histone variant H2A.Z may also be a combinatorial consequence of the number of posttranslational modifications at histone tails (Bonisch and Hake 2012), an exact nucleosome position relative to a positive regulatory DNA sequence (Marques et al. 2010), or an availability of other chromatin modifiers. For instance, the H2A.Z histone variant was found to act together with 16 different histone modifications at more than 3000 genes in the human genome (Wang et al. 2008). Taking into account a strikingly high conservation of the H2A.Z histone variant among different species (almost 80 % interspecies identity; Bonisch and Hake

2012), one can hypothesize that H2A.Z maintains the unique and specific functions that probably cannot be performed by other histones. Indeed, in budding yeast, the *htz1Δ* mutant which lacks the ability to deposit the corresponding yeast histone variant H2A.Z demonstrates a significant correlation of transcriptome profile with that in heat-stressed wild-type yeast (Kumar and Wigge 2010).

Recent studies in Zilberman and Henikoff laboratories have provided a persuasive proof that in *Arabidopsis*, the H2A.Z histone variant and DNA methylation reside in different genomic regions and mutually exclude each other from specific loci (Pearson's $r = -0.81$ for the quantitative distribution of DNA methylation and H2A.Z accumulation) (Zilberman et al. 2008). Thus, we can hypothesize that the H2A.Z histone variant can be one of the direct epigenetic receptors of the ambient environmental conditions that trigger downstream alterations in the epigenetic landscape. The evidence for the aforementioned hypothesis can be obtained from the comparison of the epigenetic profile (DNA methylation, posttranslational histone modifications, small RNA expression, etc.) before and after stress with respect to the H2A.Z distribution. Additionally, transgenerational stress experiments on mutant lines of *Arabidopsis PHOTOPERIOD-INDEPENDENT EARLY FLOWERING 1* (PIE1) and *arp6* which are defective in H2A.Z deposition at specific loci can reveal the role of H2A.Z in the development of epigenetic memory.

Unfortunately, there is no evidence that other histones or their variants can contribute to the perception of environmental cues and transmit signals to the epigenetic machinery. An exploration of this area will definitely benefit to understanding the cause of fluctuations in DNA methylation and posttranslational histone modification (PTM) profiles in response to stress.

4 DNA Methylation and Plant Stress Response

DNA methylation is a heritable epigenetic mark which involves a reversible chemical modification of cytosine residues with a methyl group that is covalently added to the C-5 position (Kovalchuk and Kovalchuk 2012). Being discovered almost a century ago (Johnson and Coghill 1925), DNA methylation has become a substantial focus of research on a number of species including bacteria, fungi, worms, insects, plants, and mammals (Kovalchuk and Kovalchuk 2012).

Due to the sessile nature of plants and their inability to escape from unfavorable environmental cues that can jeopardize homeostasis, vascular plants have reached a pick of evolution in terms of the utilization, distribution, and maintenance of DNA methylation patterns. The importance of DNA methylation in plants has been shown in a number of genetic functions, including transcription, replication, recombination, transposition, cell development, and differentiation (Mirouze et al. 2012; Kovalchuk and Kovalchuk 2012). Cytosine methylation is catalyzed by enzymes known as DNA methyltransferases (MTases) which utilize *S*-adenosylmethionine as a primary methyl donor (Kovalchuk and Kovalchuk 2012). In mammals, symmetric CpG sites are usually preferred as targets for methylation, whereas in the plant genomes, the occurrence of methylated cytosines appears to arise virtually at any sequence, including symmetric methylation at both CpG and CpHpG sites (where H = A, C, or T) and asymmetric methylation at CpHpH sites (Kovalchuk and Kovalchuk 2012). Consequently, only 2–8 % of mammalian DNA is methylated, compared to up to 50 % DNA methylation in higher plants (Zhu 2009).

De novo DNA methylation at asymmetric CpHpH sites in plants is catalyzed by methyltransferases DOMAINS REARRANGED METHYLASE 1 and 2 (DRM1 and DRM2, respectively) through the RNA-dependent DNA methylation pathway (RdDM) (Chinnusamy and Zhu 2009), whereas de novo and maintenance methylation of the symmetrical sequences CpG and CpHpG is performed by METHYLTRANSFERASE 1 (MET1) and CHROMOMETHYLASE 3 (CMT3), respectively (Henderson and Jacobsen 2007). Noteworthy, recent studies suggested a high level of redundancy between DNA methylases that demonstrate the ability of MET1 and CMT3 to establish de novo methylation, whereas the maintenance of symmetrical methylation can be performed by DRM1 and DRM2 (Lister et al. 2008).

Plants also possess enzymes that counteract the activity of DNA methylases named "DNA demethylases" (see Table 1). DNA demethylases such as DEMETER (DME), REPRESSOR OF SILENCING1 (ROS1), and DEMETER-LIKE (DML) PROTEINS DML2 AND DML3 belong to a small family of DNA glycosylases involved in DNA base excision repair (BER) and are the main enzymes involved in active locus-specific and global DNA demethylation (Saze et al. 2012).

Genome-wide analysis of the distribution of methylated cytosine (meC) in a number of plant species indicates the enrichment of methylated DNA predominantly at repeats and transposons (on average, 90 % of all sequences are methylated) where a transcriptionally repressed chromatin state is maintained (Lippman et al. 2004; Vaughn et al. 2007; Li et al. 2008; Wang et al. 2009). A comparison of methylome and transcriptome data in Arabidopsis plants revealed that moderately expressed genes exhibit DNA methylation at the transcribed coding region, while both high and low expressed genes demonstrate a significantly lower level of methylation (Zilberman et al. 2007; Cokus et al. 2008; Lister et al. 2008). An increase in methylation levels in gene bodies of actively transcribed genes is thought to be the outcome of small interfering RNA-mediated (siRNA) suppression of the expression of noncanonical promoters that reside within the coding region (Lauria and Rossi 2011). On the contrary, the occurrence of meC at the 5′ region of the gene (including the promoter and part of the transcribed sequence) and at the 3′ region (including part of the coding sequence and the 3′ UTR) is negatively correlated with gene expression and may be involved in tissue-specific gene expression or even pathogen response (Gehring et al. 2009; Zemach et al. 2010; Zilberman et al. 2007; Dowen et al. 2012).

An additional level of transcriptome regulation orchestrated by DNA methylation is achieved through the modulation of alternative splicing sites (Zhou et al. 2012). Recently, it has been speculated that DNA methylation may contribute to the definition of exon–intron boundaries with preferably higher methylation levels of

Table 1 Plant DNA methyltransferases and demethylases

Gene name	Target sequence	Effects on chromatin/ transcription	Effects of mutation
METHYLTRANSFERASE1 (*MET1*)	CpG	Maintains the global methylation of symmetrical CpG sites; involved in the RdDM pathway/ Repression	Inability to establish CpG methylation; a passive decrease of DNA methylation throughout generations
CHROMOMETHYLASE3 (*CMT3*)	Primarily CpHpG	Targets centromeric repeats and transposons; partially contributes to the establishment of DNA methylation at the CpG and CpHpG contexts/Repression	Loss of CpHpG methylation
DOMAIN REARRANGED METHYLTRANSFERASES (*DRM1*, *DRM2*)	CpG, CpHpG, and CpHpH	De novo methylation of asymmetric sites; DRM2 is involved in de novo methylation of CpG sequences in the RdDM pathway/ Repression	Loss of de novo DNA methylation
DEMETER (*DME*)	CpG, CpHpG, and CpHpH	Demethylation of silenced promoter sequences through nucleotide excision repair (NER) pathway/Activation	Inability to activate imprinted genes; seed abortion
REPRESSOR OF SILENCING1 (*ROS1*)	CpG, CpHpG, and CpHpH	Demethylation activity on methylated promoter sequences/ Activation	Local hypermethylation and transcriptional gene silencing; reduced tolerance to genotoxic agents
DEMETER-LIKE (*DML*) *PROTEINS*: *DML2 AND DML3*	CpG, CpHpG, and CpHpH	Demethylation of genes at the 5′ and 3′ regions, which leads to the reduced accumulation of both methylation at or near genes and a decrease in the number of stable epialleles/mostly unchanged	Hypermethylation of genes at the 5′ and 3′ regions

exons compared to introns, and the enrichment of meC in DNA sequences wrapped around histone octamers rather than linker sequences (Chodavarapu et al. 2010). Using an immunoprecipitation technique, Chodavarapu et al. (2010) have demonstrated the higher enrichment of RNA polymerase II (Pol II) in the exon regions compared to the intron regions. A preferential deposition of nucleosomes at the exon regions leads to the Pol II stalling at the intron–exon and exon–intron boundaries which leads to a precise mRNA splicing (Chodavarapu et al. 2010). Taking into account that at least 42 % of the intron-containing genes in Arabidopsis are alternatively spliced (Filichkin et al. 2010), DNA methylation accompanied by other chromatin marks undoubtedly provides a global and fine-tuned mechanism of gene expression regulation.

Stress-regulated alternative splicing in plants has been previously documented (Ali and Reddy 2008); nevertheless, it still remains to be elucidated how the epigenetic machinery contributes to this phenomenon. A general overview of DNA methylation alterations in response to stress indicates stress-dependent changes of methylation at specific loci. For instance, osmotic stress triggers transient DNA hypermethylation at the repetitive heterochromatic loci in tobacco cell suspension culture (Kovarik et al. 1997), while aluminum, salt, paraquat, and cold stresses initiate a decrease in CpG methylation in the coding region of the GLYCEROPHOSPHODIESTERASE-LIKE gene (NtGPDL) in tobacco plants (Choi and Sano 2007).

It appears that global demethylation leading to the activation of gene expression is an immediate response to stress that is more common in plants. Some examples include global DNA hypomethylation after abiotic stress in a number of species: *Broynia dioica* (Galaud et al. 1993), maize (Steward et al. 2002), rice (Wang et al. 2011), *Trifolium repens L.*, and *Cannabis sativa* L. (Aina et al. 2004); similarly, *Pseudomonas syringae*-challenged Arabidopsis plants (Pavet et al. 2006) as well as virus-infected tomato plants (Mason et al. 2008) that exhibit DNA hypomethylation at centromeric repeats and in several genomic regions involved in defense and stress responses, respectively. At the same time, *Mesembryanthemum crystallinum* plants exposed to salt stress demonstrated a twofold increase in CpHpG methylation (Dyachenko et al. 2006). Also, an age-dependent increase in global DNA methylation was correlated with the development of resistance in adult plants to the blight pathogen *Xanthomonas oryzae* in rice (Sha et al. 2005).

Unfortunately, the link between direct exposure to stress and global or sequence-specific DNA demethylation has not been established yet. The most plausible explanation of this phenomenon manifests in the studies showing the generation of ROS upon exposure to abiotic stress that results in oxidation of guanosine residues, which subsequently leads to the formation of 8-hydroxyguanosine (Dionisio-Sese and Tobita 1998). The occurrence of 8-hydroxyguanosine in the CpG sequences strongly suppresses methylation of adjacent cytosine residues (Cerda and Weitzman 1997). Nevertheless, this hypothesis of passive DNA demethylation does not explain a rapid demethylation at the *NtGPDL* genomic locus that has been reported to occur as soon as 1 h after stress exposure (Choi and Sano 2007). The involvement of one type of DNA demethylases, DML3, in the active DNA demethylation process can also be excluded because in Arabidopsis seedlings, DML3 is negatively regulated by miR402 which is induced upon salt, dehydration, or cold stress (Kim et al. 2010). However, recently it has been suggested that DNA demethylation may be guided to specific loci by ROS3, an RNA recognition motif containing protein which uses small RNAs as probes. ROS3 can interact with a different DNA demethylase ROS1 and possibly guide it to target loci for demethylation (Zheng et al. 2008), thus suggesting an intriguing and dynamic interplay between DNA methylation and demethylation pathways.

A decrease in DNA methylation in mutants impaired in CpG maintenance methylation (*met1*) and non-CpG methylation (*drm1/drm2/cmt3* triple mutant) has been shown to result in a significantly higher resistance against bacterial pathogen *Pseudomonas syringae* pv. *tomato* DC3000 (*Pst*) compared to that in wild-type Arabidopsis plants, thus suggesting that DNA methylation, at least to some extent, suppresses plant defense mechanisms (Dowen et al. 2012). A detailed analysis of the methelome profile in response to pathogen attack has revealed a number of TEs, pathogen-responsive and protein-coding genes that were differentially methylated at 5 days postinfection in wild-type plants (Dowen et al. 2012). Moreover, altered methylation of genes encoding TEs was linked to their own expression and/or the expression of neighboring genes as well as the accumulation of TE-associated 21-nt long siRNAs (Dowen et al. 2012). The authors speculated that a pool of 21-nt long siRNAs can be used as non-cell-autonomous messengers that arise upon stress exposure in somatic tissues and eventually can shape the epigenome of gametes, which results in trans-generational stress memory. Taking into account the possible dual function of small RNAs in the process of DNA methylation, this is an intriguing hypothesis. Future research should address this interesting phenomenon and reveal a link between the target demethylation pathway and plant response upon stress exposure.

In recent studies, the strong relationship has been established between DNA hypermethylation at particular loci in response to environmental cues and epigenetic factors involved (Baek et al. 2011; Tricker et al. 2012). For instance, the tandem repeat upstream of the start codon of a sodium transporter gene in Arabidopsis has been shown to be a putative small RNA-mediated target for methylation associated with salt stress tolerance if it is fully methylated at CpHpG and CpHpH contexts (Baek et al. 2011). A plausible relationship has been also reported between the transcriptional repression of two genes that control stomata development, the RNA-dependent DNA methylation (RdDM) pathway, and a decreased number of stomata in Arabidopsis plants cultivated under low humidity (Tricker et al. 2012).

Despite the undeniable role of DNA methylation in plant stress response, an increasing body of evidence indicates the involvement of histone occupation, octamer positioning, and PTMs as the primary stress receptive elements that can act either in cooperation with (Baubec et al. 2010; Bilichak et al. 2012) or apart from DNA methylation (Lang-Mladek et al. 2010; Tittel-Elmer et al. 2010).

Table 2 Posttranslational histone modifications

Enzyme category	Residue	Type of modification	Effect on the transcription
Histone acetyltransferase	Lysine	Acetylation	Activation
Histone deacetylases	Lysine	Deacetylation	Inhibition
Histone methyltransferases	Lysine	Methylation	Depends on the histone residue
Histone demethylases	Lysine	Demethylation	Depends on the histone residue
Ubiquitin ligase	Lysine	Ubiquitination	Activation
Ubiquitin protease	Lysine	Deubiquitination	Repression
Kinase	Serine/ Threonine	Phosphorylation	Activation
Phosphatase	Serine/ Threonine	Dephosphorylation	Inhibition
Arginine methyltransferases	Arginine	Methylation	Can be activating or repressive
Deiminase	Arginine	Demethylation	Can be activating or repressive

5 Chromatin Fluctuations and Plant Stress Response

The regulation of gene expression is not a one-dimensional process; it typically involves a tightly interwoven complex of chromatin modifiers that is connected to PTM, ATP-dependent chromatin remodeling, and histone variant replacement. In contrast to DNA methylation, histone modifications on the amino-terminal tails are highly variable and multifarious, but their role in chromatin regulation and gene expression is sometimes not completely obvious (see Table 2). The high complexity of encoded information carried by histone epigenetic marks manifests in a large number of possible posttranslational modifications combined with different histone variants that together form a histone code of a cell (Kovalchuk and Kovalchuk 2012; Chinnusamy and Zhu 2009). Protruding from the globular nucleosome core, histone tails can undergo different PTMs such as acetylation, methylation, phosphorylation, ubiquitination, sumoylation, biotinilation, carbonylation, glycosylation, and ADP ribosylation catalyzed by a plethora of enzymes (Lauria and Rossi 2011; Tariq and Paszkowski 2004; Liu et al. 2010; Kouzarides 2007; Berr et al. 2012).

A wide range of histone modifications has been connected to the gene activity. Elevated gene expression correlates with the enrichment of acetylation, certain phosphorylation, and ubiquitination (Sridhar et al. 2007; Zhang et al. 2007a) in histone N-terminal regions, while downregulation is linked to biotinylation and sumoylation (Camporeale et al. 2007; Chen et al. 2010). Histone methylation plays a dual role in the modulation of gene activity since both lysine and arginine residues are substrates for histone methyltransferases, with up to three methyl groups being attached to each lysine residue. A genome-wide analysis of histone methylation marks in plants manifests that trimethylation of histone 3 lysine 4 and di-/trimethylation of histone 3 lysine 36 (H3K4me3 and H3K36me2/me3, respectively) are enriched in actively transcribed gene sequences, whereas H3K27me3 and H3K9me2 are the main gene silencing markers (Wang et al. 2009; Zhang et al. 2007b). A few histone methylation modifications such as H3K27me1, H3K27me2, and H4K20me1 have been found to be accumulated in both transposon regions and constitutive heterochromatin regions (Roudier et al. 2011).

Increasing evidence indicates that a range of histone epigenetic marks co-interact with each other and form combinatorial clusters of gene expression regulation (Wang et al. 2008; Berger 2007; Strahl and Allis 2000; Roudier et al. 2011). For instance, a genome-wide analysis of the distribution of 39 histone modifications in $CD4^{+}T$ human cells revealed a common module consisting of 17 histone PTMs that were significantly enriched at 3,286 promoters of actively transcribed genes (Wang et al. 2008). A similar histone module recently has been described in Arabidopsis with 12 histone marks which have been collectively found in about 90 % of the genome (Roudier et al. 2011). In silico predictions of the combinatorial cluster distribution allowed the authors to define four main chromatin states in the Arabidopsis nucleus that mainly encompass active genes, repressed genes, silent repeat elements, and intergenic regions. Furthermore, a strong association has been reported between the distribution of particular histone modifications and different categories of tissue-specific alternative splicing patterns (Zhou et al. 2012). Thus, it is obvious that uniting an immense number of histone PTMs into combinatorial clusters will benefit the analysis of gene activities at the chromatin level.

The association between chromatin marks and their fluctuations in response to stress is becoming a primary focus of plant epigenetic research. A growing body of evidence indicates that histone modifications are at the forefront of stress signal perception, modulations of gene activities, and stress response. Nevertheless, such modifications at the histone tails may or may not be truly epigenetic in nature since the mechanism of propagation of histone codes during DNA replication remains obscure, and it is thought that regardless of the circumstances, not all PTMs of histones are faithfully transmitted to daughter cells after vanishing of the maintenance signal (Chinnusamy and Zhu 2009; Bonasio et al. 2010). Furthermore, to date, there is only a limited evidence of chromatin modifications in systemic tissues of stressed plants; therefore in our review, we will mostly focus on reports indicating changes in directly exposed cells at the histone level.

5.1 Histone Acetylation and Plant Stress Response

Environmental and endogenous cues can affect gene activity through alterations at the histone acetylation level. For instance, in Arabidopsis, histone deacetylases (HDAs), namely HDA6 and HDA19, catalyze deacetylation at several loci in response to abiotic and biotic stresses. Both genes are receptive to jasmonic acid (JA), and in the case of *HDA19*, they possibly mediate the plant pathogen response through the JA-regulated *PATHOGENESIS-RELATED* (PR) genes (Zhou et al. 2005; Wu et al. 2008). Furthermore, studies on knockouts in Arabidopsis revealed the involvement of the histone H4deacetylase (for HIGH EXPRESSION OF OSMOTICALLY RESPONSIVE GENES (HOS15)) in cold stress response (Zheng et al. 2008). Interestingly, *hos15* plants demonstrated constitutive expression of stress-related genes, such as *COLD REGULATED 15A* and *ALCOHOL DEHYDROGENASE 1*, but they were unable to cope with cold stress unlike wild-type plants.

Overexpression of a histone deacetylase from Arabidopsis, *AtHD2C*, which is a member of the plant-specific HD2 family of HDA, resulted in activating the abscisic acid (ABA) responsive genes and elevated salt and drought tolerance compared to that in wild-type plants (Sridha and Wu 2006), thus suggesting that either histone acetylation suppresses plant stress tolerance, at least in the aforementioned examples, or HDAs demonstrate a locus-specific activity upon stress exposure. In support of the last statement, using the yeast *Saccharomyces cerevisiae* as a model, it was shown that the stress-responsive mitogen-activated protein kinase Hog1 guides Rpd3 HDAC to the promoters of stress-related genes, which eventually leads to histone deacetylation, the entry of Pol II, and initiation of gene expression (De Nadal et al. 2004).

It is a known fact that in Arabidopsis plants, transcription factors involved in activating stress-responsive genes interact with the histone acetyltransferase (HAT) (Stockinger et al. 2001). But is there a proof of such collaboration for HDAs? The study manifests evidence of the involvement of WRKY38 and WRKY62 transcription factors and HDA19 in plant basal defense (Kim et al. 2008a). HDA19 acts as a direct negative regulator of transcription factors and induces plant stress tolerance against *Pst*DC3000 in *HDA19*-overexpressing transgenic lines. These and other observations provide a link between the locus-specific HAT activity and HDA activities, apparently suggesting that the regulation of stress-related gene expression by HATs and HDAs does not occur strictly at the level of histone acetylation/deacetylation but through the abolishment of either the transcriptional repressor or activator functions (Courey and Jia 2001; Glass and Rosenfeld 2000). Future research is absolutely needed to provide an evidence of the collaboration between other HATs and HDAs and transcriptional factors involved in abiotic stress response.

5.2 *Histone Methylation and Plant Stress Response*

Previously, we have mentioned the existence of a combinatorial histone modification module in Arabidopsis plants which virtually splits chromatin into four distinct states associated with the origin and transcriptional activity of DNA sequences (see Sect. 5) (Roudier et al. 2011). The available data on the genome-wide histone modification distribution in Arabidopsis allowed combining certain histone acetylation (H3K56ac), methylation (H3K4me2 and 3; H3K9me2 and 3; H3K27me1, 2, and 3; H3K36me3; and H4K20me1), and ubiquitinylation (H2Bub) marks into one module, thus providing an explicit proof of a tight interplay between a number of histone modifications. Therefore, one can expect that alterations in certain chromatin marks upon stress exposure would bring a range of other modifications at the same locus. A typical example includes the accumulation of two repressive chromatin marks, H3K9me2 and H3K27me3, at the potent floral repressor, the *FLOWERING LOCUS C* (FLC) gene, after vernalization (Kim and Sung 2012). Moreover, recently it has been shown that the activation of the *PR1* gene in response to salicylic acid (SA) treatment or pathogen attack is correlated with the enrichment of permissive chromatin marks H3K4me2, H3K4me3, and H3ac in the promoter region (Mosher et al. 2006; De-La-Pena et al. 2012).

In a screen of genes required for the *accelerated cell death 11* (*acd11*) mutant phenotype which exhibits a constant autoimmune response regardless of pathogen perception, a histone lysine methyltransferase SET (Su(var)3-9, E(z), and the Trithorax-conserved) DOMAIN GROUP 8 (SDG8) gene was revealed (Palma et al. 2010). The Arabidopsis SDG8 protein associated with methylations at H3K36 implements a regulatory function on the *lazarus* 5 resistance (*R*) gene which modulates strong defense responses upon pathogen attack. *sdg8* plants failed to develop full resistance to different strains of virulent *Pseudomonas* pathogens, indicating that SDG8 apparently targets a subset of *R* genes (Palma et al. 2010).

Recently, histone posttranslational modifications have been also shown to be involved in priming defense genes that permits genes to respond faster and with a higher extent to biotic stresses (Jaskiewicz et al. 2011). The same study also provides an evidence of systemic epigenetic responses at the histone level in bystander leaves. Following the localized foliar infection by the pathogen *Pseudomonas syringae* pv. *maculicola*, the authors examined histone epigenetic marks at the promoter regions of stress-responsive transcription factors of the WRKY family proteins in distal untreated leaves. Three known *WRKY* promoters (*WRKY29*, *WRKY6*, and *WRKY53*) demonstrated a significant accumulation of permissive chromatin marks: H3K4me2 and 3, H4K5ac, H4K8ac, and H4K12ac, but they failed to show an increase in transcription from the same genes without the direct stress application. These data allowed the authors to speculate that histone marks set a transcriptionally competent state of stress-related genes in systemic tissues which allows for the rapid initiation of transcription and, apparently, mediates SAR in response to pathogen attacks.

5.3 *Histone Phosphorylation/Ubiquitination and Plant Stress Response*

In plants, phosphorylation of serine and threonine residues of histone tails is catalyzed by a wide range of kinases, the most prominent of which are Aurora and NIMA kinases that phosphorylate histone H3 at serine 10 and haspin-like kinase (a haploid germ cell-specific nuclear protein) with the phosphorylation activity at T3 of histone H3 (Houben et al. 2007). The removal of phosphate groups from histone tails is catalyzed by the protein phosphatase 1 (PP1) family enzymes.

Similar to other histone epigenetic marks, histone phosphorylation has been linked to the modulation of gene activity, DNA damage repair, chromatin structure, and apoptosis (Loury and Sassone-Corsi 2004; Houben et al. 2007). For instance, phosphorylation of H2B and histone variant H2A.X at serine 14 was linked to the onset of apoptotic chromatin condensation and DNA fragmentation, respectively (Cheung et al. 2003; Thiriet and Hayes 2005). Phosphorylation of the H2A.X histone variant at serine 4 residues of the C-terminal tail generates a phosphorylated form known as γ-H2A.X (Redon et al. 2002). The C-terminal tail domain projects out towards the front of the nucleosome and interacts with the linker DNA entering the nucleosome, thus making the C-terminus relatively accessible for diffusible factors, and its phosphorylation is believed to be a hallmark of DNA double-strand breaks. Indeed, the accumulation of γ-H2A.X initiates the accumulation of other components involved in DNA double-strand break repair and transcription (Thiriet and Hayes 2005; Lang et al. 2012).

Histone monoubiquitination in Arabidopsis plants occurs at lysine 143 of histone H2B and lysine 121 of the histone variant H2A.1 with the help of the ubiquitin E3 ligase, HISTONE MONOUBIQUITINATION1 (HUB1) and polycomb group (PcG) proteins, and the PRC1 RING-finger homologs AtBMI1A and AtBMI1B, respectively (Bratzel et al. 2010; Weake and Workman 2008; Himanen et al. 2012).

Overall, phosphorylation at Ser10 and Ser28 residues of histone H3 and monoubiquitination at H2B have been correlated with transcription activation (Khorasanizadeh 2004). For instance, global enrichment of H3 phosphorylated at Ser-10 and phosphoacetylated histone H3 was sufficient to upregulate stress-related genes in response to high salinity, cold, and the exogenous ABA application in tobacco and Arabidopsis cell suspension cultures (Sokol et al. 2007), while inducing monoubiquitination of histone H2B in Arabidopsis plants led to elevated plant tolerance to necrotrophic fungi (Dhawan et al. 2009).

5.4 ATP-Dependent Chromatin Remodeling and Plant Stress Response

Chromatin remodeling which utilizes the energy of ATP molecules for altering histone–DNA interactions was found to be an additional dynamic and vital process which modulates gene expression in response to stress (Gutzat and Mittelsten Scheid 2012). There are three main classes of the ATP-dependent chromatin remodeling complexes that are known to exist in plants: the imitation switch (ISWI) ATPases, the SWI/SNF ATPases, and the chromo-domain and helicase-like domain (CHD) ATPases (Walley et al. 2008; Kwon and Wagner 2007).

Thus far, only a handful of the ATP-dependent chromatin remodeling proteins have been implemented in the plant stress response and development. For instance, the SWI/SNF class chromatin remodeling ATPase SPLAYED (SYD) was documented to be involved in the regulation of a range of stress-related and developmental processes (Walley et al. 2008; Bezhani et al. 2007). A different member of the same class—SWI3B—is able to interact with the ABA co-receptor, HYPERSENSITIVE TO ABA1 (HAB1), and is a positive regulator of the ABA-mediated response (Saez et al. 2008). Also, the SNF2/Brahma-type chromatin-remodeling protein AtCHR12 was suggested to play a role in mediating the temporary growth arrest of Arabidopsis under stress conditions (Luo et al. 2012).

The involvement of chromatin remodeling apart from either DNA methylation or certain histone epigenetic marks in the activation of repetitive elements upon heat shock has been recently reported in Arabidopsis (Tittel-Elmer et al. 2010; Pecinka et al. 2010). Nucleosome loading was significantly but reversibly reduced at TE loci in response to the prolonged heat stress (Pecinka et al. 2010). The participation of nucleosome occupancy rather than DNA or histone methylation in transcriptional regulation of TEs was further suggested by delayed re-silencing of heat stress-activated retrotransposons (TSI and ATHILA-related) in CHROMATIN ASSEMBLY FACTOR 1 (CAF-1) mutants (Pecinka and Mittelsten Scheid 2012). An additional level of restricted retrotransposon mobility can also be achieved through the PTGS pathway, with smRNAs being the main players. Moreover, new evidence indicates that the metabolism of smRNAs plays a vital role in plant development, stress response, and epigenetic landscape reshaping in the offspring of stressed plants.

6 The Metabolism of Small RNAs and Plant Stress Response

A novel type of noncoding RNAs, so-called small RNAs (smRNAs), has recently emerged and complemented the epigenetic mechanism of gene expression regulation. Ranging from 20 to 27 nt in length, smRNAs are vital regulators of global epigenome alterations during plant ontogenesis and periconception, which covers gametogenesis, fertilization, and early zygotic development (Slotkin et al. 2009; Bourc'his and Voinnet 2010). Additionally, apart from being dynamically involved in promoting the formation of long-term memory, smRNAs also preserve genome integrity from the effects of potentially harmful genomic parasites like transposons. RNA silencing is a major mechanism of smRNA action by triggering transcriptional (through DNA methylation) or posttranscriptional (through RNA degradation) gene silencing of complementary DNA or RNA, respectively. smRNAs are used as the probes that direct effector proteins to the target nucleic acid molecules through base-pairing interactions (Carthew and Sontheimer 2009).

There are two major classes of smRNAs known in plants as small interfering RNAs (siRNAs) and micro RNAs (miRNAs). Whereas the former ones are processed from long double-stranded (dsRNA) or single-stranded RNAs (ssRNAs) with substantially perfect or near-perfect hairpins, the latter ones are generated from single-stranded stem-loop-like structures of precursor miRNAs that are imperfectly folded—pre-miRNAs—through a two-step or sometimes multistep process (Ramachandran and Chen 2008). The second strand of dsRNA molecules can be synthesized by either of six Arabidopsis RNA-dependent RNA polymerases (RDRs) that recognize aberrant decapped mRNAs (Brosnan et al. 2007).

Massive amounts of data produced by the next-generation sequencing technologies revealed a variety of siRNAs expressed from endogenous loci. Three main classes of endogeneous siRNAs have been put together regarding the loci from which they are generated: natural-antisense transcript-derived siRNAs (nat-siRNAs), ncRNAs produced from two overlapping and partially converging coding transcripts (Borsani et al. 2005); heterochromatic or repeat-associated siRNAs (hc- or ra-siRNAs, respectively), dsRNAs generated from heterochromatin and DNA repeat loci (Guleria et al. 2011); and *trans*-acting siRNAs (ta-siRNAs), miRNA-guided cleavage products of mRNA which are recognized and converted into dsRNAs by RDRs (Grativol et al. 2012; Borsani et al. 2005; Sunkar et al. 2007).

dsRNA processing named "dicing" is performed by one or more of the four Dicer-like (DCL) proteins with the ribonuclease III activity. DCL1 generates 18–21-nt long smRNAs, while DCL2, DCL3, and DCL4 produce 22-nt, 24-nt, and 21-nt long smRNAs, respectively (Ruiz-Ferrer and Voinnet 2009). Following dicing, smRNAs with 3′ overhangs ends are 2-O-methylated by the methyltransferase HUA ENHANCER 1 (HEN1) that protects them from degradation. Later, smRNA duplexes can be either retained in the nucleus for the TGS pathway or exported to the cytoplasm, possibly through the exportin-5 homolog HASTY (HST) for PTGS. In the former scenario, smRNAs are picked by one of the PAZ and PIWI domains containing enzymes with endonucleolytic activities, ARGONAUTE 4 (AGO4), and incorporated into the RNA-induced transcriptional silencing (RITS) complex to promote sequence-specific DNA methylation. Besides AGO4 and siRNA, the RITS complex includes DOMAINS REARRANGED METHYLTRANSFERASE 2 (DRM2), the DDR complex composed of DEFECTIVE IN RNA-DIRECTED DNA METHYLATION 1 (DRD1), DEFECTIVE IN MERISTEM SILENCING 3 (DMS3), RNA-DIRECTED DNA METHYLATION 1 (RDM1), and Pol V (Kovalchuk and Kovalchuk 2012; Law et al. 2010).

When siRNAs are guided from the nucleus, they are primarily picked by AGO6 and incorporated into the RNA-induced silencing complex (RISC) that scans the cell for complementary nucleic acids to execute its silencing function (Ramachandran and Chen 2008). The Arabidopsis genome encodes ten AGO proteins with the RNA-binding PAZ and RNase H-like PIWI domains that belong to three phylogenetic clades. Nevertheless, only a handful of AGOs has been ascribed to the pathways in which they modulate gene activity (Vaucheret 2008).

6.1 Natural Antisense Transcript-Derived siRNAs and Plant Stress Response

Natural antisense transcript-derived siRNAs are a class of endogenous ncRNAs 21–24 nt in length that fall into two groups regarding the origin of transcripts from which they are generated. nat-siRNAs produced from mRNAs which are transcribed from opposite strands at the same locus are called *cis*-nat-siRNAs, while those generated from distinct genomic regions are named *trans*-nat-siRNAs. The members of both groups were found to regulate gene activity at the posttranscriptional level by guiding mRNA cleavage (Kovalchuk and Kovalchuk 2012).

cis-nat-siRNAs generated from convergently transcribed RNAs of *DELTA-1-PYRROLINE-5-CARBOXYLATE DEHYDROGENASE* (*P5CDH*) and *SIMILAR TO RCD ONE 5* (*SRO5*) genes were the first discovered nat-siRNAs in plants expressed in response to salt stress (Borsani et al. 2005). Whereas the former gene encodes a constitutively expressed enzyme involved in proline metabolism, the latter one is activated only in response to salt stress. When overlapping transcripts are generated, DCL2 and DICER RNA BINDING FACTOR (DRB) partners cleave the dsRNA duplex. Later, truncated *P5CDH*mRNA can become a substrate for the RNA-DEPENDENT RNA POLYMERASE6 (RDR6) together with SUPRESSOR OF GENE SILENCING3 (SGS3), which results in the generation of longer RNA duplexes. The formation of nat-siRNAs promotes the degradation of *P5CDH* mRNA, thus leading to proline accumulation, an important metabolite involved in developing salt stress tolerance (Borsani et al. 2005).

Infection of Arabidopsis plants with the bacterial pathogen *Pseudomonas syringae* carrying effector avrRpt2 results in the generation of another *cis*-nat-siRNA—nat-siRNAATGB2 derived from the overlapping region of a Rab2-like small GTP-binding protein gene (*ATGB2*) and a *PENTATRICOPEPTIDE REPEATS* (*PPR*) protein-like gene (*PPRL*) (Katiyar-Agarwal et al. 2006). The authors have shown that the production of this *cis*-nat-siRNA requires activities of HYL1, HEN1, RDR6, SGS3, and Pol IVa. The specific induction of 22-nt long nat-siRNAATGB2 leads to silencing of the antisense gene *PPRL*, which apparently is a negative regulator of the pathogen signaling pathway. Therefore, the downregulation of *PPRL* by nat-siRNAATGB2 plays a positive role in plant resistance against bacterial pathogens (Katiyar-Agarwal et al. 2006).

The most recent genome-wide analysis of plant *cis*-nat-siRNAs has identified 17,141 and 56,209 unique nat-siRNA sequences in Arabidopsis plants exposed to biotic and abiotic stresses and abiotic stress-challenged rice, respectively (Zhang et al. 2012). The biogenesis analysis revealed that the generation of 20–22-nt-long *cis*-nat-siRNAs is dependent on DCL1, whereas siRNAs ranging from 23 to 28 nt in length are produced by DCL3. Interestingly, the authors have revealed that many of the 21-nt-long nat-siRNAs were able to downregulate their target transcripts, whereas the members of the 24-nt-long siRNA class did not demonstrate a significant silencing activity.

The possible discrimination between the pathways in which smRNAs modulate gene activity comes from ten AGO proteins which demonstrate their preference to different siRNAs regarding their length and 5′ terminal nucleotide (Mallory and Vaucheret 2010). For instance, the majority of AGO1-associated smRNAs are either 21 nt or 22 nt in length and carry 5′U, while most of the smRNAs associated with AGO2 are 21-nt long and have 5′A. Furthermore, AGO4, AGO6, and AGO9 mostly precipitate with smRNAs that are 24-nt long and have 5′A, whereas AGO5 accumulates 24-nt long smRNAs with 5′C (Mallory and Vaucheret 2010). Since AGO proteins have been found to act selectively in the distinct silencing pathways, smRNAs of different sizes and with distinct 5′ terminal nucleotides can repress gene expression either at the posttranscriptional level via PTGS or at the chromatin level through TGS depending on the AGO protein with which they interact. An example of siRNAs acting in the TGS pathway includes hc- and ra-siRNAs.

6.2 Heterochromatic and Repeat-Associated siRNAs and Plant Stress Response

Heterochromatic and repeat-associated siRNAs (hc- and ra-siRNAs, respectively) are produced from transcripts generated at the heterochromatic regions and repetitive elements, respectively. Hc-siRNAs-mediated TGS plays an important role in defense against the proliferation of endogenous transposons and restriction of undesirable gene expression through DNA methylation and histone modifications.

A reinforcing heterochromatic loop starts from the generation of transcripts by Pol II, Pol III, and Pol IV. Later, RNA from heterochromatic loci or repetitive elements is targeted by RDR2 to produce long dsRNAs followed by DCL3/DRB cleavage of dsRNA into 24-nt-long duplexes which are then methylated by HEN1. These 24-nt-long siRNAs are picked by AGO4 and incorporated into RITS which eventually mediates cytosine methylation of complementary DNA sequences through the RNA-dependent DNA methylation (RdDM) pathway (Zhang et al. 2007c; Kovalchuk and Kovalchuk 2012; Haag and Pikaard 2011). The strand with the weaker hydrogen bonded 5′ end is preferentially selected as siRNA, while the opposite strand is degraded (Haag and Pikaard 2011). Then, the complete siRNA-guided RITS complex possibly interacts with another plant-specific polymerase—Pol V. Apparently, Pol V transcripts and the Pol V largest subunit bring the RITS complex into the proximity of chromatin to be modified, which eventually allows recruiting specific chromatin modifiers (Haag and Pikaard 2011).

In plants, smRNA-directed cytosine methylation occurs within any sequence regions and is thought to be primarily performed by de novo methyltransferase DRM2 armed with the members of the DDR complex. Silencing at the targeted locus can be further reinforced by the accumulation of repressive chromatin marks. For instance, the removal of permissive chromatin marks (histone acetylation and H3K4me3) and the enrichment of repressive chromatin marks (H3K9me2 and

H3K27me) contribute to transcriptional silencing at sites which are subject to Pol IV- and Pol V-mediated RdDM (Haag and Pikaard 2011).

The contribution of hc- and ra-siRNA to the immobilization of retrotransposons was further supported in the stress experiments that involved mutants impaired in the siRNA pathway (Ito et al. 2011). The authors have demonstrated that heat-stressed Arabidopsis mutants *nrpdl*, *nrpd2*, and *rdr2* significantly accumulate transcripts of a copia-type retrotransposon ONSEN which also generates extrachromosomal DNA copies. Following stress exposure, ONSEN transcripts and the extrachromosomal DNA copies gradually vanished; new ONSEN insertions were not detected in the genomic DNA of either wild-type or *nrpdl* plants. Enigmatically, high-frequency retrotransposition was observed in the next generation of stressed mutant plants that were compromised in the siRNA's metabolism. These results can suggest either the occurrence of stress memory that was maintained throughout the development of mutant plants or an insufficient reinforcement of repressive chromatin marks that usually follows the Pol IV-mediated transposon transcription.

Interestingly, two recent reports in Drosophila and Arabidopsis have shown that smRNAs produced from transposon regions can affect the expression of endogens, thus bridging TE and gene regulation networks. In the Drosophila early embryo, TE-originated PIWI-interacting smRNAs (piRNAs) can silence the *nanos* mRNA which is essential for a proper segmentation (Rouget et al. 2010), while in Arabidopsis, *Athila*-derived smRNAs directly target *OLIGOURIDYLATE BINDING PROTEIN 1b(UBP1b)* mRNA—a component of stress granules involved in responses to certain abiotic stresses (McCue et al. 2012).

6.3 Trans-*Acting siRNAs and Plant Stress Response*

ta-siRNAs are a class of plant siRNAs produced by the interaction between miRNA and siRNA pathways. The generation of ta-siRNAs starts with transcription of miRNA precursors (pre-miRNA) by Pol II. These pre-miRNAs contain miRNA sequences within a stem of a long imperfect RNA hairpin which is processed by DCL1 in the nucleus, resulting in an imperfect RNA duplex with the 2-nucleotide 3′ overhangs on each strand. Most miRNAs are found to mainly target protein-coding mRNAs through the PTGS pathway; however, some miRNAs guide the cleavage of nonprotein-coding primary transcripts of ta-siRNA (*TAS*) genes directing the formation of truncated RNA (Krasnikova et al. 2009). The resulting cleavage product of *TAS* RNAs becomes a substrate for RDR6/SGS3, leading to the generation of double-stranded *TAS* RNAs. SGS3 might stabilize the cleaved RNAs from degradation, while RDR6 synthesizes the second strand (Allen and Howell 2010). Subsequently, long double-stranded precursor ta-siRNAs (pre-ta-siRNAs) are cleaved by the cooperative action of DCL4 and DICER RNA-binding factor4 (DRB4) at 21-nt-long increments relative to the original cleavage site on both strands—a process called "phasing." Apparently, the generation of the uniform

21-nt-long siRNA occurs due to a precise slicing activity of DCL4 starting at the miRNA cleavage site. The exact sequence of miRNA-guided cleavage sets the entry point for DCL4 and thus for the phase of ta-siRNA. ds-ta-siRNAs are methylated by HEN1, and the RDR6-template strand is then loaded into the RISC with one of the AGO proteins (Kovalchuk and Kovalchuk 2012).

Most pre-ta-siRNAs have only a single miRNA target motif which is cleaved by miRNA-guided AGO1. However, there is a curious exception of these observations. For instance, mRNA transcribed from the *TAS3* gene contains two binding sites for miR390 within its sequence. One miR390 guides the AGO7-mediated cleavage of the 3′ side of *TAS3* RNA, whereas the second one interacts in a non-cleavage mode at a site near the 5′ terminus and is important for the production of ds-pre-ta-siRNA (Krasnikova et al. 2009; Montgomery et al. 2008). Subsequently, ta-siRNAs act in *trans* to reduce the expression of unrelated loci from which they are produced (Hsieh et al. 2009).

Arabidopsis ta-siRNAs are derived from eight loci that fall into four families: *TAS1*, *TAS2*, *TAS3*, and *TAS4*. *TAS1*, *TAS2*, and *TAS4* ta-siRNA biogenesis starts with either the miR173- (*TAS1*and *TAS2*) or miR828-guided (*TAS4*) cleavage at the 5′ terminus of the ta-siRNA-generating region. *TAS1/2* loci are transcribed by Pol II generating typical polyadenylated and capped transcripts. Enigmatically, *TAS* nonprotein-coding transcripts are recognized by the silencing machinery as aberrant transcripts and are targeted for degradation. The *TAS1* family which consists of transcripts from three loci (*TAS1a*, *TAS1b*, and *TAS1c*) codes for multiple ta-siRNAs with very similar sequences that are predicted to target the same mRNAs coding for unknown proteins. Whereas ta-siRNA—siR1511 produced from the *TAS2* transcript targets PPRmRNAs, ta-siRNAs sliced from the *TAS3* transcript target the Auxin Response Factor (ARF) family members ARF1, 2, 3, or 4 involved in the juvenile-to-adult transition in leaf development. Interestingly, miR828 which sets a cleavage point for *TAS4* transcripts is specifically involved in the direct regulation of the MYB transcription factor *MYB113*. In turn, the negative regulation of *MYB113* expression and related family members via miR828 can be further amplified by *TAS4*-originated ta-siRNAs (Allen and Howell 2010). Since ta-siRNAs are mobile and can perform short-distance journeys across multiple cells, they can create a gradient of suppression activity in the proximate cells (Schwab et al. 2009).

Recently, ta-siRNAs have been implemented for better adaptation to phosphate (Pi) deficiency and cold stress in *Arabidopsis* and thermosensitive genic male sterile (TGMS) lines of wheat (*Triticum aestivum*), respectively (Tang et al. 2012; Hsieh et al. 2009). Pi-deficient shoots of Arabidopsis accumulate *TAS4*-siR81(−) which is involved in the autoregulatory mechanism of PAP1/MYB75 and the biosynthesis of anthocyanin. In wheat, the *TAS3*-derived ta-siRNA-ARF was significantly repressed at an early stage of spike development during cold stress that was correlated with the upregulation of one of the *ARF* genes. Since in Arabidopsis, *TAS3a* ta-siRNAs have been shown to modulate *ARF3* gene expression which regulates late stages of flower development in a number of plants, the authors suggested that during cold treatment, an abnormal decline of ta-siRNA-*ARF* levels contributes to male sterility in the TGMS line through the negative regulation of *ARFs*. Nevertheless, it still remains to be elucidated how the expression of this and other ta-siRNAs is regulated during stress exposure.

6.4 Micro RNAs and Plant Stress Response

Micro RNAs are typically 21-nt-long single-stranded RNAs which are generated by DCL1 from endogenous transcripts containing local hairpin structures. Compared to animals, the majority of miRNA genes in plants are located separately and are transcribed by Pol II into long pri-miRNAs that contain a 5′ cap and a 3′ poly (A) tail. Processing of pri-miRNAs is believed to take place shortly after a nascent transcript folds into the secondary hairpin-like structure that might involve some of the machineries responsible for capping, splicing and polyadenylation of protein-coding transcripts (Xie et al. 2010). For instance, Arabidopsis mutants *cbp80/abh1* and *cbp20* of the cap-binding complex (CBC) accumulate elevated levels of pri-miRNAs concomitant with a decreased level of mature miRNAs (Kim et al. 2008b; Laubinger et al. 2008).

The dsRNA arm of a hairpin loop is further recognized by DCL1, and in cooperation with the zinc finger-containing protein SERRATE (SE) and the dsRNA-binding protein HYPONASTIC LEAVES1 (HYL1), mature miRNA/miRNA* duplexes are excised from the stem of the hairpin with a two-nucleotide overhang on each strand (Vazquez 2006). Subsequently, the mature miRNA/miRNA* duplexes are stabilized by methylation at the 2′-OH of the 3′ ends by HEN1 and are possibly exported to the cytoplasm through HASTY.

The last step of miRNA maturation involves a selective incorporation of the miRNA strand into the AGO1-containing miRNA–RISC. In the case of siRNA-strand selection, there are two known requirements for the miRNA-guided strand (1) whichever strand is less stably paired at its 5′ end is incorporated into RISC; (2) bearing a 5′-terminal uridine to be preferentially incorporated into AGO1 (Xie et al. 2010). Subsequently, being part of RISC, plant miRNAs target transcripts through the perfect (or near-perfect) pairing between miRNA and the mRNA transcript. Guided to the complementary transcript, AGO1 generates a single cut of the target mRNA phosphodiester backbone. Eventually, the truncated transcripts are either degraded by exonucleases or become a substrate for RDR enzymes (Kovalchuk and Kovalchuk 2012).

An emerging body of evidence suggests that miRNAs significantly contribute to plant stress response. For instance, Arabidopsis mutants *hen1* and *dcl1* that are partially compromised in miRNA metabolism are less stress tolerant compared to wild-type plants (Sunkar and Zhu 2004). Recent studies further support the contribution of miRNAs to stress response in plants. Photosynthesis results in the production of superoxide radicals which need to be scavenged as soon as they are generated to limit the production of more damaging hydroxyl radicals. Furthermore, a variety of environmental cues, such as drought, salinity, high light, cold, and heavy metals, lead to the elevated accumulation of ROS. Therefore, plants have developed a highly sophisticated and fast-acting antioxidant system which involves the miRNA-mediated regulation of ROS scavenging enzymes (Sunkar et al. 2007). CU-ZN SUPEROXIDE DISMUTASE1 and 2 enzymes (*CSD1* and *CSD2*) which participate in detoxifying superoxide radicals are upregulated in response to oxidative stress. Curiously, despite an increase in their activity in response to stress, the nuclear run-on assay revealed that mRNA levels in these genes remain unchanged. Exposure to oxidative stress decreases miR398 transcription that normally guides the cleavage of cytosolic (*CSD1*) and plastidic (*CSD2*) gene mRNAs, resulting in the fast accumulation of CSD1 and CSD2 transcripts and allowing plants to cope with the burst in ROS production (Sunkar et al. 2006).

Many miRNAs have been found to be regulated in response to UV, cold, drought, and salt stress; nevertheless, only a few smRNAs have been related to the pathways in which they modulate resistance or acclimation (Sunkar et al. 2007). For instance, miR399 (a–f) and miR395 have been shown to modulate the activities of genes involved in phosphate and sulfate homeostasis, respectively. miR399 (a–f) is induced in response to phosphate starvation and upregulates a number of genes involved in phosphate metabolism *intrans* by targeting their suppressor—a ubiquitin-conjugating enzyme PHO2. The rates of sulfate translocation and accumulation during sulfur starvation are modulated by the miR395-mediated suppressing action on ATP sulfurylases (*APS1*, *APS3*, and *APS4*) and a low-affinity sulfate transporter (*AST68*).

7 The Maintenance of the Stress-Induced Epigenetic Landscape and Memory Transmission to the Progeny

Epigenetic components that modulate the transcriptome profile upon stress exposure can be both permanent and/or reversible in their nature. However, by definition, the subjects of epigenetic research are only chromatin alterations that are stable and faithfully inherited throughout mitosis and sometimes also meiosis. Being flexible in its origin, epigenetic regulation of gene expression across generations is an attractive alternative mechanism of transgenerational plant response to stress, compared to genetic changes such as point mutations, deletions, insertions, and gross chromosomal rearrangements (Boyko and Kovalchuk 2011). An increasing evidence indicates that the propagation of the stress-induced epigenetic landscape to the next sexual generation named "transgenerational epigenetic inheritance" is a cross-kingdom phenomenon of plant adaptation and acclimation to unfavorable environmental cues (Koturbash et al. 2006; Molinier et al. 2006; Pembrey et al. 2006). The examples include an enhanced tolerance to NaCl and to a DNA-methylating agent—methyl methane sulfonate (MMS) in the offspring of Arabidopsis salt-stressed plants (Boyko et al. 2010; Rahavi et al. 2011); an elevated tolerance to heavy metals in the progeny of heavy metal-treated Arabidopsis and

Oryza sativa plants (Ou et al. 2012; Rahavi et al. 2011); an increase in tolerance to several stressors in timberline plants as a result of adaptation to UV-B radiation (Turunen and Latola 2005); an elevated tolerance to chilling or freezing stresses in the progeny of cold-treated Arabidopsis plants (Blodner et al. 2007); and the natural trans-generational adaptive plasticity to the maternal light environment in the monocarpic herb *Campanulastrum americanum* (Galloway and Etterson 2007). Trans-generational stress adaptation is not only limited to abiotic stressors. Previously, we have shown that the offspring of *Nicotiana tabacum* plants challenged with *Tobacco mosaic virus* (TMV) developed a higher resistance not only against the same pathogen but also against *Pseudomonas syringae* and *Phytophthora nicotianae*. Furthermore, most of the ascribed transgenerational effects were accompanied by alterations in DNA methylation, posttranslational histone modifications, and an increased frequency of homologous recombination events (Bilichak et al. 2012; Boyko et al. 2010; Kathiria et al. 2010). Unfortunately, a mechanical link between the acquired transgenerational stress tolerance and molecular/epigenetic factors that lead to this intriguing phenomenon has not been established yet.

It can be envisaged that SES plays an important role in the transgenerational inheritance in plants since plants do not set aside a predetermined germline linage in early development as animals do. Instead, angiosperms possess an undifferentiated state of spore mother cells (SMCs) developed from a subepidermal cell only late during ontogenesis (Drews and Koltunow 2011; Dickinson and Grant-Downton 2009). In Arabidopsis, the archespore, which is the first cell of the reproductive lineage, differentiates directly into the megaspore mother cell (MMC). Through meiosis, MMC gives rise to a tetrad of haploid megaspores, one of which survives (a functional megaspore), while the others degenerate. The functional megaspore develops into the haploid embryo sac (the female gametophyte) through three rounds of mitosis followed by cellularization. The complete embryo sac includes two gametes (the haploid egg cell and the homo-diploid central cell), two synergids, and three antipodals (Schmidt et al. 2011).

The development of the paternal germ line begins from a pollen mother cell (PMC) that undergoes meiosis in the anthers, resulting in four haploid microspores. Each microspore gives rise to a larger vegetative cell and a smaller generative cell through an asymmetric division. The generative cell which represents the male germ line undergoes a further symmetric partition to produce two identical sperm cells surrounded by vegetative cells (Calarco et al. 2012).

Most of the studies demonstrating the stress-induced transgenerational epigenetic inheritance in *Arabidopsis* were performed at the principal growth stage 1 (Boyes et al. 2001) before inflorescence emergence and germline differentiation (Molinier et al. 2006; Pecinka et al. 2009; Boyko et al. 2010; Bilichak et al. 2012). Therefore, the epigenetic memory engraved in chromatin of somatic cells upon stress exposure has to be systemically transmitted throughout a number of mitotic divisions to gametophyte initials and later survive meiosis and fertilization for being propagated into the next generation.

To date, there is no evidence confirming that epigenetic marks are reset at the gametophyte stage in plants, unlike in mammals where DNA undergoes several rounds of methylation and demethylation during germ cell proliferation and postfertilization (Feng et al. 2010). Moreover, recent advances in elucidating the molecular mechanisms underlying the formation of the epigenetic landscape in gametes revealed that both the sperm cell and the egg cell in angiosperms maintain a quiescent state concomitant with the enrichment of repressive chromatin marks at the euchromatin regions (H3K9me2, H3K27me3, and DNA hypermethylation) (Calarco et al. 2012). Subsequently, the quiescence in both gametes with the respective chromatin marks is propagated into a zygote followed by the occurrence of the silent chromatin state in the embryo (Baroux et al. 2011). In stark contrast, the global epigenetic activation occurs in non-germline reproductive cells such as a vegetative nucleus in pollen and a central cell in the embryo sac that do not contribute genetic material to the progeny but generate smRNAs that are thought to reinforce silencing of transposable elements and imprinted genes in germ cells and the embryo (Baroux et al. 2011). These observations lead us to the question: why do gamete nuclei require VN-derived smRNAs for the suppression of the already silenced TEs? We can speculate that either gametes do not possess the complete silencing machinery to suppress undesired transposition events compared to somatic cells, or smRNAs contribute to the adjustment of epigenetic landscapes later after fertilization occurs.

In the recent study, by using bisulfite sequencing of genomic DNA from Arabidopsis microspores and from their derivative sperm and vegetative cells, the authors demonstrated that symmetric CpG and CpHpG methylation was largely retained in Arabidopsis sperm cells, whereas CpHpH methylation vanished from at least 1,500 TEs. This was concomitant with the downregulation of the RdDM methyltransferase DRM2 in sperm cells that is required for de novo CpHpH methylation guided by 24-nt-long smRNAs (Calarco et al. 2012; Cao and Jacobsen 2002). Nevertheless, it has been shown that stripping off CpHpH methylation still does not activate transposons in sperm cells (Slotkin et al. 2009), albeit homozygous mutants that are compromised in the RdDM pathway (*nrpd1*, *nrpd2*, and *rdr2*) demonstrate the mobility of transgenerational retrotransposons when exposed to elevated temperature (Ito et al. 2011). This brings us to the conclusion that smRNAs derived from the transcriptionally active VN do not strictly suppress transposon activities in the sperm cells but rather modulate DNA methylation at the stage that follows periconception. Since there are no data confirming epigenetic memory resetting in the sperm and egg cells in plants, this hypothesis can possibly explain the mystery of drastic fluctuations in global DNA methylation observed immediately after stress compared to that in the progeny of stressed plants.

The global DNA demethylation followed by transcription activation is the most common immediate plant stress response (see Sect. 4). In stark contrast, 10–12 % global hypermethylation was observed in the untreated offspring of salt-stressed plants concomitant with the enrichment of repressive chromatin marks and transcription downregulation (Bilichak et al. 2012; Boyko et al. 2010). We can hypothesize that such drastic fluctuations in the global DNA methylation profile is the consequence of an increase in stress-induced accumulation of smRNAs in the parental plants which eventually direct RdDM after periconception. Indeed, in the recent study, we demonstrated that the establishment of the transgenerational memory requires the functional siRNA biogenesis pathway because both *dcl2* and *dcl3*Arabidopsis mutants failed to maintain stress-induced epigenetic inheritance (Boyko et al. 2010). The involvement of the smRNAs and RdDM pathway in the transgenerational epigenetic memory is further supported by changes in the distribution of DNA methylation in gene bodies and exon/intron regions in the progeny of salt-stressed Arabidopsis plants compared to that in the control unexposed progeny. A drastic increase in DNA methylation was observed at the 5′ and 3′ ends (up to a 1.8-fold difference) compared to that in the central part of the gene in the progeny of salt-stressed plants. Similarly, the exon regions demonstrated an increase (on average, a 1.2-fold difference) in DNA methylation compared to introns. Curiously, a significant hypermethylation either at the promoter or gene body regions was observed at the gene regions encoding global regulators of chromatin structure and DNA repair, such as *SUVH2*, *5*, *6*, *8*, *ROS1*, *UBIQUITIN-SPECIFIC PROTEASE 26* (*UBP26*), *MUTS HOMOLOG 6* (*MSH6*), and DNA repair/chromatin binding (*UVH3*) homolog. DNA hypermethylation was further reinforced with the accumulation of the transcriptionally repressive histone modification mark H3K9me2 and the depletion of the permissive mark—H3K9ac. The highest level of correlation between repressive chromatin marks and mRNA levels was observed in the *MSH6* gene ($r = -0.95$ on average) that encodes a mismatch repair protein. MSH6, together with MSH2, are involved in the initial recognition of DNA errors; therefore, the reduced expression of mismatch repair genes followed by lower levels of mismatch repair activities may result in a higher frequency of point mutations and, possibly, other genomic rearrangements in the progeny of stressed plants (Bilichak et al. 2012). Indeed, we have previously documented an increase in the number of homologous recombination events as well as in the frequency of point mutations and microsatellite instability in the progeny of plants exposed to various stresses (Kathiria et al. 2010; Yao and Kovalchuk 2011). These observations led us to hypothesize that plants may utilize epigenetic pathways to trigger locus-specific genome rearrangements, thereby forcing a rapid evolution of targeted sequences and associated phenotypes (Boyko and Kovalchuk 2011). Further support for this hypothesis has been found before (Meyers et al. 2005), where it was suggested that the evolution of plant *R*-genes involved gene duplication and recombination events.

Overall, a rapid transgenerational adaptability of plants to new growth conditions cannot be explained by the stochastic heritable variability as it is generally suggested by Darwin's theory of evolution (Boyko and Kovalchuk 2011). The disproof of the random nature of genome variability in living organisms is manifested in multiple examples of prokaryotic systems as well as the existence of hypervariable loci in humans (Foster 2007; Bjedov et al. 2003; Nolin et al. 2003). In plants, studies of DeBolt confirmed that biotic and abiotic stresses could trigger gene copy number variations (CNV) in a non-stochastic way (DeBolt 2010). CNV initiation sites were most frequently found within stress response genes and transposable elements in the offspring exposed to the same stress over multiple generations, thereby supporting the nonrandom occurrence of rearrangements. The future analysis of the epigenome and genome profiles in the progeny of plants subsequently stressed over multiple generations will possibly reveal the contribution of epigenetic components to the microevolution in plants.

8 Conclusions

Overall, the study of the exact mechanisms and players of systemic chromatin remodeling upon stress exposure is still in its infancy. Nevertheless, it becomes apparent that epigenetic components play a key role in the immediate plant stress response and possibly guide microevolution of adaptation and acclimation processes. Recently, epigenetic components have been linked to genome rearrangements in plants (Melamed-Bessudo and Levy 2012), suggesting that dynamic chromatin fluctuations can lead to the occurrence of new sequences, which eventually can result in different gene products and phenotypes. The development of next-generation sequencing technologies will definitely provide a higher resolution and understanding of alterations in the epigenetic landscape in plants in response to stress and will shed a new light on their role in adaptive plant evolution.

References

Aina R, Sgorbati S, Santagostino A, Labra M, Ghiani A, Citterio S (2004) Specific hypomethylation of DNA is induced by heavy metals in white clover and industrial hemp. Physiol Plant 121:472–480

Ali GS, Reddy AS (2008) Regulation of alternative splicing of pre-mRNAs by stresses. Curr Top Microbiol Immunol 326:257–275

Allen E, Howell MD (2010) miRNAs in the biogenesis of trans-acting siRNAs in higher plants. Semin Cell Dev Biol 21:798–804

Antoniades CA, Watts C (2013) Huntington's disease and cell therapies: past, present and future. In: Baluska F (ed) Long-distance systemic signaling and communication in plants, Signaling and communication in plants. Springer, Heidelberg

Baek D, Jiang J, Chung JS, Wang B, Chen J, Xin Z, Shi H (2011) Regulated AtHKT1 gene expression by a distal enhancer element and DNA methylation in the promoter plays an important role in salt tolerance. Plant Cell Physiol 52:149–161

Baroux C, Raissig MT, Grossniklaus U (2011) Epigenetic regulation and reprogramming during gamete formation in plants. Curr Opin Genet Dev 21:124–133

Baubec T, Dinh HQ, Pecinka A, Rakic B, Rozhon W, Wohlrab B, von Haeseler A, Scheid OM (2010) Cooperation of multiple chromatin modifications can generate unanticipated stability of epigenetic states in Arabidopsis. Plant Cell 22:34–47

Berger SL (2007) The complex language of chromatin regulation during transcription. Nature 447:407–412

Berr A, Menard R, Heitz T, Shen WH (2012) Chromatin modification and remodelling: a regulatory landscape for the control of Arabidopsis defence responses upon pathogen attack. Cell Microbiol 14:829–839

Bezhani S, Winter C, Hershman S, Wagner JD, Kennedy JF, Kwon CS, Pfluger J, Su Y, Wagner D (2007) Unique, shared, and redundant roles for the Arabidopsis SWI/SNF chromatin remodeling ATPases BRAHMA and SPLAYED. Plant Cell 19:403–416

Bilichak A, Ilnystkyy Y, Hollunder J, Kovalchuk I (2012) The progeny of Arabidopsis thaliana plants exposed to salt exhibit changes in DNA methylation, histone modifications and gene expression. PloS One 7:e30515

Bird A (2007) Perceptions of epigenetics. Nature 447:396–398

Bjedov I, Tenaillon O, Gerard B, Souza V, Denamur E, Radman M, Taddei F, Matic I (2003) Stress-induced mutagenesis in bacteria. Science 300:1404–1409

Blodner C, Goebel C, Feussner I, Gatz C, Polle A (2007) Warm and cold parental reproductive environments affect seed properties, fitness, and cold responsiveness in *Arabidopsis thaliana* progenies. Plant Cell Environ 30:165–175

Bonasio R, Tu S, Reinberg D (2010) Molecular signals of epigenetic states. Science 330:612–616

Bonisch C, Hake SB (2012) Histone H2A variants in nucleosomes and chromatin: more or less stable? Nucleic Acids Res 40(21):10719–10741

Borsani O, Zhu J, Verslues PE, Sunkar R, Zhu JK (2005) Endogenous siRNAs derived from a pair of natural cis-antisense transcripts regulate salt tolerance in Arabidopsis. Cell 123:1279–1291

Bourc'his D, Voinnet O (2010) A small-RNA perspective on gametogenesis, fertilization, and early zygotic development. Science 330:617–622

Boyes DC, Zayed AM, Ascenzi R, McCaskill AJ, Hoffman NE, Davis KR, Gorlach J (2001) Growth stage-based phenotypic analysis of Arabidopsis: a model for high throughput functional genomics in plants. Plant Cell 13:1499–1510

Boyko A, Kovalchuk I (2011) Genome instability and epigenetic modification – heritable responses to environmental stress? Curr Opin Plant Biol 14:260–266

Boyko A, Blevins T, Yao Y, Golubov A, Bilichak A, Ilnytskyy Y, Hollunder J, Meins F Jr, Kovalchuk I (2010) Transgenerational adaptation of Arabidopsis to stress requires DNA methylation and the function of Dicer-like proteins. PloS One 5:e9514

Bratzel F, Lopez-Torrejon G, Koch M, Del Pozo JC, Calonje M (2010) Keeping cell identity in Arabidopsis requires PRC1 RING-finger homologs that catalyze H2A monoubiquitination. Curr Biol 20:1853–1859

Brosnan CA, Mitter N, Christie M, Smith NA, Waterhouse PM, Carroll BJ (2007) Nuclear gene silencing directs reception of long-distance mRNA silencing in Arabidopsis. Proc Natl Acad Sci USA 104:14741–14746

Calarco JP, Borges F, Donoghue MT, Van Ex F, Jullien PE, Lopes T, Gardner R, Berger F, Feijo JA, Becker JD, Martienssen RA (2012) Reprogramming of DNA methylation in pollen guides epigenetic inheritance via small RNA. Cell 151:194–205

Camporeale G, Oommen AM, Griffin JB, Sarath G, Zempleni J (2007) K12-biotinylated histone H4 marks heterochromatin in human lymphoblastoma cells. J Nutr Biochem 18:760–768

Cao X, Jacobsen SE (2002) Role of the arabidopsis DRM methyltransferases in de novo DNA methylation and gene silencing. Curr Biol 12:1138–1144

Carthew RW, Sontheimer EJ (2009) Origins and mechanisms of miRNAs and siRNAs. Cell 136:642–655

Cerda S, Weitzman SA (1997) Influence of oxygen radical injury on DNA methylation. Mutat Res 386:141–152

Chen M, Lv S, Meng Y (2010) Epigenetic performers in plants. Dev Growth Differ 52:555–566

Cheung WL, Ajiro K, Samejima K, Kloc M, Cheung P, Mizzen CA, Beeser A, Etkin LD, Chernoff J, Earnshaw WC, Allis CD (2003) Apoptotic phosphorylation of histone H2B is mediated by mammalian sterile twenty kinase. Cell 113:507–517

Chinnusamy V, Zhu JK (2009) Epigenetic regulation of stress responses in plants. Curr Opin Plant Biol 12:133–139

Chodavarapu RK, Feng S, Bernatavichute YV, Chen P-Y, Stroud H, Yu Y, Hetzel JA, Kuo F, Kim J, Cokus SJ, Casero D, Bernal M, Huijser P, Clark AT, Kramer U, Merchant SS, Zhang X, Jacobsen SE, Pellegrini M (2010) Relationship between nucleosome positioning and DNA methylation. Nature 466:388–392

Choi CS, Sano H (2007) Abiotic-stress induces demethylation and transcriptional activation of a gene encoding a glycerophosphodiesterase-like protein in tobacco plants. Mol Genet Genomics 277:589–600

Cokus SJ, Feng S, Zhang X, Chen Z, Merriman B, Haudenschild CD, Pradhan S, Nelson SF, Pellegrini M, Jacobsen SE (2008) Shotgun bisulphite sequencing of the Arabidopsis genome reveals DNA methylation patterning. Nature 452:215–219

Courey AJ, Jia S (2001) Transcriptional repression: the long and the short of it. Genes Dev 15:2786–2796

de la Serna IL, Ohkawa Y, Imbalzano AN (2006) Chromatin remodelling in mammalian differentiation: lessons from ATP-dependent remodellers. Nat Rev Genet 7:461–473

De Nadal E, Zapater M, Alepuz PM, Sumoy L, Mas G, Posas F (2004) The MAPK Hog1 recruits Rpd3 histone deacetylase to activate osmoresponsive genes. Nature 427:370–374

Deal RB, Topp CN, McKinney EC, Meagher RB (2007) Repression of flowering in Arabidopsis requires activation of FLOWERING LOCUS C expression by the histone variant H2A.Z. Plant Cell 19:74–83

DeBolt S (2010) Copy number variation shapes genome diversity in Arabidopsis over immediate family generational scales. Genome Biol Evol 2:441–453

De-La-Pena C, Rangel-Cano A, Alvarez-Venegas R (2012) Regulation of disease-responsive genes mediated by epigenetic factors: interaction of Arabidopsis-Pseudomonas. Mol Plant Pathol 13:388–398

Dhawan R, Luo H, Foerster AM, Abuqamar S, Du HN, Briggs SD, Mittelsten Scheid O, Mengiste T (2009) HISTONE MONOUBIQUITINATION1 interacts with a subunit of the mediator complex and regulates defense against necrotrophic fungal pathogens in Arabidopsis. Plant Cell 21:1000–1019

Dickinson HG, Grant-Downton R (2009) Bridging the generation gap: flowering plant gametophytes and animal germlines reveal unexpected similarities. Biol Rev Camb Philos Soc 84:589–615

Dionisio-Sese ML, Tobita S (1998) Antioxidant responses of rice seedlings to salinity stress. Plant Sci 135:1–9

Dowen RH, Pelizzola M, Schmitz RJ, Lister R, Dowen JM, Nery JR, Dixon JE, Ecker JR (2012) Widespread dynamic DNA methylation in response to biotic stress. Proc Natl Acad Sci USA 109:E2183–E2191

Drews GN, Koltunow AM (2011) The female gametophyte. The Arabidopsis Book 9:e0155

Dunoyer P, Himber C, Voinnet O (2005) DICER-LIKE 4 is required for RNA interference and produces the 21-nucleotide small interfering RNA component of the plant cell-to-cell silencing signal. Nat Genet 37:1356–1360

Dyachenko OV, Zakharchenko NS, Shevchuk TV, Bohnert HJ, Cushman JC, Buryanov YI (2006) Effect of hypermethylation of CCWGG sequences in DNA of *Mesembryanthemum crystallinum* plants on their adaptation to salt stress. Biochem Biokhim 71:461–465

Feng S, Jacobsen SE, Reik W (2010) Epigenetic reprogramming in plant and animal development. Science 330:622–627

Filichkin SA, Priest HD, Givan SA, Shen R, Bryant DW, Fox SE, Wong WK, Mockler TC (2010) Genome-wide mapping of alternative splicing in Arabidopsis thaliana. Genome Res 20:45–58

Foster PL (2007) Stress-induced mutagenesis in bacteria. Crit Rev Biochem Mol Biol 42:373–397

Galaud J-P, Gaspar T, Boyer N (1993) Inhibition of internode growth due to mechanical stress in *Bryonia dioica*: relationship between changes in DNA methylation and ethylene metabolism. Physiol Plant 87:25–30

Galloway L, Etterson J (2007) Transgenerational plasticity is adaptive in the wild. Science 318:1134–1136

Gehring M, Bubb KL, Henikoff S (2009) Extensive demethylation of repetitive elements during seed development underlies gene imprinting. Science 324:1447–1451
Glass CK, Rosenfeld MG (2000) The coregulator exchange in transcriptional functions of nuclear receptors. Genes Dev 14:121–141
Grativol C, Hemerly AS, Ferreira PC (2012) Genetic and epigenetic regulation of stress responses in natural plant populations. Biochim Biophys Acta 1819:176–185
Guillemette B, Bataille AR, Gevry N, Adam M, Blanchette M, Robert F, Gaudreau L (2005) Variant histone H2A.Z is globally localized to the promoters of inactive yeast genes and regulates nucleosome positioning. PLoS Biol 3:2100–2110
Guleria P, Mahajan M, Bhardwaj J, Yadav SK (2011) Plant small RNAs: biogenesis, mode of action and their roles in abiotic stresses. Genomics Proteomics Bioinformatics 9:183–199
Gutzat R, Mittelsten Scheid O (2012) Epigenetic responses to stress: triple defense? Curr Opin Plant Biol 15(5):568–573
Haag JR, Pikaard CS (2011) Multisubunit RNA polymerases IV and V: purveyors of non-coding RNA for plant gene silencing. Nat Rev Mol Cell Biol 12:483–492
Hamilton A, Voinnet O, Chappell L, Baulcombe D (2002) Two classes of short interfering RNA in RNA silencing. EMBO J 21:4671–4679
Henderson IR, Jacobsen SE (2007) Epigenetic inheritance in plants. Nature 447:418–424
Himanen K, Woloszynska M, Boccardi TM, De Groeve S, Nelissen H, Bruno L, Vuylsteke M, Van Lijsebettens M (2012) Histone H2B monoubiquitination is required to reach maximal transcript levels of circadian clock genes in Arabidopsis. Plant J 72:249–260
Himber C, Dunoyer P, Moissiard G, Ritzenthaler C, Voinnet O (2003) Transitivity-dependent and -independent cell-to-cell movement of RNA silencing. EMBO J 22:4523–4533
Houben A, Demidov D, Caperta AD, Karimi R, Agueci F, Vlasenko L (2007) Phosphorylation of histone H3 in plants–a dynamic affair. Biochim Biophys Acta 1769:308–315
Hsieh LC, Lin SI, Shih AC, Chen JW, Lin WY, Tseng CY, Li WH, Chiou TJ (2009) Uncovering small RNA-mediated responses to phosphate deficiency in Arabidopsis by deep sequencing. Plant Physiol 151:2120–2132
Ito H, Gaubert H, Bucher E, Mirouze M, Vaillant I, Paszkowski J (2011) An siRNA pathway prevents transgenerational retrotransposition in plants subjected to stress. Nature 472:115–119
Jaskiewicz M, Conrath U, Peterhansel C (2011) Chromatin modification acts as a memory for systemic acquired resistance in the plant stress response. EMBO Rep 12:50–55
Johnson TB, Coghill RD (1925) Researches on pyrimidines: C111. The discovery of 5-methyl-cytosine in tuberculinic acid, the nucleic acid of the tubercle bacillus1. J Am Chem Soc 47:2838–2844
Kathiria P, Sidler C, Golubov A, Kalischuk M, Kawchuk LM, Kovalchuk I (2010) Tobacco mosaic virus infection results in an increase in recombination frequency and resistance to viral, bacterial, and fungal pathogens in the progeny of infected tobacco plants. Plant Physiol 153:1859–1870
Katiyar-Agarwal S, Morgan R, Dahlbeck D, Borsani O, Villegas A Jr, Zhu JK, Staskawicz BJ, Jin H (2006) A pathogen-inducible endogenous siRNA in plant immunity. Proc Natl Acad Sci USA 103:18002–18007
Khorasanizadeh S (2004) The nucleosome: from genomic organization to genomic regulation. Cell 116:259–272
Kim DH, Sung S (2012) Environmentally coordinated epigenetic silencing of FLC by protein and long noncoding RNA components. Curr Opin Plant Biol 15:51–56
Kim KC, Lai Z, Fan B, Chen Z (2008a) Arabidopsis WRKY38 and WRKY62 transcription factors interact with histone deacetylase 19 in basal defense. Plant Cell 20:2357–2371
Kim S, Yang JY, Xu J, Jang IC, Prigge MJ, Chua NH (2008b) Two cap-binding proteins CBP20 and CBP80 are involved in processing primary MicroRNAs. Plant Cell Physiol 49:1634–1644
Kim JY, Kwak KJ, Jung HJ, Lee HJ, Kang H (2010) MicroRNA402 affects seed germination of *Arabidopsis thaliana* under stress conditions via targeting DEMETER-LIKE Protein3 mRNA. Plant Cell Physiol 51:1079–1083

Kobor MS, Venkatasubrahmanyam S, Meneghini MD, Gin JW, Jennings JL, Link AJ, Madhani HD, Rine J (2004) A protein complex containing the conserved Swi2/Snf2-related ATPase Swr1p deposits histone variant H2A.Z into euchromatin. PLoS Biol 2:E131
Koturbash I, Baker M, Loree J, Kutanzi K, Hudson D, Pogribny I, Sedelnikova O, Bonner W, Kovalchuk O (2006) Epigenetic dysregulation underlies radiation-induced transgenerational genome instability in vivo. Int J Radiat Oncol Biol Phys 66:327–330
Kouzarides T (2007) Chromatin modifications and their function. Cell 128:693–705
Kovalchuk I, Kovalchuk O (2012) Epigenetics in health and disease, 1st edn. FT Press, Upper Saddle River, NJ
Kovarik A, Koukalová B, Bezdek M, Opatrny Z (1997) Hypermethylation of tobacco heterochromatic loci in response to osmotic stress. Theor Appl Genet 95:301–306
Krasnikova MS, Milyutina IA, Bobrova VK, Ozerova LV, Troitsky AV, Solovyev AG, Morozov SY (2009) Novel miR390-dependent transacting siRNA precursors in plants revealed by a PCR-based experimental approach and database analysis. J Biomed Biotechnol 2009:952304
Kumar SV, Wigge PA (2010) H2A.Z-containing nucleosomes mediate the thermosensory response in Arabidopsis. Cell 140:136–147
Kwon CS, Wagner D (2007) Unwinding chromatin for development and growth: a few genes at a time. Trends Genet 23:403–412
Lang J, Smetana O, Sanchez-Calderon L, Lincker F, Genestier J, Schmit A-C, Houlné G, Chabouté M-E (2012) Plant γH2AX foci are required for proper DNA DSB repair responses and colocalize with E2F factors. New Phytol 194:353–363
Lang-Mladek C, Popova O, Kiok K, Berlinger M, Rakic B, Aufsatz W, Jonak C, Hauser MT, Luschnig C (2010) Transgenerational inheritance and resetting of stress-induced loss of epigenetic gene silencing in Arabidopsis. Mol Plant 3:594–602
Laubinger S, Sachsenberg T, Zeller G, Busch W, Lohmann JU, Ratsch G, Weigel D (2008) Dual roles of the nuclear cap-binding complex and SERRATE in pre-mRNA splicing and microRNA processing in *Arabidopsis thaliana*. Proc Natl Acad Sci USA 105:8795–8800
Lauria M, Rossi V (2011) Epigenetic control of gene regulation in plants. Biochim Biophysica Acta 1809:369–378
Law JA, Ausin I, Johnson LM, Vashisht AA, Zhu JK, Wohlschlegel JA, Jacobsen SE (2010) A protein complex required for polymerase V transcripts and RNA- directed DNA methylation in Arabidopsis. Curr Biol 20:951–956
Li X, Wang X, He K, Ma Y, Su N, He H, Stolc V, Tongprasit W, Jin W, Jiang J, Terzaghi W, Li S, Deng XW (2008) High-resolution mapping of epigenetic modifications of the rice genome uncovers interplay between DNA methylation, histone methylation, and gene expression. Plant Cell 20:259–276
Lippman Z, Gendrel AV, Black M, Vaughn MW, Dedhia N, McCombie WR, Lavine K, Mittal V, May B, Kasschau KD, Carrington JC, Doerge RW, Colot V, Martienssen R (2004) Role of transposable elements in heterochromatin and epigenetic control. Nature 430:471–476
Lister R, O'Malley RC, Tonti-Filippini J, Gregory BD, Berry CC, Millar AH, Ecker JR (2008) Highly integrated single-base resolution maps of the epigenome in Arabidopsis. Cell 133:523–536
Liu C, Lu F, Cui X, Cao X (2010) Histone methylation in higher plants. Annu Rev Plant Biol 61:395–420
Loury R, Sassone-Corsi P (2004) Analysis of histone phosphorylation: coupling intracellular signaling to chromatin remodeling. Methods Enzymol 377:197–212
Luo M, Liu X, Singh P, Cui Y, Zimmerli L, Wu K (2012) Chromatin modifications and remodeling in plant abiotic stress responses. Biochim Biophysica Acta 1819:129–136
Mallory A, Vaucheret H (2010) Form, function, and regulation of ARGONAUTE proteins. Plant Cell 22:3879–3889
March-Diaz R, Garcia-Dominguez M, Lozano-Juste J, Leon J, Florencio FJ, Reyes JC (2008) Histone H2A.Z and homologues of components of the SWR1 complex are required to control immunity in Arabidopsis. Plant J 53:475–487

Marques M, Laflamme L, Gervais AL, Gaudreau L (2010) Reconciling the positive and negative roles of histone H2A.Z in gene transcription. Epigenetics 5:267–272

Mason G, Noris E, Lanteri S, Acquadro A, Accotto G, Portis E (2008) Potentiality of methylation-sensitive amplification polymorphism (MSAP) in identifying genes involved in tomato response to tomato yellow leaf curl Sardinia Virus. Plant Mol Biol Rep 26:156–173

McCue AD, Nuthikattu S, Reeder SH, Slotkin RK (2012) Gene expression and stress response mediated by the epigenetic regulation of a transposable element small RNA. PLoS Genet 8: e1002474

Melamed-Bessudo C, Levy AA (2012) Deficiency in DNA methylation increases meiotic cross-over rates in euchromatic but not in heterochromatic regions in Arabidopsis. Proc Natl Acad Sci USA 109:E981–E988

Meyers BC, Kaushik S, Nandety RS (2005) Evolving disease resistance genes. Curr Opin Plant Biol 8:129–134

Mirouze M, Paszkowski J (2011) Epigenetic contribution to stress adaptation in plants. Curr Opin Plant Biol 14:267–274

Mirouze M, Lieberman-Lazarovich M, Aversano R, Bucher E, Nicolet J, Reinders J, Paszkowski J (2012) Loss of DNA methylation affects the recombination landscape in Arabidopsis. Proc Natl Acad Sci USA 109:5880–5885

Molinier J, Ries G, Zipfel C, Hohn B (2006) Transgeneration memory of stress in plants. Nature 442:1046–1049

Molnar A, Melnyk C, Baulcombe DC (2011) Silencing signals in plants: a long journey for small RNAs. Genome Biol 12:215

Montgomery TA, Howell MD, Cuperus JT, Li D, Hansen JE, Alexander AL, Chapman EJ, Fahlgren N, Allen E, Carrington JC (2008) Specificity of ARGONAUTE7-miR390 interaction and dual functionality in TAS3 trans-acting siRNA formation. Cell 133:128–141

Mosher RA, Durrant WE, Wang D, Song J, Dong X (2006) A comprehensive structure-function analysis of Arabidopsis SNI1 defines essential regions and transcriptional repressor activity. Plant Cell 18:1750–1765

Nolin SL, Brown WT, Glicksman A, Houck GE Jr, Gargano AD, Sullivan A, Biancalana V, Brondum-Nielsen K, Hjalgrim H, Holinski-Feder E, Kooy F, Longshore J, Macpherson J, Mandel JL, Matthijs G, Rousseau F, Steinbach P, Vaisanen ML, von Koskull H, Sherman SL (2003) Expansion of the fragile X CGG repeat in females with premutation or intermediate alleles. Am J Hum Genet 72:454–464

Ou X, Zhang Y, Xu C, Lin X, Zang Q, Zhuang T, Jiang L, von Wettstein D, Liu B (2012) Transgenerational inheritance of modified DNA methylation patterns and enhanced tolerance induced by heavy metal stress in rice (*Oryza sativa* L.). PloS One 7:e41143

Palauqui JC, Balzergue S (1999) Activation of systemic acquired silencing by localised introduction of DNA. Curr Biol 9:59–66

Palauqui JC, Elmayan T, Pollien JM, Vaucheret H (1997) Systemic acquired silencing: transgene-specific post-transcriptional silencing is transmitted by grafting from silenced stocks to non-silenced scions. EMBO J 16:4738–4745

Palma K, Thorgrimsen S, Malinovsky FG, Fiil BK, Nielsen HB, Brodersen P, Hofius D, Petersen M, Mundy J (2010) Autoimmunity in Arabidopsis acd11 is mediated by epigenetic regulation of an immune receptor. PLoS Pathog 6:e1001137

Pavet V, Quintero C, Cecchini NM, Rosa AL, Alvarez ME (2006) Arabidopsis displays centromeric DNA hypomethylation and cytological alterations of heterochromatin upon attack by *Pseudomonas syringae*. Mol Plant-Microbe Interact 19:577–587

Pecinka A, Mittelsten Scheid O (2012) Stress-induced chromatin changes: a critical view on their heritability. Plant Cell Physiol 53:801–808

Pecinka A, Rosa M, Schikora A, Berlinger M, Hirt H, Luschnig C, Scheid OM (2009) Transgenerational stress memory is not a general response in Arabidopsis. PloS One 4:e5202

Pecinka A, Dinh HQ, Baubec T, Rosa M, Lettner N, Scheid OM (2010) Epigenetic regulation of repetitive elements is attenuated by prolonged heat stress in Arabidopsis. Plant Cell 22:3118–3129

Pembrey ME, Bygren LO, Kaati G, Edvinsson S, Northstone K, Sjostrom M, Golding J (2006) Sex-specific, male-line transgenerational responses in humans. Eur J Hum Genet 14:159–166

Rahavi MR, Migicovsky Z, Titov V, Kovalchuk I (2011) Transgenerational adaptation to heavy metal salts in Arabidopsis. Front Plant Sci 2:91

Ramachandran V, Chen X (2008) Small RNA metabolism in Arabidopsis. Trends Plant Sci 13:368–374

Redon C, Pilch D, Rogakou E, Sedelnikova O, Newrock K, Bonner W (2002) Histone H2A variants H2AX and H2AZ. Curr Opin Genet Dev 12:162–169

Roudier F, Ahmed I, Berard C, Sarazin A, Mary-Huard T, Cortijo S, Bouyer D, Caillieux E, Duvernois-Berthet E, Al-Shikhley L, Giraut L, Despres B, Drevensek S, Barneche F, Derozier S, Brunaud V, Aubourg S, Schnittger A, Bowler C, Martin-Magniette ML, Robin S, Caboche M, Colot V (2011) Integrative epigenomic mapping defines four main chromatin states in Arabidopsis. EMBO J 30:1928–1938

Rouget C, Papin C, Boureux A, Meunier AC, Franco B, Robine N, Lai EC, Pelisson A, Simonelig M (2010) Maternal mRNA deadenylation and decay by the piRNA pathway in the early Drosophila embryo. Nature 467:1128–1132

Ruiz-Ferrer V, Voinnet O (2009) Roles of plant small RNAs in biotic stress responses. Annu Rev Plant Biol 60:485–510

Saez A, Rodrigues A, Santiago J, Rubio S, Rodriguez PL (2008) HAB1–SWI3B interaction reveals a link between abscisic acid signaling and putative SWI/SNF chromatin-remodeling complexes in Arabidopsis. Plant Cell 20:2972–2988

Saze H, Tsugane K, Kanno T, Nishimura T (2012) DNA methylation in plants: relationship to small RNAs and histone modifications, and functions in transposon inactivation. Plant Cell Physiol 53:766–784

Schmidt A, Wuest SE, Vijverberg K, Baroux C, Kleen D, Grossniklaus U (2011) Transcriptome analysis of the Arabidopsis megaspore mother cell uncovers the importance of RNA helicases for plant germline development. PLoS Biol 9:e1001155

Schwab R, Maizel A, Ruiz-Ferrer V, Garcia D, Bayer M, Crespi M, Voinnet O, Martienssen RA (2009) Endogenous TasiRNAs mediate non-cell autonomous effects on gene regulation in *Arabidopsis thaliana*. PloS One 4:e5980

Sha AH, Lin XH, Huang JB, Zhang DP (2005) Analysis of DNA methylation related to rice adult plant resistance to bacterial blight based on methylation-sensitive AFLP (MSAP) analysis. Mol Genet Genomics 273:484–490

Slotkin RK, Vaughn M, Borges F, Tanurdzic M, Becker JD, Feijo JA, Martienssen RA (2009) Epigenetic reprogramming and small RNA silencing of transposable elements in pollen. Cell 136:461–472

Smith AP, Jain A, Deal RB, Nagarajan VK, Poling MD, Raghothama KG, Meagher RB (2010) Histone H2A.Z regulates the expression of several classes of phosphate starvation response genes but not as a transcriptional activator. Plant Physiol 152:217–225

Sokol A, Kwiatkowska A, Jerzmanowski A, Prymakowska-Bosak M (2007) Up-regulation of stress-inducible genes in tobacco and Arabidopsis cells in response to abiotic stresses and ABA treatment correlates with dynamic changes in histone H3 and H4 modifications. Planta 227:245–254

Sridha S, Wu K (2006) Identification of AtHD2C as a novel regulator of abscisic acid responses in Arabidopsis. Plant J 46:124–133

Sridhar VV, Kapoor A, Zhang K, Zhu J, Zhou T, Hasegawa PM, Bressan RA, Zhu JK (2007) Control of DNA methylation and heterochromatic silencing by histone H2B deubiquitination. Nature 447:735–738

Steward N, Ito M, Yamaguchi Y, Koizumi N, Sano H (2002) Periodic DNA methylation in maize nucleosomes and demethylation by environmental stress. J Biol Chem 277:37741–37746

Stockinger EJ, Mao Y, Regier MK, Triezenberg SJ, Thomashow MF (2001) Transcriptional adaptor and histone acetyltransferase proteins in Arabidopsis and their interactions with CBF1, a transcriptional activator involved in cold-regulated gene expression. Nucleic Acids Res 29:1524–1533

Strahl BD, Allis CD (2000) The language of covalent histone modifications. Nature 403:41–45

Sunkar R, Zhu JK (2004) Novel and stress-regulated microRNAs and other small RNAs from Arabidopsis. Plant Cell 16:2001–2019

Sunkar R, Kapoor A, Zhu JK (2006) Posttranscriptional induction of two Cu/Zn superoxide dismutase genes in Arabidopsis is mediated by downregulation of miR398 and important for oxidative stress tolerance. Plant Cell 18:2051–2065

Sunkar R, Chinnusamy V, Zhu J, Zhu J-K (2007) Small RNAs as big players in plant abiotic stress responses and nutrient deprivation. Trends Plant Sci 12:301–309

Tang Z, Zhang L, Xu C, Yuan S, Zhang F, Zheng Y, Zhao C (2012) Uncovering small RNA-mediated responses to cold stress in a wheat thermosensitive genic male-sterile line by deep sequencing. Plant Physiol 159:721–738

Tariq M, Paszkowski J (2004) DNA and histone methylation in plants. Trends Genet 6:244–251

Thiriet C, Hayes JJ (2005) Chromatin in need of a fix: phosphorylation of H2AX connects chromatin to DNA repair. Mol Cell 18:617–622

Tittel-Elmer M, Bucher E, Broger L, Mathieu O, Paszkowski J, Vaillant I (2010) Stress-induced activation of heterochromatic transcription. PLoS Genet 6:e1001175

Tricker PJ, Gibbings JG, Rodriguez Lopez CM, Hadley P, Wilkinson MJ (2012) Low relative humidity triggers RNA-directed de novo DNA methylation and suppression of genes controlling stomatal development. J Exp Bot 63:3799–3813

Turunen M, Latola K (2005) UV-B radiation and acclimation in timberline plants. Environ Pollut 137:390–403

van den Burg HA, Takken FLW (2009) Does chromatin remodeling mark systemic acquired resistance? Trends Plant Sci 14:286–294

Vance V, Vaucheret H (2001) RNA silencing in plants – defense and counterdefense. Science 292:2277–2280

Vaucheret H (2008) Plant ARGONAUTES. Trends Plant Sci 13:350–358

Vaughn MW, Tanurdzic M, Lippman Z, Jiang H, Carrasquillo R, Rabinowicz PD, Dedhia N, McCombie WR, Agier N, Bulski A, Colot V, Doerge RW, Martienssen RA (2007) Epigenetic natural variation in *Arabidopsis thaliana*. PLoS Biol 5:e174

Vazquez F (2006) Arabidopsis endogenous small RNAs: highways and byways. Trends Plant Sci 11:460–468

Voinnet O, Baulcombe DC (1997) Systemic signalling in gene silencing. Nature 389(6651):553

Voinnet O, Vain P, Angell S, Baulcombe DC (1998) Systemic spread of sequence-specific transgene RNA degradation in plants is initiated by localized introduction of ectopic promoterless DNA. Cell 95:177–187

Walley JW, Rowe HC, Xiao Y, Chehab EW, Kliebenstein DJ, Wagner D, Dehesh K (2008) The chromatin remodeler SPLAYED regulates specific stress signaling pathways. PLoS Pathog 4: e1000237

Wang Z, Zang C, Rosenfeld JA, Schones DE, Barski A, Cuddapah S, Cui K, Roh TY, Peng W, Zhang MQ, Zhao K (2008) Combinatorial patterns of histone acetylations and methylations in the human genome. Nat Genet 40:897–903

Wang X, Elling AA, Li X, Li N, Peng Z, He G, Sun H, Qi Y, Liu XS, Deng XW (2009) Genome-wide and organ-specific landscapes of epigenetic modifications and their relationships to mRNA and small RNA transcriptomes in maize. Plant Cell 21:1053–1069

Wang WS, Pan YJ, Zhao XQ, Dwivedi D, Zhu LH, Ali J, Fu BY, Li ZK (2011) Drought-induced site-specific DNA methylation and its association with drought tolerance in rice (*Oryza sativa* L.). J Exp Bot 62:1951–1960

Weake VM, Workman JL (2008) Histone ubiquitination: triggering gene activity. Mol Cell 29:653–663

Wingard SA (1928) Hosts and symptoms of ring spot, a virus disease of plants. J Agric Res 37:127–153

Wu K, Zhang L, Zhou C, Yu C-W, Chaikam V (2008) HDA6 is required for jasmonate response, senescence and flowering in Arabidopsis. J Exp Bot 59:225–234

Xie Z, Khanna K, Ruan S (2010) Expression of microRNAs and its regulation in plants. Semin Cell Dev Biol 21:790–797

Yaegashi H, Tamura A, Isogai M, Yoshikawa N (2008) Inhibition of long-distance movement of RNA silencing signals in *Nicotiana benthamiana* by Apple chlorotic leaf spot virus 50 kDa movement protein. Virology 382:199–206

Yao Y, Kovalchuk I (2011) Abiotic stress leads to somatic and heritable changes in homologous recombination frequency, point mutation frequency and microsatellite stability in Arabidopsis plants. Mutat Res 707:61–66

Youngson NA, Whitelaw E (2008) Transgenerational epigenetic effects. Annu Rev Genomics Hum Genet 9:233–257

Zemach A, Kim MY, Silva P, Rodrigues JA, Dotson B, Brooks MD, Zilberman D (2010) Local DNA hypomethylation activates genes in rice endosperm. Proc Natl Acad Sci USA 107:18729–18734

Zhang K, Sridhar VV, Zhu J, Kapoor A, Zhu JK (2007a) Distinctive core histone post-translational modification patterns in Arabidopsis thaliana. PloS One 2:e1210

Zhang X, Clarenz O, Cokus S, Bernatavichute YV, Pellegrini M, Goodrich J, Jacobsen SE (2007b) Whole-genome analysis of histone H3 lysine 27 trimethylation in Arabidopsis. PLoS Biol 5: e129

Zhang X, Henderson IR, Lu C, Green PJ, Jacobsen SE (2007c) Role of RNA polymerase IV in plant small RNA metabolism. Proc Natl Acad Sci USA 104:4536–4541

Zhang X, Xia J, Lii YE, Barrera-Figueroa BE, Zhou X, Gao S, Lu L, Niu D, Chen Z, Leung C, Wong T, Zhang H, Guo J, Li Y, Liu R, Liang W, Zhu JK, Zhang W, Jin H (2012) Genome-wide analysis of plant nat-siRNAs reveals insights into their distribution, biogenesis and function. Genome Biol 13:R20

Zheng X, Pontes O, Zhu J, Miki D, Zhang F, Li W-X, Iida K, Kapoor A, Pikaard CS, Zhu J-K (2008) ROS3 is an RNA-binding protein required for DNA demethylation in Arabidopsis. Nature 455:1259–1262

Zhou C, Zhang L, Duan J, Miki B, Wu K (2005) HISTONE DEACETYLASE19 is involved in jasmonic acid and ethylene signaling of pathogen response in Arabidopsis. Plant Cell 17:1196–1204

Zhou Y, Lu Y, Tian W (2012) Epigenetic features are significantly associated with alternative splicing. BMC Genomics 13:123

Zhu JK (2009) Active DNA demethylation mediated by DNA glycosylases. Annu Rev Genet 43:43–166

Zilberman D, Gehring M, Tran RK, Ballinger T, Henikoff S (2007) Genome-wide analysis of *Arabidopsis thaliana* DNA methylation uncovers an interdependence between methylation and transcription. Nat Genet 39:61–69

Zilberman D, Coleman-Derr D, Ballinger T, Henikoff S (2008) Histone H2A.Z and DNA methylation are mutually antagonistic chromatin marks. Nature 456:125–129

Long-Distance Signals Produced by Water-Stressed Roots

Jason Q.D. Goodger

Abstract Roots can sense small changes in soil water status and rapidly communicate this over long distances throughout the plant. This sets in motion numerous response mechanisms for water conservation and drought tolerance, largely facilitated by the hormone ABA. Despite impressive advances in the molecular mechanisms by which ABA mediates such plant responses, long-distance signaling of soil water status remains relatively poorly understood. Recent results refute the long-held hypothesis of ABA biosynthesis in roots as the primary signal, at least in the initial stage of water stress communication. This chapter examines the involvement of leaf ABA biosynthesis, pH-mediated ABA redistribution, and ABA conjugate catabolism in communicating soil water status. In addition, the chapter presents current knowledge on other xylem-borne signaling compounds such as cytokinins, 1-aminocyclopropane-1-carboxylic acid, inorganic ions, and organic acids and their possible interactions with ABA in long-distance signaling of water stress.

1 Introduction

Soil water deficit or drought is the greatest abiotic stress influencing crop productivity and survival. As soil water potential decreases, roots are increasingly unable to acquire sufficient water to replace that lost via leaf transpiration, and reductions in plant water status eventuate. Under prolonged periods of drought, decreasing plant water status can induce wilting, cell membrane damage, cell death, and ultimately result in plant mortality. Under less severe water stress, temporary interruptions to shoot growth and reductions in photosynthesis due to stomata-induced CO_2 restrictions have flow on effects on crop productivity and harvestable

J.Q.D. Goodger (✉)
School of Botany, The University of Melbourne, Parkville, VIC 3010, Australia
e-mail: jgoodger@unimelb.edu.au

F. Baluška (ed.), *Long-Distance Systemic Signaling and Communication in Plants*, Signaling and Communication in Plants 19, DOI 10.1007/978-3-642-36470-9_5,

yields. Indeed, periodic droughts can be as detrimental to reproductive crop productivity as all other environmental factors combined (Boyer and Westgate 2004). Temperature rises associated with global climate change are predicted to alter precipitation patterns resulting in increased frequency and intensity of periodic droughts (and flooding) in almost all regions of the earth (Kharin et al. 2007). Therefore, the future impact of drought on crop growth and productivity is likely to be even more dramatic than currently observed. As water resources for agricultural use become more limiting, the development of drought-tolerant crop lines will become increasingly important if we are to feed the world's burgeoning population.

Plants have evolved numerous mechanisms to avoid or tolerate periodic drought episodes, and such traits can be targeted in crop improvement strategies. The immediate ability of plants to maintain plant water status under conditions of reduced soil water availability involves some general responses, commonly observed throughout the plant kingdom. Foremost of these occur in leaves and involve reduction of transpirational water loss by stomatal closure and conservation of water by reduction or cessation of leaf growth (Bogoslavsky and Neumann 1998; Liu et al. 2001; Neumann et al. 1997). Leaf responses are thought to be concomitant with root responses including the maintenance or even acceleration of root growth to explore enlarged soil volumes for water (Vartanian et al. 1994; Passioura 1996; Sharp et al. 2004), and increases in root hydraulic conductance (Hose et al. 2000; Parent et al. 2009; Kudoyarova et al. 2011). A number of relatively less well-characterized responses are also observed in plants exposed to soil water deficits and are better known from species adapted to arid environments. These responses include synthesis of antioxidant proteins, accumulation of osmotically active solutes involved in the maintenance of cell turgor, and remobilization of assimilates from vegetative to reproductive structures or from older to younger leaves, followed by induction of leaf senescence (Chaves et al. 2003; Neumann 2008). The various stress management responses are coordinated in a manner to enable most plant species to survive transient periods of soil water stress and those capable of more specialized drought responses to survive severe and persistent drought conditions.

Before plants can initiate mechanisms to limit water loss, conserve water reserves, and search out new sources of soil water, they must sense small reductions in soil water potential and then rapidly signal this change in status to the appropriate response targets in leaves and roots. The sensitivity and rapidity of this sensing–signaling–response mechanism is such that a large proportion of crop yield losses due to drought occur at the wet end of the soil water deficit spectrum, well before changes in plant water status occur (Davies et al. 2005). This common observation is an argument against simple changes in plant water status acting as a hydraulic signal in the early stages of the drought response mechanism (but see Christmann et al. (2007) and Tardieu et al. (2010) for arguments to the contrary). Moreover, the rapidity of leaf responses to reductions in soil water suggests that root cells are capable of fine-scale sensing of changes in soil water potential and produce signals capable of rapid transport over long distances throughout the plant (Davies et al. 2002; Wilkinson and Davies 2002). The most probable candidates for these signals are xylem sap-soluble chemicals released by root cells. These chemicals are then assumed to be transported extracellularly through the xylem conduit system to targets such as receptors in guard cells of leaf stomata and induce tolerance responses.

The hormone abscisic acid (ABA) is now widely accepted as the key chemical signal mediating water stress responses, with fewer arguments to the contrary being published in recent years compared with the persistent controversy of previous decades (see Davies et al. (2005); Jia and Zhang (2008); Schachtman and Goodger (2008) and references therein). Nonetheless, ABA may not be acting as the sole or initial water stress signal, and indeed the signaling cascade is likely to involve a number of other signals and interactions. This chapter will focus on our current understanding of ABA as the long-distance water stress signal, and highlight the involvement of other signals, the complexity of which we are only beginning to fully appreciate.

2 ABA Is the Key Water Stress Signal Acting on Stomata

The molecular and physiological basis of ABA-stomatal signaling has made an attractive target for research in the past couple of decades. The cellular signaling pathway for ABA-induced stomatal closure is complex and involves an array of multiple signaling compounds [see Li et al. (2006)]. Recently significant molecular breakthroughs have been made in elucidating the key interactions between ABA and proteins in the signal transduction pathway using the model species Arabidopsis. A number of reviews have extensively covered this rapidly advancing field (Hirayama and Shinozaki 2010; Ben-Ari 2012; Hauser et al. 2011; Klingler et al. 2010; Sheard and Zheng 2009), and the key aspects of these recent discoveries will only be briefly covered here.

In 2009, two independent groups discovered that members of a new family of proteins called PYR/PYL/RCAR START proteins (RCAR) act as archetypal ABA receptors (Ma et al. 2009; Park et al. 2009). ABA binds to RCAR proteins and this complex then inhibits the activity of type 2C plant protein phosphatase enzymes (PP2C), the inhibition of which results in plant stress responses. These paradigm-shifting discoveries were followed by several crystallographic examinations (e.g., Nishimura et al. 2009; plus see Santiago et al. 2012) and mutational studies (e.g., Mosquna et al. 2011), elucidating the intricate interactions between ABA, RCAR receptors, and PP2C phosphatases. These interactions are summarized in a simplified form as follows and presented schematically in Fig. 1.

In the absence of ABA, PP2C acts as a constitutive negative regulator of a family of kinases (SnRK2; Fig. 1a) whose autophosphorylation is required for regulation of ion channels in stomatal guard cell membranes, e.g., RbohF, SLAC1, and KAT1, and activation of ABA-responsive elements (ABRE) in the promoter of ABA-responsive genes in the nucleus via transcription factors (ABF). When ABA is present, it binds to the RCAR receptor by initiating an allosteric open-to-close

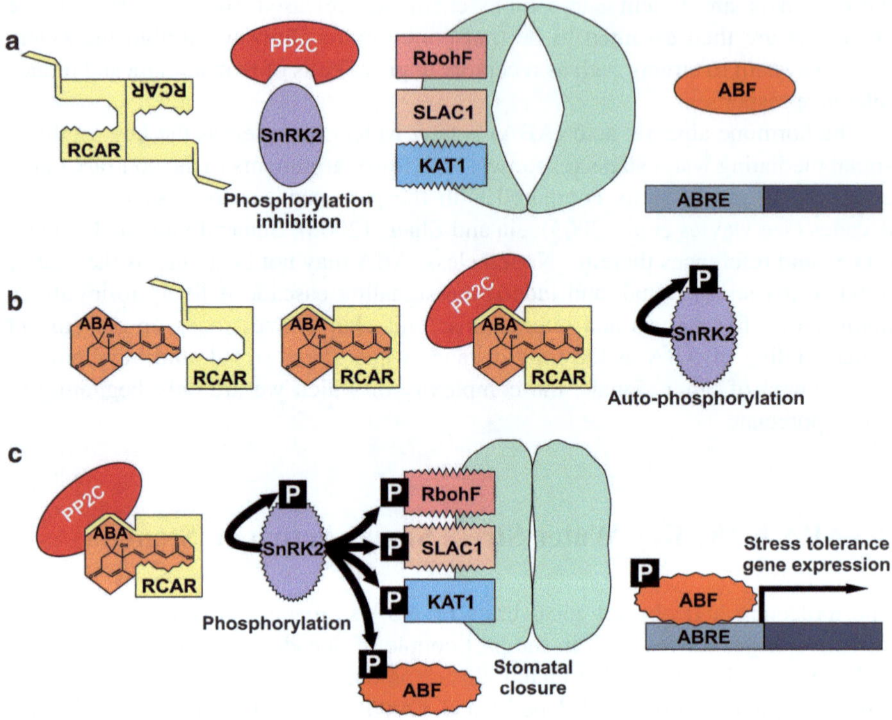

Fig. 1 Simplified ABA signal transduction pathway largely derived from molecular work on Arabidopsis. (**a**) In the absence of the hormone ABA, RCAR receptor occurs as a homodimer, PP2C phosphatase inhibits autophosphorylation of SnRK2 kinase, stomata are open, and ABF transcription factors are inactive towards ABA-responsive gene promoter elements. (**b**) When ABA is present, it binds to RCAR receptor via open-to-close transition of a gating loop, exposing hydrophobic region for PP2C to bind. This complexation represses inhibition of SnRK2 which is free to autophosphorylate. (**c**) SnRK2 phosphorylation regulates ion channels at the guard cell membrane resulting in stomatal closure and activates transcription factors such as ABF, thereby enabling transcription of ABA-responsive genes in the nucleus

transition of a gating loop on the protein (Fig. 1b). Closure of the gating loop complexes ABA with the receptor and exposes a hydrophobic binding site enabling the RCAR-ABA complex to subsequently bind to PP2C. The binding covers the phosphatase active site in PPC2, thereby repressing its negative regulation of SnRK2. SnRK2 is then free to autoactivate via autophosphorylation (Fig. 1b). Activated SnRK2 can regulate ion channels at the plasma membrane via phosphorylation, e.g., by activating RbohF and SLAC1 and inhibiting KAT1, resulting in stomatal closure and therefore reduced transpirational water loss (Fig. 1c). Activated SnRK2 is also free to activate downstream transcription factors such as ABF via phosphorylation, thereby facilitating transcription of ABA-responsive genes (ABRE) in the nucleus (Fig. 1d). ABRE gene products can be involved in a water stress feedback mechanism acting to maintain stomatal closure or curtail it when the stress has passed, and can also result in enhanced stress tolerance capabilities in the long term [see Krasensky and Jonak (2012)].

3 Is ABA the Long-Distance Water Stress Signal?

The key role of ABA in the complex molecular interactions that culminate in stress-induced stomatal closure and ABRE gene transcription is now well characterized. In contrast, the mechanism by which the hormone reaches receptors in leaf cells, purportedly from water-stressed roots, remains relatively poorly characterized (Jia and Zhang 2008; Schachtman and Goodger 2008). It has long been hypothesized that roots sensing soil water deficit synthesize ABA which is then transported to leaves via xylem, thereby providing plants with a mechanism for relatively rapidly signal transmission over long distances from roots to leaves (Jones and Mansfield 1971; Walton et al. 1976). Indeed numerous studies have shown enhanced accumulation of root, xylem, and leaf ABA, and concomitant reduction in transpiration and or growth in response to soil water reduction (Davies and Zhang 1991; Dry and Loveys 1999; Zhang and Davies 1989; Liang and Zhang 1997; Ernst et al. 2010; Goodger et al. 2005). Moreover, the results of grafting experiments on tobacco show that ABA is transported from roots under conditions of soil water deficit, although it was unclear whether this amount was sufficient to induce a response (Borel et al. 2001).

3.1 Root ABA Biosynthesis in Response to Water Stress

If root-sourced ABA is the key long-distance signal involved in root-to-shoot signaling of soil water deficit, then the hormone must be rapidly biosynthesized in root tissues that have detected soil drying. The means by which root cells detect reductions in soil water potential is poorly understood, and similarly, relatively little is known about the biosynthesis of ABA in roots under any conditions and particularly as a rapid response to soil water stress. For instance, the precise location of ABA biosynthesis in roots is still unclear. Such information is important as it may influence how roots perceive and monitor soil water availability. It has been reported that synthesis of ABA in maize roots occurs in root tips (Zhang and Tardieu 1996), whereas an increase in expression of genes in the ABA biosynthetic pathway was recently observed 1–6 cm distal to the maize root tip, rather than the proximal 1 cm of root tip (Ernst et al. 2010). In a similar manner to the latter study, results of an experiment using detached roots of pea and Asiatic dayflower suggested ABA biosynthesis occurred somewhere between the root tip and 3 cm distal to the tip (Zhang et al. 1987). In general, the question of the site of ABA biosynthesis has mainly been addressed by measuring ABA content, which does not take into account ABA production and subsequent translocation to other root or plant tissues (Freundl et al. 2000; Schraut et al. 2005). More work examining the expression of the genes for ABA biosynthesis and biosynthetic enzyme activity is required to clarify where ABA is produced in roots. In addition, the rate of biosynthesis needs to be quantified, to determine if it can occur rapidly enough under only slight soil water reductions, for root-synthesized ABA to be the primary water stress signal transported to leaves.

3.2 Leaf ABA Biosynthesis in Response to Water Stress

Interestingly, there have been greater advances in understanding ABA biosynthesis in leaves rather than roots in recent years. ABA can be synthesized in both roots and leaves (Thompson et al. 2007), and recent studies have suggested leaf biosynthesis as an alternative or additional source to root-derived ABA acting to facilitate leaf water stress responses. Indeed a number of studies have shown that ABA-induced stomatal closure is not dependent on ABA release from roots (Christmann et al. 2007; Holbrook et al. 2002). Moreover, ABA can be produced in greater amounts or at an earlier stage in leaves relative to roots, particularly in response to water stress (Qin and Zeevaart 1999; Christmann et al. 2005; Ikegami et al. 2009). For example, Ikegami et al. (2009) found that ABA accumulated in roots when intact whole plants were subjected to water stress, but not when detached roots were exposed to the same stress. Furthermore, after the application of ^{13}C-labeled ABA to leaves of whole plants in well-watered conditions, the label was detected in roots and the amount increased after imposition of water stress to the roots (Ikegami et al. 2009), suggesting transport of ABA from leaves to roots, rather than the other way around.

The hypothesis that ABA acting on stomata is produced by leaves of plants under water stress was further supported with the use of reciprocal grafts between roots and shoots of wild-type or ABA-deficient Arabidopsis (Christmann et al. 2007) and tomato (Holbrook et al. 2002) mutants, where subsequent measurement of leaf responses to water stress indicated that stomatal closure was not dependent on ABA production by the roots. Other studies have shown the ABA biosynthetic enzymes AtNCED3, AtABA2, and AAO3 to be localized to leaf vascular tissues in Arabidopsis (Cheng et al. 2002; Koiwai et al. 2004; Endo et al. 2008). It has been proposed that ABA biosynthesized in leaf vasculature is transported to stomata to directly induce stomatal closure and can also be translocated to phloem and transported down to roots to induce water uptake from soil and expression of stress-resistant genes in roots (Ikegami et al. 2009).

The timing of ABA biosynthesis in roots also casts doubt on the ability of root-synthesized ABA to be the initial signal produced by roots sensing early reductions in soil water potential. For example, expression of ABA biosynthetic genes in roots of water-stressed maize was observed to significantly increase only after leaves had responded with decreased stomatal conductance (Ernst et al. 2010). Indeed in that study, increased expression of root ABA biosynthetic genes and significant increases in xylem sap ABA concentration were only observed under extended water stress and coincided with a significant reduction in plant water status. It is known that such hydraulic changes can influence cell wall and/or cell wall–plasma membrane interactions mediated by integrin-like proteins resulting in the induction of ABA biosynthesis in root cells (Lü et al. 2007; Jia et al. 2001) and in leaves (Christmann et al. 2007). The results of these studies suggest that the ABA acting on stomata in water-stressed plants may be produced in leaves and, if so, other root-sourced signals may be transported to leaves to induce this biosynthesis.

3.3 Redistribution of ABA Under Water Stress

Arguably a more rapid response than synthesizing ABA in roots sensing soil water deficit is to redistribute existing pools of the hormone from plant sink tissues. Indeed, early reports on plant water stress responses suggested redistribution of ABA within leaf tissues may be a source of the hormone facilitating rapid leaf responses (Zeevart and Boyer 1984; Loveys 1977). Since then, it has been proposed that increases in ABA in particular tissues are not only dependent upon de novo ABA biosynthesis but can also be enhanced by ABA redistribution (Wilkinson and Davies 2002). More specifically, water stress has been observed to alkalize xylem sap in several species, and this can cause ABA to accumulate in the sap and/or move from leaf and stem symplast to apoplast resulting in increased delivery of ABA to stomata (Bacon et al. 1998; Wilkinson et al. 1998; Wilkinson and Davies 1997).

Alkalization of xylem sap pH under water stress has been observed in a variety of species including sunflower (Gollan et al. 1992), runner bean (Hartung et al. 1998), and Asiatic dayflower (Wilkinson and Davies 1997). The potential effects of sap pH increases have been outlined previously (Wilkinson 1999) and include (1) changes in ABA metabolism resulting in increased leaf ABA concentration, (2) changes in leaf water status which can directly alter guard cell turgor or increase cell sensitivity to ABA concentrations, (3) direct effects on guard cell membrane ion fluxes, and (4) altered distribution of leaf ABA increasing the concentration in the apoplast outside guard cells. In isolation or combination, these effects can enhance stomatal closure and even reduce leaf growth in certain plant species as xylem sap becomes more alkaline under drought conditions.

In particular, an increase in apoplastic ABA around guard cells due to leaf redistribution appears the most likely effect of increasing alkalinity. ABA is a weak acid and may be protonated or deprotonated due to pH changes in the apoplast of leaves. These apoplast pH changes can result from changes in xylem sap pH. When water stress induces an increase in xylem sap pH, any ABA carried in the xylem will remain deprotonated under the more alkaline conditions and will not be taken up passively by surrounding mesophyll cells. Experiments in the early 1980s on mesophyll protoplasts documented high rates of ABA uptake at acidic pH and almost no uptake at the more alkaline pH of 7.5 (Kaiser and Hartung 1981). The result is that less ABA is transported into the mesophyll cells, but more accumulates in the apoplastic space and leads to enhanced stomatal closure. In contrast, under well-watered conditions, the apoplastic space is more acidic and ABA passively enters the symplast, and apoplastic concentrations do not increase as rapidly. Therefore, the likely effect of pH is indirect in that it changes the accumulation of ABA in the apoplast with flow on effects on guard cells (Wilkinson 1999).

A number of studies using mutants deficient in ABA biosynthesis have shown that pH alone does not induce stomatal closure, but acts to enhance the ABA signal (Bacon et al. 1998; Wilkinson et al. 1998). Interestingly, the use of the tomato mutant flacca, which is deficient in ABA biosynthesis, has shown that increases of artificial sap pH can result in increased transpiration in the absence of ABA, suggesting that the hormone is not only necessary for stomatal closure but also for preventing stomatal opening (Wilkinson et al. 1998). Similarly, a study comparing Asiatic dayflower and Arabidopsis found that the increased pH of external solutions led to a 27 % increase in stomatal opening in Asiatic dayflower, but had no effect on Arabidopsis (Prokic et al. 2006). Both species, however, responded similarly with decreased conductance in the presence of ABA and only changes in the kinetics of the response were observed.

The interaction between alkalinity and ABA under water stress is relatively well documented; nonetheless changes in sap pH under water stress do not appear to be consistent in all species or even in all experiments using the same species. For example, a study showed pH of xylem sap in sunflower and Asiatic dayflower did not change significantly as soil dried, whereas in tomato the sap pH significantly increased (Jia and Davies 2007). In addition, the timing of sap pH changes does not consistently support alkalinity-induced ABA redistribution as a part of the rapid signal response in the early stages of water stress. If pH changes are to form part of a rapid response of plants to soil water deficit, then sap pH increases should be observed in the earliest stages of water stress. Refuting this notion, a study on soybean found xylem sap pH was not correlated with decreased stomatal conductance in the initial stages of water stress (Liu et al. 2003). In that study, the xylem sap did become more alkaline during extended water stress, but this alkalization occurred well after a measured decrease in stomatal conductance. Similarly, studies on maize have detected no change in sap pH at early (Goodger et al. 2005) or even moderate stages of water stress (Ernst et al. 2010), but as with soybean, there was significant alkalinization under extended water stress (Ernst et al. 2010; Bahrun et al. 2002). ABA redistribution mediated by pH changes is likely to be involved in plant water stress responses, but at least in some species, this may be a mechanism of enhancing responses under more severe stress conditions, rather than the initial signal produced by roots sensing soil water deficits (Goodger and Schachtman 2010b).

3.4 Production of ABA via Catabolism of Conjugates

Another potentially rapid means of ABA production in roots sensing soil water deficit is the catabolism of conjugated forms of the hormone such as ABA glucose esters (ABA-GE). Six different structural forms of ABA-GE have been found in xylem sap from both well-watered and water-stressed sunflower (Hansen and Dörffling 1999), and cell wall glucosidase enzymes can catabolize these conjugates to free ABA. For example, a β-glucosidase was found to be involved in the release of free ABA from ABA-GE in the cortical apoplast of roots (Sauter and Hartung 2000) and in the xylem vessel apoplast of leaves (Dietz et al. 2000), especially when de novo ABA biosynthesis was inhibited. Moreover, water stress has been shown to induce polymerization of β-glucosidase, thereby activating the enzyme to release ABA (Lee et al. 2006). Nonetheless, the mechanism of ABA-GE catabolism in the apoplast and its importance in response to water stress are still poorly understood. Furthermore, it has been argued that the amount of ABA-GE in roots is too small for ABA release from this conjugate to contribute significantly to the overall increase in ABA during water stress (Priest et al. 2006), although such low amounts do not necessarily preclude this mechanism from being an early signal that initiates potential secondary signaling components such as ABA redistribution and biosynthesis. Interestingly the ABA-GE conjugate itself has also been considered a potential candidate as a chemical stress signal (Munns and King 1988; Munns et al. 1993). Indeed, a correlation was found between increased ABA-GE concentrations and the severity of water stress (Sauter and Hartung 2002), suggesting that it may be involved in the complex signaling of water stress in plants, even in its non-catabolized form.

4 Involvement of Other Chemicals in Long-Distance ABA Signaling

Although ABA may play a dominant role in root-to-shoot signaling under water stress, it seems likely that the signaling cascade is more complex and other substances are likely to be involved. Numerous substances have been identified in xylem sap, and many of these have been implicated to a greater or lesser degree in root-to-shoot signaling under conditions of water stress (Schachtman and Goodger 2008; Jia and Zhang 2008). These substances include hormones such as cytokinins, ethylene and jasmonic acid, inorganic ions, organic acids, sugars, proteins, peptides, and microRNAs. As found at the cellular level with stomata, the signaling pathway via xylem sap may involve these substances acting in concert with ABA or perhaps even prior to ABA as root-to-shoot signals of early soil water deficit. Some of these putative signaling candidates have been relatively well characterized with respect to water stress responses and are summarized in the following sections.

In contrast, compounds such as jasmonic acid, proteins, peptides, and microRNAs (miRNAs) have not been well characterized with respect to their involvement in plant water stress responses and will only be briefly covered here. Despite jasmonic acid being a well-characterized systemic signal (see chapters in this publication), evidence for its involvement in water stress responses has only recently been presented. A study on papaya seedlings noted changes in jasmonic acid and ABA under water stress, and the authors suggested the hormones may act synergistically to facilitate stomatal closure (Mahouachi et al. 2007). Proteins have been known to be present in xylem sap for quite some time (Biles and Abeles 1991), but the extent of their abundance and distribution within plants has only more recently been established (Buhtz et al. 2004; Alvarez et al. 2006, 2008; Kehr et al. 2005; Krishnan et al. 2011; Ligat et al. 2011). Numerous proteins have been observed to be up- or downregulated under water stress (e.g., Alvarez et al. (2006) and (2008)), but their direct involvement in signaling has not been examined. Peptides have been shown to play important signaling roles in plants and are known to move systemically through phloem (Lindsey et al. 2002; Ryan et al. 2002) and xylem (Neumann 2007), but again their role in signaling of plant water stress has not been appraised. In addition, miRNAs have been implicated as potential signal molecules that move systemically in plants (Sunkar et al. 2007). Studies on miRNAs extracted from whole plants of Arabidopsis (Zhao et al. 2007) and Chinese white polar (Ren et al. 2012) have also shown that their expression can change in response to water stress, but their movement through xylem sap has yet to be examined. Notably, they have been detected in the phloem sap of oilseed rape (Buhtz et al. 2008), but much more work is required to test if they are involved in long-distance signaling of water stress.

4.1 Cytokinins

Cytokinins are largely produced in roots (Sakakibara 2006) and are known to be involved in responses to nutrient deprivation (Schachtman and Shin 2007). The hormones may also be involved in signal transmission from roots to leaves under water stress, but in contrast to ABA, cytokinin concentrations generally decrease in plants subjected to water stress (Davies et al. 2005). Mechanistic explanations for this negative relationship include ABA inducing increased expression of cytokinin oxidase genes (Brugiere et al. 2003) as well as inducing increased activity of enzymes that catalyze irreversible cytokinin degradation (Vysotskaya et al. 2009).

Increased xylem cytokinin concentrations have been shown to reduce stomatal sensitivity to xylem-delivered ABA (Radin et al. 1982), with the hormones generally thought of as antagonists of ABA action on stomata (Davies et al. 2005). Exogenous application of cytokinins has been shown to maintain stomata in an open state (Blackman and Davies 1985), and elevated levels of leaf cytokinins have been correlated with stomatal opening in some plants (Vysotskaya et al. 2004). In particular, overexpression of the isopentenyltransferase (*ipt*) gene associated with cytokinin biosynthesis resulted in increased rates of transpiration in transgenic tobacco plants (Teplova et al. 2000). Nonetheless the effects of cytokinins on stomatal conductance are less clear than those for ABA, and severe stresses are often required to significantly reduce cytokinin delivery to leaves via xylem sap (Wilkinson et al. 2012).

In contrast to the plethora of studies on ABA, there have been much fewer reports on the cytokinin content of xylem sap and how that content changes with water stress. Nonetheless, in grapevine, a 50 % reduction in zeatin (Z) and zeatin riboside (ZR) was found in plants subjected to partial root zone drying (Stoll et al. 2000). In a more recent study where Z, ZR, and zeatin nucleotide (ZN) were measured, partial root zone drying of tomato reduced the ZN content of the xylem sap (Kudoyarova et al. 2007). In at least two studies on sunflower xylem sap, Z, ZR, isopentenyladenine, and isopentenyladenosine concentrations were shown to decrease under water stress (Hansen and Dörffling 2003). A recent study on maize also found a decrease in Z and ZR concentrations in xylem sap (Alvarez et al. 2008). Surprisingly, that study also found the concentration of the aromatic cytokinin 6-benzylaminopurine (BAP) in xylem sap to increase significantly due to water stress (Alvarez et al. 2008).

The majority of available data suggests that water stress generally results in a decrease in xylem sap cytokinin concentrations, but the rapidity of this decrease and its importance in ABA signaling are unclear. It is also not clear that all plant species respond in the same way to cytokinins at the leaf and guard cell levels (Dodd et al. 2003). It has been suggested that ABA/cytokinin ratios in xylem sap are more important for facilitating stomatal responses to water stress than the absolute amounts of each hormone (Hansen and Dörffling 1999). Furthermore, the complexity of cytokinin structures is yet to be fully elucidated (Davies et al. 2005), and much more work is required to clarify the role of each compound in water stress signaling.

4.2 *Ethylene*

The volatile plant hormone ethylene may be another important factor involved in plant signaling of soil water deficit. Increased evolution of ethylene under water stress can inhibit or reduce root and leaf growth (Pierik et al. 2006; Sharp 2002), induce leaf senescence and abscission (Wilkinson et al. 2012), directly reduce photosynthesis (Rajala and Peltonen-Sainio 2001), and close stomata (Desikan et al. 2006; Vysotskaya et al. 2011). It is known that a precursor of ethylene, 1-aminocyclopropane-1-carboxylic acid (ACC), moves in the xylem from root to shoots (Else and Jackson 1998). Under water stress, ACC transported in xylem sap may result in increased evolution of ethylene in leaves. For example, the use of ACC oxidase (ACO) antisense tomato plants showed ethylene evolution was much lower under well-watered compared to water deficit conditions (Sobeih et al. 2004). In that study, the stomatal response to soil drying was the same in all plants, but a decrease in leaf growth was measured in wild type compared to ACO antisense plants (Sobeih et al. 2004). This suggests that ethylene may play a role in mediating the water stress response of decreased leaf growth.

Ostensibly contradictory to the aforementioned action of ethylene in water stress responses, the hormone can also antagonize ABA-induced stomatal closure and growth reduction (Tanaka et al. 2005; Wilkinson and Davies 2009). For example, ethylene can prevent ABA accumulation and vice versa (Spollen et al. 2000) or can modulate cellular sensitivity to ABA (Wilkinson and Davies 2010). The antagonism can produce an opposing effect of ethylene on stomatal and leaf growth when ABA accumulation is also increased (Wilkinson et al. 2012). Furthermore, a study on water-stressed maize plants showed root growth is partially maintained due to an ABA-induced decrease in ethylene synthesis (Sharp et al. 2004). In support of the ABA–ethylene antagonism theory, research on ethylene-insensitive Arabidopsis mutants have shown that the antagonism of ABA-induced stomatal closure by ethylene can be suppressed (Tanaka et al. 2005). Therefore, in a similar way to ABA/cytokinin ratios, ABA/ethylene, i.e., ABA/ACC ratios in xylem sap may be more important as components of water stress signaling than the concentrations of the hormones alone (Acharya and Assmann 2009; Wilkinson and Davies 2010).

4.3 Inorganic Ions and Organic Acids

The concentration of numerous inorganic ions and organic acids in xylem sap has been shown to change in response to water stress (Gollan et al. 1992; Goodger et al. 2005). Such changes may relate to sap charge balance maintenance (Goodger and Schachtman 2010a) or pH adjustments (Wilkinson et al. 2007), with indirect effects on root-to-shoot signaling of water stress. For example, the mechanism of pH change described in Sect. 3.3 may involve nitrate availability. As soils dry, nutrient availability is reduced due to physiochemical changes that occur. Under conditions of reduced nitrate availability, nitrate reductase activity shifts to roots which causes changes in the sap pH (Liu et al. 2005). The effects of a nitrate reductase inhibitor (sodium tungstate) have been tested (Wilkinson 2004), and it was found to prevent the alkalization of tomato sap under water stress conditions. Moreover, the addition of nitrate to water-stressed tomato plants enhanced the alkalization of xylem sap (Jia and Davies 2007). The study by Jia and Davies (2007) also determined a synergistic effect of xylem nitrate and ABA on stomatal conductance in detached leaves of Asiatic dayflower. Similarly, results of work on maize suggest that under some circumstances a combination of ABA and nitrate is required to elicit an effect on shoot physiology, where neither compound is as effective alone (Wilkinson et al. 2007).

A number of organic acids have been found in xylem sap of plants (Peterlunger et al. 1990; Senden et al. 1992), and their concentrations are known to change in response to water stress (e.g., Goodger et al. 2005). Such changes may result from the activity of nitrate reductase as described above, which is known to influence organic acid production (Scheible et al. 1997). Increased concentrations of organic acids can alter sap pH and thereby influence ABA redistribution as described in Sect. 3.3. Malate, in particular, has been shown to increase in xylem sap under water stress (Patonnier et al. 1999; Goodger et al. 2005), and a level of interdependence has been suggested between pH, malate, and cations in xylem sap of beech

(Schell 1997). It is possible that malate may also have a direct effect on plant water stress signaling. For example, stomatal opening could be prevented in ash leaves supplied with 0.5–3.0 mM malate (Patonnier et al. 1999), and elevated extracellular concentrations of malate can result in stomatal closure in broad bean (Hedrich et al. 1994).

Recently, both carbon monoxide and nitric oxide have also been implicated as potential chemical signals regulating stomatal closure in broad bean (Song et al. 2008). The xylem delivery rate of sulfate has also been observed to increase in maize plants subjected to water stress (Goodger et al. 2005; Ernst et al. 2010). Interestingly, in the study by Ernst et al. (2010), the expression of ABA biosynthetic genes in roots only increased significantly under extended water stress, whereas that of a sulfate transporter gene in roots increased from early water stress onwards. Using bioassays, that study also found an interactive effect of ABA and sulfate in decreasing maize transpiration rate, as compared to ABA alone. In support of the involvement of sulfate in ABA-mediated responses, the stomata of plant species containing inherently higher amounts of leaf ABA, e.g., peanut and tomato, have been shown to respond more rapidly to sulfur dioxide fumigation than those with inherently lower amounts, e.g., radish, perilla, and spinach (Kondo and Sugahara 1978). Similarly, a study on Argentine screwbean watered with Na_2SO_4 or NaCl found sulfate was transported from roots to leaves in greater amounts than chloride and affected gas fluxes in leaflets of treated plants, possibly through influences on membrane permeability (Reinoso et al. 2005). It has been proposed that sulfate may be a chemical signal transported from roots to leaves in the early phase of water stress sensing which can enhance the anti-transpirational effect of small amounts of ABA in leaves (Goodger and Schachtman 2010b).

5 Conclusions

Knowledge on the molecular mechanisms by which ABA facilitates stomatal closure and other water stress responses is accumulating at an impressive rate. In contrast, research on how plants sense small reductions in soil water and then communicate this to the rest of the plant is far less advanced. Such long-distance communication is likely to occur via chemical signals transported within the xylem, but there are relatively few studies that have comprehensively analyzed all xylem sap constituents transported from roots to leaves under water stress. More commonly, studies have analyzed select constituents, with a great bias towards ABA. Given the key role of ABA in regulating stomatal closure and other water stress responses, this focus is not surprising, but means potentially important interactions and components of long-distance signaling remain poorly understood. In the past, inherent analytical difficulties in accurately determining the identity and abundance of the wide variety of constituents transported in xylem sap have hindered research, but technical advances continue to alleviate such difficulties. More detailed studies are required to provide comprehensive information on changes in xylem sap composition, especially in the very early stages of soil water reductions. Such information will help clarify the contribution of different chemical signals and their potential interactions with ABA in long-distance communication of soil water deficit.

The results of detailed examinations of sap constituents are only meaningful if the experimental design is well considered. The results of many studies on changes in xylem sap constituents with water stress have produced contradictory results, and this can be due to a number of factors related to differences in experimental designs. Firstly, the choice of species is important. Studies have been conducted on a broad range of species each adapted to very different soil water regimes and therefore likely to differ in their response to water stress. Secondly, the implication of water stress is not necessarily as easy as simply withholding water, and differences in the intensities of applied water stress and the time at which sampling takes place during stress imposition can produce varying results. In particular, xylem sap sampling has often taken place at times of more advanced water stress, when it is not possible to differentiate between the initial signals sent from roots sensing soil water deficit and downstream events in the complex signaling cascade, in which ABA appears to have a greater role. Thirdly, there are many different methods used to extract xylem sap and not all accurately reflect the abundances of constituents reaching leaves from roots (Tiekstra et al. 2000; Goodger et al. 2005; Bacon et al. 1998; Goodger and Schachtman 2010a; Gollan et al. 1992). In a related manner, the way constituent data is commonly presented as concentrations rather than fluxes (Goodger and Schachtman 2010a) can make it difficult to quantify and compare what leaves are experiencing in plants. The rate of advance in this field will increase with more fundamental research using well-considered experimental designs.

Much more work is required before we fully understand the complexity of communicating water status over long distances from the rhizosphere to the guard cell membrane and the myriad of interactions that occur in between. As we increase our knowledge of the precise identity and biosynthesis of the primary root signals that culminate in plant tolerance responses, it will become possible to alter the sensitivity of crop plants to soil water deficit. This will provide new molecular breeding strategies for tailoring crops to maintain or even increase yields under the more frequent episodes of water stress that the world is predicted to experience in the coming decades. A much more detailed understanding of the complexity of long-distance plant signaling will also help reduce crop yield penalties when water is readily available—a problem commonly observed with current approaches to genetic manipulations for drought tolerance.

Acknowledgement This work was funded by a grant from the Australian Research Council (Project DP1094530).

References

Acharya B, Assmann S (2009) Hormone interactions in stomatal function. Plant Mol Biol 69:451–462

Alvarez S, Goodger JQD, Marsh EL, Chen SX, Asirvatham VS, Schachtman DP (2006) Characterization of the maize xylem sap proteome. J Proteome Res 5:963–972

Alvarez S, Marsh EL, Schroeder SG, Schachtman DP (2008) Metabolomic and proteomic changes in the xylem sap of maize under drought. Plant Cell Environ 31:325–340

Bacon MA, Wilkinson S, Davies WJ (1998) pH-regulated leaf cell expansion in droughted plants is abscisic acid dependent. Plant Physiol 118:1507–1515

Bahrun A, Jensen CR, Asch F, Mogensen VO (2002) Drought-induced changes in xylem pH, ionic composition, and ABA concentration act as early signals in field-grown maize (*Zea mays* L.). J Exp Bot 53:251–263

Ben-Ari G (2012) The ABA signal transduction mechanism in commercial crops: learning from Arabidopsis. Plant Cell Rep 31:1357–1369

Biles CL, Abeles FB (1991) Xylem sap proteins. Plant Physiol 96:597–601

Blackman PG, Davies WJ (1985) Root to shoot communication in maize plants of the effects of soil drying. J Exp Bot 36:39–48

Bogoslavsky L, Neumann PM (1998) Rapid regulation by acid pH of cell wall adjustment and leaf growth in maize plants responding to reversal of water stress. Plant Physiol 118:701–709

Borel C, Frey A, Marion-Poll A, Tardieu F, Simonneau T (2001) Does engineering abscisic acid biosynthesis in *Nicotiana plumbaginifolia* modify stomatal response to drought? Plant Cell Environ 24:477–489

Boyer JS, Westgate ME (2004) Grain yields with limited water. J Exp Bot 55:2385–2394

Brugiere N, Jiao SP, Hantke S, Zinselmeier C, Roessler JA, Niu XM, Jones RJ, Habben JE (2003) Cytokinin oxidase gene expression in maize is localized to the vasculature, and is induced by cytokinins, abscisic acid, and abiotic stress. Plant Physiol 132:1228–1240

Buhtz A, Kolasa A, Arlt K, Walz C, Kehr J (2004) Xylem sap protein composition is conserved among different plant species. Planta 219:610–618

Buhtz A, Springer F, Chappell L, Baulcombe DC, Kehr J (2008) Identification and characterization of small RNAs from the phloem of *Brassica napus*. Plant J 53:739–749

Chaves MM, Maroco JP, Pereira JS (2003) Understanding plant responses to drought – from genes to the whole plant. Funct Plant Biol 30:239–264

Cheng WH, Endo A, Zhou L, Penney J, Chen HC, Arroyo A, Leon P, Nambara E, Asami T, Seo M, Koshiba T, Sheen J (2002) A unique short-chain dehydrogenase/reductase in Arabidopsis glucose signaling and abscisic acid biosynthesis and functions. Plant Cell Environ 14:2723–2743

Christmann A, Hoffmann T, Teplova I, Grill E, Muller A (2005) Generation of active pools of abscisic acid revealed by in vivo imaging of water-stressed Arabidopsis. Plant Physiol 137:209–219

Christmann A, Weiler EW, Steudle E, Grill E (2007) A hydraulic signal in root-to-shoot signalling of water shortage. Plant J 52:167–174

Davies WJ, Zhang J (1991) Root signals and the regulation of growth and development of plants in drying soil. Annu Rev Plant Physiol Plant Mol Biol 42:55–76

Davies WJ, Wilkinson S, Loveys BR (2002) Stomatal control by chemical signalling and the exploitation of this mechanism to increase water use efficiency in agriculture. New Phytol 153:449–460

Davies WJ, Kudoyarova GR, Hartung W (2005) Long-distance ABA signaling and its relation to other signaling pathways in the detection of soil drying and the mediation of the plant's response to drought. J Plant Growth Regul 24:285–295

Desikan R, Last K, Harrett-Williams R, Taglavia C, Harter K, Hooley R, Hancock JT, Neill SJ (2006) Ethylene-induced stomatal closure in Arabidopsis occurs via AtrbohF-mediated hydrogen peroxide synthesis. Plant J 47:907–916

Dietz KJ, Sauter A, Wichert K, Messdaghi D, Hartung W (2000) Extracellular beta-glucosidase activity in barley involved in the hydrolysis of ABA glucose conjugate in leaves. J Exp Bot 51:937–944

Dodd IC, Tan LP, He J (2003) Do increases in xylem sap pH and/or ABA concentration mediate stomatal closure following nitrate deprivation? J Exp Bot 54:1281–1288

Dry PR, Loveys BR (1999) Grapevine shoot growth and stomatal conductance are reduced when part of the root system is dried. Vitis 38:151–156

Else MA, Jackson MB (1998) Transport of 1-aminocyclopropane-1-carboxylic acid (ACC) in the transpiration stream of tomato (*Lycopersicon esculentum*) in relation to foliar ethylene production and petiole epinasty. Aust J Plant Physiol 25:453–458

Endo A, Sawada Y, Takahashi H, Okamoto M, Ikegami K, Koiwai H, Seo M, Toyomasu T, Mitsuhashi W, Shinozaki K, Nakazono M, Kamiya Y, Koshiba T, Nambara E (2008) Drought induction of Arabidopsis 9-cis-epoxycarotenoid dioxygenase occurs in vascular parenchyma cells. Plant Physiol 147:1984–1993

Ernst L, Goodger JQD, Alvarez S, Marsh EL, Berla B, Lockhart E, Jung J, Li P, Bohnert HJ, Schachtman DP (2010) Sulphate as a xylem-borne chemical signal precedes the expression of ABA biosynthetic genes in maize roots. J Exp Bot 61:3395–3405

Freundl E, Steudle E, Hartung W (2000) Apoplastic transport of abscisic acid through roots of maize: effect of the exodermis. Planta 210:222–231

Gollan T, Schurr U, Schulze E-D (1992) Stomatal response to drying soil in relation to changes in the xylem sap composition of *Helianthus annus*. I. The concentration of cations, anions, amino acids in, and pH of, the xylem sap. Plant Cell Environ 15:551–559

Goodger JQD, Schachtman DP (2010a) Nitrogen source influences root to shoot signaling under drought. In: Pareek A, Sopory SK, Bohnert HJ, Govindjee (eds) Abiotic stress adaptation in plants: physiological, molecular and genomic foundation. Springer Science, Dordrecht, pp 165–173

Goodger JQD, Schachtman DP (2010b) Re-examining the role of ABA as the primary long-distance signal produced by water-stressed roots. Plant Signal Behav 5:1298–1301

Goodger JQD, Sharp RE, Marsh EL, Schachtman DP (2005) Relationships between xylem sap constituents and leaf conductance of well-watered and water-stressed maize across three xylem sap sampling techniques. J Exp Bot 56:2389–2400

Hansen H, Dörffling K (1999) Changes of free and conjugated abscisic acid and phaseic acid in xylem sap of drought-stressed sunflower plants. J Exp Bot 50:1599–1605

Hansen H, Dörffling K (2003) Root-derived trans-zeatin riboside and abscisic acid in drought-stressed and rewatered sunflower plants: interaction in the control of leaf diffusive resistance? Funct Plant Biol 30:365–375

Hartung W, Wilkinson S, Davies WJ (1998) Factors that regulate abscisic acid concentrations at the primary site of action at the guard cell. J Exp Bot 49:361–367

Hauser F, Waadt R, Schroeder JI (2011) Evolution of abscisic acid synthesis and signaling mechanisms. Curr Biol 21:R346–R355

Hedrich R, Marten I, Lohse G, Dietrich P, Winter H, Lohaus G, Heldt H-W (1994) Malate-sensitive anion channels enable guard cells to sense changes in the ambient CO_2 concentration. Plant J 6:741–748

Hirayama T, Shinozaki K (2010) Research on plant abiotic stress responses in the post-genome era: past, present and future. Plant J 61:1041–1052

Holbrook NM, Shashidhar VR, James RA, Munns R (2002) Stomatal control in tomato with ABA-deficient roots: response of grafted plants to soil drying. J Exp Bot 53:1503–1514

Hose E, Steudle E, Hartung W (2000) Abscisic acid and hydraulic conductivity of maize roots: a study using cell- and root-pressure probes. Planta 211:874–882

Ikegami K, Okamoto M, Seo M, Koshiba T (2009) Activation of abscisic acid biosynthesis in the leaves of *Arabidopsis thaliana* in response to water deficit. J Plant Res 122:235–243

Jia W, Davies WJ (2007) Modification of leaf apoplastic pH in relation to stomatal sensitivity to root-sourced abscisic acid signals. Plant Physiol 143:68–77

Jia W, Zhang J (2008) Stomatal movements and long-distance signaling in plants. Plant Signal Behav 3:772–777

Jia W, Liang J, Zhang J (2001) Initiation and regulation of water deficit-induced abscisic acid accumulation in maize leaves and roots: cellular volume and water relations. J Exp Bot 52:295–300

Jones RJ, Mansfield TA (1971) Antitranspirant activity of the methyl and phenyl esters of abscisic acid. Nature 231:331–332

Kaiser WM, Hartung W (1981) Uptake and release of abscisic acid by isolated photoautotrophic mesophyll cells, depending on pH gradients. Plant Physiol 68:202–206

Kehr J, Buhtz A, Giavalisco P (2005) Analysis of xylem sap proteins from *Brassica napus*. BMC Plant Biol 5:11

Kharin VV, Zwiers FW, Zhang X, Hegerl GC (2007) Changes in temperature and precipitation extremes in the IPCC ensemble of global coupled model simulations. J Clim 20:1419–1444

Klingler JP, Batelli G, Zhu J-K (2010) ABA receptors: the START of a new paradigm in phytohormone signalling. J Exp Bot 61:3199–3210

Koiwai H, Nakaminami K, Seo M, Mitsuhashi W, Toyomasu T, Koshiba T (2004) Tissue-specific localization of an abscisic acid biosynthetic enzyme, AAO3, in Arabidopsis. Plant Physiol 134:1697–1707

Kondo N, Sugahara K (1978) Changes in transpiration rate of SO_2-resistant and -sensitive plants with SO_2 fumigation and the participation of abscisic acid. Plant Cell Physiol 19:365–373

Krasensky J, Jonak C (2012) Drought, salt, and temperature stress-induced rearrangements and regulatory networks. J Exp Bot 63:1593–1608

Krishnan HB, Natarajan SS, Bennett JO, Sicher RC (2011) Protein and metabolite composition of xylem sap from field-grown soybeans (*Glycine max*). Planta 233:921–931

Kudoyarova GR, Vysotskaya LB, Cherkozyanova A, Dodd IC (2007) Effect of partial rootzone drying on the concentration of zeatin-like cytokinins in tomato (*Solanum lycopersicum* L.) xylem sap and leaves. J Exp Bot 58:161–168

Kudoyarova G, Veselova S, Hartung W, Farkhutdinov RG, Veselov DS, Sharipova G (2011) Involvement of root ABA and hydraulic conductivity in the control of water relations in wheat plants exposed to increased evaporative demand. Planta 233:87–94

Lee KH, Piao HL, Kim HY, Choi SM, Jiang F, Hartung W, Hwang I, Kwak JM, Lee IJ (2006) Activation of glucosidase via stress-induced polymerization rapidly increases active pools of abscisic acid. Cell 126:1109–1120

Li S, Assmann SM, Albert R (2006) Predicting essential components of signal transduction networks: a dynamic model of guard cell abscisic acid signaling. PLoS Biol 4:e312

Liang J, Zhang J (1997) Collection of xylem sap at flow rate similar to in vivo transpiration flux. Plant Cell Physiol 38:1375–1381

Ligat L, Lauber E, Albenne C, San Clemente H, Valot B, Zivy M, Pont-Lezica R, Arlat M, Jamet E (2011) Analysis of the xylem sap proteome of *Brassica oleracea* reveals a high content in secreted proteins. Proteomics 11:1798–1813

Lindsey K, Casson S, Chilley P (2002) Peptides: new signalling molecules in plants. Trends Plant Sci 7:78–83

Liu L, McDonald AJS, Stadenberg I, Davies WJ (2001) Stomatal and leaf growth responses to partial drying of root tips in willow. Tree Physiol 21:765–770

Liu FL, Jensen CR, Andersen MN (2003) Hydraulic and chemical signals in the control of leaf expansion and stomatal conductance in soybean exposed to drought stress. Funct Plant Biol 30:65–73

Liu F, Jensen CR, Andersen MN (2005) A review of drought adaptation in crop plants: changes in vegetative and reproductive physiology induced by ABA-based chemical signals. Aust J Agric Res 56:1245–1252

Loveys BR (1977) The intracellular location of abscisic acid in stressed and non-stressed leaf tissue. Physiol Plant 40:6–10

Lü B, Chen F, Gong ZH, Xie H, Zhang JH, Liang JS (2007) Intracellular localization of integrin-like protein and its roles in osmotic stress-induced abscisic acid biosynthesis in *Zea mays*. Protoplasma 232:35–43

Ma Y, Szostkiewicz I, Korte A, Moes D, Yang Y, Christmann A, Grill E (2009) Regulators of PP2C phosphatase activity function as abscisic acid sensors. Science 324:1064–1068

Mahouachi J, Arbona V, Gómez-Cadenas A (2007) Hormonal changes in papaya seedlings subjected to progressive water stress and re-watering. Plant Growth Regul 53:43–51

Mosquna A, Peterson FC, Park SY, Lozano-Juste J, Volkman BF, Cutler SR (2011) Potent and selective activation of abscisic acid receptors in vivo by mutational stabilization of their agonist-bound conformation. Proc Natl Acad Sci USA 108:20838–20843

Munns R, King RW (1988) Abscisic acid is not the only stomatal inhibitor in the transpiration stream of wheat plants. Plant Physiol 88:703–708

Munns R, Passioura JB, Milborrow BV, James RA, Close TJ (1993) Stored xylem sap from wheat and barley in drying soil contains a transpiration inhibitor with a large molecular size. Plant Cell Environ 16:867–872

Neumann PM (2007) Evidence for long distance xylem transport of signal peptide activity from tomato roots. J Exp Bot 58:2217–2223

Neumann PM (2008) Coping mechanisms for crop plants in drought-prone environments. Ann Bot 101:901–907

Neumann PM, Chazen O, Bogoslavsky L, Hartung W (1997) Role of root derived ABA in regulating early leaf growth responses to water deficits. In: Altman A, Waisel Y (eds) Biology of root formation and development. Plenum, New York, pp 147–154

Nishimura N, Hitomi K, Arvai AS, Rambo RP, Hitomis C, Cutler SR, Schroeder JI, Getzoff ED (2009) Structural mechanism of abscisic acid binding and signaling by dimeric PYR1. Science 326:1373–1379

Parent B, Hachez C, Redondo E, Simonneau T, Chaumont F, Tardieu F (2009) Drought and abscisic acid effects on aquaporin content translate into changes in hydraulic conductivity and leaf growth rate: a trans-scale approach. Plant Physiol 149:2000–2012

Park SY, Fung P, Nishimura N, Jensen DR, Fujii H, Zhao Y, Lumba S, Santiago J, Rodrigues A, Chow TF, Alfred SE, Bonetta D, Finkelstein R, Provart NJ, Desveaux D, Rodrigues PL, McCourt P, Zhu JK, Schroeder JI, Volkman BF, Cutter SR (2009) Abscisic acid inhibits type 2C protein phosphatases via the PYR/PYL family of START proteins. Science 324:1068–1071

Passioura JB (1996) Drought and drought tolerance. Plant Growth Regul 20:79–83

Patonnier M, Peltier J, Marigo G (1999) Drought-induced increase in xylem malate and mannitol concentrations and closure of *Fraxinus excelsior* L. stomata. J Exp Bot 50:1223–1231

Peterlunger E, Maragoni B, Testolin R, Vizzotto G, Costa G (1990) Carbohydrates, organic acids and mineral elements in xylem sap bleeding from kiwifruit canes. Acta Hortic 282:273–282

Pierik R, Tholen D, Poorter H, Visser EJW, Voesenek LACJ (2006) The Janus face of ethylene: growth inhibition and stimulation. Trends Plant Sci 11:176–183

Priest DM, Ambrose SJ, Vaistij FE, Elias L, Higgins GS, Ross AR, Abrams SR, Bowles DJ (2006) Use of the glucosyltransferase UGT71B6 to disturb abscisic acid homeostasis in *Arabidopsis thaliana*. Plant J 46:492–502

Prokic LJ, Jovanovic Z, McAinsh MR, Vucinic Z, Stikic R (2006) Species-dependent changes in stomatal sensitivity to abscisic acid mediated by external pH. J Exp Bot 57:675–683

Qin X, Zeevaart JAD (1999) The 9-*cis*-epoxycarotenoid cleavage reaction is the key regulatory step of abscisic acid biosynthesis in water-stressed bean. Proc Natl Acad Sci USA 96:15354–15361

Radin JW, Parker LL, Guinn G (1982) Water relations of cotton plants under nitrogen deficiency. V. Environmental control of abscisic acid accumulation and stomatal sensitivity to abscisic acid. Plant Physiol 70:1066–1070

Rajala A, Peltonen-Sainio P (2001) Plant growth regulator effects on spring cereal root and shoot growth. Agron J 93:936–943

Reinoso H, Sosa L, Reginato M, Luna V (2005) Histological alterations induced by sodium sulfate in the vegetative anatomy of *Prosopis strombulifera* (Lam.) Benth. World J Agric Sci 1:109–119

Ren Y, Chen L, Zhang Y, Kang X, Zhang Z, Wang Y (2012) Identification of novel and conserved *Populus tomentosa* microRNA as components of a response to water stress. Funct Integr Genomics 12:327–339

Ryan CA, Pearce G, Scheer J, Moura DS (2002) Polypeptide hormones. Plant Cell 14:S251–S264

Sakakibara H (2006) Cytokinins: activity, biosynthesis, and translocation. Annu Rev Plant Biol 57:431–449

Santiago J, Dupeux F, Betz K, Antoni R, Gonzalez-Guzman M, Rodriguez L, Marquez JA, Rodriguez PL (2012) Structural insights into PYR/PYL/RCAR ABA receptors and PP2Cs. Plant Sci 182:3–11

Sauter A, Hartung W (2000) Radial transport of abscisic acid conjugates in maize roots: its implication for long distance stress signals. J Exp Bot 51:929–935

Sauter A, Hartung W (2002) The contribution of internode and mesocotyl tissues to root-to-shoot signalling of abscisic acid. J Exp Bot 53:297–302

Schachtman DP, Goodger JQD (2008) Chemical root to shoot signaling under drought. Trends Plant Sci 13:281–287

Schachtman DP, Shin R (2007) Nutrient sensing and signaling: NPKS. Annu Rev Plant Biol 58:47–69

Scheible WR, Gonzalez-Fontes A, Lauerer M, Muller-Rober B, Caboche M, Stitt M (1997) Nitrate acts as a signal to induce organic acid metabolism and repress starch metabolism in tobacco. Plant Cell 9:783–798

Schell J (1997) Interdependence of pH, malate concentration, and calcium and magnesium concentrations in the xylem sap of beech roots. Tree Physiol 17:479–483

Schraut D, Heilmeier H, Hartung W (2005) Radial transport of water and abscisic acid (ABA) in roots of *Zea mays* under conditions of nutrient deficiency. J Exp Bot 56:879–886

Senden MHMN, Van der Meer AJGM, Limborgh J, Wolterbeek HT (1992) Analysis of major tomato xylem organic acids and PITC-derivatives of amino acids by RP-HPLC and UV detection. Plant Soil 142:81–89

Sharp RE (2002) Interaction with ethylene: changing views on the role of abscisic acid in root and shoot responses to water stress. Plant Cell Environ 25:211–222

Sharp RE, Poroyko V, Hejlek LG, Spollen WG, Springer GK, Bohnert HJ, Nguyen HT (2004) Root growth maintenance during water deficits: physiology to functional genomics. J Exp Bot 55:2343–2351

Sheard LB, Zheng N (2009) Signal advance for abscisic acid. Nature 462:575–576

Sobeih W, Dodd IC, Bacon MA, Grierson DC, Davies WJ (2004) Long-distance signals regulating stomatal conductance and leaf growth in tomato (*Lycopersicon esculentum*) plants subjected to partial rootzone drying. J Exp Bot 55:2353–2364

Song X, She X, Zhang B (2008) Carbon monoxide-induced stomatal closure in *Vicia faba* is dependent on nitric oxide synthesis. Physiol Plant 132:514–525

Spollen WG, LeNoble ME, Samuels TD, Bernstein N, Sharp RE (2000) Abscisic acid accumulation maintains maize primary root elongation at low water potentials by restricting ethylene production. Plant Physiol 122:967–976

Stoll M, Loveys BR, Dry PR (2000) Hormonal changes induced by partial rootzone drying of irrigated grapevine. J Exp Bot 51:1627–1634

Sunkar R, Chinnusamy V, Zhu J, Zhu J-K (2007) Small RNAs as big players in plant abiotic stress responses and nutrient deprivation. Trends Plant Sci 12:301–309

Tanaka Y, Sano T, Tamaoki M, Nakajima N, Kondo N, Hasezawa S (2005) Ethylene inhibits abscisic acid-induced stomatal closure in Arabidopsis. Plant Physiol 138:2337–2343

Tardieu F, Parent B, Simonneau T (2010) Control of leaf growth by abscisic acid: hydraulic or non-hydraulic processes? Plant Cell Environ 33:636–647

Teplova I, Farkhutdinov RG, Mitrichenko AN, Ivanov II, Veselov SY, Valcke RL, Kudoyarova GR (2000) Response of tobacco plants transformed with the *ipt* gene to elevated temperature. Russ J Plant Physl 47:367–369

Thompson AJ, Mulholland BJ, Jackson AC, McKee JMT, Hilton HW, Symonds RC, Sonneveld T, Burbidge A, Stevenson P, Taylor IB (2007) Regulation and manipulation of ABA biosynthesis in roots. Plant Cell Environ 30:67–78

Tiekstra AE, Else MA, Jackson MB (2000) External pressures based on leaf water potentials do not induce xylem sap flow at rates of whole plant transpiration from roots of flooded or well-drained tomato and maize plants. Impact of shoot hydraulic resistances. Ann Bot 86:665–674

Vartanian N, Marcotte L, Giraudat J (1994) Drought rhizogenesis in *Arabidopsis thaliana* (differential responses of hormonal mutants). Plant Physiol 104:761–767

Vysotskaya LB, Kudoyarova GR, Veselov SY, Jones HG (2004) Effect of partial root excision on transpiration, root hydraulic conductance and leaf growth in wheat seedlings. Plant Cell Environ 27:69–77

Vysotskaya LB, Korobova AV, Veselov SY, Dodd IC, Kudoyarova GR (2009) ABA mediation of shoot cytokinin oxidase activity: assessing its impacts on cytokinin status and biomass allocation of nutrient deprived durum wheat. Funct Plant Biol 36:66–72

Vysotskaya LB, Wilkinson S, Davies WJ, Arkhipova TN, Kudoyarova GR (2011) The effect of competition from neighbours on stomatal conductance in lettuce and tomato plants. Plant Cell Environ 34:729–737

Walton DC, Harrison MA, Cote P (1976) The effects of water stress on abscisic acid levels and metabolism in roots of *Phaseolus vulgaris* and other plants. Planta 131:141–144

Wilkinson S (1999) pH as a stress signal. Plant Growth Regul 29:87–99

Wilkinson S (2004) Water use efficiency and chemical signalling. In: Bacon MA (ed) Water use efficiency in plant biology. Blackwell, Oxford, pp 75–112

Wilkinson S, Davies WJ (1997) Xylem sap pH increase: a drought signal received at the apoplastic face of the guard cell that involves the suppression of saturable abscisic acid uptake by the epidermal symplast. Plant Physiol 113:559–573

Wilkinson S, Davies WJ (2002) ABA-based chemical signalling: the co-ordination of responses to stress in plants. Plant Cell Environ 25:195–210

Wilkinson S, Davies WJ (2009) Ozone suppresses soil drying- and abscisic acid (ABA)-induced stomatal closure via an ethylene-dependent mechanism. Plant Cell Environ 32:949–959

Wilkinson S, Davies WJ (2010) Drought, ozone, ABA and ethylene: new insights from cell to plant to community. Plant Cell Environ 33:510–525

Wilkinson S, Corlett JE, Oger L (1998) Effects of xylem pH on transpiration from wild-type and *flacca* tomato leaves: a vital role for abscisic acid in preventing excessive water loss even from well-watered plants. Plant Physiol 117:703–709

Wilkinson S, Bacon MA, Davies WJ (2007) Nitrate signalling to stomata and growing leaves: interactions with soil drying, ABA, and xylem sap pH in maize. J Exp Bot 58:1705–1716

Wilkinson S, Kudoyarova GR, Veselov DS, Arkhipova TN, Davies WJ (2012) Plant hormone interactions: innovative targets for crop breeding and management. J Exp Bot 63:3499–3509

Zeevart JAD, Boyer GJ (1984) Accumulation and transport of abscisic acid and its metabolites in *Ricinus* and *Xanthium*. Plant Physiol 74:934–939

Zhang J, Davies WJ (1989) Abscisic acid produced in dehydrating roots may enable the plant to measure the water status of the soil. Plant Cell Environ 12:73–81

Zhang J, Tardieu F (1996) Relative contribution of apices and mature tissues to ABA synthesis in droughted maize root systems. Plant Cell Physiol 37:598–605

Zhang J, Schurr U, Davies WJ (1987) Control of stomatal behaviour by abscisic acid which apparently originates in the roots. J Exp Bot 38:1174–1181

Zhao B, Liang R, Ge L, Li W, Xiao H, Lin H, Ruan K, Jin Y (2007) Identification of drought-induced microRNAs in rice. Biochem Biophys Res Commun 354:585–590

Oxygen Deficiency-Induced Root-to-Shoot Communication

Laura Arru, Silvia Fornaciari, and Stefano Mancuso

Abstract During anoxic condition, moving from the root which represents the primary oxygen-sensing organ, the plant effects a redefinition of the metabolism across the whole body. During such a phenotypical and physiological reshaping of the plant, the changes in a single organ may affect and determine metabolic adjustments elsewhere in order to assure plant survival. Here, we review the main mechanisms adopted in roots subject to anoxia, considering different tolerance degrees and different strategies, and how these adaptations reflect on the behaviour of the entire organism.

Keywords Abiotic stress • Flooding • Anaerobiosis • Oxygen sensing • Oxygen deficiency signalling • Physiological response

1 Introduction

The land plant progenitors, ancestral green algae, lived into aquatic environment (Graham 1996). From the marginal habitats of ponds and marshes, early plants moved onto land, emerging and beginning land conquest around 460 million of years ago (Beerling 2007). Many important changes had to be faced before plants could begin their life on the land. First of all, the danger of desiccation: in order to bypass the hazard, they evolved a thin waxy layer on the surface called cuticle.

L. Arru
Department of Life Sciences, University of Modena and Reggio Emilia, Reggio Emilia, Italy

S. Fornaciari
Interdepartmental Research Centre for Agri-food Biological Resources Improvement and Valorisation, University of Modena and Reggio Emilia, Reggio Emilia, Italy

S. Mancuso (✉)
Department of Plant, Soil and Environmental Science, University of Firenze, Florence, Italy
e-mail: stefano.mancuso@unifi.it

F. Baluška (ed.), *Long-Distance Systemic Signaling and Communication in Plants*, Signaling and Communication in Plants 19, DOI 10.1007/978-3-642-36470-9_6,

Algae are known to have no cuticle, while nearly all land plants do. At the same time it took shape the necessity of breaches in the cuticle in order to allow gas exchange and CO_2 assimilation; thus, stomata have evolved, with a progressive more and more sophisticated regulation of the opening (Manetas 2012). It has been an ever changing of characteristics due to an evolutionary pressure, during which plants began to be adapted to the new environmental conditions. Different environmental conditions did mean, as it happens for any similar change and for any life form, new kind of stresses to face. So terrestrial plants evolved stress-induced metabolic pathways and morphological adaptation that confer new adaptive abilities.

Nowadays, flooding is a major issue for plant survival in many regions of the world, and flooding stress tolerance implies a wide range of adaptive competences able to grant the survival. Between the aquatic life and the strictly terrestrial life, there is an intermediate state where plants must be able to compensate the effects of the stress due to temporary flooding and then return to a drier climate (Braendle and Crawford 1999; Jackson et al. 2009). This intermediate state involved a high number of adaptive solutions, related to different degrees of flooding stress conditions, which is—in all of these forms—one of the major stresses in many countries of the northern hemisphere (Copolovici and Niinemets 2010). The need to re-evolve a flooding tolerance induced the plants to achieve competences typically related to aquatic life once again, independently regained in a high number of cases (Jackson et al. 2009). This suggests the hypothesis that a restricted number of mutations is involved in the capacity to withstand the flooding stress (Voesenek and Pierik 2008; Colmer and Voesenek 2009), conferring this ability in many taxonomic groups even highly distant related (Jackson et al. 2009). An interesting point is that there are not only various interspecies responses but also significant intraspecies differences, suggesting a recent evolutionary origin of the adaptive traits (Jackson et al. 2009).

As highly evolved life forms, plants need to respond to external stimuli as whole organisms, particularly during stress conditions. Coherently, during low oxygen adaptation, it is a key challenge to the whole-plant response, even involving organs not directly subject to roots flooding. Sure enough it is clear that every organ of a plant is affected by a stressful condition, even if the stress initially hits only one or few organs of the entire plant. This implies a root-to-shoot signalling that enables the coordination of the adaptive response in the whole body of the plant. A clear, symptomatic and representative example is the induction of stomatal closure during roots flooding: the adaptive significance of this phenomenon seems to be that it prevents the loss of oxygen and the dehydration of the leaves during the stress. The way flooded roots communicate to aerial parts the need of a stomatal closure is still under debate. Most of the observation and subsequent consideration about plant behaviour under oxygen deficiency stress must first begin with the assumption that plant response is coordinated across the whole organism, including chemical, electrical and physical signalling that help in regulating the defence pathways of the whole plant.

2 Flooding

Different plant habitats impose different organism development, providing characteristics for the evolution of specific adaptation. Strictly aquatic plants (hydrophytes) need an obligate submergence, while semiaquatic plants present aerial parts such as leaves and stem, but require a submerged or soaking wet soil state of the roots. "Amphibious" plants can grow in both terrestrial and submerged conditions, while terrestrial plants might even tolerate periods of flooding, but usually live a dry condition. Roots and shoots of wetland species and, in some adverse conditions, dryland species may form the aerenchyma, a plant tissue containing enlarged gas spaces and channels, which allow gas exchange among the involved organs. Aerenchyma occurs in many plants and can be formed during the normal development or in response to particular stresses (i.e. hypoxia, high temperature or drought) (Evans 2004). In case of flooding, its role is fundamental since the excess of water at root level impedes the gas diffusion, leading to a drastic change in the levels of oxygen, CO_2 and ethylene (Bailey-Serres et al. 2012).

Roots flooding of semiaquatic plants that constitutively form aerenchyma commonly does not cause serious impacts, while, on the contrary, most of the crop species suffer an affection of both growth and development during submergence (Christianson et al. 2010). The main problem related to the flooding is the reduced oxygen availability (Copolovici and Niinemets 2010), but also the alteration in the soil chemical composition or the change in the pH of the soil during waterlogging might be involved in the reduced ATP production or into other flooding-related plant metabolism alteration (Bailey-Serres and Voesenek 2008; Pucciariello and Perata 2012). It is also well known that a period of adaptation to the oxygen deficiency status before the drastic oxygen shortage can induce the biosynthesis of target enzymes and trigger the metabolic shift to the anaerobic fermentation (Jackson et al. 2009; Fu et al. 2012).

Anoxia is known to cause an array of physiological stress responses such as the decrease in photosynthesis rates and the stomatal closure (Jackson et al. 2009). It is already known that plants can communicate a stressful condition to their neighbours. Plants warn their own kind for the onset of a biotic stress condition by means of volatile organic compounds (VOCs), but there are few data about communication under an abiotic stress. Anyway, it is documented as an emission of specific volatile compounds under flooding conditions. The emissions of volatile abiotic stress compounds include methanol (Holzinger et al. 2000), nitric oxide (NO) (Dat et al. 2004; Dordas et al. 2004) and chemicals typically produced during the anaerobic metabolism such as ethanol and acetaldehyde (Atkinson et al. 2008; Rottenberger et al. 2008). What is still under debate is if these molecules have a real specific signalling role towards other plants or if they slip throughout the plant while acting as internal metabolites, with a more or less known role in intra-plant root-to-shoot signalling. There is in fact a good possibility that at least some of these volatile compounds may be simple by-products of anaerobic pathways activated to survive during the flooding condition. Interestingly, in 2010,

Copolovici and Niinemets showed that the amount of these molecules varies depending on the waterlogging tolerance ability of the plant, suggesting a higher emission burst of these compounds in the less tolerant species (Copolovici and Niinemets 2010). As demonstrated in those studies based on a more integrated point of view, the effect of waterlogging is not limited to the organs directly subject to the stress: the damage can rapidly extend to aerial part, where plant metabolic activity depending on root functions encounter hindrances (Jackson et al. 2003). It has been observed that root waterlogging induces a change in the gene transcription of leaves in most of the plants (Christianson et al. 2010). On the other hand, a whole-plant approach elucidated that into plants subject to oxygen-deficient conditions, it can be promoted an adaptive response which involves many different modification through the whole plant (Colmer and Voesenek 2009).

This becomes particularly evident considering the LOQS and the LOES responses. These are the acronyms for the two main strategies for adaptation to submergence: Low Oxygen Quiescence Syndrome (LOQS) and Low Oxygen Escape Syndrome (LOES). During a flooding stress, when oxygen becomes poorly available, an LOQS plant is characterised by a reduced growth and a quiescence state in order to maintain a sufficient substrate supply to survive in the de-submergence phase; on the contrary, the LOES plant first of all invests on the rapid growth of submerged shoots, in order to reach the water surface and resume the gas exchange, before applying on other several acclimative modifications (Colmer and Voesenek 2009; Pucciariello and Perata 2012).

When the roots are the organs suffering low oxygen availability, it is the entire organism that is affected by the stress. A disturbance in the root organ leads to changes in many processes both at short metabolic level and in long-lasting morphological adaptation, with the aim of preserving the life of the entire plant. So it happens that the shoot/root growth ratio must be adjusted, and the coordination of this integrated response is often regulated by phytohormones. One adaptation of LOES plants is the formation of aerenchyma, constituted by intercellular gas-filled lacunas in both roots and shoots (Drew et al. 1979; Kawase 1981). The thickening of cell walls of external cells and the hydrophobic surface of the lacunas allows the longitudinal transport of oxygen from air to submerged organs inhibiting the oxygen escape (Rascio et al. 1994; Raven 1996; Sorrell et al. 1997; Babourina and Rengel 2010). When the diffusion of oxygen from the roots to the soil is inhibited, aerenchyma-dependent advantages for submerged plants might be even enhanced with the development of barrier to radial O_2 loss (ROL) which impedes the oxygen loss (Armstrong 1979; Visser et al. 2000; Colmer 2003) and the entrance of toxic compounds accumulated in the flooded soil (Armstrong and Armstrong 2005). Other adaptation to LOES plants is the hyponastic growth of the leaves, which may promote the reaching of the water surface and the perception of light (Colmer and Voesenek 2009); the formation of lenticels, small pores which promote gas exchange; or the formation of adventitious roots (Kozlowski 1984).

Under anaerobic conditions, as cellular respiration is inhibited, ATP production is insufficient for energy-consuming processes. In order to face this energy crisis, plants promote the optimisation of ATP consumption, i.e. limiting protein synthesis

(Branco-Price et al. 2008) and favouring the regeneration of NAD^+ (needed to sustain glycolysis) by means of the fermentation pathways (Bailey-Serres et al. 2012). Moreover, there is a strategic inhibition of the ATP-consuming processes which are not of vital importance for plant surviving. Those pathways that cannot be denied are replaced with alternative ones, less energy-consuming: an example can be the sucrose degradation. Sucrose is broken into two monosaccharides by invertase, and then hexokinase catalyses the ATP-consuming reaction that generates the first intermediate of glycolysis. This step requires two molecules of ATP. When there is a lack of oxygen, this pathway is substituted by sucrose synthase and UDP-glucose pyrophosphorylase, which does not involve any ATP-consuming step, requiring only one molecule of inorganic pyrophosphate (PPi). The energy required for this alternative step is even less revealing considering that PPi is generated by a wide range of reaction as a waste product (Geigenberger 2003; Magneschi and Perata 2009; Arru and Fornaciari 2010).

3 Anaerobic Metabolism

Under anaerobic conditions the oxidative phosphorylation pathway is inhibited, and thus the main ATP providing steps and NAD^+ regenerating reactions are suppressed. To avoid stress, plant cell needs NAD^+ to sustain glycolysis, and fermentation has the main role in providing NAD^+ to the glycolytic pathway. Two fermentative pathways, ethanolic and lactic fermentations, can regenerate NAD^+ from NADH. Perata and Alpi proposed in 1993 the sustain hypothesis according to which lactate dehydrogenase (LDH) pathway prevails in early phases of anaerobic metabolism, resulting in a fall in cytoplasmic pH which induces the activation of the pyruvate decarboxylase enzyme (PDC) and the subsequent ethanol formation by means of alcohol dehydrogenase (ADH) (Perata and Alpi 1993). In 1992, Andreev and Vartapetian noticed that it is not a rule for all plants that the ethanolic fermentation is a consequence of the lactic fermentation. Some plant species do not show lactic acid production before ethanol production (Andreev and Vartapetian 1992). Ricard and his collaborators (1994) demonstrated that ethanol, lactate and alanine can be produced simultaneously in many species under O_2 deficiency, although in different relative quantities, depending on the species and the degree of O_2 deficiency (Ricard et al. 1994). Alanine is produced by means of an alternative metabolism of pyruvate that allows the formation of an end product (alanine in itself) which does not diffuse out of the cells as ethanol does (Ricoult et al. 2006; Limami et al. 2008). This switch allows the lack of carbon loss and the formation of a co-product (2-oxoglutarate) which can be further metabolised to succinate (Bailey-Serres et al. 2012) providing ATP production. An alternative pathway may recycle the fermentation products in order to avoid the carbon loss due to ethanol and acetaldehyde emission: the root-derived ethanol can be converted back to acetaldehyde by ADH enzyme activity and further oxidised to acetate by aldehyde dehydrogenase (ALDH), reducing NAD^+ to NADH (Arru and Fornaciari 2010; Pucciariello and Perata 2012).

There is also an alternative pathway to fermentation that acts during oxygen deficiency stress regenerating NAD^+: nitrate reduction generates NO oxidising NADH to NAD^+ (Igamberdiev and Hill 2004; Igamberdiev et al. 2005). Since hypoxic stress induces the biosynthesis of a specific stress-induced class of haemoglobin with high affinity with oxygen and nitric oxide, the produced NO does not accumulate: haemoglobin participates in the reconversion of NO to nitrate by means of NAD(P)H oxidation, leading to a lower $NAD(P)H/NAD(P)^+$ ratio within the hypoxic cell (Igamberdiev and Hill 2004). The whole cycle oxidises 2.5 moles of NADH per mole of nitrate recycled, thus participating actively in the maintenance of redox and energy status in oxygen-lacking cells (Igamberdiev and Hill 2004).

4 Sensing Organ

During flooding, plants act a redefinition of the gene transcription in the whole plant (Bailey-Serres and Voesenek 2008; Kreuzwieser et al. 2009; Christianson et al. 2010). The adjustment of molecular pathways for the enzyme biosynthesis in those organs which are not submerged suggests the presence of a systemic intra-plant communication. The perception of oxygen availability is the primary step in the plant adaptation. Root demonstrates to be very sensitive as perceptive organ, acting as explorer, leading the plant into new, unknown and possibly more advantageous regions of the soil (Aiken and Smucker 1996; Dat et al. 2004), searching, choosing, escaping and recognising healthier conditions (Gruntman and Novoplansky 2004; Hodges 2009; Trewavas 2009; Baluška et al. 2010; Mugnai et al. 2012). Root apices are covered by a root cap (Barlow 2003) behind which the root might be considered as composed by different zones: a meristematic zone, a transition zone and a zone of rapid cell elongation (Baluška et al. 1990, 1996, 2001; Verbelen et al. 2006; Kieffer et al. 2009). In particular, the cells of the transition zone have unique functional and sensorial properties, showing high sensibility to a wide range of parameters such as touch (Baluška et al. 1996), extracellular calcium (Ishikawa and Evans 1992), water and salt stress (Winch and Pritchard 1999; Ober and Sharp 2003), gravity and auxin (Mancuso et al. 2007; Sobol and Kordyum 2009), aluminium (Illéš et al. 2006; Marciano et al. 2010) and hypoxia (Mancuso and Boselli 2002; Mancuso and Marras 2006; Mugnai et al. 2011).

These transition zone cells present the highest rate of vesicle recycling activity (Baluška et al. 2010), a synchronised electrical activity (Masi et al. 2009), and demonstrated to be a key zone for root hormonal activity (Baluška et al. 2010). All of the processes occurring in the transition zone cause a great intake of ATP and oxygen, much more than in the other root regions (Mugnai et al. 2012). Coherently, the transition zone seems to be the most sensitive region of the root to oxygen deprivation. Moreover, it might be the point of departure of intercellular communication. In fact, selectively imposing hypoxia to the transition zone, the cells of the transition zone emit the highest amount of nitric oxide, and the entire root shows an acclimation response (Mugnai et al. 2012). However, it must be noted that this is an intricate region of knowledge, and the precise site where the root perceives oxygen availability is still unclear, such as the way how it acclimates to hypoxia.

5 Sensing Mechanisms

Cellular oxygen status can be perceived directly or indirectly, and both these mechanisms bring to acclimation responses that prolong plant survival during oxygen deprivation (Bailey-Serres and Chang 2005). In 2011, Licausi et al. showed that part of the ubiquitin-dependent N-end rule pathway for protein degradation, which is active in both mammals and plants, has a role into the oxygen-sensing mechanism in *Arabidopsis thaliana* (Licausi et al. 2011). In particular, they studied a conserved amino-terminal sequence of a member of the subgroup VII of the ERF (ethylene response factor) transcription factor family RAP2.12, widely conserved in higher plants. This conserved domain shows to be related to oxygen-dependent post-translational modifications, which lead to the degradation of RAP2.12 under aerobic conditions. This means that, when plant suffers oxygen deficiency stress, RAP2.12 is not degraded and accumulates in the nucleus, where it activates the gene expression of those genes related to hypoxia acclimation (Licausi 2011; Licausi et al. 2011).

Other sensing mechanisms involve the perception of ATP and ADP ratio, carbohydrates and pyruvate, cytosolic acidification and cytosolic calcium or presence and amount of reactive oxygen (ROS) and nitrogen (NOS) species (Bailey-Serres et al. 2012). Among other reactive oxygen species, H_2O_2 involvement has been reported in many different kinds of programmed cell death induced by abiotic stress (De Pinto et al. 2012). For example, H_2O_2 is involved in the hypoxic stress-induced lysigenous aerenchyma formation in Arabidopsis plants (Muhlenbock et al. 2007) and Arabidopsis plantlets pretreated with H_2O_2 demonstrated to be more tolerant to anoxia (Banti et al. 2010). Nitric oxide and the hypoxia-inducible O_2-binding protein haemoglobin seem to be involved in the altered oxygen status sensing. However, it has been suggested that a key role of the hypoxia-induced haemoglobin is in scavenging H_2O_2, and in oxidising NO back to nitrate. This could help in maintaining the redox state and the energy state in the hypoxic cell (Dordas et al. 2003; Igamberdiev and Hill 2004; Borisjuk and Rolletschek 2008).

6 Signalling

Most of the molecules or the parameters involved into the indirect sensing at root level might be involved into the signalling pathways in the whole plant. Since these oxygen availability indicators are involved in the signalling of various stimuli, biotic and abiotic, it is reasonable that the co-presence of more than one of these molecules becomes a key factor for reporting an oxygen deprivation-specific condition (Pucciariello and Perata 2012).

Three different kinds of signal seem to be involved in the root-to-shoot communication: positive, negative and accumulative (Jackson 2002). It might be referred to a positive message when, responding to the root stress, the output of signal molecules from root-to-shoot is generated or increased. When the export of a signal molecule is reduced, it could be considered a negative message, and the message can be considered accumulative when the root demand for molecules coming from the shoot decreases, leading to an accumulation of these chemicals in the sources (Jackson 2002; Else et al. 2009). For example, it has been observed an increase in Ca^{2+} concentration in maize and Arabidopsis when subject to hypoxia and anoxia (Sedbrook et al. 1996; Subbaiah et al. 2000). In some cases, this increase seems to be required for the *Adh1* gene expression (Subbaiah et al. 2000). Calcium concentration increase can be distinguished into two phases: a first spike within few minutes and a subsequent abundant calcium presence (Sedbrook et al. 1996; Subbaiah et al. 2000) which is decoded by an array of proteins belonging to the calcium sensor relays and responders (Lecourieux et al. 2006).

Other hints come from the studies on the rice plant, the quintessential model plant for hypoxic and anoxic plant response. Rice (*Oryza sativa*) is one of the plants better adapted to survive in flooding conditions, even if there is considerable variation among the different ecotypes. The ability of rice to tolerate oxygen deprivation is known to also depend on a coordinated response to both oxygen and sugar deficiency. In this survival response it involved a particular protein kinase, CIPK15 (Lee et al. 2009), a calcineurin B-like (CBL)-interacting protein kinase 15. The CBL target for this interaction however still remains unknown. Briefly, CIPK15 has been recognised as an upstream regulator of SnRK1 (Snf1-related kinase), which in turn is involved in the activation of a MYBS1 factor, that finally binds specifically the sequence in the promoter region of a specific starch-degrading α-amylase isoform. This amylase represents a key element linking sugar and oxygen deficiency signalling crosstalk, being induced in anoxic rice seedlings and regulated by sugar starvation (well reviewed in Pucciariello and Perata 2012).

There is another example of convergence between sugar and energy-lack sensing during hypoxia coming from Arabidopsis. Baena-González and colleagues in 2007 observed that there is a subgroup of genes regulated by proteins belonging to the Snfl-related kinase (SnRK1) family which can be considered as plant "energy" sensor, involved in the underwater growth in response of an affected impaired sugar and energy production (Baena-González et al. 2007).

Nitric oxide is another well-documented signal molecule involved in many stress-induced adaptations in many plants (Wang and Yang 2005; Sun et al. 2007; Leterrier et al. 2012). The NO formation in response to environmental stresses has been frequently observed (Desikan et al. 2004; Besson-Bard et al. 2008; Mugnai et al. 2012). Recently, NO has been described both as a highly important molecule in the phytohormones signalling pathways (Leterrier et al. 2012) and as an important signalling molecule in itself, able to induce specific cascade of events (Hasanuzzaman et al. 2012). NO has long been known to possess a role in the regulation of the plant response. It has been shown to be involved at intracellular level, where it regulates the actin cytoskeleton, the vesicle trafficking, the endocytosis and the polarity of growing tip cells (Prado et al. 2004, 2008; Lombardo et al. 2006; Salmi et al. 2007; Kasprowicz et al. 2009; Wang et al. 2009). It has also been observed that NO plays a role among cells, participating in root growth, seed germination and stomatal closure (Lamattina et al. 2003; Neill et al. 2003, 2008; Correa-Aragunde et al. 2004; Desikan et al. 2004; Pagnussat et al. 2004; Lanteri et al. 2008).

It has been demonstrated that during oxygen deficiency at root level, the cells of the transition zone show an enhanced emission of nitric oxide. The transition zone is known to be the root zone with the highest oxygen consumption amount, due to the wide range of ATP-consuming processes there localised. The meaning of this increased emission is still not completely clear, but it has been noted an involvement of the NO emission in the decrease of oxygen requirement in the transition zone. The reduced requirement of oxygen is related to the inhibition of vesicle recycling and actin polymerisation during the stress. This helps to reduce oxygen demand in a circumstance when there is no O_2 availability. The effect seems to produce a systemic signal throughout the entire root favouring the acclimatation as well (Mugnai et al. 2012).

Low oxygen availability causes ROS production in the eukaryotic cells, both for plants and for animals, where they seem to be necessary to start hypoxia response (Bell et al. 2007; Guzy and Schumacker 2006). The rapid formation of ROS species in oxygen deficiency conditions is a well-demonstrated phenomenon, and these molecules demonstrated to have an importance in the cell-to-cell communication in plant responses to diverse stress situations (Baxter-Burrell et al. 2002; Chang et al. 2012; Pucciariello et al. 2012).

In plant sensitive to hypoxia like pea (*Pisum sativum*) and soybean (*Glycine max*), hypoxia enhanced markedly the production of hydroperoxides, superoxide anion radical and, most of all, hydrogen peroxide (Ershova et al. 2011); in pigeon pea (*Cajanus cajan*) after 6 days of waterlogging, superoxide radical, hydrogen peroxide and the enzymes involved in antioxidant synthesis gradually increased (Kumutha et al. 2009). In the apoplast of root meristems of hypoxic wheat (*Triticum aestivum*), it has been demonstrated that H_2O_2 accumulates in response to anoxia (Blokhina et al. 2003), as well as in Arabidopsis seedlings, where H_2O_2 levels increase in response to O_2 deprivation (Baxter-Burrell et al. 2002). Santosa and colleagues (2006) demonstrated that ethane, a product of membrane peroxidation by ROS, evolves from submerged rice seedlings in a closed system as levels of O_2

fall to 1 %, providing evidence that ROS form as O_2 levels decline (Santosa et al 2006).

Even if ROS are dangerous and unwanted by-products of anaerobic processes, however, they demonstrated to have all those characteristics necessary for a signal molecule to be effective: rapid production, specific-response induction and quick removal when no more required (Pauly et al. 2006). Thus, no surprise that evolution did favour response strategies involving ROS-dependent signalling pathways. Several studies reported significant increases in mRNAs encoding enzymes involved in ROS signalling (Lasanthi-Kudahettige et al. 2007; Liu et al. 2005; Loreti et al. 2005). In particular, it has been suggested that an RHO-like small G protein (ROP) is activated during oxygen deprivation, inducing an accumulation of H_2O_2 and a subsequent switch to anaerobic fermentation and other stress-adaptive pathways (Baxter-Burrell et al. 2002; Yang and Fu 2007).

Among all the H_2O_2-responsive proteins, the heat-shock transcription factors (HSFs) and the heat-shock proteins (HSPs) are both related to heat stress and oxygen deprivation stress (Vandenbroucke et al. 2008; Mustroph et al. 2010). In effect, in nature rarely a single stress factor isolately occurs, and plants under stress have been evolved to simultaneously respond to multiple combined stressful events. Interestingly, HSP transcripts increase during low oxygen stress (Mustroph et al. 2010), and members of the HSFs are supposed to be H_2O_2 molecular sensors (Miller and Mittler 2006). This links up again to the calcium-mediated signalling, since there are evidences about a role of the ROS in the activation of an ROS-mediated plasma membrane calcium channel (Lecourieux et al. 2006), and one more time calcium is involved in plant signal transduction in low oxygen deficiency stress as explained above.

pH change is a very general signal (Felle 2010): it is already well known that most of the stresses acidify the cytoplasm and alkalise the apoplast (Wilkinson 1999; Felle et al. 2004, 2005). In response to waterlogging, the lowering of the cytosolic pH and the subsequent acidification of xylem sap might be a root signal in itself. Changes in pH originate in anoxic roots and can be transmitted by means of the xylem sap to the leaves, which lead to stomatal closure (see below Sect. 7). Anaerobic conditions may impair the working of proton pumps inside the membrane of the root cells, causing a pH lowering within the same cells (Ratcliffe 1997). This acidification can be extended from cell to cell, simplistically, until the xylem sap. This is how in stressed anoxic roots the pH change acts as a messenger for the whole plant (Jia and Davies 2007). On the other hand, there are evidences correlating the abscisic acid (ABA)-dependent stomatal closure and the pH changes during oxygen deficiency stress. pH seems to show a role in quantifying the sufficient ABA concentration necessary in the guard cells for inducing the closing of the stomata (Wilkinson 1999). This hypothesis however is not the only possible explanation for all the plant species: studies on citrus plants demonstrated that the amount of ABA rising from the xylem and entering the leaves does not have a significant interaction with the pH, being the ABA entering the guard cells too low to achieve a physiologically active level, at least at the beginning of the stress (Rodríguez-Gamir et al. 2011). A hint comes from a supported hypothesis, showing an indirect relationship between changes in pH and stomatal closure. Acidosis is known to reduce hydraulic conductivity of the roots by affecting the activity of aquaporins (Tournaire-Roux et al. 2003; Ehlert et al. 2009; Bramley and Tyerman 2010); this in the end could induce the stomatal closure (Rodríguez-Gamir et al. 2011). Moreover, acidosis and redox balance changes induced by anoxic conditions are also supposed to promote the nitrate reductase activity, increasing the emission of NO (Dordas et al. 2004).

It is well known that hormones have a fundamental role in mediating signal transduction and inducing physiological responses for both biotic and abiotic stresses. The main hormones involved into the oxygen deprivation stress demonstrated to be ethylene, ABA and gibberellins (Gas), acting as coordinated factors (Kende et al. 1998; Peeters et al. 2002; Bailey-Serres and Voesenek 2008).

Ethylene seems to act as key signal for both submerged and aerial parts during waterlogging, and it is involved into the formation of aerenchyma (Sairam et al. 2008). In hypoxic roots of maize, application of ethylene demonstrated to induce the aerenchyma formation, while ethylene antagonists inhibit its development (Drew et al. 1981; Konings 1982; Jackson et al. 1985). More recently, a higher concentration of ACC (1-aminocyclopropane-1-carboxylic acid), the ethylene precursor, and ACC synthase has been found in roots of hypoxic maize, with respect to normoxic roots (He et al. 1994; Geisler-Lee et al. 2010). It has been suggested that ethylene may also have a role in the induction of adventitious roots. This induction depends on a balance between ethylene and auxin signalling (Clark et al. 1999; Negi et al. 2010), where ethylene could affect the formation of adventitious roots induced by auxin, as seen in tomato (Vidoz et al. 2010). In rice ethylene plays a role regulating ethylene response factor (ERFs) transcription factors differently depending on the strategy, in particular regulating Sub1A during a quiescent strategy (Fukao et al. 2006; Xu et al. 2006) and SK1/SK2 during the escape strategy (Hattori et al. 2009). Ethylene induces, in submerged organs of rice, a hormonal signalling cascade which regulates the balance between ABA and GAs, influencing cell elongation (Bailey-Serres et al. 2012). The reduced elongation of the underwater petiole is influenced by an increased level of ABA and a reduced level of GA also in *R. palustris* and *R. acetosa* (Benschop et al. 2005; Chen et al. 2010).

Gibberellins and cytokinins are supposed to have a role in root-to-shoot signalling during waterlogging, since it has been shown a reduced level in the sap of plants during the stress (Burrows and Carr 1969; Reid et al. 1969). It has been suggested that both GAs and cytokinins may promote stomatal opening, antagonising the ABA effect (Pospíšilová 2003; Kumar et al. 2004). Moreover, it has been observed an evident relationship between the enhanced levels of NO during oxygen deprivation and the decreased levels of GAs (Christianson et al. 2010), and it has been speculated that this increase in NO production influences stomatal closure (Christianson et al. 2010).

Increased levels of ABA are known to be involved in stomatal closure in diverse species (Jackson and Hall 1987; Neuman and Smith 1991; Zhang and Zhang 1994; Ahmed et al. 2006), but evidences have shown that more than one signal generated by submerged roots is probably responsible for stomatal closure (Jackson 2002; Christianson et al. 2010; Rodríguez-Gamir et al. 2011).

7 A Case Under Study: The Stomatal Closure

Among other effects, the reduction in the stomatal conductance (Jackson et al. 2009) due to root anoxic conditions is a critical response, symptomatic of the behaviour of a whole organism entirely covered by signalling pathways.

One of the first hypotheses was that early closure of stomata is induced by increased levels of ABA. Nowadays, the comprehension of the mechanism and the actors involved is ampliated (Else et al. 2009). During oxygen deficiency stress, the impedance of water supply is one of the earliest responses; it can occur within few hours (Jackson et al. 2003), since the lack of oxygen, combined with an increase in CO_2 amounts at soil levels, might induce this decrease in root hydraulic conductance (Kramer 1940; Everard and Drew 1989; Birner and Steudel 1993). It has also been suggested an involvement of aquaporins. The lowest cytosolic pH due to anoxic stress causes a malfunction in the aquaporins, with a consequent decrease of the root hydraulic conductance compared to well-drained plants (Tournaire-Roux et al. 2003). The resultant loss of leaf hydration brings to a consequent stomata closure in order to limit the water loss (Andersen et al. 1984; Else et al. 2001; Jackson et al. 2003). At the same time, stomata closure induces a reduction in the CO_2 uptake (Rodríguez-Gamir et al. 2011), causing a decline in the CO_2 assimilation. In tomato and in mung bean it has been demonstrated that flooding can affect photosynthesis by limiting the quantum efficiency of PSII (Janowiak et al. 2002; Ahmed et al. 2006) and by damaging the photosynthetic apparatus (Chirkova et al. 1995; Building 2001).

This complex behaviour that means a response which reflects on the aerial parts of the plant when the lacking of oxygen is imposed at root level suggests the presence of a root-to-shoot signalling (Sairam et al. 2008; Ehlert et al. 2009). There are many different studies where it is shown that waterlogging does not induce a loss of leaf turgor, meaning that the stomatal closure is not due to the dehydration of the leaves (Pereira and Kozlowski 1977; Jackson et al. 1978; Bradford and Hsiao 1982; Jackson and Hall 1987; Else et al. 1996; Blanke and Cooke 2004; Garcia-Sanchez et al. 2007). So the sequence of events between root waterlogging and stomatal closure is based on cause–effect relations which are not simple and clear: the hypothesis that ABA would be the main signal involved in the change of stomatal conductance is confirmed only in some cases (Jackson and Hall 1987; Neuman and Smith 1991; Zhang and Zhang 1994; Ahmed et al. 2006), but not in other ones (Jia and Davies 2007; Arbona and Gomez-Cadenas 2008; Rodríguez-Gamir et al. 2011).

In citrus seedling, waterlogging induces a rapid decrease in the transpiration rate due to the stomatal closure, but the increased ABA levels into leaves appear three times later with respect to the stomatal closure (Rodríguez-Gamir et al. 2011).

Moreover, ABA level in waterlogged roots, as well as the transport of ABA through the xylem sap, is lower in stressed plants compared to the control (Arbona and Gomez-Cadenas 2008; Rodríguez-Gamir et al. 2011).

According to these studies, it has been shown that waterlogging does not induce an increase in ABA production in roots (Jackson et al. 1988; Zhang and Zhang 1994), while a significant amount of ABA seems to be due to the early senescence of old leaves when plant is subject to anoxic stress. ABA is produced in the old leaves and then transported to the young leaves (Zhang and Zhang 1994; Rodríguez-Gamir et al. 2011).

Else and colleagues described how ABA may interact with ethylene in the induction of stomatal closure: in tomato plants ABA concentration inside leaves slightly increases, probably as a consequence of impeded export and internal redistribution (Else et al. 1996, 2006). In these plants, stomatal closing may be due to an interaction between this ABA and the ethylene derived from the ACC given off from the roots to the shoots, which significantly increases in the xylem sap of submerged plants (Bradford and Hsiao 1982; Else and Jackson 1998; Xu et al. 2007; Else et al. 2009). Also GAs and cytokinins have been suggested to be involved in stomatal closure, and their role as stomatal opening promoter is downregulated by a strongly reduced delivery in the transpiration stream (Else et al. 2009).

A hypothetic but controversial signal for stomatal closure might be the acidification of the xylem sap in itself (Jia and Davies 2007): experiments with acidic buffer supplied to the roots and brought to the whole plant cause a reduction of hydraulic conductance and the closing of the stomata (Rodríguez-Gamir et al. 2011). However, experiments on detached leaves seem to not confirm this hypothesis, even if caution might be used when results obtained from an isolated organ such as a leaf need to be extrapolated to the behaviour in the continuum of a whole plant.

8 Genes

The number of genes selectively expressed during oxygen deprivation stress is significantly different between flooding-sensitive plants and flooding-tolerant ones: more than 5,000 genes show altered expression levels in hypoxic stressed poplar tissues, a flooding-tolerant plant, while similar conditions did change much less the transcript abundance into Arabidopsis plants, the flooding-sensitive, with about 150 genes involved in stress-dependent regulation (Klok et al. 2002; Liu et al. 2005; Arru and Fornaciari 2010). In spite of this difference in the genes amount, for both these species a high percentage of altered genes are transcriptional factors. The presence of a differential transcriptional factor expression is indicative of the existence of a more complex transcriptional network, which sophisticated regulation contributes to the stress adaptation in response to the lack of oxygen (Kreuzwieser et al. 2009).

Arabidopsis and rice plants have been deeper studied in their response to hypoxia and anoxia, so much that nowadays there are many microarray datasets providing a considerable amount of information (Branco-Price et al. 2005; Liu et al. 2005; Loreti et al. 2005; Lasanthi-Kudahettige et al. 2007; van Dongen et al. 2008; Banti et al. 2010; Jung et al. 2010; Licausi et al. 2010; Mustroph et al. 2010; Lee et al. 2011).

It has been shown that there are common genes which are regulated into the whole plant as well as specific genes regulated differently among the various organs of the plant (Pucciariello and Perata 2012). In particular, ethylene demonstrated to have a central role in the regulation/induction of genes related to oxygen deprivation response: family VII of the ethylene response factors (ERF VII) is involved in the stress adaptation responding to the increased ethylene production (Nakano et al. 2006). Both LOES and LOQS mechanisms involve genes belonging to this family (Pucciariello and Perata 2012). In mature plants of rice acting the quiescent strategy, it has been demonstrated that the ERF submergence 1A gene (SUB1A) plays an important role (Fukao et al. 2006; Xu et al. 2006). In order to conserve energy for the re-oxygenation phase, the submergence-induced allele Sub1A-1 activates two specific genes: the Slender-Rice 1 (SLR1) and the Slender-Rice like 1 (SLRL1) which inhibit GA-induced plant elongation (Bailey-Serres and Voesenek 2008; Pucciariello and Perata 2012). On the opposite, during the escape strategy, underwater elongation is mediated by two different genes also belonging to the ERF family VII: Snorkel 1 (SK1) and Snorkel 2 (SK2), which promote fast elongation in order to reach the water surface and re-establish gas exchange (Hattori et al. 2009).

Arabidopsis plants subject to low oxygen conditions have shown a strong upregulation of genes also belonging to the ERF family VII, such as the Hypoxia Responsive ERF (HRE)1 and the HRE2 (Licausi et al. 2010; Yang et al. 2011). The importance of these genes is confirmed by experiments where transgenic plants overexpressing HRE1 and HRE2 showed an enhanced tolerance to anoxia, while plants with HRE1 and HRE2 knockout showed a reduced expression of many hypoxic-related downstream genes (Pucciariello and Perata 2012). Even if Arabidopsis plants under oxygen deprivation increased the expression of hypoxic genes, including the ERF, and subsequently enhance low oxygen tolerance (Hinz et al. 2010), the overexpression of the ERF genes in plants under normoxia does not enhance the expression of hypoxic genes (Licausi 2011).

There are genes related to the heat shock which are involved also in the anoxia response; in particular the heat-shock transcription factors (HSFs) and the heat-shock proteins (HSPs) are both related to heat stress and oxygen deprivation stress (Vandenbroucke et al. 2008; Mustroph et al. 2010). Arabidopsis seedlings overexpressing HsfA2 demonstrated to be more tolerant to oxygen deprivation (Banti et al. 2010) and in particular to anoxia, where the adaptation mechanism seems to be more complex (Licausi et al. 2011).

9 Conclusions

Oxygen sensing and signalling in plants is a topic of major importance, for our knowledge and for the many possible applications in crop cultivation. However, even if it is clear that a mechanism for low oxygen sensing must exist, only indirect sensing and signalling mechanisms have been proposed until now. A real understanding of plant behaviour during biotic and abiotic stresses implies the assumption that plant is an organism into which all the single parts communicate each other and respond together and coherently to stimuli.

It would not be believable a so extended presence of the plants (in terms of time and space) on the land without a sophisticated intercellular communication that can grant survival and growth under a so widespread array of conditions. In order to understand how the single parts take place in this internal signalling system during plant responses to environment, it would be easy and immediate trying to find analogies with the animal mechanisms. Even if sometimes evolution steps have been similar for certain metabolic pathways or the nature of the molecules involved, plants must obviously respond to external stimuli in different ways, considering the absence of moving ability and differences in sensing mechanisms. During submergence roots are obliged without chance to a hypoxic or anoxic condition, and the whole plant must be able to withstand the stress imposed by the lack of oxygen. In order to achieve this ability, plants have evolved in different ways, and describing and understanding the metabolic consequences of the stress represent a major step for the characterisation of plant behaviour during adaptation and survival to flooding.

Studies on oxygen deprivation stress have underlined the presence of flooding-sensitive and flooding-tolerant species and even differences between adaptive strategies belonging to the same tolerant species (i.e. rice ecotypes). Evidences have shown that chemical, electrical and hydraulic signals are involved in the response of flooded plants. So the topic of a deeper study would include knowledge from molecular biology, biochemistry and physics intracells, intercells and interorgans, for better understanding the mechanisms through which the whole plant can withstand oxygen deficiency stress.

References

Ahmed S, Nawata E, Sakuratani T (2006) Changes of endogenous ABA and ACC, and their correlations to photosynthesis and water relations in mungbean (*Vigna radiata* (L.) Wilczak cv. KPS1) during waterlogging. Environ Exp Bot 57:278–284

Aiken RM, Smucker AJM (1996) Root system regulation of whole plant growth. Annu Rev Phytopathol 34:325–346

Andersen PC, Lombard PB, Westwood MN (1984) Effect of root anaerobiosis on the water relations of several Pyrus species. Physiol Plant 62:245–252

Andreev VY, Vartapetian BB (1992) Induction of alcoholic and lactic fermentation in the early stages of anaerobic incubation of higher plants. Phytochemistry 31:1859–1861

Arbona V, Gomez-Cadenas A (2008) Hormonal modulation of citrus responses to flooding. J Plant Growth Regul 27:241–250
Armstrong W (1979) Aeration in higher plants. Adv Bot Res 7:225–332
Armstrong J, Armstrong W (2005) Rice: sulphide-induced barriers to root radial oxygen loss, Fe^{2+} and water uptake, and lateral root emergence. Ann Bot 96:625–638
Arru L, Fornaciari S (2010) Root oxygen deprivation and leaf biochemistry in trees. In: Mancuso S, Shabala S (eds) Waterlogging signaling and tolerance in plants. Springer, Heidelberg, pp 181–195
Atkinson CJ, Harrison-Murray RS, Taylor JM (2008) Rapid flood-induced stomatal closure accompanies xylem sap transportation of root-derived acetaldehyde and ethanol in *Forsythia*. Environ Exp Bot 64:196–205
Babourina O, Rengel Z (2010) Ion transport in aquatic plants. In: Mancuso S, Shabala S (eds) Waterlogging signaling and tolerance in plants. Springer, Heidelberg, pp 221–238
Baena-González E, Rolland F, Thevelein JM, Sheen J (2007) A central integrator of transcription networks in plant stress and energy signaling. Nature 448:938–942
Bailey-Serres J, Chang R (2005) Sensing and signaling in response to oxygen deprivation in plants and other organisms. Ann Bot 96:507–518
Bailey-Serres J, Voesenek LACJ (2008) Flooding stress: acclimations and genetic diversity. Annu Rev Plant Biol 59:313–339
Bailey-Serres J, Fukao T, Gibbs DJ, Holdsworth MJ, Lee SC, Licausi F, Perata P, Voesenek LACJ, van Dongen JT (2012) Making sense of low oxygen sensing. Trends Plant Sci 17:129–138
Baluška F, Kubica Š, Hauskrecht M (1990) Postmitotic 'isodiametric' cell growth in the maize root apex. Planta 181:269–274
Baluška F, Volkmann D, Barlow PW (1996) Specialized zones of development in roots: view from the cellular level. Plant Physiol 112:3–4
Baluška F, Volkmann D, Barlow PW (2001) A polarity crossroad in the transition growth zone of maize root apices: cytoskeletal and developmental implications. J Plant Growth Regul 20:170–181
Baluška F, Mancuso S, Volkmann D, Barlow PW (2010) Root apex transition zone: a signalin-g–response nexus in the root. Trends Plant Sci 15:402–408
Banti V, Mafessoni F, Loreti E, Alpi A, Perata P (2010) The heat-inducible transcription factor HsfA2 enhances anoxia tolerance in *Arabidopsis*. Plant Physiol 152:1471–1483
Barlow PW (2003) The root cap: cell dynamics, cell differentiation and cap function. J Plant Growth Regul 21:261–286
Baxter-Burrell A, Yang Z, Springer PS, Bailey-Serres J (2002) RopGAP4- dependent Rop GTPase rheostat control of Arabidopsis oxygen deprivation tolerance. Science 296:2026–2028
Beerling D (2007) The emerald planet. How plants changed Earth's history. Oxford University Press, New York
Bell EL, Klimova TA, Eisenbart J, Moraes CT, Murphy MP, Budinger GRS, Chandel NS (2007) The Qo site of the mitochondrial complex III is required for the transduction of hypoxic signaling via reactive oxygen species production. J Cell Biol 177:1029–1036
Benschop JJ, Jackson MB, Gühl K, Vreeburg RAM, Croker SJ, Peeters AJM, Voesenek LACJ (2005) Contrasting interactions between ethylene and abscisic acid in *Rumex* species differing in submergence tolerance. Plant J 44:756–768
Besson-Bard A, Pugin A, Wendehenne D (2008) New insights into nitric oxide signaling in plants. Annu Rev Plant Biol 59:21–39
Birner PT, Steudel E (1993) Effects of anaerobic conditions on water and solute relations and on active transport in roots of maize (*Zea mays* L.). Planta 190:474–483
Blanke MM, Cooke DT (2004) Effects of flooding and drought on stomatal activity, transpiration, photosynthesis, water potential and water channel activity in strawberry stolons and leaves. Plant Growth Regul 42:153–160
Blokhina O, Virolainen E, Fagerstedt KV (2003) Antioxidants, oxidative damage and oxygen deprivation stress: a review. Ann Bot 91:179–194

Borisjuk L, Rolletschek H (2008) Nitric oxide is a versatile sensor of low oxygen stress in plants. Plant Signal Behav 3:391–393

Bradford KJ, Hsiao TC (1982) Stomatal behaviour and water relations of waterlogged tomato plants. Plant Physiol 70:1508–1513

Braendle R, Crawford RMM (1999) Plants as amphibians. Perspect Plant Ecol Evol Syst 2:56–78

Bramley H, Tyerman S (2010) Root water transport under waterlogged conditions and the roles of aquaporins. In: Mancuso S, Shabala S (eds) Waterlogging signaling and tolerance in plants. Springer, Heidelberg, pp 151–180

Branco-Price C, Kawaguchi R, Ferreira RB, Bailey-Serres J (2005) Genomewide analysis of transcript abundance and translation in Arabidopsis seedlings subjected to oxygen deprivation. Ann Bot 96:647–660

Branco-Price C, Kaiser KA, Jang CJH, Larive CK, Bailey-Serres J (2008) Selective mRNA translation coordinates energetic and metabolic adjustments to cellular oxygen deprivation and reoxygenation in *Arabidopsis thaliana*. Plant J 56:743–755

Building MH (2001) Long-term anoxia tolerance in leaves of *Acorus calamus* L. and *Iris pseudacorus* L. J Exp Bot 52:2213–2225

Burrows WJ, Carr DJ (1969) Effects of flooding root system of sunflower plants on cytokinin content in xylem sap. Physiol Plant 22:1105–1112

Chang R, Jang CJH, Branco-Price C, Nghiem P, Bailey-Serres J (2012) Transient MPK6 activation in response to oxygen deprivation and reoxygenation is mediated by mitochondria and aids seedling survival in Arabidopsis. Plant Mol Biol 78:109–122

Chen X, Pierik R, Peeters AJM, Poorter H, Visser EJ, Huber H, de Kroon H, Voesenek LACJ (2010) Endogenous ABA as a key switch for natural variation in flooding-induced shoot elongation. Plant Physiol 154:969–977

Chirkova TV, Leffler S, Novitskaya LO (1995) Some peculiarities of chloroplast and mitochondria state in leaves of wheat and rice seedlings under conditions of anoxia and long-term darkness. Fiziol Rast (Moscow) 42:368–376 (Russ J Plant Physiol Eng Transl)

Christianson JA, Llewellyn DJ, Dennis ES, Wilson IW (2010) Global gene expression responses to waterlogging in roots and leaves of Cotton (*Gossypium hirsutum* L.). Plant Cell Physiol 51:21–37

Clark DG, Gubrium EK, Barrett JE, Nell TA, Klee HJ (1999) Root formation in ethylene-insensitive plants. Plant Physiol 121:53–60

Colmer TD (2003) Aerenchyma and an inducible barrier to radial oxygen loss facilitate root aeration in upland, paddy and deep-water rice (*Oryza sativa* L.). Ann Bot 91:301–309

Colmer TD, Voesenek LACJ (2009) Flooding tolerance: suites of plant traits in variable environments. Funct Plant Biol 36:665–681

Copolovici L, Niinemets U (2010) Flooding induced emissions of volatile signaling compounds in three tree species with differing waterlogging tolerance. Plant Cell Environ 33:1582–1594

Correa-Aragunde N, Graziano M, Lamattina L (2004) Nitric oxide plays a central role in determining lateral root development in tomato. Planta 218:900–905

Dat JF, Capelli N, Folzer H, Bourgeade P, Badot PM (2004) Sensing and signaling during plant flooding. Plant Physiol Biochem 42:273–282

De Pinto MC, Locato V, De Gara L (2012) Redox regulation in plant programmed cell death. Plant Cell Environ 35:234–244

Desikan R, Cheung MK, Bright J, Henson D, Hancock JT, Neill SJ (2004) ABA, hydrogen peroxide, and nitric oxide signaling in stomatal guard cells. J Exp Bot 55:205–212

Dordas C, Rivoal J, Hill RD (2003) Plant haemoglobins, nitric oxide and hypoxic stress. Ann Bot 91:173–178

Dordas C, Hasinoff BB, Rivoal J, Hill RD (2004) Class I hemoglobins, nitrate and NO levels in hypoxic maize cell suspension cultures. Planta 219:66–72

Drew MC, Jackson MB, Giffard S (1979) Ethylene-promoted adventitious rooting and the development of cortical air spaces (aerenchyma) in roots may be adaptive responses to flooding in *Zea mays* L. Planta 147:83–88

Drew MC, Jackson MB, Giffard SC, Campbell R (1981) Inhibition by silver ions of gas space (aerenchyma) formation in adventitious roots of *Zea mays* L. subjected to exogenous ethylene or to oxygen deficiency. Planta 153:217–224

Ehlert C, Maurel C, Tardieu F, Simonneau T (2009) Aquaporin-mediated reduction in maize root hydraulic conductivity impacts cell turgor and leaf elongation even without changing transpiration. Plant Physiol 150:1093–1104

Else MA, Jackson MB (1998) Transport of 1-aminocyclopropane-1- carboxylic acid (ACC) in the transpiration stream of tomato (*Lycopersicon esculentum*) in relation to foliar ethylene production and petiole epinasty. Aust J Plant Physiol 25:453–458

Else MA, Tiekstra AE, Croker SJ, Davies WJ, Jackson MB (1996) Stomatal closurein flooded tomato plants involves abscisic acid and a chemically unidentified anti-transpirant in xylem sap. Plant Physiol 112:239–247

Else MA, Coupland D, Dutton L, Jackson MB (2001) Decreased root hydraulic conductivity reduces leaf water potential, initiates stomatal closure and slows leaf expansion in flooded plants of castor oil (*Ricinus communis*) despite diminished delivery of ABA from the roots to shoots in xylem sap. Physiol Plant 111:46–54

Else MA, Taylor JM, Atkinson CJ (2006) Anti-transpirant activity in xylem sap from flooded tomato (*Lycopersicon esculentum* Mill.) plants is not due to pH-mediated redistributions of root- or shoot-sourced ABA. J Exp Bot 57:3349–3357

Else MA, Janowiak F, Atkinson CJ, Jackson MB (2009) Root signals and stomatal closure in relation to photosynthesis, chlorophyll a fl uorescence and adventitious rooting of flooded tomato plants. Ann Bot 103:313–323

Ershova A, Popova N, Berdnikova O (2011) Production of reactive oxygen species and antioxidant enzymes of pea and soybean plants under hypoxia and high CO_2 concentration in medium. Russ J Plant Physiol 58:982–990

Evans DE (2004) Aerenchyma formation. New Phytol 161:35–49

Everard JD, Drew MC (1989) Water relations of sunflower (*Helianthus annuus* L.) shoots during exposure of the root system to oxygen deficiency. J Exp Bot 40:1255–1264

Felle HH (2010) pH signaling during anoxia. In: Mancuso S, Shabala S (eds) Waterlogging signaling and tolerance in plants. Springer, Heidelberg, pp 79–98

Felle HH, Herrmann A, Hanstein S, Hückelhoven R, Kogel KH (2004) Apoplastic pH signaling in barley leaves attacked by the powdery mildew fungus *Blumeria graminis* f. sp. *hordei*. Mol Plant-Microbe Interact 17:118–123

Felle HH, Herrman A, Hueckelhoven R, Kogel K-H (2005) Root-to-shoot signaling: apoplastic alkalinization, a general stress response and defence factor in barley (*Hordeum vulgare*). Protoplasma 227:17–24

Fu XY, Peng SX, Yang S, Chen YH, Zhang JY, Mo WP, Zhu JY, Ye YX, Huang XM (2012) Effects of flooding on grafted annona plants of different scion/rootstock combinations. Agric Sci 3:249–256

Fukao T, Xu K, Ronald PC, Bailey-Serres J (2006) A variable cluster of ethylene response factorlike genes regulates metabolic and developmental acclimation responses to submergence in rice. Plant Cell 18:2021–2034

Garcia-Sanchez F, Syvertsen JP, Gimeno V, Botia P, Perez-Perez JG (2007) Responses to flooding and drought stress by two citrus rootstock seedlings with different water-use efficiency. Physiol Plant 130:532–542

Geigenberger P (2003) Response of plant metabolism to too little oxygen. Curr Opin Plant Biol 6:247–256

Geisler-Lee J, Caldwell C, Gallie DR (2010) Expression of the ethylene biosynthetic machinery in maize roots is regulated in response to hypoxia. J Exp Bot 61:857–871

Graham LG (1996) Green algae to land plants: an evolutionary transition. J Plant Res 109:7737–7742

Gruntman M, Novoplansky A (2004) Physiologically mediated self/non-self discrimination in roots. Proc Natl Acad Sci USA 101:3863–3867

Guzy RD, Schumacker PT (2006) Oxygen sensing by mitochondria at complex III: the paradox of increased reactive oxygen species during hypoxia. Exp Physiol 91:807–819

Hasanuzzaman M, Gill SS, Fujita M (2012) Physiological role of nitric oxide in plants grown under adverse environmental conditions. In: Tuteja N, Gill SS (eds) Plant acclimation to environmental stress, Chap 11. Springer, New York

Hattori Y, Nagai K, Furukawa S, Song XJ, Kawano R, Sakakibara H, Wu J, Matsumoto T, Yoshimura A, Kitano H, Matsuoka M, Mori H, Ashikari M (2009) The ethylene response factors SNORKEL1 and SNORKEL2 allow rice to adapt to deep water. Nature 460:1026–1030

He CJ, Morgan PW, Drew MC (1994) Induction of enzymes associated with lysigenous aerenchyma formation in roots of *Zea mays* during hypoxia or nitrogen starvation. Plant Physiol 105:861–865

Hinz M, Wilson IW, Yang J, Buerstenbinder K, Llewellyn D, Dennis ES, Sauter M, Dolferus R (2010) *Arabidopsis* RAP2.2: an ethylene response transcription factor that is important for hypoxia survival. Plant Physiol 144:218–231

Hodges A (2009) Root decisions. Plant Cell Environ 32:628–640

Holzinger R, Sandoval-Soto L, Rottenberger S, Crutzen PJ, Kesselmeier J (2000) Emissions of volatile organic compounds from *Quercus ilex* L. measured by proton transfer reaction mass spectrometry under different environmental conditions. J Geophys Res 105:20573–20579

Igamberdiev AU, Hill RD (2004) Nitrate, NO and haemoglobin in plant adaptation to hypoxia: an alternative to classic fermentation pathways. J Exp Bot 55:2473–2482

Igamberdiev AU, Baron K, Manać̌h-Little N, Stoimenova M, Hill RD (2005) The haemoglobin/nitric oxide cycle: involvement in flooding stress and effects on hormone signaling. Ann Bot 96:557–564

Illéš P, Schlicht M, Pavlovkin J, Lichtscheidl I, Baluška F, Ovečka M (2006) Aluminium toxicity in plants: internalisation of aluminium into cells of the transition zone in Arabidopsis root apices relates to changes in plasma membrane potential, endosomal behaviour, and nitric oxide production. J Exp Bot 57:4201–4213

Ishikawa H, Evans ML (1992) Induction of curvature in maize roots by calcium or by thigmostimulation. Role of postmitotic isodiametric growth zone. Plant Physiol 100:762–768

Jackson MB (2002) Long-distance signaling from roots to shoots assessed: the flooding story. J Exp Bot 53:175–181

Jackson MB, Hall KC (1987) Early stomatal closure in waterlogged pea plants is mediated by abscisic acid in the absence of foliar water deficits. Plant Cell Environ 10:121–130

Jackson MB, Gales K, Campbell DJ (1978) Effect of waterlogged soil conditions on the production of ethylene and on water relationships in tomato plants. J Exp Bot 29:183–193

Jackson MB, Fenning TM, Drew MC, Saker LR (1985) Stimulation of ethylene production and gas-space (aerenchyma) formation in adventitious roots of *Zea mays* L. by small partial pressures of oxygen. Planta 165:486–492

Jackson MB, Young S, Hall KC (1988) Are roots a source of abscisic acid for the shoots of flooded pea plants? J Exp Bot 39:1631–1637

Jackson MB, Saker LR, Crisp CM, Else MA, Janowiak F (2003) Ionic and pH signaling from roots to shoots of flooded tomato plants in relation to stomatal closure. Plant Soil 253:103–113

Jackson MB, Ishizawa K, Ito O (2009) Evolution and mechanisms of plant tolerance to flooding stress. Ann Bot 103:137–142

Janowiak F, Else MA, Jackson MB (2002) A loss of photosynthetic efficiency does not explain stomatal closure in flooded tomato plants. In: Advances of agricultural sciences problem issues, Issue 481. Ecophysiological aspects of plants responses to stress factors. Polish Academy of Sciences, Warsaw, pp 229–234

Jia WS, Davies WJ (2007) Modification of leaf apoplastic pH in relation to stomatal sensitivity to root-sourced ABA signals. Plant Physiol 143:68–77

Jung KH, Seo YS, Walia H, Cao P, Fukao T, Canlas PE, Amonpant F, Bailey-Serres J, Ronald PC (2010) The submergence tolerance regulator *Sub1A* mediates stress-responsive expression of *AP2/ERF* transcription factors. Plant Physiol 152:1674–1692

Kasprowicz A, Szuba A, Volkmann D, Baluška F, Wojtaszek F (2009) Nitric oxide modulates dynamic actin cytoskeleton and vesicle trafficking in a cell type-specific manner in root apices. J Exp Bot 60:1605–1617
Kawase M (1981) Anatomical and morphological adaptation of plants to waterlogging. Hortic Sci 16:30–34
Kende H, van der Knaap E, Cho H-T (1998) Deepwater rice: a model plant to study stem elongation. Plant Physiol 118:1105–1110
Kieffer M, Neve J, Kepinski S (2009) Defining auxin response contexts in plant development. Curr Opin Plant Biol 13:12–20
Klok EJ, Wilson IW, Wilson D, Chapman SC, Ewing RM, Somerville SC, Peacock WJ, Dolferus R, Dennis ES (2002) Expression profile analysis of the low-oxygen response in Arabidopsis root cultures. Plant Cell 14:2481–2494
Konings H (1982) Ethylene-promoted formation of aerenchyma in seedling roots of *Zea mays* L. under aerated and non-aerated conditions. Physiol Plant 54:119–124
Kozlowski TT (1984) Responses of woody plants to flooding. In: Kozlowski TT (ed) Flooding and plant growth. Academic, New York, pp 129–163
Kramer PJ (1940) Causes of decreased absorption of water by plants in poorly aerated media. Am J Bot 27:216–220
Kreuzwieser J, Hauberg J, Howell KA, Carroll A, Rennenberg H, Millar AH, Whelan J (2009) Differential response of gray poplar leaves and roots underpins stress adaptation during hypoxia. Plant Physiol 149:461–473
Kumar B, Pandey DM, Goswami CL, Jain S (2004) Effect of growth regulators on photosynthesis, transpiration and related parameters in water stressed cotton. Biol Plant 44:475–478
Kumutha D, Ezhilmathi K, Sairam RK, Srivastava GC, Deshmukh PS, Meena RC (2009) Waterlogging induced oxidative stress and antioxidant activity in pigeonpea genotypes. Biol Plant 53:75–84
Lamattina L, Garcia-Mata C, Graziano M, Pagnussat G (2003) Nitric oxide: the versatility of an extensive signal molecule. Annu Rev Plant Biol 54:109–136
Lanteri ML, Laxalt AM, Lamattina L (2008) Nitric oxide triggers phosphatidic acid accumulation via phospholipase D during auxin-induced adventitious root formation in cucumber. Plant Physiol 147:188–198
Lasanthi-Kudahettige R, Magneschi L, Loreti E, Gonzali S, Licausi F, Novi G, Beretta O, Vitulli F, Alpi A, Perata P (2007) Transcript profiling of the anoxic rice coleoptile. Plant Physiol 144:218–231
Lecourieux D, Ranjeva R, Pugin A (2006) Calcium in plant defence signaling pathways. New Phytol 171:249–269
Lee KW, Chen PW, Lu CA, Chen S, Ho THD, Yu SM (2009) Coordinated responses to oxygen and sugar deficiency allow rice seedlings to tolerate flooding. Sci Signal 2:ra61
Lee SC, Mustroph A, Sasidharan R, Vashisht D, Pedersen O, Oosumi T, Voesenek LACJ, Bailey-Serres J (2011) Molecular characterization of the submergence response of the *Arabidopsis thaliana* ecotype Columbia. New Phytol 190:457–471
Leterrier M, Valderrama R, Chaki M, Airaki M, Palma JM, Barroso JB, Corpas FJ (2012) Function of nitric oxide under environmental stress conditions. In: Khan NA, Nazar R, Iqbal N, Anjum NA (eds) Phytohormones and abiotic stress tolerance in plants. Springer, Berlin, pp 99–113
Licausi F (2011) Regulation of the molecular response to oxygen limitations in plants. New Phytol 190:550–555
Licausi F, van Dongen JT, Giuntoli B, Novi G, Santaniello A, Geigenberger P, Perata P (2010) *HRE1* and *HRE2*, two hypoxia-inducible ethylene response factors, affect anaerobic responses in *Arabidopsis thaliana*. Plant J 62:302–315
Licausi F, Kosmacz M, Weits DA, Giuntoli B, Giorgi FM, Voesenek LACJ, Perata P, van Dongen JT (2011) Oxygen sensing in plants is mediated by an N-end rule pathway for protein destabilization. Nature 479:419–422

Limami AM, Glévarec G, Ricoult C, Cliquet JB, Planchet E (2008) Concerted modulation of alanine and glutamate metabolism in young Medicago truncatula seedlings under hypoxic stress. J Exp Bot 59:2325–2335

Liu F, VanToai T, Moy LP, Bock G, Linford LD, Quackenbush J (2005) Global transcription profiling reveals comprehensive insights into hypoxic response in Arabidopsis. Plant Physiol 137:1115–1129

Lombardo MC, Graziano M, Polacco JC, Lamattina L (2006) Nitric oxide functions as a positive regulator of root hair development. Plant Signal Behav 1:28–33

Loreti E, Poggi A, Novi G, Alpi A, Perata P (2005) A genome-wide analysis of the effects of sucrose on gene expression in *Arabidopsis* seedlings under anoxia. Plant Physiol 137:1130–1138

Magneschi L, Perata P (2009) Rice germination and seedling growth in the absence of oxygen. Ann Bot 103:181–196

Mancuso S, Boselli M (2002) Characterisation of the oxygen fluxes in the division, elongation and mature zones of Vitis roots: influence of oxygen availability. Planta 214:767–774

Mancuso S, Marras AM (2006) Adaptative response of Vitis root to anoxia. Plant Cell Physiol 47:401–409

Mancuso S, Marras AM, Mugnai S, Schlicht M, Žársky V, Li G, Song L, Xue HW, Baluška F (2007) Phospholipase Dz2 drives vesicular secretion of auxin for its polar cell-cell transport in the transition zone of the root apex. Plant Signal Behav 2:240–244

Manetas Y (2012) Short evolutionary history of plants. In: Manetas Y (ed) Alice in the land of plants: biology of plants and their importance for planet earth. Springer, Berlin, pp 113–160

Marciano DPRO, Toledo-Ramos F, Neiva-Alvim M, Magalhaes JR, Costa Franca MG (2010) Nitric oxide reduces the stress effects of aluminum on the process of germination and early root growth of rice. J Plant Nutr Soil Sci 173:885–891

Masi E, Ciszak M, Stefano G, Renna L, Azzarello E, Pandolfi C, Mugnai S, Baluška F, Arecchi FT, Mancuso S (2009) Spatio-temporal dynamics of the electrical network activity in the root apex. Proc Natl Acad Sci USA 106:4048–4053

Miller G, Mittler R (2006) Could heat shock transcription factors function as hydrogen peroxide sensors in plants? Ann Bot 98:279–288

Mugnai S, Marras AM, Mancuso S (2011) Effect of hypoxic acclimation on anoxia tolerance in Vitis roots: response of metabolic activity and K+ fluxes. Plant Cell Physiol 52:1107–1116

Mugnai S, Azzarello E, Baluška F, Mancuso S (2012) Local root apex hypoxia induces no-mediated hypoxic acclimation of the entire root. Plant Cell Physiol 53:912–920

Muhlenbock P, Plaszczyca M, Mellerowicz E, Karpinski S (2007) Lysigenous aerenchyma formation in *Arabidopsis* is controlled by Lesion Simulating Disease1. Plant Cell 19:3819–3830

Mustroph A, Lee SC, Oosumi T, Zanetti ME, Yang H, Ma K, Yaghoubi-Masihi A, Fukao T, Bailey-Serres J (2010) Cross-kingdom comparison of transcriptomic adjustments to low-oxygen stress highlights conserved and plant-specific responses. Plant Physiol 152:1484–1500

Nakano T, Suzuki K, Fujimura T, Shinshi H (2006) Genome-wide analysis of the ERF gene family in *Arabidopsis* and rice. Plant Physiol 140:411–432

Negi S, Sukumar P, Liu X, Cohen JD, Muday GK (2010) Genetic dissection of the role of ethylene in regulating auxin-dependent lateral and adventitious root formation in tomato. Plant J 61:3–15

Neill SJ, Desikan R, Hancock JT (2003) Nitric oxide signaling in plants. New Phytol 159:11–35

Neill S, Barros R, Bright J, Desikan R, Hancock J, Harrison J, Morris P, Ribeiro D, Wilson I (2008) Nitric oxide, stomatal closure, and abiotic stress. J Exp Bot 59:165–176

Neuman DS, Smith BA (1991) The influence of leaf water status and ABA on leaf growth and stomata of Phaseolus seedlings with hypoxic roots. J Exp Bot 42:1499–1506

Ober ES, Sharp RE (2003) Electrophysiological responses of maize roots to low water potentials: relationship to growth and ABA accumulation. J Exp Bot 54:813–824

Pagnussat GC, Lanteri ML, Lombardo MC, Lamattina L (2004) Nitric oxide mediates the indole acetic acid induction activation of a mitogen-activated protein kinase cascade involved in adventitious root development. Plant Physiol 135:279–286

Pauly N, Pucciariello C, Mandon K, Innocenti G, Jamet A, Baudouin E, Herouart D, Frendo P, Puppo A (2006) Reactive oxygen and nitrogen species and glutathione: key players in the legume- *Rhizobium* symbiosis. J Exp Bot 57:1769–1776

Peeters AJM, Cox MCH, Benschop JJ, Vreeburg RAM, Bou J, Voesenek LACJ (2002) Submergence research using *Rumex palustris* as a model; looking back and going forward. J Exp Bot 53:391–398

Perata P, Alpi A (1993) Plant responses to anaerobiosis. Plant Sci 93:1–17

Pereira JS, Kozlowski TT (1977) Variations among woody angiosperms in response to flooding. Physiol Plant 45:184–192

Pospíšilová J (2003) Participation of phytohormones in the stomatal regulation of gas exchange during water stress. Biol Plant 46:491–506

Prado AM, Porterfield DM, Feijo JA (2004) Nitric oxide is involved in growth regulation and re-orientation of pollen tubes. Development 131:2707–2714

Prado AM, Colaco R, Moreno N, Silva AC, Feijo JA (2008) Targeting of pollen tubes to ovules is dependent on nitric oxide(NO) signaling. Mol Plant 1:703–714

Pucciariello C, Perata P (2012) Flooding tolerance in plants. In: Shabala S (ed) Plant stress physiology. CAB International, Oxford, pp 148–171

Pucciariello C, Parlanti S, Banti V, Novi G, Perata P (2012) Reactive oxygen species-driven transcription in Arabidopsis under oxygen deprivation. Plant Physiol 159:184–196

Rascio N, Mariani P, Dalla Vecchia F, Zanchin A, Pool A, Larcher W (1994) Ultrastructural and photosynthetic features of leaves and stems of *Elodea canadensis*. J Plant Physiol 144:314–323

Ratcliffe RG (1997) In vivo NMR studies of the metabolic response of plant tissue to anoxia. Ann Bot 79:39–48

Raven JA (1996) Into the voids: the distribution, function, development and maintenance of gas spaces in plants. Ann Bot 78:137–142

Reid DM, Crozier A, Harvey BMR (1969) Effects of flooding on export of gibberellins from root to shoot. Planta 89:376–379

Ricard B, Couee I, Raymond P, Saglio PH, Saint-Ges V, Pradet A (1994) Plant metabolism under hypoxia and anoxia. Plant Physiol Biochem 32:1–10

Ricoult C, Echeverria LO, Cliquet JB, Limami AM (2006) Characterization of alanine aminotransferase (AlaAT) multigene family and hypoxic response in young seedlings of the model legume *Medicago truncatula*. J Exp Bot 57:3079–3089

Rodríguez-Gamir J, Ancillo G, González-Mas MC, Primo-Millo E, Iglesias DJ, Forner-Giner MA (2011) Root signaling and modulation of stomatal closure in flooded citrus seedlings. Plant Physiol Biochem 49:636–645

Rottenberger S, Kleiss B, Kuhn U, Wolf A, Piedade MTF, Junk W, Kesselmeier J (2008) The effect of flooding on the exchange of the volatile C2-compounds ethanol, acetaldehyde and acetic acid between leaves of Amazonian floodplain tree species and the atmosphere. Biogeosciences 5:1085–1100

Sairam RK, Kumutha D, Ezhilmathi K, Deshmukh PS, Srivastava GC (2008) Physiology and biochemistry of waterlogging tolerance in plants. Biol Plant 52:401–412

Salmi ML, Morris KE, Roux SJ, Porterfield DM (2007) Nitric oxide and cGMP signaling in calcium-dependent development of cell polarity in Ceratopteris richardii. Plant Physiol 144:94–104

Santosa IE, Ram PC, Boamfa EI, Laarhoven LJ, Reuss J, Jackson MB, Harren FJM (2006) Patterns of peroxidative ethane emission from submerged rice seedlings indicate that damage from reactive oxygen species takes place during submergence and is not necessarily a postanoxic phenomenon. Planta 226:193–202

Sedbrook JC, Kronebush PJ, Borisy GG, Trewavas AJ, Masson P (1996) Transgenic aequorin reveals organ specific cytosolic Ca^{2+} responses to anoxia in *Arabidopsis thaliana* seedlings. Plant Physiol 111:243–257

Sobol M, Kordyum E (2009) Distribution of calcium ions in cells of the root distal elongation zone under clinorotation. Microgravity Sci Technol 21:179–185

Sorrell BK, Brix H, Orr PT (1997) *Eleocharis sphacelata*: internal gas transport pathways and modelling of aeration by pressurized flow and diffusion. New Phytol 136:433–442

Subbaiah CC, Bush DS, Sachs MM (2000) Elevation of cytosolic calcium precedes anoxic gene expression in maize suspension-cultured cells. Plant Cell 6:1747–1762

Sun BT, Jing Y, Chen KM, Song LL, Chen FJ, Zhang LX (2007) Protective effect of nitric oxide on iron deficiency-induced oxidative stress in maize (Zea mays). J Plant Physiol 164:536–543

Tournaire-Roux C, Sutka M, Javot H, Gout E, Gerbeau P, Luu DT, Bligny R, Maurel C (2003) Cytosolic pH regulates root water transport during anoxic stress through gating of aquaporins. Nature 425:393–397

Trewavas A (2009) What is plant behavior? Plant Cell Environ 32:606–616

van Dongen JT, Frohlich A, Ramirez-Aguilar SJ, Schauer N, Fernie AR, Erban A, Kopka J, Clark J, Langer A, Geigenberger P (2008) Transcript and metabolite profiling of the adaptive response to mild decreases in oxygen concentration in the roots of Arabidopsis plants. Ann Bot 103:269–280

Vandenbroucke K, Robbens S, Vandepoele K, Inzé D, Van de Peer Y, Van Breusegem F (2008) Hydrogen peroxide-induced gene expression across kingdoms: a comparative analysis. Mol Biol Evol 25:507–516

Verbelen J, Cnodder T, Le J, Vissenberg K, Baluška F (2006) Root apex of *Arabidopsis thaliana* consists of four distinct zones of growth activities: meristematic zone, transition zone, fast elongation zone, and growth terminating zone. Plant Signal Behav 1:296–304

Vidoz ML, Loreti E, Mensuali A, Alpi A, Perata P (2010) Hormonal interplay during adventitious root formation in flooded tomato plants. Plant J 63:551–562

Visser EJW, Colmer TD, Blom CWPM, Voesenek LACJ (2000) Changes in growth, porosity, and radial oxygen loss from adventitious roots of selected mono- and dicotyledonous wetland species with contrasting types of aerenchyma. Plant Cell Environ 23:1237–1245

Voesenek LACJ, Pierik R (2008) Plant stress profiles. Science 320:880–881

Wang YS, Yang ZM (2005) Nitric oxide reduces aluminum toxicity by preventing oxidative stress in the roots of *Cassiatora* L. Plant Cell Physiol 46:1915–1923

Wang Y, Chen T, Zhang C, Hao H, Liu P, Zheng M, Baluška F, Šamaj J, Lin J (2009) Nitric oxide modulates the influx of extracellular Ca^{2+} and actin filament organization during cell wall construction in Pinus bungeana pollen tubes. New Phytol 182:851–862

Wilkinson S (1999) pH as a stress signal. Plant Growth Regul 29:87–99

Winch S, Pritchard J (1999) Acid-induced cell wall loosening is confined to the accelerating region of the root growing zone. J Exp Bot 50:1481–1487

Xu K, Xu X, Fukao T, Canlas P, Maghirang-Rodriguez R, Heuer S, Ismail AM, Bailey-Serres J, Ronald PC, Mackill DJ (2006) Sub1A is an ethylene-response-factor-like gene that confers submergence tolerance to rice. Nature 442:705–708

Xu ZS, Xia LQ, Chen M, Cheng XG, Zhang RY, Li LC, Zhao YX, Lu Y, Ni ZY, Liu L, Qiu ZG, Ma YZ (2007) Isolation and molecular characterization of the *Triticum aestivum* L. ethylene-responsive factor 1 (TaERF1) that increases multiple stress tolerance. Plant Mol Biol 65:719–732

Yang Z, Fu Y (2007) ROP/RAC GTPase signaling. Curr Opin Plant Biol 10:490–494

Yang CY, Hsu FC, Li JP, Wang NN, Shih MC (2011) The AP2/ERF transcription factor AtERF73/HRE1 modulates ethylene responses during hypoxia in *Arabidopsis*. Plant Physiol 156:202–212

Zhang J, Zhang X (1994) Can early wilting of old leaves account for much of the ABA accumulation in flooded leaves? J Exp Bot 45:1335–1342

Systemic Signaling in the Maintenance of Phosphate Homeostasis

Aaron P. Smith

Abstract Inorganic phosphate (Pi) availability in soils is often low and heterogeneous. Therefore, plants have evolved numerous adaptations to maintain Pi homeostasis that are regulated by both local and systemic signaling pathways. The level of sophistication in Pi signaling is exemplified by the presence of a key systemic signaling circuit that has been elucidated in the past decade. This circuit appears conserved among angiosperms and involves a phloem-mobile microRNA that targets a root-specific E2 conjugase. Recent advances regarding this circuit and its regulators, as well as additional systemic signaling factors, are the focus of this chapter.

Keywords Phosphorus • Phosphate starvation response • microRNA • PHR1 • Sucrose • Noncoding RNA • Hormones

1 Introduction

Phosphorus levels in soil can be high, but plant-available inorganic phosphate (Pi) is often present at low levels (Marschner 1995; Raghothama 1999). As a result, plants have evolved numerous mechanisms for maintaining optimal Pi acquisition and distribution when Pi availability is low. Local sensing of low Pi primarily initiates root morphological changes aimed at enhancing Pi foraging capabilities, such as cessation of primary root growth and increased proliferation of root hairs and lateral roots (also adventitious and cluster/proteoid roots) (Lynch 2011; Peret et al. 2011; Vance et al. 2003). In addition to local signaling pathways, plants exhibit systemic, or long-distance, regulation of Pi starvation responses (PSR). The demonstration of systemic signaling pathways was revealed by "split-root" studies in

A.P. Smith (✉)
Department of Biological Sciences, Louisiana State University, A759 Life Sciences Annex, Baton Rouge, LA 70803, USA
e-mail: apsmith@lsu.edu

F. Baluška (ed.), *Long-Distance Systemic Signaling and Communication in Plants*, Signaling and Communication in Plants 19, DOI 10.1007/978-3-642-36470-9_7,

which the root systems of individual plants were separated into two compartments at different (i.e., high and low) Pi regimes. In these studies, the induction of various Pi starvation-induced genes was attenuated in low Pi compartments, apparently by mobile, "high Pi" signals originating in roots from the Pi-replete compartments (Burleigh and Harrison 1999; Franco-Zorrilla et al. 2005; Liu et al. 1998). More recently, Thibaud et al. (2010) demonstrated the presence of systemic signaling on a global scale and placed numerous PSR genes into several local or systemic categories based on the degree of influence from systemic signals. In contrast to local Pi-signaling pathways, systemic signaling serves to facilitate Pi acquisition, mobilization, and redistribution (Chiou and Lin 2011; Thibaud et al. 2010; Yang and Finnegan 2010).

Although two distinct networks are recognized in Pi signaling, coordination between local and systemic pathways is essential for proper maintenance of Pi homeostasis. Indeed, this sophisticated level of signaling integration has led to difficulty in teasing apart the various layers of signaling and regulatory components. Nonetheless, great progress has been made in recent years in identifying a key, systemic regulatory circuit that modulates Pi acquisition and distribution, and the components that regulate or fine-tune it. Herein, recent advances regarding this conserved circuit as well as additional systemic Pi-signaling components and regulators are presented.

2 Shoot-to-Root Signals

2.1 miR399

Arabidopsis microRNA399 (miR399) was the first miRNA demonstrated to be specifically and highly induced in response to Pi deficiency (Fujii et al. 2005). The Arabidopsis genome encodes six primary miR399 genes, which are all induced to varying degrees by Pi deficiency. Transgenic Arabidopsis or rice, each overexpressing their respective miR399, accumulated high Pi levels in shoots, which led to Pi toxicity (Aung et al. 2006; Bari et al. 2006; Chiou et al. 2006; Fujii et al. 2005; Hu et al. 2011). Overexpression of Arabidopsis miR399 in tomato also resulted in increased Pi accumulation (Gao et al. 2010). Collectively, these reports demonstrate that miR399 induces Pi acquisition and translocation.

Early searches for Arabidopsis miRNA targets predicted that miR399 targets a ubiquitin E2 conjugase (UBC24) (Allen et al. 2005; Sunkar and Zhu 2004). A *ubc24* T-DNA knockout exhibited phenotypes similar to those observed in miR399-overexpressing plants (Chiou et al. 2006), which were also similar to the previously characterized *pho2* mutant (Delhaize and Randall 1995). Subsequently, two groups independently demonstrated that mutation of *UBC24* was the source of the *pho2* phenotype (Aung et al. 2006; Bari et al. 2006). *UBC24*/*PHO2* (*PHO2* hereafter) is highly expressed in Pi-replete roots, but is downregulated in response to Pi deficiency, whereas miR399 expression shows the reciprocal response (Bari et al. 2006; Chiou et al. 2006; Fujii et al. 2005).

Via grafting experiments between *pho2* mutants and wild-type Arabidopsis, it was shown that loss of *PHO2* in roots, but not shoots, was sufficient to confer Pi overaccumulation in shoots (Bari et al. 2006). Additional grafting experiments showed that miR399-overexpression scions grafted onto wild-type rootstocks led to mature miR399 in roots under Pi-replete conditions, despite undetectable root levels of miR399 primary transcripts (Lin et al. 2008; Pant et al. 2008). miR399 genes are expressed in the vasculature (Aung et al. 2006), and miR399 accumulates in phloem sap, particularly under Pi-deficiency conditions (Buhtz et al. 2008; Pant et al. 2008). Together these observations provide strong support for a model in which miR399, acting as a systemic signal, is induced in shoots in response to Pi deficiency and travels to roots where it targets *PHO2* mRNA for degradation, thereby regulating Pi acquisition and translocation. Orthologs of both miR399 and PHO2 have been identified in many diverse angiosperm species and when examined have been shown to have the same responses to Pi deficiency as the Arabidopsis components (Kuo and Chiou 2011). This demonstrates that the miR399-PHO2 pathway is a highly conserved and integral Pi regulatory circuit.

Until recently, downstream targets of PHO2 remained elusive. Initial reports indicated that transcript levels for the high-affinity phosphate transporter family (Pht1) genes *Pht1;8* and *Pht1;9* were elevated in *pho2* mutant roots, suggesting that upregulation of the encoded transporters resulted in the Pi overaccumulation phenotype (Aung et al. 2006; Bari et al. 2006). Although RNA interference-mediated knockdown of *Pht1;8* in *pho2* was reported to decrease leaf Pi accumulation to wild-type levels (Bari et al. 2006), T-DNA knockout of *Pht1;8* and/or *Pht1;9* in *pho2* did not affect Pi accumulation (Kuo and Chiou 2011). This apparent discrepancy warrants further investigation. In expression studies of *pho2* mutants in both Arabidopsis and rice, the transcript levels of a number of genes linked to Pi transport, mobilization, or distribution have been shown to be upregulated (Bari et al. 2006; Hu et al. 2011; Liu et al. 2010a). However, these transcriptional responses may be indirect, particularly in light of PHO2 being a ubiquitin E2 conjugase, and thus likely responsible for protein degradation as a mechanism for modulating Pi responses.

Notably, a recent study by Liu et al. (2012) has identified a bona fide, direct target of PHO2, the SPX-EXS family member, PHO1 (Hamburger et al. 2002; Wang et al. 2004). Originally characterized as components of Pi sensing and transport in yeast, SPX domain proteins have more recently been implicated in Pi signaling and transport in plants (Duan et al. 2008; Hamburger et al. 2002; Lenburg and O'Shea 1996). In contrast to *PHO2*, mutation of *PHO1* disrupts Pi loading into the xylem, resulting in low Pi shoot levels (Poirier et al. 1991). In a *pho2* suppressor screen, Liu et al. (2012) identified two suppressors that each carried mutations in *PHO1*. Also, transient co-expression of PHO1 and PHO2 in tobacco leaves resulted in a reduction in *PHO1* expression in a PHO2 dosage-dependent manner that required the ubiquitin-conjugating activity of PHO2 (Liu et al. 2012). Further, bimolecular fluorescence and split-ubiquitin complementation assays indicated that PHO1 and PHO2 physically interact. Together these results indicate that PHO2 targets PHO1 for protein degradation and places PHO1 as a key downstream component in the miR399-PHO2 Pi signaling circuit. Importantly, overexpression of *PHO1* does not completely mimic the *pho2* phenotype (i.e., Pi overaccumulation in leaves), indicating that PHO2 targets additional components involved in modulating Pi translocation and/or acquisition (Liu et al. 2012).

2.2 Other miRNAs

Through a variety of approaches, a number of miRNAs in addition to miR399 have been shown to be responsive to Pi deficiency (Kuo and Chiou 2011). Of these miRNA families, those that are Pi responsive in multiple plant species include miR156, miR159, miR166, miR319, miR395, miR398, miR447, and miR827. It is likely that, similar to miR399, these miRNAs comprise regulatory components of Pi-signaling pathways, possibly as systemic signals. Indeed, miR169, miR827, and miR2111 were detected in the phloem of rapeseed, and their abundance was responsive to Pi deficiency, consistent with a role in modulating systemic Pi signaling (Pant et al. 2008).

Of particular note, miR827, which is highly and specifically induced in response to Pi deficiency, has been shown to target genes encoding SPX domain-containing proteins in Arabidopsis and rice (Hsieh et al. 2009; Lin et al. 2010; Lundmark et al. 2010; Pant et al. 2009). In Arabidopsis, miR827 targets an SPX protein that also contains a C-terminal RING domain and exhibits ubiquitin E3 ligase activity (Hsieh et al. 2009). This protein, NITROGEN LIMITATION ADAPTATION (NLA), also known as BENZOIC ACID HYPERSENSITIVE1 (BH1) (Yaeno and Iba 2008), was originally identified as being involved in growth responses to nitrogen starvation (Peng et al. 2007). Mutation of *NLA* suppressed N-deficiency induced anthocyanin accumulation and led to early senescence (Peng et al. 2007, 2008). Interestingly, this early senescence phenotype was shown to be linked to Pi overaccumulation, and knockout of either the *Pht1;1* Pi transporter or *PHF1*, which is required for proper processing of Pht1 proteins (Gonzalez et al. 2005), suppressed this phenotype (Kant et al. 2011). These observations demonstrated that NLA plays an important role in maintaining Pi homeostasis. Notably, rice miR827 also targets SPX-containing proteins, but not from the same subclass as NLA. These targets, OsSPX-MFS1 and OsSPX-MFS2, each contain a major facilitator superfamily (MFS) domain, which is typically involved in cellular transport (Lin et al. 2010). Both of these proteins are negatively regulated by miR827, particularly in shoots (Lin et al. 2010). Consistently, overexpression of miR827 or reduced expression of *OsSPX-MFS1* resulted in increased Pi accumulation in leaves and defective translocation of Pi from mature to young leaves (Secco et al. 2012). Interestingly, Arabidopsis miR2111 targets a root-specific E3 ligase (Pant et al. 2009). Therefore, miRNA-directed regulation of proteolytic components appears to be a common mechanism used by plants to maintain Pi homeostasis.

2.3 Sucrose

Many studies have suggested a role for sugar (particularly sucrose) signaling in modulating PSR (Hammond and White 2011). Pi deficiency both induces shoot-derived carbohydrate levels (Cakmak et al. 1994; Ciereszko et al. 1996; Foyer and Spencer 1986), and activates expression of sugar-responsive genes (Ciereszko et al. 2001; Nielsen et al. 1998; Sadka et al. 1994). Reciprocally, exogenous application of sucrose induces many PSR genes (Franco-Zorrilla et al. 2005; Karthikeyan et al. 2007; Liu et al. 2005; Muller et al. 2007). A typical response to Pi starvation is biased allocation of carbon to roots to optimize Pi foraging capacity. This is achieved largely through shoot-to-root translocation of sucrose (Cakmak et al. 1994; Hermans et al. 2006). The *pho3* mutant was identified as having attenuated responses to Pi deficiency, including decreased acid phosphatase activity, and decreased Pi accumulation (Zakhleniuk et al. 2001). The *pho3* phenotype was found to result from a mutation in *SUC2*, a sucrose transporter involved in phloem loading (Lloyd and Zakhleniuk 2004). As a result, *pho3/suc2* mutant plants cannot adequately translocate carbohydrates to roots and thus have elevated shoot carbohydrate concentrations (Gottwald et al. 2000; Lloyd and Zakhleniuk 2004). This lack of carbon allocation to roots correlates with attenuated PSR (Liu et al. 2005; Zakhleniuk et al. 2001). Recently, overexpression of SUC2 was shown to result in increased sucrose concentrations in roots and shoots and exaggerated PSR under Pi-replete conditions (Lei et al. 2011b). Transcriptomic analysis revealed that 73 % of genes induced during Pi deficiency in wild type are also induced in *SUC2*-overexpressors under Pi-replete conditions (Lei et al. 2011b). Studies with white lupin (Liu et al. 2005) and common bean (Liu et al. 2010b) have demonstrated the importance of maintaining phloem transport in mounting a full response to Pi starvation. Reduced photosynthesis, dark treatment, or stem girdling, all attenuated Pi starvation-induced gene expression in roots (Liu et al. 2005, 2010b). However, photosynthates (likely sucrose) were shown to act upstream of miR399 induction in response to Pi deficiency (Liu et al. 2010b). Therefore, it is possible that limited phloem movement of miR399, and not sucrose, results in diminished PSR gene induction in roots. In this capacity, sucrose may act more as a local signaling molecule that induces miR399 expression in shoots, rather than a systemic signaler. Nevertheless, the importance of sucrose as a systemic signal in young Arabidopsis seedlings to control root growth was recently identified (Kircher and Schopfer 2012). Also, sugar signaling has been implicated in the long-distance control of responses to nitrogen deficiency (Coruzzi and Zhou 2001; Lejay et al. 2003). It is of interest to further investigate the nutritive and signaling roles of sucrose in Pi signaling.

2.4 Auxin

Many phytohormones have been implicated in Pi-signaling pathways (Chiou and Lin 2011). The majority of studies have focused on the impact of hormone action on root growth responses to Pi deficiency. Many of these interactions may be due to local, rather than systemic, signaling pathways (Thibaud et al. 2010). Nevertheless, hormones may also act as systemic signals to modulate responses to Pi deficiency and/or as regulators of systemic signals.

A number of studies have demonstrated a link between auxin synthesis/signaling and the regulation of PSR, particularly with regard to Pi-dependent changes in root system architecture (Chiou and Lin 2011). For example, Pi deficiency was shown to induce expression of *TRANSPORT INHIBITOR RESPONSE1* (*TIR1*), which encodes an auxin receptor. TIR1 is required for proper lateral root development in response to Pi limitation (Perez-Torres et al. 2008). Although direct evidence for auxin as a systemic regulator of Pi-signaling pathways is lacking, it is known that shoot-derived auxin impacts root development (Leyser 2011). Also, auxin responses are linked with those of strigolactones and cytokinins, which may be involved in root-to-shoot Pi signaling (see below). Together these observations suggest a role for auxin in systemic Pi-signaling pathways.

3 Root-to-Shoot Signals

3.1 Pi

Prior to the initiation of shoot-derived Pi systemic signaling pathways, a low Pi signal must be transmitted from roots to shoots. However, the source of this signal and the associated mechanisms remain unclear but may involve Pi itself, cytokinins, strigolactones, or a β-carotene-related compound. Experiments probing the impact of the Pi analog, phosphite (Phi), on PSR have supported the notion that Pi itself is a signaling component. Phi is taken up by plants through Pi transporters but cannot be metabolized *in planta* (Carswell et al. 1996; Ouimette and Coffey 1990; Sukarno et al. 1993). During Pi deficiency, exogenous application of Phi attenuates a number of PSR (Carswell et al. 1996, 1997; Kobayashi et al. 2006; Ticconi et al. 2001; Varadarajan et al. 2002). This indicates that Phi is adequate to signal a Pi-sufficient state, even though it cannot serve as a nutritive source of phosphorus. However, whether Pi acts systemically to modulate Pi-signaling pathways remains unclear. Induction of the *Medicago truncatula* Pi starvation-induced gene, *Mt4*, was attenuated in the Pi-depleted portion of a split-root system but was not accompanied by an increase in endogenous Pi levels (Burleigh and Harrison 1999). This suggests that Pi does not act as a systemic signal to regulate *Mt4*. In plants with attenuated *PHO1* expression, shoot Pi levels were low, but PSR remained suppressed (Rouached et al. 2011). This suggests that PHO1 may be involved in the translocation of a signaling molecule other than Pi from roots to shoots. Recent evidence revealing PHO1 as a direct target of PHO2 highlights PHO1 as a key component of both root-to-shoot and shoot-to-root systemic signaling (Liu et al. 2012).

3.2 Cytokinin

Pi deficiency is known to cause reductions in cytokinin concentrations and sensitivity (Franco-Zorrilla et al. 2002; Kuiper et al. 1988). Reciprocally, cytokinin negatively regulates many PSR genes (Martin et al. 2000; Wang et al. 2006). Translocation of cytokinins from roots to shoots via the xylem has been demonstrated and has also been linked to the sensing of nitrogen availability (Kudo et al. 2010; Takei et al. 2001). This together with the observation that cytokinin affects expression of PSR genes, including members of the Mt4/TPSI1 family (see below), suggests a role for cytokinin in systemic Pi signaling (Martin et al. 2000). However, in split-root experiments, cytokinin-dependent repression of PSR genes was limited to the local portion of roots receiving cytokinin treatment (Franco-Zorrilla et al. 2005). It is possible that cytokinin contributes to relaying a Pi-deficiency signal from root to shoot that precedes shoot-to-root Pi-signaling pathways but plays a more prominent role in PSR gene regulation locally.

3.3 Strigolactone

Strigolactones have been implicated in both local and systemic Pi-signaling pathways. In rice and Arabidopsis, Pi deficiency led to increased strigolactone levels in roots and suppression of tiller or lateral bud outgrowth in wild-type shoots, but not in those of mutants deficient in strigolactone biosynthesis or signaling (Kohlen et al. 2011; Umehara et al. 2010). Also, a decrease in root strigolactone levels preceded the induction of tiller bud outgrowth in Pi-starved wild-type rice seedlings after transfer to Pi-replete media (Umehara et al. 2010). Moreover, direct analysis of xylem sap in Arabidopsis plants identified strigolactones induced in roots in response to Pi deficiency that were subsequently transported to shoots (Kohlen et al. 2011). Together these results suggest that strigolactones act as a systemic root-to-shoot signal that modulates Pi-signaling pathways. Interestingly, a recent study revealed that a number of systemically controlled PSR genes are misregulated in an Arabidopsis strigolactone-signaling mutant (*max2-1*) in response to Pi deficiency (Mayzlish-Gati et al. 2012). Notably, *PHO2* transcript levels were 125-fold higher in *max2-1* than in wild type. This may indicate that loss of the MAX2 strigolactone-signaling component interferes with induction of miR399 in shoots and subsequent transport of mature miR399 to roots.

Strigolactones also promote arbuscular mycorrhizal (AM) associations. These symbioses result in enhanced Pi acquisition for the plant and enhanced carbohydrate acquisition for the AM fungus (Smith and Read 2008). Ironically, a lot of what is known regarding Pi-signaling pathways in plants was determined through work on Arabidopsis, which does not associate with AM fungi. Nevertheless, it has become apparent that the central miR399/PHO2 circuit is conserved in AM plants and, more strikingly, is involved in regulating AM symbioses. Higher levels of low Pi-induced miR399 primary transcripts were observed in leaves of mycorrhizal *Medicago truncatula* plants as compared to non-mycorrhizal plants (Branscheid et al. 2010). Also, consistent with phloem translocation from shoots to roots, mature miR399 levels were higher in roots of mycorrhizal plants relative to non-mycorrhizal plants. Further, the elevated miR399 levels in roots of these plants correlated with low *PHO2* transcript levels despite an increase in Pi acquisition by colonized roots. Together these results support a model in which a systemic signal from mycorrhizal roots travels to the shoot where it induces miR399 expression. This would lead to increased accumulation of mature miR399 levels in roots via phloem that would suppress PHO2. It is possible that PHO2-dependent, high Pi signals are involved in attenuating AM symbiosis in Pi-replete plants. Therefore, suppressing PHO2 activity via increased miR399 levels may be important for sustaining AM symbiosis. It is tempting to speculate that strigolactones serve as the associated systemic root-to-shoot signal. This warrants further investigation.

3.4 BYPASS1-Related Carotenoid

Characterization of the Arabidopsis *bypass1* (*bps1*) mutant has indicated the presence of a novel carotenoid-derived, systemic signal in roots that regulates leaf growth. Grafting experiments showed that a *bps1* rootstock was sufficient to induce the premature arrest of leaf growth phenotype in a wild-type scion, whereas a wild-type rootstock restored the wild-type phenotype in a *bps1* scion (Van Norman et al. 2004). *BPS1* encodes a plant-specific protein of unknown function, and the nature of the BPS1-associated signal has yet to be determined. Growth of *bps1* plants in the presence of the carotenoid biosynthesis inhibitor fluridone (Bartels and Watson 1978) resulted in restoration of the wild-type phenotype, consistent with the signal being carotenoid-derived (Van Norman et al. 2004). Importantly, abscisic acid (ABA) and strigolactones are carotenoid-derived hormones (Raghavendra et al. 2010; Rameau 2010), but experiments with double mutants between *bps1* and ABA or strigolactone biosynthesis mutants indicated that the BPS1 signal is not dependent on these two hormones (Van Norman and Sieburth 2007). Although the BPS1 signal has not been implicated in Pi signaling, another phenotype of *bps1* plants is leaf necrosis, which is a known phenotype of Pi overaccumulation (Delhaize and Randall 1995). It is of interest to probe the potential disruption of PSR in *bps1*.

4 Regulators of Systemic Pi Signals

4.1 *At4/IPS1 Noncoding RNAs*

More than a decade ago, two homologous, highly Pi starvation-inducible genes, coined *Medicago truncatula 4* (*Mt4*) and *tomato phosphate starvation-induced gene* (*TPSI1*), were independently identified (Burleigh and Harrison 1997; Liu et al. 1997). These genes shared only a short region of homology and neither appeared to encode a long open reading frame. Orthologs in Arabidopsis were later identified and termed *At4* and *induced by Pi starvation 1* (*IPS1*) (Burleigh and Harrison 1999; Martin et al. 2000). Hence, Mt4/TPSI1 and At4/IPS1 are recognized family names for this group of genes.

It is now clear that At4/IPS1 genes play an important role in regulating systemic Pi signaling. During Pi deficiency, mutation of *At4* resulted in decreased Pi translocation to roots and Pi overaccumulation in shoots, whereas overexpression of *IPS1* led to decreased shoot Pi levels and higher *PHO2* transcript abundance (Franco-Zorrilla et al. 2007; Shin et al. 2006). Interestingly, the short region of homology among the At4/IPS1 family members is complementary to miR399, except for two to three bases in the middle of the transcript (Franco-Zorrilla et al. 2007; Shin et al. 2006). Notably, these mismatches lie in the region necessary for miRNA-guided cleavage of target transcripts. As a result, At4/IPS1 transcripts bind miR399 but are not degraded. In contrast, a modified *IPS1* transcript completely complimentary to miR399 was shown to be degraded (Franco-Zorrilla et al. 2007). This novel regulatory paradigm, in which At4/IPS1 transcripts imperfectly base pair with miR399 to dampen its impact on *PHO2* mRNA, was coined target mimicry (Franco-Zorrilla et al. 2007). This mechanism provides an important layer of regulation that fine-tunes plant responses to Pi deficiency.

4.2 *PHR1*

Sensing of low external Pi concentrations results in major transcriptional changes in plant tissues (Nilsson et al. 2010). A number of transcription factors involved in modulating PSR have been characterized (Jain et al. 2012). The first identified and most studied plant transcription factor is the Arabidopsis MYB protein PHOSPHATE STARVATION RESPONSE 1 (PHR1) (Rubio et al. 2001). PHR1 and its homolog PHR1-like1 (PHL1) have been shown to have a global impact on mediating PSR (Bustos et al. 2010). Upon perception of a low-Pi signal from roots to shoots, PHR1/PHL1 activate the expression of numerous downstream loci via the PHR1-binding site (P1BS) *cis*-regulatory motif, including miR399, miR827, and At4/IPS1genes (Bari et al. 2006; Bustos et al. 2010; Nilsson et al. 2007; Rubio et al. 2001). Hence, PHR1/PHL1 are key activators of systemic Pi signaling pathways. As yet, the molecular mechanism that triggers PHR1 activation is unknown. Neither *PHR1* nor its rice ortholog, *OsPHR2*, exhibit significant low-Pi induction at the transcript level (Nilsson et al. 2007; Rubio et al. 2001; Zhou et al. 2008). PHR1 has been shown to be a target of SUMOylation, consistent with it being regulated post-translationally (Miura et al. 2005).

4.3 Other Transcriptional Regulators

In addition to PHR1, a number of transcriptional regulators of PSR have been characterized, and these have been the focus of recent reviews (Jain et al. 2012; Nilsson et al. 2010). Herein, the regulators with the most potential to modulate systemic Pi-signaling pathways will be summarized. Four classes of transcriptional regulators have been implicated in systemic Pi signaling: class I SPX domain-containing proteins, as well as MYB, WRKY, and bHLH transcription factors. Both the rice SPX1 and Arabidopsis SPX3 proteins have been shown to negatively regulate responses to Pi deficiency, including expression of PSR genes. Both are induced by PHR1 family proteins and antagonize PHR1-dependent induction of PSR genes, including At4/IPS1 genes and miR399 (Duan et al. 2008; Liu et al. 2010a). In addition to PHR1, other MYB proteins have been shown to regulate At4/IPS1 family genes and/or miR399. Overexpression of Arabidopsis *MYB62* suppressed *IPS1* induction, despite lower shoot Pi levels (Devaiah et al. 2009). In contrast, under both Pi-deplete and replete conditions, overexpression of the rice MYB protein, *OsMYB2-1*, increased *IPS1* and miR399 transcript abundance, whereas RNAi knockdown of *OsMYB2-1* decreased expression of these genes (Dai et al. 2012). WRKY proteins comprise a large class of stress-responsive transcription factors, and some members of this family regulate Pi-signaling pathways. WRKY6 has been shown to negatively regulate *PHO1* expression in a Pi-dependent manner (Chen et al. 2009). Further, WRKY6 and WRKY42 were both shown to bind WRKY box *cis*-elements present in the *PHO1* promoter (Chen et al. 2009). In contrast to WRKY6 (and possibly WRKY42), WRKY75 appears to be a positive regulator of PSR in light of the observation that RNAi knockdown of *WRKY75* decreased *At4* and *IPS1* transcript abundance (Devaiah et al. 2007). Finally, the rice bHLH transcription factor, OsPTF1 plays a role in the regulation of many Pi-related phenotypes. OsPTF1 appears to regulate PSR genes via *cis*-elements distinct from the P1BS element (Yi et al. 2005). Hence, this bHLH transcription factor may impact miR399-independent, systemic signaling pathways.

4.4 Calcium

Calcium ion (Ca^{2+}) is a secondary messenger involved in multiple cellular responses (Kudla et al. 2010). Recent evidence suggests a role for Ca^{2+} in regulating systemic Pi signaling. CAX1 and CAX3 functionally overlap as vacuolar

Ca^{2+}/H^+ antiporters that maintain proper subcellular compartmentalization of Ca^{2+} (Cheng et al. 2003, 2005). Recent work has shown that mutation of both *CAX1* and *CAX3* causes increased transcript abundance of several PSR genes in shoots (Liu et al. 2011). Also, the *cax1cax3* double mutant overaccumulates Pi in shoots (Liu et al. 2011). Interestingly, grafting experiments with *cax1cax3* scion and wild-type rootstock demonstrated that a signal from *cax1cax3* shoots led to increased Pi acquisition in roots, possibly through upregulation of the Pht1;1 Pi transporter (Liu et al. 2011). Notably, miR399 levels were unaffected in *cax1cax3*, indicating that the systemic effect from mutation of *CAX1* and *CAX3* was independent of the miR399-PHO2 pathway.

4.5 Hormones

As mentioned above, auxin, cytokinin, and strigolactone may act as Pi-related systemic signals and/or regulators. In addition to these hormones, ethylene and gibberellin (GA) also appear to regulate systemic Pi-signaling pathways. Pi starvation induces ethylene biosynthesis and sensitivity in shoots and roots, at least partially through the PHR1 transcriptional activator (Nagarajan and Smith 2012). Recent studies on several *hypersensitive to Pi starvation* (*hps*) mutants further support a role for ethylene in regulating PSR (Lei et al. 2011a; Wang et al. 2012a, b; Yu et al. 2012). These mutants, which exhibit mis-regulated expression of a number of PSR genes, including miR399 and At4/IPS1 genes, were shown to be allelic to components associated with ethylene biosynthesis or signaling, namely, CTR1 (Kieber et al. 1993), ETO1 (Guzman and Ecker 1990), and SABRE (Aeschbacher et al. 1995). In contrast to ethylene, GA levels are decreased by Pi deficiency (Jiang et al. 2007). This appears to result, at least in part, by repression of GA biosynthetic genes by the MYB62 transcription factor (Devaiah et al. 2009). Because *MYB62*-overexpression also suppresses many PSR genes and affects root system architecture, Pi acquisition, and acid phosphatase activity, it is possible that GA is involved in MYB62-dependent regulation of PSR.

Plants treated with ABA exhibit some phenotypes similar to those of Pi-deficient plants, including altered root-to-shoot biomass ratios and root-hair proliferation (Trull et al. 1997). However, examination of ABA-deficient and insensitive mutants revealed an essentially normal response to Pi starvation, suggesting ABA does not play a major role in Pi-signaling pathways (Trull et al. 1997). Nonetheless, miR399 induction was detected in ABA-treated plants (Sunkar and Zhu 2004), and some At4/IPS1 genes were shown to be repressed by ABA treatment (Shin et al. 2006). These observations suggest modulation of miR399-related systemic Pi-signaling pathways by ABA. In a recent Arabidopsis transcriptome analysis, PSR genes known to be responsive to other factors shared the most overlap with ABA as compared to other hormones (Woo et al. 2012). This may indicate that ABA preferentially regulates Pi-responsive genes that are not highly specific to Pi deficiency. Recently, Pi and PHO1 were shown to be necessary for stomatal responses to ABA, establishing a potentially novel link between Pi signaling and ABA (Zimmerli et al. 2012).

5 Perspectives

Phosphorus is an essential nutrient required for many cellular components and proper control of enzymatic reactions and signal transduction. Accordingly, the maintenance of optimal *in planta* Pi levels is essential for plant growth and development. Because plants acquire Pi from soils, most of which are Pi-deficient, complex regulatory networks have evolved to control Pi acquisition and distribution. Particularly elegant are systemic Pi-signaling pathways that facilitate the long-distance communication needed to integrate Pi sensing with plant development and other environmental signals.

Future work on systemic Pi signaling should continue to focus on uncovering the actual long-distance signals, with particular attention paid to miRNAs, which are emerging as crucial components of Pi signaling. Also, it is of interest to determine the extent to which Pi-signaling pathways are conserved among species. Finally, current and forthcoming "systems" technologies should be harnessed for uncovering points of cross talk among Pi and other signaling pathways, such as those for sugars, hormones, and other nutrients. An obvious future goal of this field of work is to improve the phosphorus use-efficiency of crop plants, particularly in the face of the imminent disappearance of readily available phosphate fertilizer sources.

References

Aeschbacher RA, Hauser MT, Feldmann KA, Benfey PN (1995) The SABRE gene is required for normal cell expansion in Arabidopsis. Gene Dev 9:330–340

Allen E, Xie ZX, Gustafson AM, Carrington JC (2005) microRNA-directed phasing during trans-acting siRNA biogenesis in plants. Cell 121:207–221

Aung K, Lin SI, Wu CC, Huang YT, Su CL, Chiou TJ (2006) pho2, a phosphate overaccumulator, is caused by a nonsense mutation in a microRNA399 target gene. Plant Physiol 141:1000–1011

Bari R, Pant BD, Stitt M, Scheible WR (2006) PHO2, microRNA399, and PHR1 define a phosphate-signaling pathway in plants. Plant Physiol 141:988–999

Bartels PG, Watson CW (1978) Inhibition of carotenoid synthesis by fluridone and norflurazon. Weed Sci 26:198–203

Branscheid A, Sieh D, Pant BD, May P, Devers EA, Elkrog A, Schauser L, Scheible WR, Krajinski F (2010) Expression pattern suggests a role of miR399 in the regulation of the cellular response to local Pi increase during arbuscular mycorrhizal symbiosis. Mol Plant-Microbe Interact 23:915–926

Buhtz A, Springer F, Chappell L, Baulcombe DC, Kehr J (2008) Identification and characterization of small RNAs from the phloem of *Brassica napus*. Plant J 53:739–749

Burleigh SH, Harrison MJ (1997) A novel gene whose expression in *Medicago truncatula* roots is suppressed in response to colonization by vesicular-arbuscular mycorrhizal (VAM) fungi and to phosphate nutrition. Plant Mol Biol 34:199–208

Burleigh SH, Harrison MJ (1999) The down-regulation of *Mt4*-like genes by phosphate fertilization occurs systemically and involves phosphate translocation to the shoots. Plant Physiol 119:241–248

Bustos R, Castrillo G, Linhares F, Puga MI, Rubio V, Perez-Perez J, Solano R, Leyva A, Paz-Ares J (2010) A central regulatory system largely controls transcriptional activation and repression responses to phosphate starvation in Arabidopsis. PLoS Genet 6(9):e1001102

Cakmak I, Hengeler C, Marschner H (1994) Changes in phloem export of sucrose in leaves in response to phosphorus, potassium and magnesium deficiency in bean plants. J Exp Bot 45:1251–1257

Carswell C, Grant BR, Theodorou ME, Harris L, Niere JO, Plaxton WC (1996) The fungicide phosphonate disrupts the phosphate-starvation response in *Brassica nigra* seedlings. Plant Physiol 110:105–110

Carswell MC, Grant BR, Plaxton WC (1997) Disruption of the phosphate-starvation response of oilseed rape suspension cells by the fungicide phosphonate. Planta 203:67–74

Chen YF, Li LQ, Xu Q, Kong YH, Wang H, Wu WH (2009) The WRKY6 transcription factor modulates PHOSPHATE1 expression in response to low Pi stress in Arabidopsis. Plant Cell 21:3554–3566

Cheng NH, Pittman JK, Barkla BJ, Shigaki T, Hirschi KD (2003) The Arabidopsis *cax1* mutant exhibits impaired ion homeostasis, development, and hormonal responses and reveals interplay among vacuolar transporters. Plant Cell 15:347–364

Cheng NH, Pittman JK, Shigaki T, Lachmansingh J, LeClere S, Lahner B, Salt DE, Hirschi KD (2005) Functional association of Arabidopsis CAX1 and CAX3 is required for normal growth and ion homeostasis. Plant Physiol 138:2048–2060

Chiou TJ, Lin SI (2011) Signaling network in sensing phosphate availability in plants. Annu Rev Plant Biol 62:185–206

Chiou TJ, Aung K, Lin SI, Wu CC, Chiang SF, Su CL (2006) Regulation of phosphate homeostasis by microRNA in Arabidopsis. Plant Cell 18:412–421

Ciereszko I, Gniazdowska A, Mikulska M, Rychter AM (1996) Assimilate translocation in bean plants (*Phaseolus vulgaris* L) during phosphate deficiency. J Plant Physiol 149:343–348

Ciereszko I, Johansson H, Hurry V, Kleczkowski LA (2001) Phosphate status affects the gene expression, protein content and enzymatic activity of UDP-glucose pyrophosphorylase in wild-type and pho mutants of Arabidopsis. Planta 212:598–605

Coruzzi GM, Zhou L (2001) Carbon and nitrogen sensing and signaling in plants: emerging 'matrix effects'. Curr Opin Plant Biol 4:247–253

Dai XY, Wang YY, Yang A, Zhang WH (2012) OsMYB2P-1, an R2R3 MYB transcription factor, is involved in the regulation of phosphate-starvation responses and root architecture in rice. Plant Physiol 159:169–183

Delhaize E, Randall PJ (1995) Characterization of a phosphate-accumulator mutant of *Arabidopsis thaliana*. Plant Physiol 107:207–213

Devaiah BN, Karthikeyan AS, Raghothama KG (2007) WRKY75 transcription factor is a modulator of phosphate acquisition and root development in Arabidopsis. Plant Physiol 143:1789–1801

Devaiah BN, Madhuvanthi R, Karthikeyan AS, Raghothama KG (2009) Phosphate starvation responses and gibberellic acid biosynthesis are regulated by the MYB62 transcription factor in Arabidopsis. Mol Plant 2:43–58

Duan K, Yi K, Dang L, Huang H, Wu W, Wu P (2008) Characterization of a sub-family of Arabidopsis genes with the SPX domain reveals their diverse functions in plant tolerance to phosphorus starvation. Plant J 54:965–975

Foyer C, Spencer C (1986) The relationship between phosphate status and photosynthesis in leaves – effects on intracellular ortho-phosphate distribution, photosynthesis and assimilate partitioning. Planta 167:369–375

Franco-Zorrilla JM, Martin AC, Solano R, Rubio V, Leyva A, Paz-Ares J (2002) Mutations at CRE1 impair cytokinin-induced repression of phosphate starvation responses in Arabidopsis. Plant J 32:353–360

Franco-Zorrilla JM, Martin AC, Leyva A, Paz-Ares J (2005) Interaction between phosphate-starvation, sugar, and cytokinin signaling in Arabidopsis and the roles of cytokinin receptors CRE1/AHK4 and AHK3. Plant Physiol 138:847–857

Franco-Zorrilla JM, Valli A, Todesco M, Mateos I, Puga MI, Rubio-Somoza I, Leyva A, Weigel D, Garcia JA, Paz-Ares J (2007) Target mimicry provides a new mechanism for regulation of microRNA activity. Nat Genet 39:1033–1037

Fujii H, Chiou TJ, Lin SI, Aung K, Zhu JK (2005) A miRNA involved in phosphate-starvation response in Arabidopsis. Curr Biol 15:2038–2043

Gao N, Su YH, Min J, Shen WS, Shi WM (2010) Transgenic tomato overexpressing ath-miR399d has enhanced phosphorus accumulation through increased acid phosphatase and proton secretion as well as phosphate transporters. Plant Soil 334:123–136

Gonzalez E, Solano R, Rubio V, Leyva A, Paz-Ares J (2005) PHOSPHATE TRANSPORTER TRAFFIC FACILITATOR1 is a plant-specific SEC12-related protein that enables the endoplasmic reticulum exit of a high-affinity phosphate transporter in Arabidopsis. Plant Cell 17:3500–3512

Gottwald JR, Krysan PJ, Young JC, Evert RF, Sussman MR (2000) Genetic evidence for the in planta role of phloem-specific plasma membrane sucrose transporters. Proc Natl Acad Sci USA 97:13979–13984

Guzman P, Ecker JR (1990) Exploiting the triple response of Arabidopsis to identify ethylene-related mutants. Plant Cell 2:513–523

Hamburger D, Rezzonico E, Petetot JMC, Somerville C, Poirier Y (2002) Identification and characterization of the Arabidopsis PHO1 gene involved in phosphate loading to the xylem. Plant Cell 14:889–902

Hammond JP, White PJ (2011) Sugar signaling in root responses to low phosphorus availability. Plant Physiol 156:1033–1040

Hermans C, Hammond JP, White PJ, Verbruggen N (2006) How do plants respond to nutrient shortage by biomass allocation? Trends Plant Sci 11:610–617

Hsieh LC, Lin SI, Shih ACC, Chen JW, Lin WY, Tseng CY, Li WH, Chiou TJ (2009) Uncovering small RNA-mediated responses to phosphate deficiency in Arabidopsis by deep sequencing. Plant Physiol 151:2120–2132

Hu B, Zhu CG, Li F, Tang JY, Wang YQ, Lin AH, Liu LC, Che RH, Chu CC (2011) LEAF TIP NECROSIS1 plays a pivotal role in the regulation of multiple phosphate starvation responses in rice. Plant Physiol 156:1101–1115

Jain A, Nagarajan VK, Raghothama KG (2012) Transcriptional regulation of phosphate acquisition by higher plants. Cell Mol Life Sci 69:3207–3224

Jiang CF, Gao XH, Liao L, Harberd NP, Fu XD (2007) Phosphate starvation root architecture and anthocyanin accumulation responses are modulated by the gibberellin-DELLA signaling pathway in Arabidopsis. Plant Physiol 145:1460–1470

Kant S, Peng MS, Rothstein SJ (2011) Genetic regulation by NLA and microRNA827 for maintaining nitrate-dependent phosphate homeostasis in Arabidopsis. Plos Genet 7(3): e1002021

Karthikeyan AS, Varadarajan DK, Jain A, Held MA, Carpita NC, Raghothama KG (2007) Phosphate starvation responses are mediated by sugar signaling in Arabidopsis. Planta 225:907–918

Kieber JJ, Rothenberg M, Roman G, Feldmann KA, Ecker JR (1993) CTR1, a negative regulator of the ethylene response pathway in Arabidopsis, encodes a member of the raf family of protein kinases. Cell 72:427–441

Kircher S, Schopfer P (2012) Photosynthetic sucrose acts as cotyledon-derived long-distance signal to control root growth during early seedling development in Arabidopsis. Proc Natl Acad Sci USA 109:11217–11221

Kobayashi K, Masuda T, Takamiya KI, Ohta H (2006) Membrane lipid alteration during phosphate starvation is regulated by phosphate signaling and auxin/cytokinin cross-talk. Plant J 47:238–248

Kohlen W, Charnikhova T, Liu Q, Bours R, Domagalska MA, Beguerie S, Verstappen F, Leyser O, Bouwmeester H, Ruyter-Spira C (2011) Strigolactones are transported through the xylem and play a key role in shoot architectural response to phosphate deficiency in nonarbuscular mycorrhizal host Arabidopsis. Plant Physiol 155:974–987

Kudla J, Batistic O, Hashimoto K (2010) Calcium signals: the lead currency of plant information processing. Plant Cell 22:541–563

Kudo T, Kiba T, Sakakibara H (2010) Metabolism and long-distance translocation of cytokinins. J Integr Plant Biol 52:53–60

Kuiper D, Schuit J, Kuiper PJC (1988) Effects of internal and external cytokinin concentrations on root growth and shoot to root ratio of *Plantago major* ssp *Pleiosperma* at different nutrient conditions. Plant Soil 111:231–236

Kuo HF, Chiou TJ (2011) The role of microRNAs in phosphorus deficiency signaling. Plant Physiol 156:1016–1024

Lei M, Zhu C, Liu Y, Karthikeyan AS, Bressan RA, Raghothama KG, Liu D (2011a) Ethylene signalling is involved in regulation of phosphate starvation-induced gene expression and production of acid phosphatases and anthocyanin in Arabidopsis. New Phytol 189:1084–1095

Lei MG, Liu YD, Zhang BC, Zhao YT, Wang XJ, Zhou YH, Raghothama KG, Liu D (2011b) Genetic and genomic evidence that sucrose is a global regulator of plant responses to phosphate starvation in Arabidopsis. Plant Physiol 156:1116–1130

Lejay L, Gansel X, Cerezo M, Tillard P, Muller C, Krapp A, von Wiren N, Daniel-Vedele F, Gojon A (2003) Regulation of root ion transporters by photosynthesis: functional importance and relation with hexokinase. Plant Cell 15:2218–2232

Lenburg ME, O'Shea EK (1996) Signaling phosphate starvation. Trends Biochem Sci 21:383–387

Leyser O (2011) Auxin, self-organisation, and the colonial nature of plants. Curr Biol 21: R331–R337

Lin SI, Chiang SF, Lin WY, Chen JW, Tseng CY, Wu PC, Chiou TJ (2008) Regulatory network of microRNA399 and PHO2 by systemic signaling. Plant Physiol 147:732–746

Lin SI, Santi C, Jobet E, Lacut E, El Kholti N, Karlowski WM, Verdeil JL, Breitler JC, Perin C, Ko SS, Guiderdoni E, Chiou TJ, Echeverria M (2010) Complex regulation of two target genes encoding SPX-MFS proteins by rice miR827 in response to phosphate starvation. Plant Cell Physiol 51:2119–2131

Liu CM, Muchhal US, Raghothama KG (1997) Differential expression of TPS11, a phosphate starvation-induced gene in tomato. Plant Mol Biol 33:867–874

Liu C, Muchhal US, Uthappa M, Kononowicz AK, Raghothama KG (1998) Tomato phosphate transporter genes are differentially regulated in plant tissues by phosphorus. Plant Physiol 116:91–99

Liu JQ, Samac DA, Bucciarelli B, Allan DL, Vance CP (2005) Signaling of phosphorus deficiency-induced gene expression in white lupin requires sugar and phloem transport. Plant J 41:257–268

Liu F, Wang ZY, Ren HY, Shen CJ, Li Y, Ling HQ, Wu CY, Lian XM, Wu P (2010a) OsSPX1 suppresses the function of OsPHR2 in the regulation of expression of OsPT2 and phosphate homeostasis in shoots of rice. Plant J 62:508–517

Liu JQ, Allan DL, Vance CP (2010b) Systemic signaling and local sensing of phosphate in common bean: cross-talk between photosynthate and microRNA399. Mol Plant 3:428–437

Liu TY, Aung K, Tseng CY, Chang TY, Chen YS, Chiou TJ (2011) Vacuolar Ca2+/H+ transport activity is required for systemic phosphate homeostasis involving shoot-to-root signaling in Arabidopsis. Plant Physiol 156:1176–1189

Liu TY, Huang TK, Tseng CY, Lai YS, Lin SI, Lin WY, Chen JW, Chiou TJ (2012) PHO2-dependent degradation of PHO1 modulates phosphate homeostasis in Arabidopsis. Plant Cell 24:2168–2183

Lloyd JC, Zakhleniuk OV (2004) Responses of primary and secondary metabolism to sugar accumulation revealed by microarray expression analysis of the Arabidopsis mutant, pho3. J Exp Bot 55:1221–1230

Lundmark M, Korner CJ, Nielsen TH (2010) Global analysis of microRNA in Arabidopsis in response to phosphate starvation as studied by locked nucleic acid-based microarrays. Physiol Plant 140:57–68

Lynch JP (2011) Root phenes for enhanced soil exploration and phosphorus acquisition: tools for future crops. Plant Physiol 156:1041–1049

Marschner H (1995) Mineral nutrition of higher plants, 2nd edn. Academic, Boston, MA

Martin AC, del Pozo JC, Iglesias J, Rubio V, Solano R, de la Pena A, Leyva A, Paz-Ares J (2000) Influence of cytokinins on the expression of phosphate starvation responsive genes in Arabidopsis. Plant J 24:559–567

Mayzlish-Gati E, De Cuyper C, Goormachtig S, Beeckman T, Vuylsteke M, Brewer P, Beveridge C, Yermiyahu U, Kaplan Y, Enzer Y, Wininger S, Resnick N, Cohen M, Kapulnik Y, Koltai H (2012) Strigolactones are involved in root response to low phosphate conditions in Arabidopsis. Plant Physiol 160(3):1329–1341

Miura K, Rus A, Sharkhuu A, Yokoi S, Karthikeyan AS, Raghothama KG, Baek D, Koo YD, Jin JB, Bressan RA, Yun DJ, Hasegawa PM (2005) The Arabidopsis SUMO E3 ligase SIZ1 controls phosphate deficiency responses. Proc Natl Acad Sci USA 102:7760–7765

Muller R, Morant M, Jarmer H, Nilsson L, Nielsen TH (2007) Genome-wide analysis of the Arabidopsis leaf transcriptome reveals interaction of phosphate and sugar metabolism. Plant Physiol 143:156–171

Nagarajan VK, Smith AP (2012) Ethylene's role in phosphate starvation signaling: more than just a root growth regulator. Plant Cell Physiol 53:277–286

Nielsen TH, Krapp A, Roper-Schwarz U, Stitt M (1998) The sugar-mediated regulation of genes encoding the small subunit of Rubisco and the regulatory subunit of ADP glucose pyrophosphorylase is modified by phosphate and nitrogen. Plant Cell Environ 21:443–454

Nilsson L, Muller R, Nielsen TH (2007) Increased expression of the MYB-related transcription factor, PHR1, leads to enhanced phosphate uptake in *Arabidopsis thaliana*. Plant Cell Environ 30:1499–1512

Nilsson L, Muller R, Nielsen TH (2010) Dissecting the plant transcriptome and the regulatory responses to phosphate deprivation. Physiol Plant 139:129–143

Ouimette DG, Coffey MD (1990) Symplastic entry and phloem translocation of phosphonate. Pestic Biochem Physiol 38:18–25

Pant BD, Buhtz A, Kehr J, Scheible WR (2008) MicroRNA399 is a long-distance signal for the regulation of plant phosphate homeostasis. Plant J 53:731–738

Pant BD, Musialak-Lange M, Nuc P, May P, Buhtz A, Kehr J, Walther D, Scheible WR (2009) Identification of nutrient-responsive Arabidopsis and rapeseed microRNAs by comprehensive real-time polymerase chain reaction profiling and small RNA sequencing. Plant Physiol 150:1541–1555

Peng MS, Hannam C, Gu HL, Bi YM, Rothstein SJ (2007) A mutation in NLA, which encodes a RING-type ubiquitin ligase, disrupts the adaptability of Arabidopsis to nitrogen limitation. Plant J 50:320–337

Peng MS, Hudson D, Schofield A, Tsao R, Yang R, Gu HL, Bi YM, Rothstein SJ (2008) Adaptation of Arabidopsis to nitrogen limitation involves induction of anthocyanin synthesis which is controlled by the NLA gene. J Exp Bot 59:2933–2944

Peret B, Clement M, Nussaume L, Desnos T (2011) Root developmental adaptation to phosphate starvation: better safe than sorry. Trends Plant Sci 16:442–450

Perez-Torres CA, Lopez-Bucio J, Cruz-Ramirez A, Ibarra-Laclette E, Dharmasiri S, Estelle M, Herrera-Estrella L (2008) Phosphate availability alters lateral root development in Arabidopsis by modulating auxin sensitivity via a mechanism involving the TIR1 auxin receptor. Plant Cell 20:3258–3272

Poirier Y, Thoma S, Somerville C, Schiefelbein J (1991) A mutant of Arabidopsis deficient in xylem loading of phosphate. Plant Physiol 97:1087–1093

Raghavendra AS, Gonugunta VK, Christmann A, Grill E (2010) ABA perception and signalling. Trends Plant Sci 15:395–401

Raghothama KG (1999) Phosphate acquisition. Annu Rev Plant Physiol Plant Mol Biol 50:665–693

Rameau C (2010) Strigolactones, a novel class of plant hormone controlling shoot branching. Curr Biol 333:344–349

Rouached H, Stefanovic A, Secco D, Arpat AB, Gout E, Bligny R, Poirier Y (2011) Uncoupling phosphate deficiency from its major effects on growth and transcriptome via PHO1 expression in Arabidopsis. Plant J 65:557–570

Rubio V, Linhares F, Solano R, Martin AC, Iglesias J, Leyva A, Paz-Ares J (2001) A conserved MYB transcription factor involved in phosphate starvation signaling both in vascular plants and in unicellular algae. Genes Dev 15:2122–2133

Sadka A, Dewald DB, May GD, Park WD, Mullet JE (1994) Phosphate modulates transcription of soybean Vspb and other sugar-inducible genes. Plant Cell 6:737–749

Secco D, Wang C, Arpat BA, Wang ZY, Poirier Y, Tyerman SD, Wu P, Shou HX, Whelan J (2012) The emerging importance of the SPX domain-containing proteins in phosphate homeostasis. New Phytol 193:842–851

Shin H, Shin HS, Chen R, Harrison MJ (2006) Loss of At4 function impacts phosphate distribution between the roots and the shoots during phosphate starvation. Plant J 45:712–726

Smith SE, Read DJ (2008) Mycorrhizal symbiosis. Academic, New York

Sukarno N, Smith SE, Scott ES (1993) The effect of fungicides on vesicular-arbuscular mycorrhizal symbiosis. 1. The effects on vesicular-arbuscular mycorrhizal fungi and plant-growth. New Phytol 125:139–147

Sunkar R, Zhu JK (2004) Novel and stress-regulated microRNAs and other small RNAs from Arabidopsis. Plant Cell 16:2001–2019

Takei K, Sakakibara H, Taniguchi M, Sugiyama T (2001) Nitrogen-dependent accumulation of cytokinins in root and the translocation to leaf: implication of cytokinin species that induces gene expression of maize response regulator. Plant Cell Physiol 42:85–93

Thibaud MC, Arrighi JF, Bayle V, Chiarenza S, Creff A, Bustos R, Paz-Ares J, Poirier Y, Nussaume L (2010) Dissection of local and systemic transcriptional responses to phosphate starvation in Arabidopsis. Plant J 64:775–789

Ticconi CA, Delatorre CA, Abel S (2001) Attenuation of phosphate starvation responses by phosphite in Arabidopsis. Plant Physiol 127:963–972

Trull MC, Guiltinan MJ, Lynch JP, Deikman J (1997) The responses of wild-type and ABA mutant *Arabidopsis thaliana* plants to phosphorus starvation. Plant Cell Environ 20:85–92

Umehara M, Hanada A, Magome H, Takeda-Kamiya N, Yamaguchi S (2010) Contribution of strigolactones to the inhibition of tiller bud outgrowth under phosphate deficiency in rice. Plant Cell Physiol 51:1118–1126

Van Norman JM, Sieburth LE (2007) Dissecting the biosynthetic pathway for the bypass1 root-derived signal. Plant J 49:619–628

Van Norman JM, Frederick RL, Sieburth LE (2004) BYPASS1 negatively regulates a root-derived signal that controls plant architecture. Curr Biol 14:1739–1746

Vance CP, Uhde-Stone C, Allan DL (2003) Phosphorus acquisition and use: critical adaptations by plants for securing a nonrenewable resource. New Phytol 157:423–447

Varadarajan DK, Karthikeyan AS, Matilda PD, Raghothama KG (2002) Phosphite, an analog of phosphate, suppresses the coordinated expression of genes under phosphate starvation. Plant Physiol 129:1232–1240

Wang Y, Ribot C, Rezzonico E, Poirier Y (2004) Structure and expression profile of the Arabidopsis PHO1 gene family indicates a broad role in inorganic phosphate homeostasis. Plant Physiol 135:400–411

Wang XM, Yi KK, Tao Y, Wang F, Wu ZC, Jiang D, Chen X, Zhu LH, Wu P (2006) Cytokinin represses phosphate-starvation response through increasing of intracellular phosphate level. Plant Cell Environ 29:1924–1935

Wang C, Huang W, Ying Y, Li S, Secco D, Tyerman S, Whelan J, Shou H (2012a) Functional characterization of the rice SPX-MFS family reveals a key role of OsSPX-MFS1 in controlling phosphate homeostasis in leaves. New Phytol 196:139–148

Wang LS, Dong JS, Gao ZY, Liu D (2012b) The Arabidopsis gene HYPERSENSITIVE TO PHOSPHATE STARVATION 3 encodes ETHYLENE OVERPRODUCTION 1. Plant Cell Physiol 53:1093–1105

Woo J, MacPherson CR, Liu J, Wang H, Kiba T, Hannah M, Wang XJ, Bajic VB, Chua NH (2012) The response and recovery of the *Arabidopsis thaliana* transcriptome to phosphate starvation. BMC Plant Biol 12:62

Yaeno T, Iba K (2008) BAH1/NLA, a RING-type ubiquitin E3 ligase, regulates the accumulation of salicylic acid and immune responses to *Pseudomonas syringae* DC3000. Plant Physiol 148:1032–1104

Yang XJ, Finnegan PM (2010) Regulation of phosphate starvation responses in higher plants. Ann Bot 105:513–526

Yi K, Wu Z, Zhou J, Du L, Guo L, Wu Y, Wu P (2005) OsPTF1, a novel transcription factor involved in tolerance to phosphate starvation in rice. Plant Physiol 138:2087–2096

Yu HL, Luo N, Sun LC, Liu D (2012) HPS4/SABRE regulates plant responses to phosphate starvation through antagonistic interaction with ethylene signalling. J Exp Bot 63:4527–4538

Zakhleniuk OV, Raines CA, Lloyd JC (2001) pho3: a phosphorus-deficient mutant of *Arabidopsis thaliana* (L.) Heynh. Planta 212:529–534

Zhou J, Jiao FC, Wu ZC, Li YY, Wang XM, He XW, Zhong WQ, Wu P (2008) OsPHR2 is involved in phosphate-starvation signaling and excessive phosphate accumulation in shoots of plants. Plant Physiol 146:1673–1686

Zimmerli C, Ribot C, Vavasseur A, Bauer H, Hedrich R, Poirier Y (2012) PHO1 expression in guard cells mediates the stomatal response to abscisic acid in Arabidopsis. Plant J 72:199–211

Long-Distance Signaling of Iron Deficiency in Plants

Yusuke Enomoto and Fumiyuki Goto

Abstract Iron is an essential nutrient used for many physiological reactions in a whole plant body. A large amount of iron exists in the Earth's crust, but plants cannot uptake iron from roots efficiently because of the low iron solubility. The uptake and translocation of iron from roots to shoots are strictly controlled in order to maintain homeostasis of cytosol. It has been suggested that long-distance signaling from shoots to roots is involved in the regulation mechanisms of iron uptake. The identification of genes related to iron uptake was made possible because of the rapid development of molecular biology since the 1990s.

In this chapter, we describe the history of the discovery of the iron uptake genes and their regulation factors, and explain the interaction of these factors. Furthermore, we show some models of long-distance signaling which consistently explain the relationship between the phenotype of some mutants and the gene functions involved in iron uptake.

Keywords Iron • Nicotianamine • Transporter • Long-distance signal • Iron deficiency

1 Introduction

Most plants acquire nutrients from soil. Recently, transporters have been identified for the uptake of several essential and nonessential nutrients. The elucidation of the mechanisms of uptake and effective application would contribute to human health

Y. Enomoto
Hiroo Gakuen Junior & Senior High School, Minami Azabu 5-1-14, Minato-ku, Tokyo 106-0047, Japan
e-mail: y-enomoto@hiroogakuen.ed.jp

F. Goto (✉)
Biotechnology Sector, Central Research Institute of Electric Power Industry (CRIEPI), 1646 Abiko, Abiko-shi 270-1194, Japan
e-mail: gotoh@criepi.denken.or.jp

F. Baluška (ed.), *Long-Distance Systemic Signaling and Communication in Plants*, Signaling and Communication in Plants 19, DOI 10.1007/978-3-642-36470-9_8,

as well as evolution of plant science (White and Brown 2010). 1.5–2 billion people of the world have iron deficiency anemia and the development of agricultural science is thought to be the way to overcome the disease (Lynch 2011). Iron is an essential nutrient for plant growth and is abundant in soil. However growth inhibition caused by iron deficiency is observed in about 30 % of cultivated land in the world (Mori 1999). The low availability of iron in soil is caused by the chemical form of iron. Iron in the surface of the soil is oxidized and its chemical form is iron (III)-hydroxide (Fe $(OH)_3$), which is insoluble and of low availability for plants. Most plants need soluble iron (10^{-8} M to 10^{-4} M) for normal growth, but the solubility of iron at pH $= 7$ is approximately 10^{-17} M. Plants have developed an iron uptake system at the root surface in order to solve the rhizospheric iron problem. However, even then, iron deficiency is induced by the limitation of iron in alkaline soil where iron is insoluble (Staiger 2002). Large amounts of iron exist in the soil; nevertheless iron deficiency is mainly counteracted by spreading fertilizer over the field. It is thought reasonable to enhance the expression of genes which regulate the iron uptake system of plants rather than to use iron fertilizer (Staiger 2002). In recent years, many findings about the absorption and translocation of iron and its regulation factors have been reported (Curie and Briat 2003; Kobayashi and Nishizawa 2012). Taking advantage of these findings will dramatically increase the use of poor agricultural lands and will also increase biomass generation.

It has been known that signaling of the iron absorption mechanism in roots occurs when the concentration of iron in the cytoplasm in shoots is reduced, but the signaling pathway is unknown. Finding upstream regulators such as the sensor for iron concentration or the signaling molecules provides important clues to elucidate the mechanism by which higher plants maintain iron homeostasis. Recent studies using genetics and molecular biology have identified iron uptake genes, for example, genes coding transporter, oxidoreductase, synthase of phytosiderophore, and transcription factor. The iron uptake system in roots is thought to be controlled by systemic long-distance signals (Schmidt 2003). In this chapter, we describe the iron uptake system tightly regulated by iron concentration and the circumstantial evidence indicating existence of the long-distance signals.

2 Iron Uptake in Roots

2.1 Maintenance of Iron Concentration in Cytoplasm

Iron is harmful in high concentrations but is also an essential micronutrient for higher plants. Iron has roles in respiration and photosynthesis (Mori 1999). Iron exists as the iron–sulfur cluster and heme iron in each molecule, so that iron atoms work as an active center for electron transfer reactions. Plants absorb iron through roots from rhizosphere and systemically distribute it. Decrease of distribution of iron for leaves causes chlorosis symptoms along the leaf veins because the synthesis

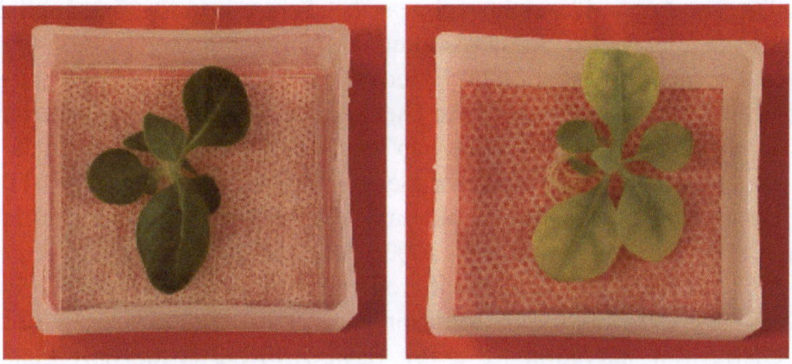

Fig. 1 Two-week-old tobacco seedlings were treated under a normal growing condition (*left*) and an iron-deficient condition (*right*) for 2 weeks (Yoshihara et al. 2006)

of chlorophyll and activity of photosynthesis are inhibited by iron deficiency (Fig. 1). On the other hand, excess iron causes severe growth inhibition by reactive oxygen species generated by Fenton reaction ($Fe^{2+} + H_2O_2 \rightarrow Fe^{3+} + HO^- + HO^-$) due to the existence of free iron in cytosol. In order to avoid the symptoms derived from excess or deficient iron, plant cells have developed mechanisms to keep iron at a constant concentration (Staiger 2002). Iron taken into roots is transported to immature tissues through a xylem in the form combined with phytosiderophore such as citric acid and nicotianamine (Rellán 2010). In general, Fenton reaction is inhibited because the iron concentration in the cytoplasm is kept constant by the reaction molecules such as ferritin, which capture the excess iron, and Nramp transporter, which isolates iron to the vacuole (Harrison 1996; Thomine 2003; Curie et al. 2000). Aconitase, RNA binding proteins, plays a key role in the regulation of iron homeostasis in animals; however, their homologues do not function as a regulator for iron uptake in plants (Arnaud et al. 2007). In contrast to aconitase, ferritin, iron-storage protein, exists in both animal cells and plant cells (Lawson et al. 1989). The function of plant ferritin is similar to the animal one though the regulation of the gene expression is different. In order to understand the system of the plant iron homeostasis, we have to reveal the genes related to sensors of iron status in a cell, the signal transduction system bearing the information of iron status, and iron uptake from the soil.

2.2 *Genes Involved in Iron Uptake in Dicot Plants*

In the usual cultivation except for the foliar spray, plants acquire enough amounts of iron for the whole body from their roots. Plants absorb iron through roots from rhizosphere and translocate iron from roots to the whole body. Further, iron is released from vacuoles to cytoplasm under the iron-deficient condition (Curie and

Briat 2003). The iron oxide in the soil near the surface is mostly insolubilized, and plants have developed two known strategies for iron uptake. In this section, we describe the mechanism of strategy I used by dicotyledonous plants (strategy II is mentioned in the next section). Ferric chelate reductase (FRO) reduces ferric iron which slightly dissolves in rhizosphere to produce soluble ferrous iron. The ferrous iron is transported to the cytosol via the iron regulated transporter (IRT) in the plasma membrane of the root's epidermis (Mori 1999; Curie and Briat 2003; Clemens 2001; Hall and Williams 2003; Schmidt 2003) (Fig. 2). *Arabidopsis* mutant *frd1* has very low activity of iron reduction derived from the defect of *AtFRO2* (Yi and Guerinot 1996; Robinson et al. 1999). The expression of *AtFRO2* is induced by iron starvation for a few days. *AtFRO2* is the only gene whose transcriptional product has the activity of iron reduction for the absorption of iron in roots though seven genes of the *FRO* gene family have been isolated from *Arabidopsis* (Wu et al. 2005). The homologous genes of *AtFRO2* were isolated in tobacco (*Nicotiana tabacum*), tomato (*Lycopersicon esculentum*), pea (*Pisum sativum*), and others. These genes are considered to be one of the most important factors for the iron uptake system in non-graminaceous plants (Hodoshima et al. 2007; Li et al. 2004; Waters et al. 2002).

IRT1 classified in *ZIP* family was cloned from *Arabidopsis* cDNA library as a gene which is capable of recovering the iron transport ability of yeast mutant *fet3fet4* (defective in both high- and low-affinity iron uptake). Vert et al. (2002) screened *irt1-1* mutant of *Arabidopsis*, which is a lack of *IRT1*, from T-DNA-tagged lines. The mutant was defective in the ability of iron absorption. Interestingly, excess Ni and Cd induce the expression of *IRT1*, and *AtIRT1* is able to transport these metals (Vert et al. 2002; Yoshihara et al. 2006; Nishida et al. 2011). Connolly et al. (2002, 2003) showed the part of the control mechanism of *AtIRT1* and *AtFRO2* using transgenic plants which overly express these genes. Although *AtFRO2* and *AtIRT1* were driven by the constitutive promoter of CaMV 35S, the accumulation of *AtFRO2* mRNA and that of IRT1 protein were limited in the roots under the iron-deficient condition. These posttranscriptional regulations of the iron uptake genes are thought to be a mechanism to avoid an excess iron symptom. The strict controls of many steps have probably been developed for safety absorption of iron (Fig. 3).

2.3 Comparison of Iron Uptake Strategy Between Dicot Plants and Graminaceous Plants

In the previous section, we described strategy I: the iron absorption mechanism of non-graminaceous plants including dicotyledoneae. In this section, strategy II is described. Takagi et al. (1984) first reported the unique strategy of iron absorption in graminaceous plants (strategy II plants). Strategy II plants emit mugineic acids, chelating agents, into the rhizosphere to solubilize the iron. Continuous studies after

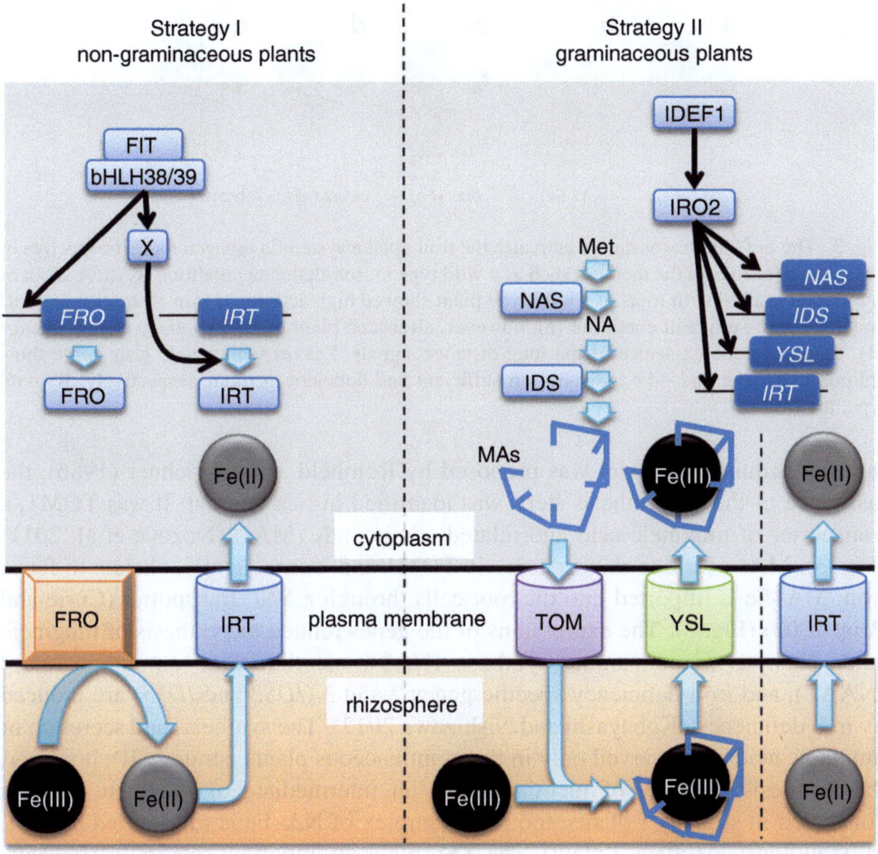

Fig. 2 Schematic illustration indicates two strategies of the iron uptake system. The most different point between the two strategies is that strategy II plants absorb chelating iron; consequently, the iron uptake system in strategy II plants is more complicated than that of strategy I plants. Transcription factors including FIT, bHLH38/39, IDEF1, and IRO2 control the expressions of downstream genes in each strategy. *Abbreviation*: *Fe(III)* ferrous iron, *Fe(II)* ferric iron, *FRO* Ferric chelate reductoxidase, *IRT* Iron-related transporter, *FIT* Fe deficiency Induced Transcription Factor, *bHLH* basic helix–loop–helix, *X* unknown factor, *TOM* transporter of mugineic acid family phytosiderophores 1, *YSL* Yellow stripe like, *IDEF1* IDE-binding factor 1, *IRO2* iron-related transcription factor 2, *NAS* Nicotianamine synthases, *IDS* Iron deficiency-specific clone, *Met* Methionine, *NA* Nicotianamine, *MAs* Mugineic acid family

the discovery of strategy II have revealed the biosynthesis pathway of mugineic acids [deoxymugineic acid (DMA), mugineic acid (MA), epihydroxy-deoxymugineic acid (epiHDMA), hydroxymugineic acid (HMA), and epihydroxymugineic acid (epiHMA)]. *S*-adenosylmethionine in a methionine cycle is the first material of the pathway (Mori 1999). Furthermore, genes encoding an enzyme required for each biosynthesis step of these compounds have been identified (Nakanishi et al. 2000; Bashir et al. 2006). In 2011, a quarter of a century after

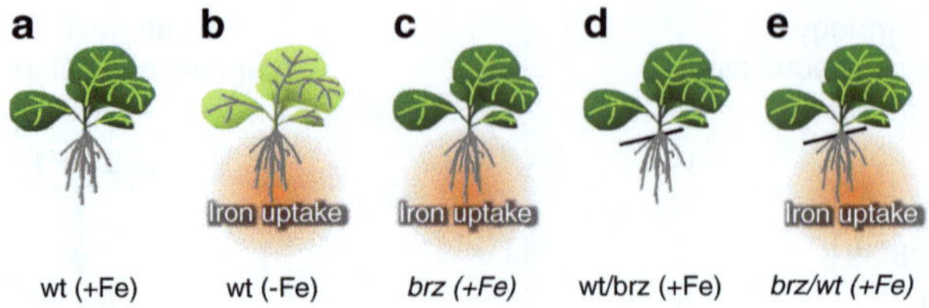

Fig. 3 The *brz* mutant constantly activates the iron uptake system in tobacco roots irrespective of iron concentrations in the medium such as a wild type in iron-deficient condition (**c**, **a**). A shoot of *brz* mutant grafted with roots of a wild-type plant showed high activity of iron absorption in roots under the iron-sufficient condition (**e**); however, a tobacco plant grafted contrary had no change (**d**), suggesting the existence of the long-distance signals. Leaves with a *pale gray color* show chlorosis (**b**). +Fe and −Fe indicate iron sufficient and deficient medium, respectively. *Wt* wild type, *brz* bronze

the iron acquisition model was proposed by Römheld and Marschner (1986), the last piece of the biosynthesis steps was identified in rice and oat. It was TOM1, a transporter of mugineic acid and related compounds (MAs) (Nozoye et al. 2011) (Fig. 3). MAs exude to rhizosphere via TOM1 and, consequently, chelate to ferric iron; MA–Fe is imported into the root cells through a YSL transporter (Curie and Briat 2003) (Fig. 3). The expressions of the genes related to synthesis of mugineic acids such as nicotianamine synthase (NAS), nicotianamine aminotransferase (NAAT), and iron-deficiency specific genes 2 and 3 (*IDS2* and *IDS3*) are induced by iron deficiency (Kobayashi and Nishizawa 2012). The synthesis and secretion of mugineic acid are observed only in the graminaceous plants (strategy II); however, the synthesis pathway of nicotianamine, an intermediate of mugineic acid, is conserved in strategy I plants, too. The complex of NA–Fe is transported by YSL in *Arabidopsis* (strategy I plant). The YSL plays an important role in the transportation of iron from source tissues to sink tissues through xylem and phloem (DiDonato et al. 2004; Schaaf et al. 2005; Le Jean et al. 2005; Schuler et al. 2012).

In addition, it is suggested that strategy II plants have the absorbing mechanism of strategy I. *OsIRT1*, homologue of *AtIRT1*, iron transporters, exists in rice (strategy II plant) and expresses under the iron-deficient condition. IRT I is essential for the absorption of iron in strategy I plants (Bughio et al. 2002). Vasconcelos et al. (2004) suggested that ferrous iron is taken directly from the rice roots via OsIRT1 because large amounts of ferrous iron exist in paddy fields in the reduced state. However, there are no homologues of *AtFRO2* in rice. *AtFRO2* overexpressed in rice does not work. These results indicate that the expression of *AtFRO2* is tightly regulated at post-transcription level in Arabidopsis (strategy I plant). In contrast, there is no regulation mechanism of *AtFRO2* in rice (strategy II plant) (Connolly et al. 2003). Ishimaru et al. (2007) introduced *refre1/372*, iron reductase gene of yeast into rice plants in order to improve the efficiency of the absorption of ferrous iron through the IRT1, and showed the transgenic rice plants grown in calcareous alkaline soil acquired the tolerance to withstand the stress of iron deficiency. We mentioned the function of strategy I genes, such as *IRT1* and *FRO2*, in strategy II plants as shown above. On the contrary, some reports showed that the strategy II genes were introduced into strategy I plants. The expression of *GUS* reporter gene under the control of a promoter of *HvIDS2*, which is related to synthesis of mugineic acid in barley, was induced by iron deficiency in transgenic tobacco plants (Yoshihara et al. 2003). The *GUS* expression suggests that tobacco (strategy I plant) has the similar mechanism for induction of iron deficiency signaling as barley (strategy II plant). The iron deficiency responsible elements, IDE1 and IDE2, were identified from the promoter of *HvIDS2*. IDE1 and IDE2 are conserved in several genes in both strategy plants, such as *HvNAAT* (nicotianamine aminotransferase), *HvNAS1* (nicotianamine synthase), *HvIDS3* (iron deficiency-specific clone 3) in barley, *OsNAS1* and *OsIRT1* in rice, and further *AtIRT1* and *AtFRO2* in *Arabidopsis* (Kobayashi et al. 2003a, b). To sum up, it is considered that the controlled manner of gene expressions and the function of proteins, such as OsIRT1 and AtIRT1, are conserved among strategy I and II plants (Ishimaru et al. 2006). Therefore it is expected that comparison of the two mechanisms may shed light on unknown areas of agriculture.

3 Long-Distance Signals Associated with the Regulation of Iron Uptake

3.1 Historical Background of Studies on Long-Distance Signals

We show here some examples where the inhibition of iron translocation from roots to shoots was caused by various reasons and display the evidence to suggest the presence of long-distance signals. Many researchers have attempted to clarify what the signal is since Grusak and Pezeshgi proposed, “long-distance signals for iron absorption,” based on the genetics of the pea in 1996. Pea mutants *brz* constantly had high activity of iron reductase in roots. When the shoots of mutants were grafted onto the rootstock of wild-type peas, the iron reductase activity in the roots of the wild-type peas was enhanced. On the other hand, when the shoots of wild-type peas were grafted onto the rootstock of the mutants, the iron reductase activity in the roots of the mutants was not enhanced (Grusak and Pezeshgi 1996) (Fig. 3). These results strongly suggested that the iron reductase activity in the roots of the wild-type was enhanced by something in the shoots. Consequently, the model of the long-distance signals from shoots to roots for the activation of the system for iron absorption was proposed; however, the mechanism of the signaling is still unknown.

The shoots of tomato mutants, *chloronerva*, grown in an iron-sufficient condition exhibit severe chlorosis and iron reductase activity in the roots increased, whereas the iron concentration of shoots of the mutant increased higher than that of the wild type (Herbik et al. 1996) (Fig. 4). Ling et al. (1999) showed that the phenotype of *chloronerva* was caused by the mutation of the nicotianamine

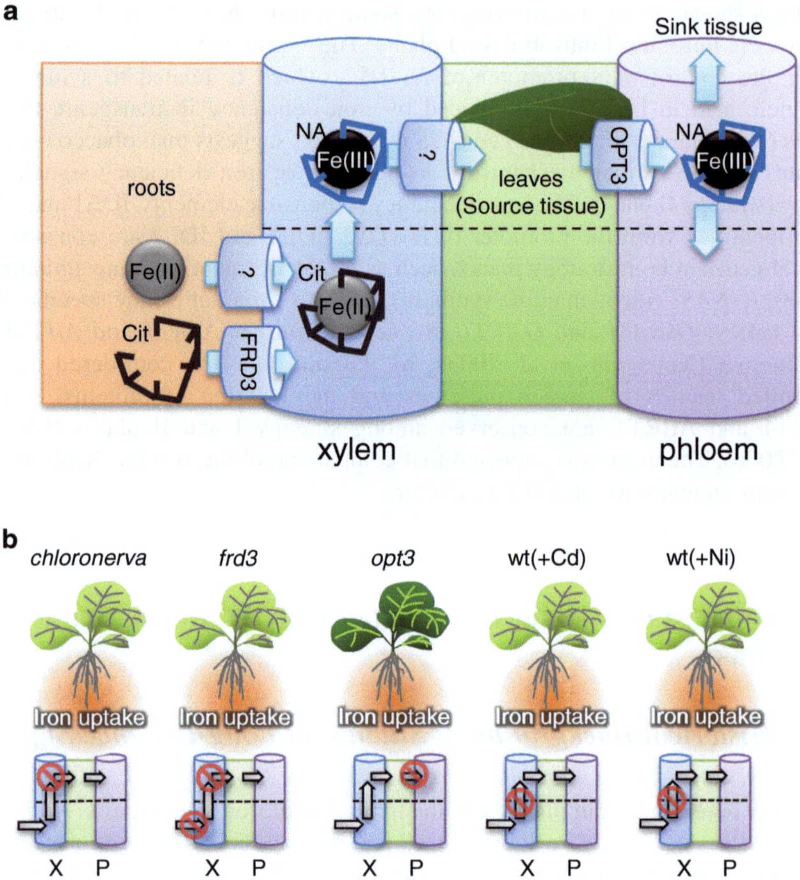

Fig. 4 (**a**) The schematic representation of the transportation of iron from root cells to sink tissues via specific transporters. Transportation of iron among tissues needs phytosiderophores (nicotianamine and citrate acid) and their transporters. (**b**) The *crosses* in simplified illustrations of an *upper panel* show the defective parts of the iron transportation pathway in mutants. The *pale gray* of leaves indicates chlorosis. *Chloronerva* is a defective mutant of nicotianamine synthases. This mutant is not able to form the nicotianamine–iron complex, resulting in iron not being transported from xylem to leaves. The *frd3* is a mutant deficient in the gene encoding a citrate transporter in roots. This mutant does not form the citrate acid–iron complex in xylem; consequently, iron is not transferred from the citrate acid–iron complex to nicotianamine. The *opt3* is a defective mutant of the oligopeptide transporter which is classified in a *YSL* gene family. No chlorosis is observed; however, the iron absorption in roots is activated. The sink tissues become iron deficient because iron is not transported from leaves to phloem. Excess Cd and Ni inhibit iron transport from roots to leaves and the leaves show chlorosis. *Abbreviation*: *Fe(III)* ferrous iron, *Fe (II)* ferric iron, *Cit* Citrate, *NA* Nicotianamine, *FRO3* Ferric chelate reductoxidase 3, *X* xylem, *P* phloem, *frd3* ferric reductase defective, *opt3* oligopeptide transporter, *wt* wild type, *Cd* cadmium, *Ni* nickel

synthase (NAS) gene. Nicotianamine (NA) is a molecule that plays an important role in the translocation of iron in the whole body. NA bonded to iron moves among tissues through the xylem or phloem (von Wiren et al. 2000). Hence lack of NA in the mutants leads to lack of iron in the cells of leaves even though iron is absorbed in the roots.

Arabidopsis mutant *frd3* shows chlorosis of leaves although it shows high activity of iron reduction in the roots irrespective of the iron concentration in the medium; consequently, iron accumulates in leaves and roots. FRD3 protein functions in roots for the secretion of citrate to xylem in wild types (Fig. 4). Chlorosis of *frd3* is recovered by the addition of citric acid into a medium (Durrett et al. 2007). Therefore, *frd3* probably accumulates iron as a useless form in xylem because frd3 lacks citric acid to capture iron as an available form (Durrett et al. 2007). Lack of MATE (Multidrug And Toxic Compound Extrusion) gene causes low activity of transportation of citric acid (Rogers and Guerinot 2002; Green and Rogers 2004) (Fig. 4). Citrate–Fe is necessary to transport iron from xylem to the shoot cells. The YSL family transporters translocate iron chelated with NA, citrate, or other chelators between different organs such as xylem-to-shoot cells and xylem-to-seeds (Schaaf et al. 2005; Le Jean et al. 2005). The distribution of iron to new leaves and old leaves is carried out using the specificity of the transport of iron chelate via OsYSL16 in rice (Kakei et al. 2012). Both of *chloronerva* and *frd3* are the mutants that lack the ability to transport iron as an available form to the appropriate tissues. Accordingly, no abnormality is observed in the signal transduction pathways that convey the state of iron deficiency to roots (Fig. 4).

Opt3-2 mutant of *Arabidopsis* lacks a small part of *OPT3* gene that encodes an oligopeptide transporter. The *Opt3-2* accumulates iron in shoots; however, iron transportation to seeds is simultaneously reduced (Stacey et al. 2007). The iron transportation to seeds is likely not in the path of the signal transduction of iron deficiency because the response to iron deficiency is observed during the vegetative growth phase as well as the reproductive growth phase. Therefore, it is thought that the defect of the mutant occurred in two steps: iron transportation system from source to sink and signal transduction system to promote iron acquisition (Stacey et al. 2007) (Fig. 4).

There are some reports on the iron deficiency responses mediated by indirect response to the exposure of heavy metals. In tobacco plants exposed to cadmium, iron concentration in shoots decreased and iron-responsible genes expressed in roots (Yoshihara et al. 2006; Hodoshima et al. 2007) (Fig. 4). In Arabidopsis exposed to nickel, iron concentration in shoots decreased and the expression of *AtIRT1* increased in roots (Nishida et al. 2011) (Fig. 4). These results can be explained as follows. First, iron transport from roots to shoots is inhibited competitively with other heavy metals; second, leaves have a lack of iron; third, iron-deficient signals are sent from shoots to roots; fourth, the expression of *IRT* is induced in roots; fifth, Cd or Ni including Fe is taken up into roots; sixth, Fe accumulates in leaves (Hodoshima et al. 2007; Nishida et al. 2011). In short, the fact that the response to iron deficiency is enhanced in roots of the mutants or under the

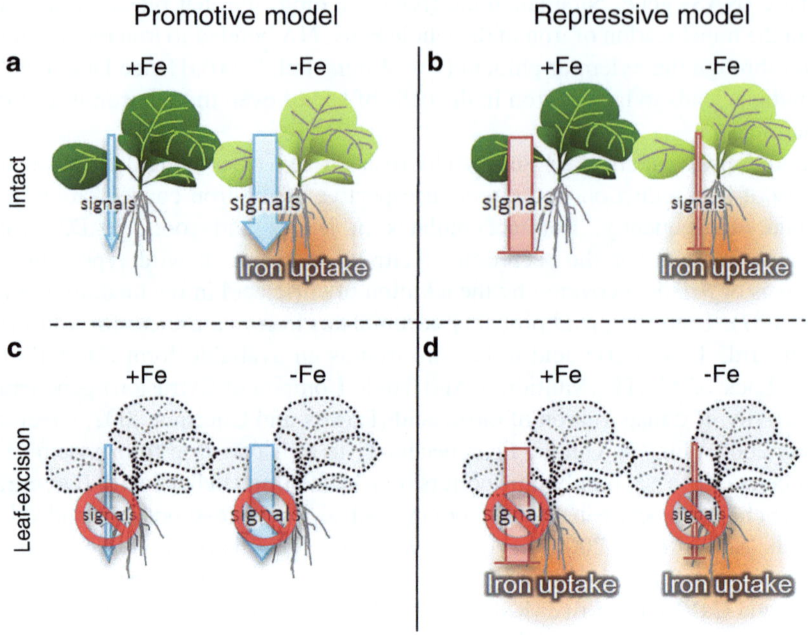

Fig. 5 The schematic representation indicates the models of iron deficiency signaling shown by Vert et al. (2003) (**a**, **b**) and the leaf-excision experiment carried out by Enomoto et al. (2007) to clarify which model is reasonable to explain the signaling (**c**, **d**). (**a**) Promotive model: the signals are not transferred from the shoots to the roots under the iron-sufficient condition; however, signals are transferred under the iron-deficient condition to take iron up from the roots. (**b**) Repressive model: the signals continuously come from the leaves to the roots to repress the expression of genes involved in iron uptake under the iron-sufficient condition. On the other hand, the signals are not transferred when the plant is iron deficiency. (**c**) If the promotive model is correct, the expressions of genes such as IRT and FRO would not be observed. (**d**) If the repressive model is correct, these genes likely would express in both iron sufficient and deficient condition

heavy metal exposure conditions demonstrates the presence of the long-distance signals bearing the information of iron deficiency of shoots.

3.2 Approaches for Elucidation of the Long-Distance Signal Molecules

Vert et al. (2003) proposed two signal models relating to the iron absorption: promotive and repressive models (Fig. 5). In the promotive model, signals are transmitted from iron-deficient leaves to the roots and genes involved in iron absorption are expressed (Fig. 5a). No signals are transmitted from iron-sufficient leaves. On the other hand, in the repressive model, signals are transmitted from iron-sufficient leaves to roots and the expressions of genes are repressed (Fig. 5b).

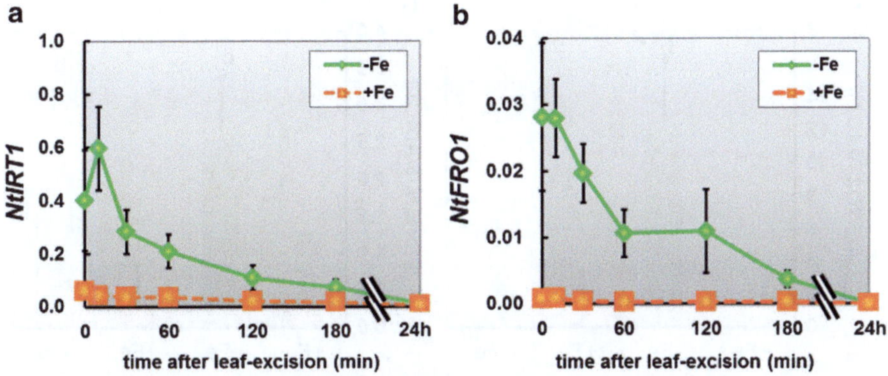

Fig. 6 Time course of the *NtIRT1* and *NtFRO1* expressions in roots of plants whose leaves were excised. The experimental scheme is shown in Fig. 2c, d. The relative expressions of genes *NtIRT1* (**a**) and *NtFRO1* (**b**) against *NtACT* were measured with the semiquantitative PCR analysis after the leaf excision. Error bars indicate S.D. with 5 plants. *Abbreviations*: +Fe, iron sufficient in the medium; −Fe, lack of iron in the medium; NtIRT1, *Nicotiana tabacum* iron regulated transporter; NtFRO1, ferric reduction oxidase 1; NtACT, actin (Enomoto et al. 2007)

Reduction of the repressive signals in the iron-deficient condition results in inducing the gene expressions. There were few clues which model was reasonable when these models were proposed because of only a few data. Enomoto et al. (2007) supported the promotive signals which was better than the repressive model to explain signal transduction of iron deficiency in leaves. The expressions of *NtIRT1* and *NtFRO1* increased in the iron-deficient condition in tobacco plants. However, the gene expressions were quickly reduced 3 h after excision of leaves (Fig. 6). If repressive signals control the mechanism of signals, the gene expressions must be increased by the excision of leaves. However, they did not obtain such results. Furthermore, they observed similar results with the expression of NAS1 gene in rice (Enomoto et al. 2009), suggesting that a common mechanism of long-distance signal transduction exists between strategy I and II plants (Fig. 7c).

Comprehensive analysis, such as transcriptome and proteome analysis, has been attempted to find factors involved in the mechanism of transmitting the information of the concentration of iron in the leaves to the roots (Thimm et al. 2001; Negishi et al. 2002). However, we have no idea of what the long-distance signals are: low-molecular compound and macromolecule, such as peptide, protein, and RNA. Plant hormones or reused iron were formerly thought to be the signal substances (Landsberg 1984; Maas et al. 1988; Bienfait 1989); however, these ideas have been denied due to the results of studies using mutants associated with plant hormones (Schmidt et al. 2000). Shimidt et al. found that there were no differences of the response of iron deficiency between wild-type and mutant *aux1-7*, lack of auxin transmission, and concluded that auxin was not involved in the response of iron deficiency. However, lately, Chen et al. (2010) strongly suggested that auxin and ethylene were involved in the signal of iron deficiency. They showed that auxin

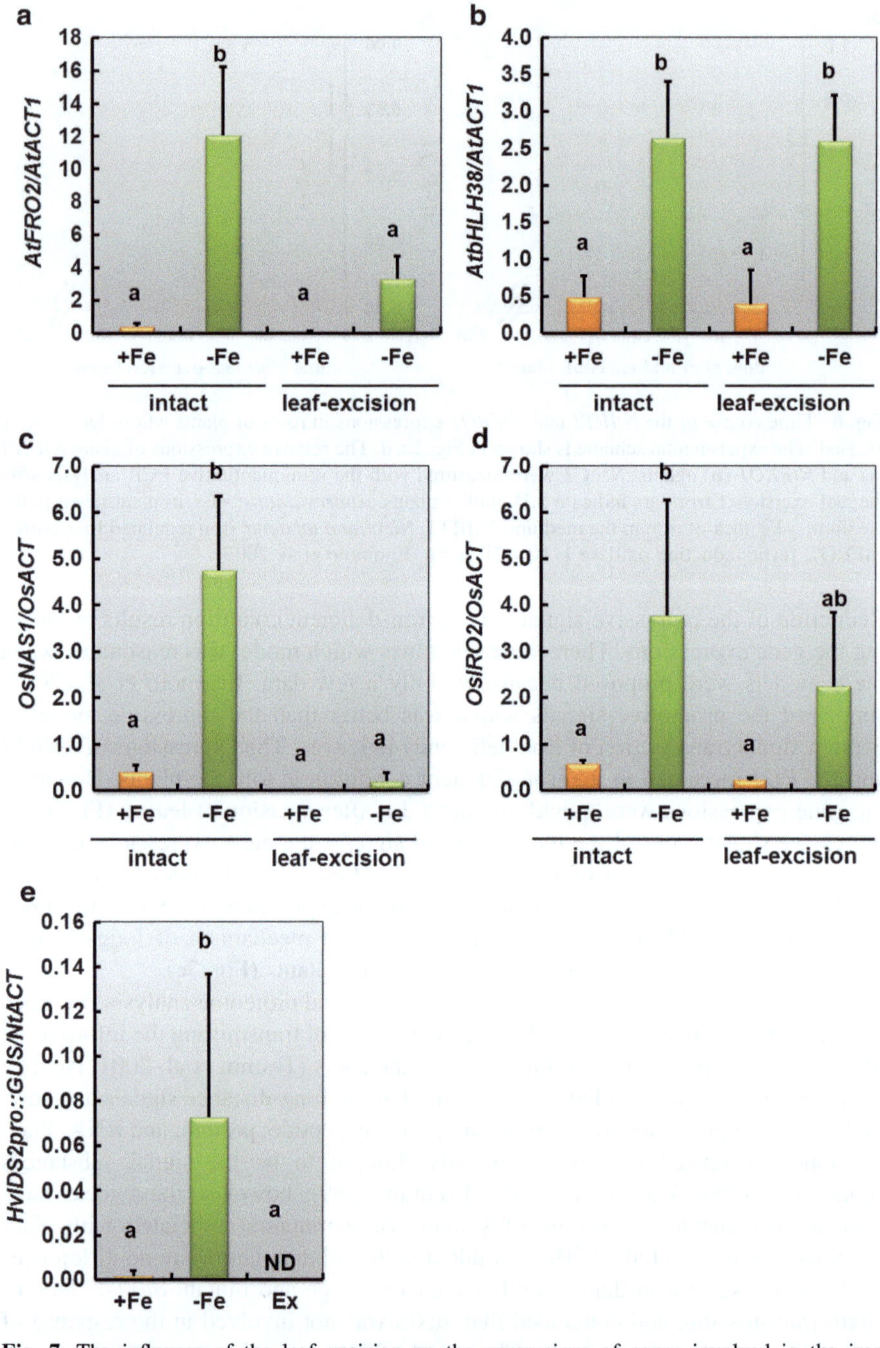

Fig. 7 The influence of the leaf excision on the expressions of genes involved in the iron absorption of *Arabidopsis*, tobacco, and rice plants. The experimental scheme is shown in Fig. 5c, d. The relative expressions of genes *AtFRO2*, *AtbHLH38* (**a**, **b**), *OsNAS1*, *OsIRO2*

involved in the expression of *FRO* in auxin hyperaccumulation mutant of *yucca* and NO worked as a signal transmission substance in the downstream of the auxin signaling pathway. Interestingly, supply of NO or NAA results in the expression of *AtFRO2* irrespective of iron presence. These results suggest that the safety mechanism works to prevent excess iron disorders, which is similar to the posttranscriptional regulation shown in the former section.

4 What Is the Signal?

4.1 Upstream Regulators for the Iron Uptake Genes

In order to examine how genes related to iron absorption (*FRO* and *IRT* in strategy I and *YSL* and *TOM* in strategy II) on the cell membrane of the epidermis are activated through the signal pathway, some approaches have been attempted. In the last 10 years, many transcriptional factors have been obtained and analyzed (Fig. 2). The breakthrough was the discovery of *FIT* (*FER/FRU/bHLH29*) (Ling et al. 2002; Colangelo and Guerinot 2004; Jakoby et al. 2004; Yuan et al. 2005). In the loss-of-function mutant *fit*, the expression of *FRO2* disappeared though transcript of *IRT1* accumulated in the iron-deficient roots as well as the wild-type roots. However, protein of IRT1 did not accumulate. These results demonstrate that FIT is a transcription factor related to the transcriptional control of *FRO* and the posttranscriptional control of *IRT*. Further, FIT did not accumulate in tomato plants overexpressing *FIT* in the iron-sufficient condition because of the suppression of the post-transcription (Brumbarova and Bauer 2005). In the same way as the tomato, *Arabidopsis* overexpressing *FIT* transcribed FRO and IRT only in the iron-deficient roots (Colangelo and Guerinot 2004). Genes that are not able to respond to *fit1-1* have been selected from the genes in response to iron deficiency in the wild type. However, it is still unknown what genes work upstream of *AtFIT1*, *AtbHLH38/39*. Incidentally heterodimer of AtFIT1 (AtbHLH29) and AtbHLH38/39 (AtbHLH38 or 39) is involved in the transcriptional control of *AtFRO2* (Yuan et al. 2008). In yeast containing *GUS* gene driven by *AtFRO2* promoter, co-expression of *AtFIT1* and *AtbHLH38* coding transcriptional factors which control

Fig. 7 (continued) (**c**, **d**), and *HvIDS2pro::GUS* (**d**) against inner ACT of each plant were measured with the semiquantitative PCR analysis 24 h after the leaf excision. Error bars and letter above the bars indicate S.D. with 3–5 plants and significant difference in each treatment, respectively (Tukey–Cramer test, $P < -0.05$). These results suggest that *AtFRO2*, *OsNAS1*, and *HvIDS2* in the roots were regulated by the long-distance signals from the shoots; however, the expressions of *AtbHLH38* and *OsIRO2* were not influenced by the signals. *Abbreviation*: +Fe, iron sufficient in the medium; −Fe, lack of iron in the medium; *At FRO2*, *Arabidopsis thaliana* ferric reduction oxidase 2; *AtbHLH38*, basic helix loop helix 38; *OsNAS1*, *Oryza sativa* Nicotianamine synthases; *OsIRO2*, iron-related transcription factor 2; *HvIDS2*, *Hordeum vulgare* iron deficiency-specific clone 2; *GUS*, β-glucuronidase (Enomoto et al. 2009)

the expression of genes related to iron absorption in Arabidopsis activates transcriptional activity of *GUS*, indicating that the complex of AtFIT1 and AtbHLH38 controls the expression of *AtFRO2*. In transgenic Arabidopsis expressing *AtFRO2*, the activity of AtFRO2 was affected by the concentration of iron in the medium, indicating that the expression of *AtFRO2* is post-transcriptionally controlled.

In Arabidopsis overexpressing both *AtFIT1* and *AtbHLH38*, not only the expression of *AtFRO2* but also enzyme activity of *AtFRO2* increased. It probably means that the complex of AtFIT1 and AtbHLH38 controls unknown factors which post-transcriptionally regulate the expression of *AtFRO2*. It should be considered well that the overexpression of *AtFIT1* and *AtbHLH38* is a necessary condition for the activation of *AtFRO2*. However, this is not a sufficient condition because Yuan et al. (2008) used cDNA of the two transcriptional factors to produce transgenic plants. In wild-type Arabidopsis, iron-deficient responsible factors probably are involved in splicing or the control of translation of the two transcriptional factors. POPEYE is also known as another bHLH transcription factor which represses iron-responsible genes (Long et al. 2010). Interestingly, iron-reducing activity and the expression of AtIRT1 in a mutant pye-1 decrease significantly less than that of the wild types grown under the iron-deficient condition, suggesting that POPEYE is involved in induction of the expression of AtIRT1 and AtFRO2 which are activated by transmission of long-distance signals.

OsIRO2 whose expression increased in both shoots and roots under the iron-deficient condition was cloned from rice plants after microarray analysis. At first *OsIRO2* was thought to be a rice-specific transcription factor because its similarity to *FIT1* (*AtbHLH29*) of Arabidopsis was low (Ogo et al. 2007). However, lately, it has been shown that *AtbHLH38* interacts with *OsIRO2* and the similarity between *OsIRO2* and *AtbHLH38* of Arabidopsis is high. *OsIRO2* transcriptionally regulates *OsNAS1* and *OsNAATA* involved in mugineic acids synthesis, and *OsYSL15* coding a mugineic acid–iron complex transporter (Lee et al. 2009). These results suggest that there is commonality between transcription factors such as *AtIRT1* and *AtRFO2* *i*n strategy I and *OsNAS1* and *OsYSL15* in strategy II. However, note that the overexpression of a single gene among *FIT*, *bHLH38/39/100/101* in strategy I was not able to induce the expression of *AtIRT1* and *AtFRO2*. On the other hand, the overexpression of only *OsIRO2* in rice induced the expression of *OsYSL15*.

The commonality between transcription factors is also suggested by the *IDE1* element mentioned above. Kobayashi et al. (2007) cloned *IDEF1* which binds to *IDE1*, 4 years after the first proposals of *cis*-elements of *IDE1* and *2*. They showed that the expression of *IRO2* was controlled by IDEF1 (Fig. 2). However, no transcription factor corresponding to *IDEF1* has been cloned from strategy I plants. It is expected that the missing piece will be found in the near future since the control mechanism of genes related to iron absorption is common between strategy I and II plants.

4.2 The Local Signals Regulate the Expressions of bHLHs

Enomoto et al. (2009) showed that the *bHLH38*, which directly regulates the expression of *AtIRT1* and *AtFRO2*, is a transcription factor activated with no relation to the shoot-to-root signals proposed by Grusak and Pezeshgi (1996) (Fig. 7). They carried out leaf excision experiments. The expression level of *AtFRO2* in roots of iron-deficient plants after the leaf excision was significantly lower than that of intact plants (Fig. 7a). On the contrary, the expression level of *AtbHLH38* was kept irrespective of leaf excision (Fig. 7b). In addition, the expression pattern of *OsIRO2*, homologue of *AtbHLH38*, was similar to that of *AtbHLH38* (Fig. 7d). Taken together, the expressions of genes, such as *OsNAS1* of rice and *AtFRO2* of Arabidopsis, which are directly responsible for iron absorption, are likely controlled by long-distance signals. However, *bHLHs* involved in the transcription of these genes are under the control of other signals (Fig. 7c). Moreover, it is suggested that the expression of GUS gene driven by the promoter of oats (strategy II) IDS2 in tobacco (strategy I) is regulated by long-distance signals (Fig. 7e) as well as AtFRO2 and OsNAs1.

The expression of *AtbHLH38*, *39*, *100*, and *101* in both roots and shoots increased in response to iron deficiency (Wang et al. 2007). We focused on the reasons why the expression pattern of *FRO* in transgenic Arabidopsis expressing *FIT* was the same as that in the wild types. If *FIT* itself is controlled as post-transcription, FIT protein probably does not accumulate in iron-sufficient roots like the tomato. Even if *FIT* is expressed with bHLH, *FIT* likely is not able to transcript *FRO*. In short, FIT is not in control of iron-deficient signal at translation and the stability of protein. Although *AtFIT1* and *AtFRO2* root specifically express, genes of *AtbHLH38* family express in shoots under the iron-deficient condition. When *FIT* is overexpressed, mRNA of FIT accumulates in shoots. However, transcriptional activity of FRO is not observed. Since mRNA of *BHLH38* family accumulates in iron-deficient shoots, at least more than one transcription factor is needed for the expression of *AtFRO2*.

The extension of a main root depends on the iron concentration in Arabidopsis roots under the phosphate-deficient condition (Ward et al. 2008). The *frd3* mutants accumulate excess iron and express *AtIRT1* and *AtFRO2* as if the mutants lack iron. However actually, the growth of the main roots behaves like iron-sufficient response under the phosphate-deficient condition. These observations suggest the existence of another signal transduction pathway which is different from the pathway inducing the expression of *AtIRT1* and *AtFRO2*. It is thought that ethylene is involved in the formation of root hair in response to iron deficiency and the response is affected by the iron concentration in the roots (Schmidt et al. 2000). Clarification of the mechanism of the local signal transmission controlled by *bHLHs* is required for understanding the whole mechanism of iron absorption.

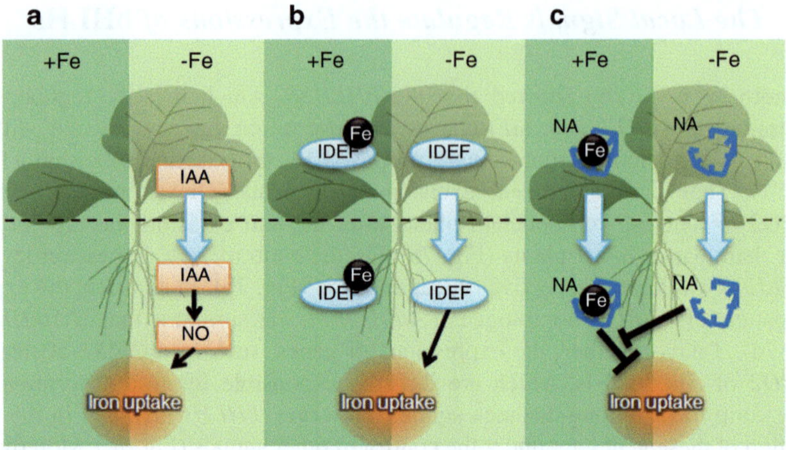

Fig. 8 The three models of the long-distance signaling from the shoots to the roots. (**a**) Auxin model; the synthesis of IAA is enhanced in the meristem in response to the iron deficiency and IAA is sent to the roots. IAA works as the promotive signal so that the genes involved in iron absorption are expressed via the NO signal transduction pathway. (**b**) Nicotianamine model; Na–Fe represses the iron absorption in the roots under the iron-sufficient condition. In the iron-deficient condition, free NA increases and NA–Fe is replaced with free NA by the antagonistic inhibition; consequently, iron is absorbed from the roots. (**c**) IDEF1 model; the transcription factor, IDEF1 moves among tissues and controls the expression of downstream genes related to iron absorption under the iron-deficient condition. When enough iron is supplied, Fe is combined with IDEF1 and Fe–IDEF1 does not induce the gene expressions

4.3 Model of the Long-Distance Signaling for Iron Uptake

We here introduce three models of the shoot-to-root signaling associated with iron deficiency, including candidate substances (Fig. 8). These models are not inconsistent with each other, and in fact, the signaling system likely consists of two or three models.

The first model is auxin which mediates systemic long-distance signals of iron-deficient response. It is suggested that auxin and ethylene are involved in the formation of root hair and the extension of lateral roots in response to the iron deficiency (Schmidt and Steinbach 2000; Schikora and Schmidt 2001; Giehl et al. 2012). Further, nitric oxide (NO) located in the downstream of the signal transduction pathway of auxin activates the transcription of *AtFIT* and *AtFRO2* (Chen et al. 2010; Wu et al. 2012). However, Schmidt et al. (2000) reported that auxin did not affect iron reduction activity in some Arabidopsis mutants with impaired auxin pathway. Further, Enomoto et al. (2007) showed that the expression of *NtIRT1* in roots was positively correlated with the amount of leaves in the experiment of leaf excision; there is no qualitative difference between mature leaves and immature leaves containing meristem in which auxisn is produced, with respect to the ability of activation of iron-deficient response. Hence auxin probably is associated with the signal transduction pathway of the iron-deficient response. However, auxin is not likely a major substance for the signal.

The second candidate is a transcription factor, IDEF1 identified in rice (Fig. 8). The concept of this model consists of the mobility of IDEF1 among tissues, similar to FT known as a signal transduction substance for flowering (Corbesier et al. 2007).

NAS2 and *IRO2* exist downstream of *IDEF1* in rice, and *IRO2* controls iron-responsible genes such as *NAS1,3*, *IRT1*, *IDS3*, and *YSL15* (MA–Fe transporter) (Ogo et al. 2006; Kobayashi et al. 2007; Lee et al. 2009) (Fig. 2). In strategy II plants, transcriptional factors similar to IDEF1 are not identified. IDEF1 has a *cis*-element, IDE1. IDE1 is probably controlled by the long-distance signals because the IDS2 promoter containing IDE1 responds to iron deficiency in transgenic tobacco and the expression of *NtIRT1* is reduced by leaf excision (Kobayashi et al. 2003a, b; Enomoto et al. 2009). IDEF1 expresses in both shoots and roots in response to iron deficiency (Kobayashi et al. 2007). IDEF1 functions as a sensor by binding to iron for controlling the expressions of iron-responsive genes because the lack of an iron-binding element in IDEF1 induces the expression of the genes (Kobayashi et al. 2012). IDEF1 works as a sensor to control the genes of downstream of IDEF1 by binding to iron. Because the expression of IDEF1 is promoted in the whole body in response to iron deficiency, the expressions of iron-responsive genes are enhanced if the iron-binding site of IDEF1 is removed (Kobayashi et al. 2007, 2012). It follows from what has been said that the IDEF1 moves among the tissues, resulting in transmission of the information of iron status. However, it should be noticed that IDEF1-like transcription factors have not been identified in strategy I plants. Therefore, we should expect to identify a transcription factor upstream of FIT and bHLH38 in strategy I plants in order to understand the long-distance signal pathway of iron deficiency.

The last is NA which has been proposed as a sensor molecule (Fig. 8) for more than 10 years (Ling et al. 1999). In general NA is known as a scavenger of excess iron or an essential molecule for iron transferring between tissues (Pich et al. 2001). It is thought that NA is not only a sensor for iron concentration in a cell but also is a substance of long-distance signal transduction (Curie and Briat 2003). The suggested model is as follows; an unidentified receptor binding to NA–Fe represses the induction of expressions of *AtFRO2*, although the receptor binding to NA does not repress it (Fig. 8). The expression of *FRO2* is not regulated in mutants which lack the ability to synthesize NA, such as *chloronelva* and Arabidopsis mutant with quadruple mutation of *NAS* genes (Klatte et al. 2009). These mutants likely form no NA–Fe because of lack of NA synthesis, resulting in no control of *FRO2* expression. On the other hand, transgenic Arabidopsis expressing *NAS* of *Thlaspi caerulescens* showed elevation of the expression of *FRO2*, irrespective of iron concentration in the roots (Cassin et al. 2009). In the condition of high concentration of NA in a cell, for example, the transgenic Arabidopsis expressing *NAS* and wild types suffering from iron deficiency, relatively high amount of free NA comparing with NA–Fe probably stops the depression of the expression of *FRO2*. Consequently, *FRO2* starts to express.

It is necessary that NA moves between shoots and roots or among tissues, for NA to work as a substance of signal transduction. In rice, NA is an essential molecule for iron transportation among tissues and the presence of NA is confirmed in both xylem and phloem (Takahashi et al. 2003; Kakei et al. 2012). In Arabidopsis, NA is necessary for iron transportation among tissues as well, for example, iron is transferred to seeds via AtYSL1 (Le Jean et al. 2005). OPT3 mentioned in Sect. 3.1 is also a transporter related to iron transportation among tissues (Fig. 4). Stacey et al. (2007) discussed that mutation of OPT3 reduced iron transportation to seeds and disconnected an unknown signal transduction pathway. If NA–Fe works as a long-distance signal, the phenotype can be explained as one step: NA–Fe is not transported from leaves to phloem. The restricted supply of NA–Fe results in the lack of iron in seeds and upregulation of the genes involved in the iron absorption. The NA model is suitable for explanation of long-distance signal transduction without inconsistency among behaviors in many mutants.

Acknowledgment We thank David Johnson for critical reading. We thank Springer for permission to use figures shown in Figs. 1 and 6.

References

Arnaud N, Ravet K, Borlotti A, Touraine B, Boucherez J, Fizames C, Briat JF, Cellier F, Gaymard F (2007) The IRE/IRP1-cytosolic aconitase iron regulatory switch does not operate in plants. Biochem J 405:523–531

Bashir K, Inoue H, Nagasaka S, Takahashi M, Nakanishi H, Mori S, Nishizawa NK (2006) Cloning and characterization of deoxymugineic acid synthase genes from graminaceous plants. J Biol Chem 281:32395–32402

Bienfait HF (1989) Prevention of stress in iron metabolism of plants. Acta Bot Neerl 38:105–129

Brumbarova T, Bauer P (2005) Iron-mediated control of the basic helix-loop-helix protein FER, a regulator of uptake in tomato. Plant Physiol 137:1018–1026

Bughio N, Yamaguchi H, Nishizawa NK, Nakanishi H, Mori S (2002) Cloning an iron-regulated metal transporter from rice. J Exp Bot 53:1677–1682

Cassin G, Mari S, Curie C, Briat JF, Czernic P (2009) Increased sensitivity to iron deficiency in Arabidopsis thaliana overaccumulating nicotianamine. J Exp Bot 60:1249–1259

Chen WW, Yang JL, Qin C, Jin CW, Mo JH, Ye T, Zheng SJ (2010) Nitric oxide acts downstream of auxin to trigger root ferric-chelate reductase activity in response to iron deficiency in Arabidopsis. Pant Physiol 154:810–819

Clemens S (2001) Molecular mechanism of plant metal tolerance and homeostasis. Planta 212:475–486

Colangelo EP, Guerinot ML (2004) The essential basic helix-loop-helix protein FIT1 is required for the iron deficiency response. Plant Cell 12:3400–3412

Connolly EL, Fett JP, Guerinot ML (2002) Expression of the IRT1 metal transporter is controlled by metals at the levels of transcript and protein accumulation. Plant Cell 14:1347–1357

Connolly EL, Campbell NH, Grotz N, Prichard CL, Guerinot ML (2003) Overexpression of the FRO2 ferric chelate reductase confers tolerance to growth on low iron and uncovers posttranscriptional control. Plant Physiol 133:1102–1110

Corbesier L, Vincent C, Jang S, Fornara F, Fan Q, Searle I, Giakountis A, Farrona S, Gissot L, Turnbull C, Coupland G (2007) FT protein movement contributes to long-distance signaling in floral induction of Arabidopsis. Science 316:1030–1033

Curie C, Briat JF (2003) Iron transport and signaling in plants. Plant Biol 54:183–206

Curie C, Alonso JM, Le Jean M, Ecker JR, Briat JF (2000) Involvement of NRAMP1 from *Arabidopsis thaliana* in iron transport. Biochem J 347:749–755

DiDonato RJ Jr, Roberts LA, Sanderson T, Eisley RB, Walker EL (2004) Arabidopsis Yellow Stripe-Like2 (YSL2): a metal-regulated gene encoding a plasma membrane transporter of nicotianamine–metal complexes. Plant J 39:403–414

Durrett TP, Gassmann W, Rogers EE (2007) The FRD3-mediated efflux of citrate into the root vasculature is necessary for efficient iron translocation. Plant Physiol 144:197–205

Enomoto Y, Hodoshima H, Shimada H, Shoji K, Yoshihara T, Goto F (2007) Long-distance signals positively regulate the expression of iron uptake genes in tobacco roots. Planta 227:81–89

Enomoto Y, Hashida S, Shoji K, Shimada H, Yoshihara T, Goto F (2009) Expressions of iron uptake genes in roots are affected by long-distance signals both in non-graminaceous and in graminaceous plants. In: IPNC XVI paper:1209

Giehl RF, Lima JE, von Wirén N (2012) Localized iron supply triggers lateral root elongation in Arabidopsis by altering the AUX1-mediated auxin distribution. Plant Cell 24:33–49

Green LS, Rogers EE (2004) FRD3 controls iron localization in Arabidopsis. Plant Physiol 136:2523–2531

Grusak MA, Pezeshgi S (1996) Shoot-to-root signal transmission regulates root Fe(III) reductase activity in the dgl mutant of pea. Plant Physiol 110:329–334

Hall JL, Williams LE (2003) Transition metal transporters in plants. J Exp Bot 54:2601–2613

Herbik A, Giritch A, Horstmann C, Becker R, Balzer HJ, Baumlein H, Stephan UW (1996) Iron and copper nutrition-dependent changes in protein expression in a tomato wild type and the nicotianamine-free mutant *chloronerva*. Plant Physiol 111:533–540

Hodoshima H, Enomoto Y, Shoji K, Shimada H, Goto F, Yoshihara T (2007) Differential regulation of cadmium-inducible expression of iron-deficiency-responsive genes in tobacco and barley. Physiol Plant 129:622–634

Ishimaru Y, Suzuki M, Tsukamoto T, Suzuki K, Nakazono M, Kobayashi T, Wada Y, Watanabe S, Matsuhashi S, Takahashi M, Nakanishi H, Mori S, Nishizawa NK (2006) Rice plants take up iron as an Fe3+-phytosiderophore and as Fe^{2+}. Plant J 45:335–346

Ishimaru Y, Kim S, Tsukamoto T, Oki H, Kobayashi T, Watanabe S, Matsuhashi S, Takahashi M, Nakanishi H, Mori S, Nishizawa NK (2007) Mutational reconstructed ferric chelate reductase confers enhanced tolerance in rice to iron deficiency in calcareous soil. Proc Natl Acad Sci USA 104:7373–7378

Jakoby M, Wang HY, Reidt W, Weisshaar B, Bauer P (2004) FRU (BHLH029) is required for induction of iron mobilization genes in Arabidopsis thaliana. FEBS Lett 577:528–534

Kakei Y, Ishimaru Y, Kobayashi T, Yamakawa T, Nakanishi H, Nishizawa NK (2012) OsYSL16 plays a role in the allocation of iron. Plant Mol Biol 79:583–594

Klatte M, Schuler M, Wirtz M, Fink-Straube C, Hell R, Bauer P (2009) The analysis of Arabidopsis nicotianamine synthase mutants reveals functions for nicotianamine in seed iron loading and iron deficiency responses. Plant Physiol 150:257–271

Kobayashi T, Nishizawa NK (2012) Iron uptake, translocation, and regulation in higher plants. Annu Rev Plant Biol 63:131–152

Kobayashi T, Nakayama Y, Itai RN, Nakanishi H, Yoshihara T, Mori S, Nishizawa NK (2003a) Identification of novel cis-acting elements, IDE1 and IDE2, of the barley IDS2 gene promoter conferring iron-deficiency-inducible, root-specific expression in heterogeneous tobacco plants. Plant J 36:780–793

Kobayashi T, Yoshihara T, Jiang T, Goto F, Nakanishi H, Mori S, Nishizawa NK (2003b) Combined deficiency of iron and other divalent cations mitigates the symptoms of iron deficiency in tobacco plants. Physiol Plant 119:400–408

Kobayashi T, Ogo Y, Itai RN, Nakanishi H, Takahashi M, Mori S, Nishizawa NK (2007) The transcription factor IDEF1 regulates the response to and tolerance of iron deficiency in plants. Proc Natl Acad Sci USA 104:19150–19155

Kobayashi T, Itai RN, Aung MS, Senoura T, Nakanishi H, Nishizawa NK (2012) The rice transcription factor IDEF1 directly binds to iron and other divalent metals for sensing cellular iron status. Plant J 69:81–91

Landsberg EC (1984) Regulation of iron-stress-response by whole-plant activity. J Plant Nutr 7:609–621

Lawson DM, Treffry A, Artymiuk PJ, Harrison PM, Yewdall SJ, Luzzago A, Cesareni G, Levi S, Arosio P (1989) Identification of the ferroxidase centre in ferritin. FEBS Lett 254:207–210

Le Jean M, Schikora A, Mari S, Briat JF, Curie C (2005) A loss-of-function mutation in AtYSL1 reveals its role in iron and nicotianamine seed loading. Plant J 44:769–782

Lee S, Chiecko JC, Kim SA, Walker EL, Lee Y, Guerinot ML, An G (2009) Disruption of OsYSL15 leads to iron inefficiency in rice plants. Plant Physiol 150:786–800

Li L, Cheng X, Ling HQ (2004) Isolation and characterization of Fe(III)-chelate reductase gene LeFRO1 in tomato. Plant Mol Biol 54:125–136

Ling HQ, Koch G, Baumlein H, Ganal MW (1999) Map-based cloning of chloronerva, a gene involved in iron uptake of higher plants encoding nicotianamine synthase. Proc Natl Acad Sci USA 96:7098–7103

Ling HQ, Bauer P, Bereczky Z, Keller B, Ganal M (2002) The tomato fer gene encoding a bHLH protein controls iron-uptake responses in roots. Proc Natl Acad Sci USA 99:13938–13943

Long TA, Tsukagoshi H, Busch W, Lahner B, Salt DE, Benfey PN (2010) The bHLH transcription factor POPEYE regulates response to iron deficiency in Arabidopsis roots. Plant Cell 22:2219–2236

Lynch SR (2011) Why nutritional iron deficiency persists as a worldwide problem. J Nutr 141:763S–768S

Maas FM, van de Wetering DA, van Beusichem ML, Bienfait HF (1988) Characterization of phloem iron and its possible role in the regulation of Fe-efficiency reactions. Plant Physiol 87:167–171

Mori S (1999) Iron acquisition by plants. Curr Opin Plant Biol 2:250–253

Nakanishi H, Yamaguchi H, Sasakuma T, Nishizawa NK, Mori S (2000) Two dioxygenase genes, Ids3 and Ids2, from Hordeum vulgare are involved in the biosynthesis of mugineic acid family phytosiderophores. Plant Mol Biol 44:199–207

Negishi T, Nakanishi H, Yazaki J, Kishimoto N, Fujii F, Shimbo K, Yamamoto K, Sakata K, Sasaki T, Kikuchi S, Mori S, Nishizawa NK (2002) cDNA microarray analysis of gene expression during Fe-deficiency stress in barley suggests that polar transport of vesicles is implicated in phytosiderophore Fe-deficient barley roots. Plant J 30:83–94

Nishida S, Tsuzuki C, Kato A, Aisu A, Yoshida J, Mizuno T (2011) AtIRT1, the primary iron uptake transporter in the root, mediates excess nickel accumulation in *Arabidopsis thaliana*. Plant Cell Physiol 52:1433–1442

Nozoye T, Nagasaka S, Kobayashi T, Takahashi M, Sato Y, Sato Y, Uozumi N, Nakanishi H, Nishizawa NK (2011) Phytosiderophore efflux transporters are crucial for iron acquisition in graminaceous plants. J Biol Chem 286:5446–5454

Ogo Y, Itai RN, Nakanishi H, Inoue H, Kobayashi T, Suzuki M, Takahashi M, Mori S, Nishizawa NK (2006) Isolation and characterization of IRO2, a novel iron-regulated bHLH transcription factor in graminaceous plants. J Exp Bot 57:2867–2878

Ogo Y, Nakanishi Itai R, Nakanishi H, Kobayashi T, Takahashi M, Mori S, Nishizawa NK (2007) The rice bHLH protein OsIRO2 is an essential regulator of the genes involved in Fe uptake under Fe-deficient conditions. Plant J 51:366–377

Pich A, Manteuffel R, Hillmer S, Scholz G, Schmidt W (2001) Fe homeostasis in plant cells: does nicotianamine play multiple roles in the regulation of cytoplasmic Fe concentration? Planta 213:967–976

Rellán-Álvarez R, Giner-Martínez-Sierra J, Orduna J, Orera I, Rodríguez-Castrillón JA, García-Alonso JI, Abadía J, Alvarez-Fernández A (2010) Identification of a tri-iron(III), tri-citrate complex in the xylem sap of iron-deficient tomato resupplied with iron: new insights into plant iron long-distance transport. Plant Cell Physiol 51:91–102

Robinson NJ, Procter CM, Connolly EL, Guerinot ML (1999) A ferric-chelate reductase for iron uptake from soils. Nature 397:694–697

Rogers EE, Guerinot ML (2002) FRD3, a member of the multidrug and toxin efflux family, controls iron deficiency responses in Arabidopsis. Plant Cell 14:1787–1799

Römheld V, Marschner H (1986) Evidence for a specific uptake system for iron phytosiderophores in roots of grasses. Plant Physiol 80:175–180

Schaaf G, Schikora A, Haberle J, Vert G, Ludewig U, Briat JF, Curie C, von Wiren N (2005) A putative function for the Arabidopsis Fe-phytosiderophore transporter homolog AtYSL2 in Fe and Zn homeostasis. Plant Cell Physiol 46:762–774

Schikora A, Schmidt W (2001) Iron stress-induced changes in root epidermal cell fate are regulated independently from physiological responses to low iron availability. Plant Physiol 125:1679–1687

Schmidt W (2003) Iron solutions: acquisition strategies and signaling pathways in plants. Trends Plant Sci 8:188–193

Schmidt W, Steinbach S (2000) Sensing iron—a whole plant approach. Ann Bot 86:589–593

Schmidt W, Tittel J, Schikora A (2000) Role of hormones in the induction of iron deficiency responses in Arabidopsis roots. Plant Physiol 122:1109–1118

Schuler M, Rellán-Álvarez R, Fink-Straube C, Abadía J, Bauer P (2012) Nicotianamine functions in the phloem-based transport of iron to sink organs, in pollen development and pollen tube growth in arabidopsis. Plant Cell 24:2380–2400

Stacey MG, Patel A, McClain WE, Mathieu M, Remley M, Rogers EE, Gassmann W, Blevins DG, Stacey G (2007) The arabidopsis AtOPT3 protein functions in metal homeostasis and movement of iron to developing seeds. Plant Physiol 146:589–601

Staiger D (2002) Chemical strategies for iron acquisition in plants. Angew Chem Int Ed 41:2259–2264

Takagi S, Nomoto K, Takemoto T (1984) Physiological aspect of mugineic acid, a possible phytosiderophore of graminaceous plants. J Plant Nutr 7:469–477

Takahashi M, Terada Y, Nakai I, Nakanishi H, Yoshimura E, Mori S, Nishizawa NK (2003) Role of nicotianamine in the intracellular delivery of metals and plant reproductive development. Plant Cell 15:1263–1280

Thimm O, Essigmann B, Kloska S, Altmann T, Buckhout TJ (2001) Response of Arabidopsis to iron deficiency stress as revealed by microarray analysis. Plant Physiol 127:1030–1043

Vasconcelos M, Musetti V, Li CM, Datta SK, Grusak MA (2004) Functional analysis of transgenic rice (Oryza sativa L.) transformed with an *Arabidopsis thaliana* ferric reductase (AtFRO2). Soil Sci Plant Nutr 50:1151–1157

Vert G, Grotz N, Dedaldechamp F, Gaymard F, Guerinot ML, Briat JF, Curie C (2002) IRT1, an Arabidopsis transporter essential for iron uptake from the soil and for plant growth. Plant Cell 14:1223–1233

Vert GA, Briat JF, Curie C (2003) Dual regulation of the Arabidopsis high-affinity root iron uptake system by local and long-distance signals. Plant Physiol 132:796–804

von Wiren N, Khodr H, Hider RC (2000) Hydroxylated phytosiderophore species possess an enhanced chelate stability and affinity for iron(III). Plant Physiol 124:1149–1158

Wang HY, Klatte M, Jakoby M, Bäumlein H, Weisshaar B, Bauer P (2007) Iron deficiency-mediated stress regulation of four subgroup Ib BHLH genes in *Arabidopsis thaliana*. Planta 226:897–908

Ward JT, Lahner B, Yakubova E, Salt DE, Raghothama KG (2008) The effect of iron on the primary root elongation of Arabidopsis during phosphate deficiency. Plant Physiol 147:1181–1191

Waters BM, Blevins DG, Eide DJ (2002) Characterization of FRO1, a pea ferric-chelate reductase involved in root iron acquisition. Plant Physiol 129:85–94

White PJ, Brown PH (2010) Plant nutrition for sustainable development and global health. Ann Bot 105:1073–1080

Wu H, Li L, Du J, Yuan Y, Cheng X, Ling HQ (2005) Molecular and biochemical characterization of the Fe(III) chelate reductase gene family in *Arabidopsis thaliana*. Plant Cell Physiol 46:1505–1514

Wu T, Zhang HT, Wang Y, Jia WS, Xu XF, Zhang XZ, Han ZH (2012) Induction of root Fe(III) reductase activity and proton extrusion by iron deficiency is mediated by auxin-based systemic signalling in Malus xiaojinensis. J Exp Bot 63:859–870

Yi Y, Guerinot ML (1996) Genetic evidence that induction of root Fe(III) chelate reductase activity is necessary for iron uptake under iron deficiency. Plant J 10:835–844

Yoshihara T, Kobayashi T, Goto F, Masuda T, Higuchi K, Nakanishi H, Nishizawa NK, Mori S (2003) Regulation of the iron-deficiency responsive gene, ids2, of barley in tobacco. Plant Biotechnol 20:33–41

Yoshihara T, Hodoshima H, Miyano Y, Shoji K, Shimada H, Goto F (2006) Cadmium inducible Fe deficiency responses observed from macro and molecular views in tobacco plants. Plant Cell Rep 25:365–373

Yuan YX, Zhang J, Wang DW, Ling HQ (2005) AtbHLH29 of Arabidopsis thaliana is a functional ortholog of tomato FER involved in controlling iron acquisition in strategy I plants. Cell Res 15:613–621

Yuan Y, Wu H, Wang N, Li J, Zhao W, Du J, Wang D, Ling HQ (2008) FIT interacts with AtbHLH38 and AtbHLH39 in regulating iron uptake gene expression for iron homeostasis in Arabidopsis. Cell Res 18:385–397

Strigolactones and the Coordinated Development of Shoot and Root

Hinanit Koltai and Christine A. Beveridge

Abstract Strigolactones are plant hormones with diverse biological roles. In addition to their effect on plant communication in the rhizosphere, they act as signalling molecules in both shoot and root to regulate several aspects of plant growth and development. In this chapter we will present the role of strigolactones as regulators of development and growth of different plant parts. We will highlight some of their properties as signalling molecules, including their modes of action, their movement in the plant and their crosstalk with other plant hormones. Also, we will review evidence that strigolactones contribute to the response of shoot and root to nutrient conditions and discuss their role in the coordination of shoot and root development under different growth conditions.

Keywords Strigolactones • Shoot • Root • Lateral buds • Phosphate • Hormones • Ethylene • Cytokinin • Auxin • Root hairs • Primary root • Lateral root

1 Introduction

Strigolactones (SLs), a family of substances produced by plants, are now known to be plant hormones. SLs have diverse biological roles. They are involved with communication in the rhizosphere, as stimulators of seed germination of parasitic plant (*Striga* and *Orobanche*; Cook et al. 1966; reviewed by Xie et al. 2010) and hyphal branching of the symbiotic arbuscular mycorrhizal fungi (AMF; reviewed

H. Koltai (✉)
Institute of Plant Sciences, Agricultural Research Organization (ARO), the Volcani Center, POB 6, Bet Dagan 50250, Israel
e-mail: hkoltai@agri.gov.il

C.A. Beveridge
School of Biological Sciences, The University of Queensland, St. Lucia, Brisbane, QLD 4072, Australia
e-mail: c.beveridge@uq.edu.au

F. Baluška (ed.), *Long-Distance Systemic Signaling and Communication in Plants*, Signaling and Communication in Plants 19, DOI 10.1007/978-3-642-36470-9_9,

by Koltai et al. 2012). Recently they were shown to also regulate shoot development acting to repress lateral bud outgrowth (Gomez-Roldan et al. 2008; Umehara et al. 2008). Furthermore, SLs regulate development of other plant parts including secondary growth (Agusti et al. 2011), adventitious root formation (Rasmussen et al. 2012b), lateral root formation and root hair length (Kapulnik et al. 2011a; Ruyter-Spira et al. 2011).

SLs are found in a wide variety of plant species, including dicots, monocots and primitive plants (e.g. Xie et al. 2010; Proust et al. 2011; Delaux et al. 2012; Liang et al. 2010; Koltai et al. 2010a). SLs, terpenoid lactones derived from carotenoids (Matusova et al. 2005), are synthesised in roots and shoot although SLs identified to date are most abundant in roots. They all have a common structure consisting of two lactones connected by an enol–ether bridge (reviewed by Xie et al. 2010).

Several steps of SL biosynthesis are known (Brewer et al. 2013; Ruyter-Spira et al. 2012). These include activity of Cytochrome P450 and of two carotenoid cleavage dioxygenase (CCD) enzymes (CCD7 and CCD8) (e.g. reviewed by Dun et al. 2009a; Ruyter-Spira et al. 2012). Also, in rice, the iron-binding protein DWARF 27 (D27) was suggested to be involved in SL biosynthesis (Lin et al. 2009; Ruyter-Spira et al. 2012). Moreover, the path from β-carotene to carlactone, a SL-like plant hormone, has been deciphered in vitro. It was suggested that D27, a β-carotene isomerise, catalyses conversion of all-*trans*-β-carotene into 9-*cis*-β-carotene. The latter might be a substrate for cleavage by CCD7, whereas CCD8 may act downstream by incorporation of oxygen, producing carlactone (Alder et al. 2012). Putative regulators of the SL biosynthesis pathways in rice and *Medicago* were suggested to be the GRAS-type transcription factors NSP1 and NSP2 (Liu et al. 2011).

Dwarf 14 (D14) and MORE AXILLARY GROWTH 2 (MAX2) are likely proteins involved in SL signalling. Mutants in these genes are hyper-branching and show a reduced response to SLs (reviewed by, e.g. Smith and Waters 2012). DAD2, a petunia homologue of the rice and Arabidopsis D14 genes, has been shown to interact with PhMAX2A in vitro in the presence of GR24 (Hamiaux et al. 2012). This has yet to be demonstrated in vivo. GR24 is a SL analogue with biological activity (e.g. Umehara et al. 2008). MAX2 is an F-box protein that might be part of the ubiquitin-mediated degradation of as-yet unknown protein targets (Stirnberg et al. 2007), while D14 is a α/β hydrolase also related to proteins involved in signal transduction (Arite et al. 2009). Along with their multiple regulatory roles in plant development, SLs are thought to coordinate the development of shoot and root in response to growth conditions. Here we will present the roles SLs fulfil in shoot and root development, their suggested modes of action and feedback regulation and an example as to their activity as coordinators of plant development in response to growth conditions.

2 Strigolactones Regulate Development of Different Plant Parts

2.1 Role of Strigolactones in Axillary Shoot Development

The mutants used to identify the genetic regulation of SLs in plants were increased branching mutants. Years earlier, a novel signal was identified as causal of the branching phenotype because the mutants had altered levels of a graft-transmissible signal that suppressed shoot branching; this signal could not be attributed to altered levels of one of the established plant hormones (Beveridge et al. 1997). SLs are required to promote axillary bud "dormancy" but are not thought to be required for axillary bud formation. It should be noted that this dormancy is an active state of metabolically active buds. It is not clear whether axillary buds of SL mutant plants ever enter a phase of dormancy or whether they are always released to grow. Notably, in most species, some axillary buds of SL mutant plants do not grow out and hence other dormancy mechanisms are at play. Studies in pea indicate that SL can inhibit growing buds over quite a wide range of growth states, including large buds of up to 1 cm and can act directly in the bud itself (Dun et al. 2013). The axillary bud specific transcription factor *BRANCHED 1* (*BCR1*) is a likely direct target of SLs in pea (Braun et al. 2012; Dun et al. 2012) and is directly or indirectly involved in SL signalling in other species (Aguilar-Martínez et al. 2007; Finlayson et al. 2010; Takeda et al. 2003; Brewer et al. 2009).

2.2 Role of Strigolactones in Shoot Secondary Growth

Secondary growth is the term used to describe the lateral growth the plant axes, leading to enhanced girth. In shoots it is caused by activity of the vascular cambium leading to production of secondary vascular tissue and wood production (Miyashima et al. 2012). Whereas SLs inhibit axillary bud outgrowth, they promote secondary growth. This has been shown in Arabidopsis, pea and Eucalypts and requires the same MAX2 dependent signalling as for shoot branching inhibition (Agusti et al. 2011).

2.3 Role of Strigolactones in Root Development

SLs have been shown to affect different aspects of root development. In Arabidopsis, under conditions of sufficient phosphate nutrition, SLs negatively regulate lateral root formation (Kapulnik et al. 2011a). Mutants, deficient in either SL response (i.e. *max2*) or biosynthesis (i.e. *max3* and *max4*), had more lateral roots than the wild type (Kapulnik et al. 2011a; Ruyter-Spira et al. 2011), whereas treatment of seedlings with GR24 repressed lateral root formation. This repression by GR24 was in the wild-type and SL-synthesis mutants, but not in the SL-response mutant, suggesting that the negative effect of SLs on lateral roots formation is MAX2 dependent (Kapulnik et al. 2011a; Ruyter-Spira et al. 2011).

SLs are also involved with root hair elongation. Exogenous supplementation of GR24 led to an increase in root hair length in the wild-type and SL-deficient mutants (*max3* and *max4*), but not in *max2*, the SL-response mutant, suggesting that here too the affect of SL is mediated via MAX2 (Kapulnik et al. 2011a).

A role for SLs in primary root growth has been demonstrated under conditions of carbohydrate limitation, which usually lead to a reduction in primary root length (Jain et al. 2007). Under these conditions the SL-deficient and SL-response mutants had a shorter primary root and less primary meristem cell number than those of the wild-type plants. Accordingly, under these conditions, GR24 treatments had a positive effect on primary root elongation and meristem cell number, in a MAX2-dependent manner (Ruyter-Spira et al. 2011).

2.4 Role of Strigolactones in Adventitious Root Formation

Adventitious root formation from stems is negatively regulated by SLs in Arabidopsis and pea. Enhanced adventitious rooting was found in both species in SL-deficient and response mutants, whereas SL treatments reduced adventitious rooting in the SL biosynthesis mutant and WT, but not in the *max2* mutant (Rasmussen et al. 2012b). Moreover, based on the response of mutants to plant hormones, it was suggested that SLs and cytokinins act independently to suppress adventitious rooting, whereas at least partial dependency between SLs and auxin activity was demonstrated for this process (Rasmussen et al. 2012b). In accordance, it was found in tomato that transgenic plants with reduced *SlCCD8* expression, and thereby reduced SLs levels, displayed excessive adventitious root development (Kohlen et al. 2012), further supporting a negative role for SLs in this process. The use of SL inhibitors to enhance rooting has been shown in a few cases (Rasmussen et al. 2012a) and adds promise to the hope that we may be able to overcome the restriction of woody plants to adventitious rooting, thereby solving an important hurdle for propagation of woody plants for industry and for conservation of endangered species.

3 Strigolactones Act as Signalling Molecules

3.1 Strigolactone Movement

The main SLs discovered to date are found in highest concentrations in roots. Many grafting studies have indicated that SLs, their metabolites or other unknown secondary messengers move in the root-to-shoot direction to inhibit shoot branching (reviewed by Dun et al. 2009a). Moreover, some evidence as to the presence of the SL orobanchol in the xylem sap of Arabidopsis was found (Kohlen et al. 2011), suggesting orobanchol to comply with this model, i.e., to be produced in the root and move towards the shoot through vasculature. The suggestion that SLs themselves are moving rather than any degradation products is reinforced in vitro by the findings of Hamiaux et al. (2012). Although these are yet to be repeated in vivo, the petunia PhDAD2 was shown to interact with PhMAX2A in a GR24 concentration-dependent manner. Second, DAD2 was shown to have the ability to hydrolyse GR24, and this ability was associated with the ability of DAD2 to interact with PhMAX2A. The GR24 hydrolysis products did not modulate branching nor stimulated the DAD2–PhMAX2A protein interaction. Therefore, the working hypothesis is that DAD2 may act to bind the mobile SL signal and catalyse it during its PhMAX2A interaction, for SL signal transduction (Hamiaux et al. 2012).

Together, these results suggest that SLs themselves, rather than their hydrolysis or downstream products, are the active compounds, and that they may be actively transported to their target organs (e.g. in or near shoot buds) for their activity.

As for the mode of SL movement, at least a partial insight came from the study of an ATP-binding cassette (ABC) transporter in petunia (Kretzschmar et al. 2012). The Petunia ABC transporter PDR1 was suggested to function as a cellular SL exporter. Evidently, overexpression of the Petunia PDR1 in Arabidopsis resulted with increased tolerance to high concentrations of GR24, due probably to increased export of SLs from the roots. Also, *pdr1* mutant has an enhanced branching phenotype and reduced mycorrhiza symbiotic interactions associated with a reduction in SL exudation from their roots. PDR1 was found to be present in cells in the plasma membrane, consistent with a suggested role in secretion. It is expressed in root tissues, extensively in individual subepidermal cells of the lateral roots. In the stem it was largely restricted to the vasculature and nodal tissues adjacent to leaf axils but absent from dormant buds. It was suggested that this pattern of expression in the stem is consistent with PDR1 functioning as a SL transporter. PDR1 may confer cellular mobility between cells that might be required to deliver SLs to their site of action (Kretzschmar et al. 2012).

It is interesting to mention, in this regard, that SLs might be produced not only in roots, but also in other plant parts. Studies showed that the pea *rms1* (CCD8) is expressed in different plant tissues, in addition to roots, although to a lesser extent than in roots. These parts include mainly epicotyl tissue and internode tissue (Foo et al. 2005; Dun et al. 2009b). These findings are in line with the suggestion that SLs could act locally, as a second messenger for auxin action to directly repress bud outgrowth (Brewer et al. 2009; detailed below). Further evidence that SLs may act at the site of its own biosynthesis comes from grafting studies. Branching inhibition is greater in wild-type shoots grafted to SL-deficient roots, rather than the reciprocal combination, and hence the shoot is actually better than the root at inhibiting branching (Foo et al. 2001; Morris et al. 2005). The weaker long-distance inhibition of branching by root-derived SLs, than for shoot-derived SLs, suggests that either the compound breaks down very rapidly over distance (like in a few minutes, because xylem transport is rapid), or that the bioactive compound is more abundant in shoots, or that root-derived sources cannot be delivered as well as shoot-derived ones to the receptor sites. This, and the clear effect of SLs in roots, their main site of production, suggests that at least in some of the cases, SLs might act in the same cells in which they are produced, or very nearby. As discussed below, the dual action of auxin on the levels of cytokinin and SL also makes interpretation complex, not to mention the possibility that some actions of SL may be via auxin transport.

3.2 *Strigolactone Signalling and Crosstalk*

It is likely that for most processes, effects of SLs rely on a mutual influence between the different plant hormones, such that, for example, one might affect the biosynthesis or transport of the other. SL-related biological processes that demonstrate this most clearly are described below.

3.2.1 Shoot Branching

Shoot branching is regulated by the interplay of many hormonal signals and factors in addition to SL. The earliest studies of shoot branching focused on the observation that auxin can largely replace the shoot tip in its ability to inhibit the growth of axillary buds below (apical dominance). It was soon realised that this auxin inhibition must be indirect and indeed auxin clearly acts via two opposing signals, SL and cytokinin, which inhibit and promote axillary bud outgrowth, respectively. Auxin regulates the expression of SL and cytokinin biosynthesis genes and SLs have been shown to inhibit branching in auxin depleted plants and auxin signalling mutants (Shimizu-Sato et al. 2009; Brewer et al. 2009). Signal cross talk in regulating shoot branching is also observed after SL and cytokinin reception as *BRC1* expression is rapidly and independently modified by both of these hormones (Dun et al. 2012). This occurs to varying degrees in different species. Auxin transport is often enhanced in SL mutants, although we are yet to determine what role this has in shoot branching (Leyser 2009; Renton et al. 2012; Domagalska and Leyser 2011). As discussed below, the SL biosynthesis genes are also likely to be feedback regulated, both locally and systemically by auxin and non-auxin-related signalling.

3.2.2 Root Branching and Directional Growth

Roots form lateral roots early during seedling development. This process involves formation of dynamic gradients of auxin with maxima at the primordia tips. These gradients are formed as a result of asymmetrically localisation of PIN1, auxin efflux carriers (Benkova et al. 2003).

As indicated above, SLs are involved with lateral root formation. Arabidopsis mutants flawed in SL sensing or biosynthesis have an increased number of lateral roots early following seed germination, and GR24 treatments reduced the number of lateral roots under optimal growth conditions (Kapulnik et al. 2011a; Ruyter-Spira et al. 2011). In accordance, upon treatment of seedlings with GR24 a decrease in PIN1-GFP intensity was apparent in lateral root primordia, suggesting an involvement for PIN1 in the GR24-mediated reduction of lateral root development. Also, once auxin was exogenously applied, GR24 application induced, rather than reduced, lateral root development, whereas no reduction in PIN1-GFP intensity was observed under these conditions (Ruyter-Spira et al. 2011). Therefore, Ruyter-Spira et al. (2011) suggested that SLs modulate in roots auxin flux associated with lateral root primordia and thus alter the auxin optima necessary for lateral root formation. Under optimal growth conditions without auxin supplementation SLs reduce auxin import to the lateral root primordia and thereby lead to inhibition of lateral root development. In contrast, under high auxin levels (i.e. exogenous auxin supplementation), the SL-mediated reduction in auxin flux reduces the excess, inhibiting levels of auxin in lateral root primordia and thereby enhance lateral root development. Further evidence of the effect of SLs on auxin flux in roots came from the study of Koltai et al. (2010b). The inhibitory effect of GR24 on root hair formation was reversed by 2,4-D only, which is a synthetic auxin that is not secreted by efflux carriers, suggesting a functional involvement of SLs with auxin efflux in the root.

Another example to demonstrate this effect of SLs on auxin efflux may be the interference of SLs with root directional growth. In both tomato and Arabidopsis, GR24 treatments induced asymmetric root growth (Koltai et al. 2010b; Ruyter-Spira et al. 2011). This, again, might be explained by interference of SLs with auxin flux, leading to asymmetric auxin distribution. Ruyter-Spira et al. (2011) suggested this effect to be a result of distorted expression of the PIN auxin efflux carriers. However, SL might also systemically affect auxin sensitivity or levels, since GR24-treated plants showed decreased GUS staining from the auxin-response reporter DR5-GUS in their aerial parts (Ruyter-Spira et al. 2011). Another evidence for a SL-mediated response that involves auxin sensitivity comes from studies of SL regulation of root development under conditions of phosphate starvation (detailed below).

3.2.3 Root Hair Elongation

Root hair development, i.e., to establish and maintain root hair tip growth, requires activation of intracellular signal transduction pathways and multiple cellular components such as ion channels, cytoskeleton and cell wall materials (reviewed by Ishida et al. 2008). Moreover, root hair tip elongation was suggested to be a function of a hormonal balance in the epidermal cell layer. It was shown that ethylene, which was previously found to promote auxin biosynthesis (Swarup et al. 2007) and/or auxin transport (Ruzicka et al. 2007), directs auxin in the epidermal cell layer to promote root hair elongation while inhibiting root epidermal cell elongation (Strader et al. 2010).

As detailed above, SLs positively affect root hair elongation in young seedlings (Kapulnik et al. 2011a). Analysis of the SLs response mutant *max2* suggested that SL signalling is not necessary for the root hair elongation induced by auxin; however, auxin signalling was required, at least in part, for the positive effect of SLs on root hair elongation. This is because the auxin-receptor mutant *tir1-1* (Dharmasiri et al. 2005) had reduced sensitivity to SL relative to the wild type (Kapulnik et al. 2011b). Ethylene too was shown to be involved with the root hair-SLs response. The ethylene-signalling mutants *etr* and *ein* had significantly reduced response to GR24 and aminoethoxyvinylglycine (AVG, an ethylene-synthesis inhibitor) had a negative effect on root hair elongation in response to SLs. Also, treatment with GR24 induced transcription of the 1-aminocyclopropane-1-carboxylic acid (ACC) synthases, involved in ethylene biosynthesis (Kapulnik et al. 2011b). These results suggest that SLs exert their effect on root hair length at least partially through the ethylene pathway (Koltai 2011).

3.2.4 Shoot Secondary Growth

The vascular cambium, a stem cell-like tissue, regulates secondary growth through an auxin (and additional hormones) dependent manner (Miyashima et al. 2012). Based on studies of cell-specific activation of SL signalling, SLs positively regulate cambial activity by a local induction of the cambium-specific stem cell niche and of vascular tissue formation. Hence, it was suggested that in Arabidopsis (and other species) a local increase in SL levels is capable of stimulating secondary growth independently from an effect on shoot branching (Agusti et al. 2011). Expression of MAX2 under the control of the (pro)cambium-specific WOX4 promoter in *max2-1* mutants background was sufficient to confer secondary growth at WT-like levels, suggesting a local, cambium-specific, MAX2-dependent activity of SLs (Agusti et al. 2011). Cell-proliferating activity that leads to secondary growth is regulated over a long distance by auxin (Miyashima et al. 2012) This and the above results of SL activity suggest that SL signalling regulates the process of secondary growth in the cambium in a cell-autonomous manner, as a secondary messenger of auxin. This might be carried via an auxin-dependent stimulation of SL biosynthesis in the cambium (Agusti et al. 2011).

3.2.5 Strigolactone Feedback Regulation

As SLs regulate plant development, their level should be carefully regulated as part of a homeostatic steady state. Three groups of molecules are suggested to carry this function of SL feedback regulation. One is auxin which positively regulates SL levels in roots and stems. MAX3 (CCD7), MAX4 (CCD8) and DWARF27

(D27; Waters et al. 2012) transcripts were shown to be positively regulated by auxin in pea and Arabidopsis. In SL response and synthesis mutants *MAX3*, *MAX4* and *D27* expression is upregulated, consistent with the findings of increased auxin in the polar auxin transport stream of these mutants (Bennett et al. 2006). Also, auxin depletion treatments reduced SL biosynthesis gene expression in pea (for RMS5 and RMS1; Foo et al. 2005; Johnson et al. 2006) and Arabidopsis, in the different plant tissues (Hayward et al. 2009). In addition, based on analysis of auxin signalling mutants in this respect, it was suggested that this feedback regulation of auxin on SL biosynthesis involves auxin signalling (Hayward et al. 2009).

A second group of molecules to regulate SL levels are SLs themselves. It was shown in several plant species, including pea and Arabidopsis that in addition to SL mutants showing higher levels of SL biosynthesis gene expression and/or SL content (e.g. Foo et al. 2005; Dun et al. 2009b; Hayward et al. 2009; Umehara et al. 2010), GR24 treatments can reduce expression of SL biosynthesis genes (Mashiguchi et al. 2009; Dun et al. 2012).

Systemic feedback regulation of SL production requires a signal in addition to SLs, since SLs move upward in the plant and the feedback signalling can move basipetally from shoot to root. Perhaps a more complete model for SL feedback regulation comes from the assumption of the presence of a third molecule(s) to regulate SL levels. SL biosynthesis and response mutants in Arabidopsis and pea, with the exception of the hyper-branching mutant *rms2* in pea, have reduced levels of the major cytokinins in xylem sap in comparison to wild-type plants (Foo et al. 2005, 2007). Based on grafting experiments and double mutant analysis it was suggested that in pea the regulation of root cytokinin export is mediated by an inhibitory signal. Further studies indicate that local signalling and response to SLs might downregulate the long-distance feedback signal in pea (Foo et al. 2007). In addition to regulating expression of SL biosynthesis genes, this long-distance signal also reduces cytokinin export from root. Since this negative feedback inhibition was not detected in *rms2*, it was suggested that RMS2 is essential for long-distance feedback regulation of cytokinin export from roots (Foo et al. 2001, 2005, 2007). Whether auxin may constitute the only long-distance signal involved in SL feedback signalling may therefore be revealed by understanding the function of *RMS2*.

4 Strigolactone-Related Signalling for Coordinated Response of Root and Shoot to Phosphate Conditions

SLs clearly affect both shoot and root development; here we discuss evidence of SL association with plant responses to growth conditions. SLs are potential coordinators of shoot and root development for fine regulation of growth under a changing environment of nutrients. Indeed, SLs have a clear role in coordination of plant development, acting as a means of inter-plant communication for nutrient status.

One of the essential macronutrients is phosphorus (P). It is required by plants for growth and development, as it is forms essential components of cellular macromolecules and participates in major metabolic processes. Plants acquire P from the soil, mostly in its inorganic phosphate (Pi) form. However, Pi levels are limiting factors for development in many habitats and vary considerably (Bieleski 1973; Maathuis 2009).

Roots serve as the main organ for Pi acquisition and have developed a number of ways to enhance Pi absorption. Among these is an ability to alter root structure to increase the root surface to enhance Pi uptake from the soil. Specifically, once Pi-deprivation conditions develop, plants alter their root system architecture by inhibition of primary root growth accompanied by promotion of lateral root formation and elongation. These changes are suggested to increase the root surface for foraging of P, leading to an increased ability of Pi absorption (reviewed by López-Bucio et al. 2003; Peret et al. 2011).

Another change taking place under reduced Pi levels is the elongation of root hairs and an increase in their number; the root hair response to low-P availability was suggested to be an efficient strategy for P acquisition (Bates and Lynch 2000; reviewed by Gilroy and Jones 2000).

SL production is induced in several plant species under low Pi conditions (e.g. Yoneyama et al. 2007; López-Ráez and Bouwmeester 2008; Kohlen et al. 2011). Moreover it was suggested that there is a correlation between shoot P levels and SL exudation across plant species (Yoneyama et al. 2012). As described above, SLs were shown to regulate positively root hair elongation and negatively lateral root formation (Kapulnik et al. 2011a), suggesting they are associated with these structural features which are important for root adaptation to Pi conditions. Moreover, SLs were shown to be essential for the roots ability to sense of or respond to low Pi conditions in term of root hair density. Evidently, SL biosynthesis or response mutants are flawed in their low Pi response, observed as a reduced ability to increase root hair density under low Pi conditions, shortly after germination (Mayzlish-Gati et al. 2012). In agreement with a role for SLs in the root response to Pi deprivation, they exert a positive, rather than negative effect on lateral root formation under these conditions (Ruyter-Spira et al. 2011).

The lack of increment in root hair density in the SL mutants following germination under low Pi conditions is also accompanied by their reduced expression of several Pi transporters (Mayzlish-Gati et al. 2012). Therefore it might be that the SL mutants suffer from reduced levels of internal phosphorous (P). However, the SL-insensitive mutant *max2* plants had P concentrations similar to the wild type under both low and high Pi conditions (Mayzlish-Gati et al. 2012), suggesting that the plant can acquire P even in the absence of a SL response. These results also suggest that unlike the WT, low levels of P in the SL mutant under low Pi conditions were not sufficient to provoke an increase in root hair density. Taken together, root sensing of, or response, to low Pi may require the activity of MAX2 and wild-type levels of SLs.

The SL pathway is also important for the shoot response to low Pi conditions. This was demonstrated in both Arabidopsis and rice, in which shoots respond with a reduced branching phenotype under growth conditions of Pi deprivation. In Arabidopsis, SL (orobanchol) detected in xylem sap and was enhanced under Pi deficiency; this was in correlation with the changes in shoot architecture (Kohlen et al. 2011). In rice, under these conditions, root SL (2′-epi-5-deoxystrigol) levels increased, and tiller bud outgrowth was inhibited. Moreover, the SL-deficient or insensitive mutants were not able to suppress tiller bud outgrowth under low Pi growth conditions (Umehara et al. 2010; reviewed by Umehara 2011).

Together, these findings suggest SLs mediate the low Pi response in root and shoot by changing plant architecture: once plants encounter and sense Pi-deprivation conditions in its growth media, shoot branching is inhibited, root branching is induced and root hair length and density are stimulated. As a result the ratio of shoot to root in reduced (Ericsson 1995).

The SL-associated response to low Pi conditions may be mediated through auxin signalling. In the WT, auxin was suggested to be required for the low Pi response in roots, especially in regards to the induced formation of lateral root and arrest of primary root growth (reviewed by Lopez-Bucio et al. 2002; Chiou and Lin 2011). Moreover, increased auxin sensitivity was detected under reduced Pi availability, resulting from induction of *TIR1* transcription (Lopez-Bucio et al. 2002; Perez-Torres et al. 2008). Interestingly, the SL-response mutant, under these conditions of Pi deprivation, displayed a reduction rather than induction of *TIR1* transcription (Mayzlish-Gati et al. 2012). Also, exogenous supplementation of indole-3-acetic acid (IAA) to SL-insensitive and SL biosynthesis mutant roots resulted with complementation of the mutants' phenotypes to that of the wild type (Mayzlish-Gati et al. 2012). These results and the fact that the *tir1* mutant was flawed in its low Pi response in comparison to the wild type (Perez-Torres et al. 2008), and this deficiency could not be restored by GR24 application (Mayzlish-Gati et al. 2012), suggest that SL signalling pathway is upstream to the TIR1-mediated low Pi response.

Thus, SL regulation on its own biosynthesis (detailed above) may be carried also by SL effect on both plant structure and auxin response. On the one hand, increased SL levels under low Pi conditions might lead to increased sensitivity to shoot-derived auxin, and therefore, to increased SL biosynthesis in roots under these conditions. On the other hand, the changes confer by SLs in the root (e.g. root hair elongation and lateral root formation) may lead to increased Pi uptake (although this has not yet been tested). If so, since high Pi levels reduce SL biosynthesis (e.g. Yoneyama et al. 2007; López-Ráez and Bouwmeester 2008; Kohlen et al. 2011), the SL-associated low Pi response and the likely increase in Pi levels may lead to reduced SL levels, and as a result to lower levels of SL biosynthesis (Koltai 2012).

A carefully coordinated Pi signalling is suggested to involve both root and shoot, and to require precise communication between them (reviewed by Chiou and Lin 2011). Also, both local and systemic signalling in root and shoot are taking place during plant response to Pi deprivation. It might be that SLs act as communication molecules in this shoot–root communication in response to Pi growth conditions.

5 Concluding Remarks

It is very clear that SLs act as signalling molecules in both shoot and root to regulate several aspects of plant growth and development in complex cross talk with other plant hormones. SLs were initially discovered as molecules of plant–microbe communication. They are exuded from plants and necessary for parasitic plant seed germination, and for hyphal branching of the symbiotic, arbuscular mycorrhiza fungi (AMF; reviewed by Xie et al. 2010; Koltai et al. 2012). AMF is known to promote the plant's ability to acquire Pi (e.g. Bucher 2007). Therefore, SLs may benefit plants under Pi-deprived conditions not only by regulating plant development but also by promoting AMF hyphal branching, and, as a result, the mycorrhizal association, for increased Pi acquisition. The outcome of this is a measured whole plant response to nutrient status involving modulations of nutrient uptake and developmental responses in root and shoot.

Future challenges for SL research include understanding the mechanisms of SL action and how it differs in different processes. In comparison with other plant hormones, we have much to learn about the later stages of the biosynthetic pathway so that we can develop a better understanding of the bioactivity of SLs and the importance of differences in relative abundance. Again, in comparison with other plant hormones, the levels and identities of SLs in shoot material is poorly understood, as is the relative contribution of root-derived and shoot-derived sources. It is possible that various growth conditions not only affect the level of SL, but potentially also their structural composition and/or relative shoot verses root supply.

References

Aguilar-Martínez JA, Poza-Carrión C, Cubas P (2007) Arabidopsis BRANCHED1 acts as an integrator of branching signals within axillary buds. Plant Cell 19:458–472

Agusti J, Herold S, Schwarz M, Sanchez P, Ljung K, Dun EA, Brewer PB, Beveridge CA, Sieberer T, Sehr EM, Greb T (2011) Strigolactone signaling is required for auxin-dependent stimulation of secondary growth in plants. Proc Natl Acad Sci USA 108:20242–20247

Alder A, Jamil M, Marzorati M, Bruno M, Vermathen M, Bigler P, Ghisla S, Bouwmeester H, Beyer P, Al-Babili S (2012) The path from {beta}-carotene to carlactone, a strigolactone-like plant hormone. Science 335:1348–1351

Arite T, Umehara M, Ishikawa S, Hanada A, Maekawa M, Yamaguchi S, Kyozuka J (2009) d14, a strigolactone-insensitive mutant of rice, shows an accelerated outgrowth of tillers. Plant Cell Physiol 50:1416–1424

Bates TR, Lynch JP (2000) The efficiency of *Arabidopsis thaliana* (Brassicaceae) root hairs in phosphorus acquisition. Am J Bot 87:964–970

Benkova E, Michniewicz M, Sauer M, Teichmann T, Seifertov D, Jurgens G, Friml J (2003) Local, efflux-dependent auxin gradients as a common module for plant organ formation. Cell 115:591–602

Bennett T, Sieberer T, Willett B, Booker J, Luschnig C, Leyser O (2006) The Arabidopsis MAX pathway controls shoot branching by regulating auxin transport. Curr Biol 16:553–563

Beveridge CA, Murfet IC, Kerhoas L, Sotta B, Miginiac E, Rameau C (1997) The shoot controls zeatin riboside export from pea roots. Evidence from the branching mutant *rms4*. Plant J 11:339–345

Bieleski R (1973) Phosphate pools, phosphate transport, and phosphate availability. Annu Rev Plant Physiol 24:225–252

Braun N, de Saint GA, Pillot J-P, Sp B-M, Dalmais M, Antoniadi I, Li X, Maia-Grondard A, Le Signor C, Bouteiller N, Luo D, Bendahmane A, Turnbull C, Rameau C (2012) The pea TCP transcription factor PsBRC1 acts downstream of strigolactones to control shoot branching. Plant Physiol 158:225–238

Brewer PB, Dun EA, Ferguson BJ, Rameau C, Beveridge CA (2009) Strigolactone acts downstream of auxin to regulate bud outgrowth in pea and Arabidopsis. Plant Physiol 150:482–493

Brewer PB, Koltai H, Beveridge CA (2013) Diverse roles of strigolactones in plant development. Mol Plant 6(1):18–28

Bucher M (2007) Functional biology of plant phosphate uptake at root and mycorrhiza interfaces. New Phytol 173:11–26

Chiou T-J, Lin S-I (2011) Signaling network in sensing phosphate availability in plants. Annu Rev Plant Biol 62:185–206

Cook CE, Whichard LP, Turner B, Wall ME, Egley GH (1966) Germination of witchweed (*Striga lutea* Lour.): isolation and properties of a potent stimulant. Science 154:1189–1190

Delaux P-M, Xie X, Timme RE, Puech-Pages V, Dunand C, Lecompte E, Delwiche CF, Yoneyama K, Bécard G, Séjalon-Delmas N (2012) Origin of strigolactones in the green lineage. New Phytol 195:857–871

Dharmasiri N, Dharmasiri S, Estelle M (2005) The F-box protein TIR1 is an auxin receptor. Nature 435:441–445

Domagalska MA, Leyser O (2011) Signal integration in the control of shoot branching. Nat Rev Mol Cell Biol 12:211–221

Dun EA, Brewer PB, Beveridge CA (2009a) Strigolactones: discovery of the elusive shoot branching hormone. Trends Plant Sci 14:364–372

Dun EA, Hanan J, Beveridge CA (2009b) Computational modeling and molecular physiology experiments reveal new insights into shoot branching in pea. Plant Cell 21:3459–3472

Dun EA, de Saint Germain A, Rameau C, Beveridge CA (2012) Antagonistic action of strigolactone and cytokinin in bud outgrowth control. Plant Physiol 158:487–498

Dun EA, de Saint Germain A, Rameau C, Beveridge C (2013) Dynamics of strigolactone function and shoot branching responses in *Pisum sativum*. Mol Plant 6:128–140

Ericsson T (1995) Growth and shoot: root ratio of seedlings in relation to nutrient availability. Plant Soil 168–169:205–214

Finlayson SA, Krishnareddy SR, Kebrom TH, Casal JJ (2010) Phytochrome regulation of branching in Arabidopsis. Plant Physiol 152:1914–1927

Foo E, Turnbull CGN, Beveridge CA (2001) Long-distance signaling and the control of branching in the *rms1* mutant of pea. Plant Physiol 126:203–209

Foo E, Bullier E, Goussot M, Foucher F, Rameau C, Beveridge CA (2005) The branching gene RAMOSUS1 mediates interactions among two novel signals and auxin in pea. Plant Cell 17:464–474

Foo E, Morris SE, Parmenter K, Young N, Wang H, Jones A, Rameau C, Turnbull CGN, Beveridge CA (2007) Feedback regulation of xylem cytokinin content is conserved in pea and Arabidopsis. Plant Physiol 143:1418–1428

Gilroy S, Jones DL (2000) Through form to function: root hair development and nutrient uptake. Trends Plant Sci 5:56–60

Gomez-Roldan V, Fermas S, Brewer PB, Puech-Pages V, Dun EA, Pillot J-P, Letisse F, Matusova R, Danoun S, Portais J-C, Bouwmeester H, Becard G, Beveridge CA, Rameau C, Rochange SF (2008) Strigolactone inhibition of shoot branching. Nature 455:189–194

Hamiaux C, Drummond RSM, Janssen BJ, Ledger SE, Cooney JM, Newcomb RD, Snowden KC (2012) DAD2 is an α/β hydrolase likely to be involved in the perception of the plant branching hormone, strigolactone. Curr Biol 22:2032–2036

Hayward A, Stirnberg P, Beveridge C, Leyser O (2009) Interactions between auxin and strigolactone in shoot branching control. Plant Physiol 151:400–412

Ishida T, Kurata T, Okada K, Wada T (2008) A genetic regulatory network in the development of trichomes and root hairs. Annu Rev Plant Biol 59:365–386

Jain A, Poling MD, Karthikeyan AS, Blakeslee JJ, Peer WA, Titapiwatanakun B, Murphy AS, Raghothama KG (2007) Differential effects of sucrose and auxin on localized phosphate deficiency-induced modulation of different traits of root system architecture in Arabidopsis. Plant Physiol 144:232–247

Johnson X, Brcich T, Dun EA, Goussot M, Haurogne K, Beveridge CA, Rameau C (2006) Branching genes are conserved across species. Genes controlling a novel signal in pea are coregulated by other long-distance signals. Plant Physiol 142:1014–1026

Kapulnik Y, Delaux P-M, Resnick N, Mayzlish-Gati E, Wininger S, Bhattacharya C, Séjalon-Delmas N, Combier J-P, Bécard G, Belausov E, Beeckman T, Dor E, Hershenhorn J, Koltai H (2011a) Strigolactones affect lateral root formation and root-hair elongation in Arabidopsis. Planta 233:209–216

Kapulnik Y, Resnick N, Mayzlish-Gati E, Kaplan Y, Wininger S, Hershenhorn J, Koltai H (2011b) Strigolactones interact with ethylene and auxin in regulating root-hair elongation in Arabidopsis. J Exp Bot 62:2915–2924

Kohlen W, Charnikhova T, Liu Q, Bours R, Domagalska MA, Beguerie S, Verstappen F, Leyser O, Bouwmeester H, Ruyter-Spira C (2011) Strigolactones are transported through the xylem and play a key role in shoot architectural response to phosphate deficiency in nonarbuscular mycorrhizal host Arabidopsis. Plant Physiol 155:974–987

Kohlen W, Charnikhova T, Lammers M, Pollina T, Tóth P, Haider I, Pozo MJ, de Maagd RA, Ruyter-Spira C, Bouwmeester HJ, López-Ráez JA (2012) The tomato CAROTENOID CLEAVAGE DIOXYGENASE8 (SlCCD8) regulates rhizosphere signaling, plant architecture and affects reproductive development through strigolactone biosynthesis. New Phytol 196:535–547

Koltai H, LekKala SP, Bhattacharya C, Mayzlish-Gati E, Resnick N, Wininger S, Dor E, Yoneyama K, Yoneyama K, Hershenhorn J, Joel DM, Kapulnik Y (2010a) A tomato strigolactone-impaired mutant displays aberrant shoot morphology and plant interactions. J Exp Bot 61:1739–1749

Koltai H (2011) Strigolactones are regulators of root development. New Phytol 190:545–549

Koltai H (2012) Strigolactones activate different hormonal pathways for regulation of root development in response to phosphate growth conditions. Ann Bot. doi:10.1093/aob/mcs216

Koltai H, Dor E, Hershenhorn J, Joel D, Weininger S, Lekalla S, Shealtiel H, Bhattacharya C, Eliahu E, Resnick N, Barg R, Kapulnik Y (2010b) Strigolactones' effect on root growth and root-hair elongation may be mediated by auxin-efflux carriers. J Plant Growth Regul 29:129–136

Koltai H, Matusova R, Kapulnik Y (2012) Strigolactones in root exudates as a signal in symbiotic and parasitic interactions. In: Vivanco JM, Baluška F (eds) Secretions and exudates in biological systems, Signaling and communication in plants. Springer, Berlin, pp 49–73

Kretzschmar T, Kohlen W, Sasse J, Borghi L, Schlegel M, Bachelier JB, Reinhardt D, Bours R, Bouwmeester HJ, Martinoia E (2012) A petunia ABC protein controls strigolactone-dependent symbiotic signalling and branching. Nature 483:341–344

Leyser O (2009) The control of shoot branching: an example of plant information processing. Plant Cell Environ 32:694–703

Liang J, Zhao L, Challis R, Leyser O (2010) Strigolactone regulation of shoot branching in chrysanthemum (*Dendranthema grandiflorum*). J Exp Bot 61:3069–3078

Lin H, Wang R, Qian Q, Yan M, Meng X, Fu Z, Yan C, Jiang B, Su Z, Li J, Wang Y (2009) DWARF27, an iron-containing protein required for the biosynthesis of strigolactones, regulates rice tiller bud outgrowth. Plant Cell 21:1512–1525

Liu W, Kohlen W, Lillo A, Op den Camp R, Ivanov S, Hartog M, Limpens E, Jamil M, Smaczniak C, Kaufmann K, Yang W-C, Hooiveld GJEJ, Charnikhova T, Bouwmeester HJ, Bisseling T, Geurts R (2011) Strigolactone biosynthesis in *Medicago truncatula* and rice requires the symbiotic GRAS-type transcription factors NSP1 and NSP2. Plant Cell 23:3853–3865

Lopez-Bucio J, Hernandez-Abreu E, Sanchez-Calderon L, Nieto-Jacobo MF, Simpson J, Herrera-Estrella L (2002) Phosphate availability alters architecture and causes changes in hormone sensitivity in the Arabidopsis root system. Plant Physiol 129:244–256

López-Bucio J, Cruz-Ramírez A, Herrera-Estrella L (2003) The role of nutrient availability in regulating root architecture. Curr Opin Plant Biol 6:280–287

López-Ráez J, Bouwmeester H (2008) Fine-tuning regulation of strigolactone biosynthesis under phosphate starvation. Plant Signal Behav 3:963–965

Maathuis FJM (2009) Physiological functions of mineral macronutrients. Curr Opin Plant Biol 12:250–258

Mashiguchi K, Sasaki E, Shimada Y, Nagae M, Ueno K, Nakano T, Yoneyama K, Suzuki Y, Asami T (2009) Feedback-regulation of strigolactone biosynthetic genes and strigolactone-regulated genes in Arabidopsis. Biosci Biotechnol Biochem 73:2460–2465

Matusova R, Rani K, Verstappen FWA, Franssen MCR, Beale MH, Bouwmeester HJ (2005) The strigolactone germination stimulants of the plant-parasitic Striga and Orobanche spp. are derived from the carotenoid pathway. Plant Physiol 139:920–934

Mayzlish-Gati E, De Cuyper C, Goormachtig S, Beeckman T, Vuylsteke M, Brewer P, Beveridge C, Yermiyahu U, Kaplan Y, Enzer Y, Wininger S, Resnick N, Cohen M, Kapulnik Y, Koltai H (2012) Strigolactones are involved in root response to low phosphate conditions in Arabidopsis. Plant Physiol 160:1329–1341

Miyashima S, Sebastian J, Lee JY, Helariutta Y (2012) Stem cell function during plant vascular development. EMBO J 32(2):178–193

Morris SE, Cox MCH, Ross JJ, Krisantini S, Beveridge CA (2005) Auxin dynamics after decapitation are not correlated with the initial growth of axillary buds. Plant Physiol 138:1665–1672

Peret B, Clement M, Nussaume L, Desnos T (2011) Root developmental adaptation to phosphate starvation: better safe than sorry. Trends Plant Sci 16:442–450

Perez-Torres C-A, Lopez-Bucio J, Cruz-Ramirez A, Ibarra-Laclette E, Dharmasiri S, Estelle M, Herrera-Estrella L (2008) Phosphate availability alters lateral root development in Arabidopsis by modulating auxin sensitivity via a mechanism involving the TIR1 auxin receptor. Plant Cell 20:3258–3272

Proust H, Hoffmann B, Xie X, Yoneyama K, Schaefer DG, Yoneyama K, Nogué F, Rameau C (2011) Strigolactones regulate protonema branching and act as a quorum sensing-like signal in the moss *Physcomitrella patens*. Development 138:1531–1539

Rasmussen A, Beveridge CA, Geelen D (2012a) Inhibition of strigolactones promotes adventitious root formation. Plant Signal Behav 7:694–697

Rasmussen A, Mason M, De Cuyper C, Brewer PB, Herold S, Agusti J, Geelen DNV, Greb T, Goormachtig S, Beeckman T, Beveridge CA (2012b) Strigolactones suppress adventitious rooting in Arabidopsis and pea. Plant Physiol 158:1976–1987

Renton M, Hanan J, Ferguson BJ, Beveridge CA (2012) Models of long-distance transport: how is carrier-dependent auxin transport regulated in the stem? New Phytol 194:704–715

Ruyter-Spira C, Kohlen W, Charnikhova T, van Zeijl A, van Bezouwen L, de Ruijter N, Cardoso C, Lopez-Raez JA, Matusova R, Bours R, Verstappen F, Bouwmeester H (2011) Physiological effects of the synthetic strigolactone analog GR24 on root system architecture in Arabidopsis: another belowground role for strigolactones? Plant Physiol 155:721–734

Ruyter-Spira C, Al-Babili S, van der Krol S, Bouwmeester H (2012) The biology of strigolactones. Trends Plant Sci. doi:10.1016/j.tplants.2012.10.003

Ruzicka K, Ljung K, Vanneste S, Podhorska R, Beeckman T, Friml J, Benkova E (2007) Ethylene regulates root growth through effects on auxin biosynthesis and transport-dependent auxin distribution. Plant Cell 19:2197–2212

Shimizu-Sato S, Tanaka M, Mori H (2009) Auxin-cytokinin interactions in the control of shoot branching. Plant Mol Biol 69:429–435

Smith SM, Waters MT (2012) Strigolactones: destruction-dependent perception? Curr Biol 22: R924–R927

Stirnberg P, Furner IJ, Ottoline Leyser HM (2007) MAX2 participates in an SCF complex which acts locally at the node to suppress shoot branching. Plant J 50:80–94

Strader LC, Chen GL, Bartel B (2010) Ethylene directs auxin to control root cell expansion. Plant J 64:874–884

Swarup R, Perry P, Hagenbeek D, Van Der Straeten D, Beemster GTS, Gr S, Bhalerao R, Ljung K, Bennett MJ (2007) Ethylene upregulates auxin biosynthesis in Arabidopsis seedlings to enhance inhibition of root cell elongation. Plant Cell 19:2186–2196

Takeda T, Suwa Y, Suzuki M, Kitano H, Ueguchi-Tanaka M, Ashikari M, Matsuoka M, Ueguchi C (2003) The OsTB1 gene negatively regulates lateral branching in rice. Plant J 33:513–520

Umehara M (2011) Strigolactone, a key regulator of nutrient allocation in plants. Plant Biotechnol 28:429–437

Umehara M, Hanada A, Yoshida S, Akiyama K, Arite T, Takeda-Kamiya N, Magome H, Kamiya Y, Shirasu K, Yoneyama K, Kyozuka J, Yamaguchi S (2008) Inhibition of shoot branching by new terpenoid plant hormones. Nature 455:195–200

Umehara M, Hanada A, Magome H, Takeda-Kamiya N, Yamaguchi S (2010) Contribution of strigolactones to the inhibition of tiller bud outgrowth under phosphate deficiency in rice. Plant Cell Physiol 51:1118–1126

Waters MT, Brewer PB, Bussell JD, Smith SM, Beveridge CA (2012) The Arabidopsis ortholog of rice DWARF27 acts upstream of MAX1 in the control of plant development by strigolactones. Plant Physiol 159:1073–1085

Xie X, Yoneyama K, Yoneyama K (2010) The strigolactone story. Annu Rev Phytopathol 48:93–117

Yoneyama K, Xie X, Kusumoto D, Sekimoto H, Sugimoto Y, Takeuchi Y, Yoneyama K (2007) Nitrogen deficiency as well as phosphorus deficiency in Sorghum promotes the production and exudation of 5-deoxystrigol, the host recognition signal for arbuscular mycorrhizal fungi and root parasites. Planta 227:125–132

Yoneyama K, Xie X, Kim HI, Kisugi T, Nomura T, Sekimoto H, Yokota T, Yoneyama K (2012) How do nitrogen and phosphorus deficiencies affect strigolactone production and exudation? Planta 235:1197–1207

Auxin as Long-Distance Signal Controlling Root Architecture in Response to Nitrogen

Giel E. van Noorden and Ulrike Mathesius

Abstract Plants show extensive root plasticity in response to nitrogen availability. Lateral root initiation and emergence are affected by nitrogen concentration and distribution to coordinate the ability to capture maximum nitrogen while minimizing carbon expenditure. Legumes and actinorhizal plants have additionally evolved symbioses with nitrogen-fixing bacteria that leads to the formation of nodules. Nodule numbers are controlled by systemic autoregulation of nodulation (AON) signaling through a receptor-like kinase acting in the shoot. The AON genes also control lateral root formation in response to nitrogen to varying extents in different legumes. Auxin transport control from the shoot to the root is one of the signals affecting nodule and lateral root development, and this is under the control of the AON gene in the legume *Medicago truncatula*. Nitrogen availability modulates long-distance auxin transport and this partly requires the function of the AON gene. Thus we propose a model in which nitrogen availability is perceived in the shoot, is processed by the AON gene, and feeds back on root architecture via control of shoot-to-root auxin transport.

Keywords Auxin transport • Nitrogen • Root development • Nodulation • Autoregulation

1 Phenotypic Plasticity Responses to Nitrogen Availability

Plants have to balance the capture of nutrients by the root system with the availability of carbon from photosynthesis in the shoot. Nutrient acquisition by roots can be maximized by changes in the expression of transport proteins as well as by

G.E. van Noorden • U. Mathesius (✉)
Plant Science Division, Research School of Biology, Australian National University, Canberra, ACT 0200, Australia
e-mail: giel.vannoorden@anu.edu.au; ulrike.mathesius@anu.edu.au

F. Baluška (ed.), *Long-Distance Systemic Signaling and Communication in Plants*, Signaling and Communication in Plants 19, DOI 10.1007/978-3-642-36470-9_10,

changes in root architecture, for example, lateral root initiation and elongation change in response to nutrient concentrations and distribution in the soil; however, this carries carbon costs to form new lateral roots (Ruffel et al. 2011).

Nitrogen is required in large amounts for plant growth and plants show large developmental plasticity in response to nitrogen availability in the soil (Walch-Liu et al. 2006; Zhang et al. 1999). Nitrogen is mainly available as nitrate or ammonium from the soil, but amino acids and proteins can also be taken up (Richardson et al. 2009). Multiple studies have focused on the local responses to nitrogen and have provided an overview of the transport and assimilation of nitrogen by the plant (Krouk et al. 2010a).

Many legumes and so-called actinorhizal plants have additionally gained the ability to take up nitrogen from a symbiosis with nitrogen-fixing bacteria called rhizobia and Frankia, respectively. These symbionts initiate the development of root nodules in specific legume and actinorhizal hosts. Inside the nodules the bacteria convert atmospheric nitrogen into ammonia, which is exported to the plant host as amino acids. In return, the plant transports carbon sources, mainly organic acids, into the nodules as nitrogen fixation has high energy demands (White et al. 2007).

While determinants of nitrogen-use efficiency have been identified as parts of the nitrogen uptake, transport, and assimilation machinery (Garnett et al. 2009), much less is known about the genetic regulation of the developmental plasticity responses to nitrogen. Developmental plasticity, i.e., the ability to adapt root development to local changes in the N environment, is regulated by both local and systemic signals, and most of this work has focused on the responses to nitrate as the N source (Alvarez et al. 2012; Forde 2002; Walch-Liu et al. 2005; Zhang et al. 1999). Nitrate can act locally to alter N uptake and transport capacity, but is also likely to act as a mobile signal to alter N uptake in distal parts of the root system. Split-root experiments have shown that systemic signals communicate the N status between roots exposed to different N concentrations (e.g., Forde 2002; Ruffel et al. 2011; Salon et al. 2009). Likely systemic signals include cytokinin as a root-to-shoot signal communicating the root N status (Ruffel et al. 2011) and miRNAs that target auxin response factor (ARF) (Gifford et al. 2008) and auxin receptor (Vidal et al. 2010) targets in Arabidopsis that control lateral root development. However, it has not clearly been demonstrated if and how cytokinin and miRNAs move systemically in response to varying N environments.

Other candidate genes controlling nitrogen-related phenotypes are a class of genes regulating legume nodule number by a systemic mechanism called autoregulation (Reid et al. 2011b; please refer to (Hayashi et al. 2013) for a detailed overview of the mechanism of AON). Loss of function of the autoregulation genes causes supernodulation, and in most of these mutants this still occurs in the presence of high nitrogen concentrations, which in wild-type plants would inhibit nodulation (Streeter 1988). The autoregulation of nodulation (AON) mutants were therefore first termed *nts* (nitrate-tolerant symbiosis) mutants in soybean (Carroll et al. 1985a, b). A leucine-rich repeat receptor-like kinase (LRR-RLK) was identified as a key gene responsible for the systemic regulation of nodule number. This receptor kinase, also called NARK (Nodulation Autoregulation Receptor Kinase), has been cloned from several legumes, including the model legumes *Lotus japonicus* (Krusell et al. 2002; Nishimura et al. 2002) and *Medicago truncatula* (Schnabel et al. 2005), as well as the crops soybean (*Glycine max*; (Searle et al. 2003)) and pea (*Pisum sativum*; (Krusell et al. 2002)). AON mutants are unable to control nodule number and their roots are supernodulated with high densities of nodules along the whole root system. Grafting of mutant shoots with wild-type roots has shown that NARK acts in the shoot to regulate nodule numbers by long-distance signal(s) (Delves et al. 1986). Split-root experiments in which one root was infected with rhizobia several days prior to infection of a separate root demonstrated that infection of the first root by rhizobia induces a signal that moves to the shoot, activates NARK, and subsequently a signal is sent back from the shoot to the distant root to temporarily inhibit further nodule development (Kosslak and Bohlool 1984). The upwards moving signal is likely to involve a (group of) peptides of the CLE (CLV3/ESR-related) family, which have been implicated in meristem maintenance in general, although definitive experiments showing CLE peptide movement to the shoot are still lacking (Mortier et al. 2010, 2012; Okamoto et al. 2009; Reid et al. 2011a).

As mentioned above, most autoregulation mutants also show a so-called *nts* phenotype (Carroll et al. 1985a; Sagan et al. 1995), although AON mutants differ in their degree of nitrate tolerance. Grafting studies showed that the *nts* phenotype is caused by a disfunctional AON gene action in the shoot in soybean and *M. truncatula* (Day et al. 1989; Jeudy et al. 2010; Sagan et al. 1995). However, nitrogen supply also has a local effect on nodulation in *M. truncatula* which affects nodule size and nodule color (Jeudy et al. 2010). Therefore, it is likely that an inhibitory signal generated after nitrate perception in the root interacts with the autoregulation signal, perceived through NARK, to downregulate nodule density. One possibility is that the NARK receptor directly or indirectly perceives the C/N ratio in the shoot and subsequently triggers a systemic signal that travels to the root to control nodule numbers depending on N demand and C supply.

Interestingly, most autoregulation mutants also show phenotypes in uninoculated plants, for example, an increased density of lateral roots and a short root system in *L. japonicus* (Wopereis et al. 2000), and shortened internodes and a reduced shoot system in pea (Novak et al. 2011), suggesting that these phenotypes are also under the control of a long-distance signaling system regulated by NARK or its homologs. So far, the long-distance shoot-to-root signals that regulate nodule number or other root architecture traits have not been identified (see Hayashi et al. 2013). However, auxin is one candidate for a shoot-to-root signal controlling nodule and lateral root development and its possible role as a long-distance signal controlling root responses to N is discussed below.

2 Auxin as a Long-Distance Signal Controlling Lateral Root Development in Response to Nitrogen

Auxin is primarily synthesized in the shoot, and transported to the root both via phloem transport, depending on source–sink relation, and by active polar cell-to-cell transport through auxin transporters (Friml 2003). Auxin is a crucial regulator of lateral root development and auxin signaling is one of the first requirements for lateral root initiation (Fukaki et al. 2007). Auxin is required for several steps during lateral root development, including initiation and elongation. It has been shown that while lateral root initiation depends on auxin transport from the root tip to the lateral root initiation zone (Casimiro et al. 2001; Reed et al. 1998), long-distance auxin transport from the shoot to the root is required for subsequent lateral root elongation (Bhalerao et al. 2002).

Nitrate availability affects both lateral root initiation and elongation, and these two processes are mediated by auxin signaling. In Arabidopsis, it has been shown that local areas of high nitrate concentration trigger lateral root elongation, whereas high systemic concentrations of nitrogen in the plant inhibit lateral root emergence, presumably to minimize C investment into lateral roots if they are not needed (Zhang and Forde 1998; Zhang et al. 1999; Forde 2002).

The auxin-resistant Arabidopsis mutant *arx4* was used to show that the local stimulation of lateral root elongation in response to nitrate (1 mM) requires auxin signaling (Zhang et al. 1999). However, another study found no increase in lateral root elongation in response to a local supply of 1 mM nitrate in wild-type or *axr4* mutant plants (Linkohr et al. 2002). Other evidence for the link between auxin and nitrate signaling came from the identification of the nitrate transporter NRT1.1, which was shown to cotransport nitrate and auxin. This study suggested that nitrate sensing could be directly linked to lateral root elongation through auxin redistribution by the NRT1.1 transporter in an emerging lateral root (Krouk et al. 2010b). In addition, the outgrowth of lateral roots was inhibited by the application of a synthetic auxin transport inhibitor above the site of localized nitrate supply in Arabidopsis, suggesting that auxin transport is involved in the lateral root response to nitrate (Guo et al. 2005).

Lateral root emergence is usually inhibited in response to high (generally >10 mM) uniform nitrate treatment, while lateral root initiation is stimulated by low nitrate environments. Both conditions were shown to cause changes in auxin concentration and response in the root (Table 1). Microarray experiments showed that several auxin response and auxin transport genes are regulated in nitrate-treated Arabidopsis plants (Gifford et al. 2008; Gutierrez et al. 2007).

Table 1 Summary of studies measuring auxin response, content, or transport in response to nitrate treatment

Plant species	Nitrate treatment	"Auxin measurement"	Time after N treatment	Plant phenotype	Reference
Arabidopsis	High C (4.5 mM sucrose) and low N (0.02 mM NH_4NO_3) treatment	Reduced auxin response in root, increased auxin response (*DR5:GUS*) in hypocotyl	4–6 days	Repression of lateral root initiation	Malamy and Ryan (2001)
Arabidopsis	High N (50 mM KNO_3) treatment	Reduced expression of DR5: GUS in lateral root primordia	Not specified	Inhibition of lateral root emergence	Bao et al. (2007)
Arabidopsis	0.5 mM $Ca(NO_3)_2$ treatment in split-root setup	Auxin transport measurement from shoot to root showed reduced auxin transport in N-treated plants	1 day	Increased lateral root length; no change in lateral root density after 5 days	Liu et al. (2010)
Maize (*Zea mays* L.)	High N (5–10 mM KNO_3) treatment	Reduced auxin content in root and phloem sap	12 days (root); 5 h (phloem)	Inhibition of root growth and lateral root length but not density	Tian et al. (2008)
Maize (*Zea mays*)	1 mM KNO_3 local	Application of auxin transport inhibitor above site of lateral root emergence	Not specified	Reduced lateral root growth (length) in response to nitrate	Guo et al. (2005)
Pineapple (*Ananas comosus*)	Shift from >40 mM N to no N (nitrate plus ammonium)	Increased auxin content in root	12–16 h	Not determined	Tamaki and Mercier (2007)

(continued)

Table 1 (continued)

Plant species	Nitrate treatment	"Auxin measurement"	Time after N treatment	Plant phenotype	Reference
Arabidopsis	Shift from high (50 mM) to low (1 mM) KNO_3	Increased auxin content in roots	24 h	Enhanced lateral root emergence	Walch-Liu et al. (2006)
Soybean wild type (*Glycine max*)	8 mM KNO_3	Reduced root auxin content after N treatment of rhizobia-inoculated plants	12 days	Reduced numbers of nodules	Caba et al. (2000)
Soybean *nts382* mutant	8 mM KNO_3	No change in auxin content after N treatment	12 days	Unchanged numbers of nodules	Caba et al. (2000)
Medicago truncatula wild type	2.5 mM KNO_3	Auxin transport measurement using radiolabeled auxin showed increased auxin transport in response to N	10 days	Reduced numbers of nodules, increased numbers of lateral roots	Jin et al. (2012)
Medicago truncatula sunn-1 mutant	2.5 mM KNO_3	Unchanged auxin transport in response to N	10 days	Reduced numbers of nodules, no change in lateral root numbers	Jin et al. (2012)

The repression of lateral root initiation in Arabidopsis seedlings grown under high C (4.5 mM sucrose) and low N (0.02 mM NH_4NO_3) conditions was linked to high auxin response in the hypocotyl and low auxin response in the root, suggesting that auxin transport from the hypocotyl to the root could be blocked. The repression of lateral root initiation could be rescued by the addition of auxin to the roots, strongly suggesting a role for correct auxin transport and localization in lateral root response to C/N ratio (Malamy and Ryan 2001). Shifting roots from high to low nitrate containing medium increased the root auxin content in pineapple (Tamaki and Mercier 2007) and in Arabidopsis, where this increase in auxin content was accompanied by enhanced lateral root emergence (Walch-Liu et al. 2006). In maize, high nitrate treatment led to an inhibition of root growth and reduced auxin content in the root and in phloem sap, particularly close to the root tip (Tian et al. 2008). A study by Bao et al. (2007) showed reduced expression of the auxin-responsive reporter, *DR5:GUS*, in response to high nitrate, again suggesting that the auxin response is affected by nitrate. Measurements of auxin transport from shoot to root showed that a 0.5 mM nitrate treatment reduced auxin transport 1 day after nitrate application, and this was correlated with increased lateral root length but no change to lateral root density in maize (Liu et al. 2010). In *Arabidopsis*, application of a synthetic auxin transport inhibitor above the site of nitrate treatment prevented the elongation of lateral roots in response to nitrate, strengthening the idea that shoot-to-root auxin transport in response to N can influence lateral root length (Guo et al. 2005).

3 Auxin as a Long-Distance Signal Controlling Nodule Development in Response to Nitrogen

As for lateral root formation, correct auxin transport and localization are required for legume nodule development (Mathesius 2008). Studies utilizing auxin-responsive reporter constructs have shown that auxin responses are localized in early dividing cells of a nodule primordium in various legumes (Mathesius et al. 1998; Pacios-Bras et al. 2003; Suzaki et al. 2012; Takanashi et al. 2011; van Noorden et al. 2007) and in actinorhizal nodules in Casuarina (Perrine-Walker et al. 2010). While both lateral root and nodule organogenesis require auxin, the thresholds required appear to be different. This was demonstrated in *M. truncatula* roots in which the cell cycle regulator *CDC16* was silenced (Kuppusamy et al. 2009). Silenced roots showed reduced auxin response and lower numbers of lateral roots but increased numbers of nodules.

It is likely that local auxin transport inhibition is necessary for adjusting local auxin concentrations during nodule initiation. This is supported by several studies using radiolabeled auxin that showed that rhizobia locally inhibit auxin transport (Boot et al. 1999; Mathesius et al. 1998). In addition, application of synthetic auxin transport inhibitors to the whole root system induces pseudo-nodules in alfalfa and

M. truncatula (Hirsch et al. 1989; Rightmyer and Long 2011), although it is currently not known if the auxin transport inhibitors act locally or on shoot-to-root auxin transport.

Several studies have shown that auxin acts as a long-distance signal from the shoot to the root during nodulation and that this is involved in the regulation of auxin transport. In wild-type soybean, the auxin content of *Rhizobium*-inoculated roots was increased, but reduced in the presence of nitrate, whereas an autoregulation mutant did not respond to nitrate or inoculation with changes in auxin concentrations (Caba et al. 2000). Direct measurements of auxin movement from the shoot to the root were done in *M. truncatula*, which showed that in wild-type plants rhizobia inhibit the transport of auxin from shoot to root within 24 h, the time coinciding with the onset of autoregulation (van Noorden et al. 2006). This inhibition was not triggered by rhizobia in the AON mutant *sunn1* (*super numeric nodules-1*), suggesting that autoregulation could act through systemic control of auxin transport. Instead, this mutant showed constitutively higher auxin transport from shoot to root, correlating with significantly higher root auxin concentration and higher nodule numbers (van Noorden et al. 2006). Consistent with that idea, inhibition of shoot-to-root auxin transport by application of a synthetic auxin transport inhibitor to the root–shoot junction rescued the supernodulation phenotype in the *sunn-1* mutant (van Noorden et al. 2006). Visualization of auxin responses in the roots of *L. japonicus* and its AON mutant *har1* showed enlarged regions of auxin response in inoculated *har1* mutant roots (Suzaki et al. 2012). Expression of CLE peptides in wild-type *L. japonicus* roots, which inhibited nodule development but still allowed some initial cortical cell divisions, led to initial auxin responses in dividing cells but vestiges of diminished auxin response in the CLE-overexpressing plants. This suggests that long-distance signaling via CLE peptides could inhibit nodule development by limiting auxin transport or response (Suzaki et al. 2012).

Ethylene signaling has also been implicated in long-distance auxin transport during nodulation. The ethylene-insensitive *sickle* mutant of *M. truncatula* is hypernodulated, presumably because of reduced defense responses, although it is not defective in AON (Penmetsa and Cook 1997; Penmetsa et al. 2003). In this mutant, shoot-to-root auxin transport is similar to wild type under nitrogen-deficient uninoculated conditions, but similar to the *sunn-1* mutant, rhizobia are unable to inhibit shoot-to-root auxin transport (Prayitno et al. 2006). Shoot-to-root auxin transport was also inhibited by application of the ethylene precursor 1-aminocyclopropane-1-carboxylic acid (ACC) in the wild type but not the *sickle* mutant, even though a auxin transport inhibitor still inhibited shoot-to-root auxin transport in *sickle* (Prayitno et al. 2006). This study highlighted two points (1) there is no straight correlation between total shoot-to-root auxin transport and nodule numbers and (2) ethylene signaling is required for shoot-to-root auxin transport control. Ethylene is likely to mediate the inhibitory effect of nitrate on nodulation (Caba et al. 1998), but so far it is unknown whether ethylene modifies auxin transport in response to nitrate.

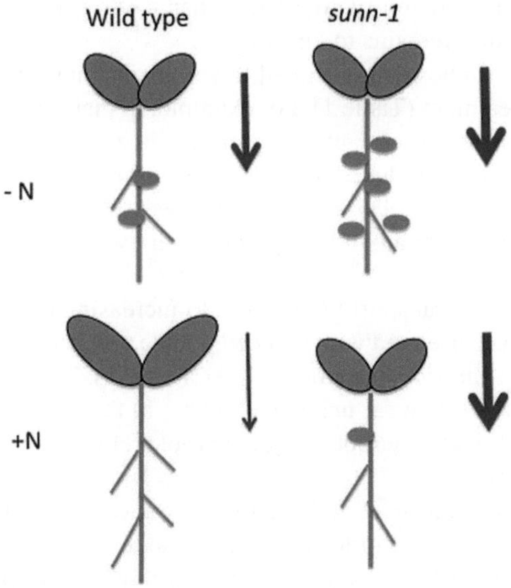

Fig. 1 Model for the control of long-distance auxin transport in response to nitrogen by the *SUNN* receptor-like kinase. In this model, SUNN perceives an N tissue signal of unknown nature and responds by altering shoot-to-root auxin transport. This has effects on lateral root and nodule density as well as shoot and root growth (represented by size of plant). However, nitrate also acts locally to inhibit nodule initiation and modulate lateral root initiation and emergence. The auxin transport capacity is indicated by an *arrow*, with the thickness of the *arrow* indicating the amount of auxin transport. Modified from Jin et al. (2012)

Jin et al. (2012) utilized the AON mutant *sunn-1* of *M. truncatula* (Schnabel et al. 2005, 2010) to test (1) whether altering C or N supply alters shoot-to-root auxin transport and (2) whether this altered auxin transport is correlated with altered nodulation and root architecture. This study showed that even though the *sunn-1* mutant had higher tissue N concentrations than the wild type after treatment with nitrate, it showed a much reduced growth response to N treatment under uninoculated conditions, suggesting that the *SUNN* gene is required for "translating" a tissue N signal into a developmental response. While nodule numbers were higher in the *sunn-1* mutant as expected, this mutant showed a similar reduction in nodule numbers in response to 2.5 mM nitrate as the wild type, similar to earlier studies characterizing the *sunn-1* mutant (Schnabel et al. 2010). This is in contrast to other supernodulation mutants that clearly show an *nts* phenotype (Carroll et al. 1985a, b).

Auxin transport measurements with radiolabeled auxin showed that in wild-type plants, increasing tissue N concentrations were significantly correlated with increased auxin transport from shoot to root in 10-day-old plants (Jin et al. 2012; Fig. 1). In contrast, plants grown under higher CO_2 concentrations showed lower auxin transport capacity. In the *sunn-1* mutant nitrogen was unable to alter auxin transport, suggesting that the SUNN gene controls long-distance auxin transport in response to nitrogen availability. Whether other AON genes from other legumes have the same function remains to be shown.

Curiously, other studies demonstrated lower auxin transport from shoot to root following nitrate treatment (Table 1). For example, in maize a low (0.5 mM) nitrate treatment reduced shoot-to-root auxin transport (Liu et al. 2010) and auxin concentration in phloem sap was reduced after a 5 or 10 mM nitrate treatment (Tian et al. 2008). Whether these differences are due to species effects, the time of treatment, or techniques used to measure auxin transport capacity will have to be investigated in future studies.

The increased auxin transport in response to increasing tissue N concentrations in *M. truncatula* was correlated with reduced nodule and lateral root densities in the wild type but not in the *sunn-1* mutant (Jin et al. 2012). This suggests that SUNN perceives an N signal (of so far unknown nature) in the shoot, and modules auxin transport from the shoot to the root to regulate root architecture according to N need and C supply.

Currently the mechanism of linking long-distance auxin transport with local accumulation of auxin at sites of lateral root and nodule development is not known. The finding by Jin et al (2012) that long-distance auxin transport was negatively correlated with lateral root density suggests that auxin concentration in the root might be superoptimal for lateral root initiation, even though external auxin application is known to increase lateral root numbers, e.g., Wightman et al. (1980). However, it agrees with detailed determinations of root auxin gradients, which show that lateral root founder cells are established at sites of auxin minima (Dubrovsky et al. 2011). The finding by Liu et al. (2010) that 0.5 mM nitrate decreased shoot-to-root auxin transport but increased lateral root elongation also suggests an inverse requirement for shoot auxin transport on lateral root development. It is also possible that it is not the absolute amount of auxin transported that acts as a long-distance signal but rather relative changes in the transport of auxin that could be perceived as a developmental signal.

There was also a negative correlation between nodule density and long-distance auxin transport in *M. truncatula* (Jin et al. 2012). Again, this was unexpected because higher auxin transport in *sunn-1* mutants is correlated with higher nodule density (van Noorden et al. 2006). Like for lateral root initiation, it might be the total auxin transport or content in roots, but the establishment of local auxin gradients that determine optimum conditions for nodule initiation. The transient establishment local auxin minima in the root before nodule initiation (Mathesius et al. 1998), which also occur in *sunn-1* mutants (van Noorden et al. 2006), might be a critical local control point for nodule initiation.

4 Conclusions

So far, little is known about the control of long-distance auxin transport and its role in regulating root architecture and very few studies have directly measured auxin transport from shoot to root. The majority of characterized genes controlling architecture responses to nitrogen and nitrogen use efficiency are involved in nitrate uptake and assimilation, rather than developmental control (Garnett et al. 2009). There is growing evidence that nitrogen availability alters long-distance auxin transport from the shoot to the root and, presumably, root auxin content as a consequence, but results vary depending on plant species and experimental system used (Table 1). The *SUNN* gene, which is controlling nodule numbers in *M. truncatula*, has been linked to the control of long-distance auxin transport and density of nodules and lateral roots (summarized in Fig. 1). Future questions remain as to the systemic N signal perceived by plants, the mechanism of N perception by SUNN, and by the molecular details of how SUNN mediates changes in long-distance auxin transport. The auxin importer AUX1 is a good candidate for an auxin transporter that could affect auxin loading from the shoot to the root (Marchant et al. 2002). Another interesting question for future research is whether the AON genes have been recruited for nodulation control from related genes that regulate lateral root architecture in response to nitrogen in nonlegumes.

Acknowledgements We thank the Australian Research Council for funding through an ARC discovery grant (DP120102970) and a Future Fellowship to UM (FT100100669).

References

Alvarez JM, Vidal EA, Gutierrez RA (2012) Integration of local and systemic signaling pathways for plant N responses. Curr Opin Plant Biol 15:185–191

Bao J, Chen F, Gu R, Wang G, Zhang F, Mi G (2007) Lateral root development of two Arabidopsis auxin transport mutants, *auxl-7* and *eirl-1*, in response to nitrate supplies. Plant Sci 173:417–425

Bhalerao RP, Eklof J, Ljung K, Marchant A, Bennett M, Sandberg G (2002) Shoot-derived auxin is essential for early lateral root emergence in Arabidopsis seedlings. Plant J 29:325–332

Boot KJM, van Brussel AAN, Tak T, Spaink HP, Kijne JW (1999) Lipochitin oligosaccharides from *Rhizobium leguminosarum* bv. *viciae* reduce auxin transport capacity in *Vicia sativa* subsp nigra roots. Mol Plant-Microbe Interact 12:839–844

Caba JM, Recalde L, Ligero F (1998) Nitrate-induced ethylene biosynthesis and the control of nodulation in alfalfa. Plant Cell Environ 21:87–93

Caba JM, Centeno ML, Fernandez B, Gresshoff PM, Ligero F (2000) Inoculation and nitrate alter phytohormone levels in soybean roots: differences between a supernodulating mutant and the wild type. Planta 211:98–104

Carroll BJ, McNeil DL, Gresshoff PM (1985a) Isolation and properties of soybean *Glycine max* (L) Merr mutants that nodulate in the presence of high nitrate concentrations. Proc Natl Acad Sci USA 82:4162–4166

Carroll BJ, McNeil DL, Gresshoff PM (1985b) A supernodulation and nitrate-tolerant symbiotic (*nts*) soybean mutant. Plant Physiol 78:34–40

Casimiro I, Marchant A, Bhalerao RP, Beeckman T, Dhooge S, Swarup R, Graham N, Inze D, Sandberg G, Casero PJ, Bennett M (2001) Auxin transport promotes Arabidopsis lateral root initiation. Plant Cell 13:843–852

Day DA, Caroll BJ, Delves AC, Gresshoff PM (1989) Relationship between autoregulation and nitrate inhibition in soybeans. Physiol Plant 75:37–42

Delves AC, Mathews A, Day DA, Carter AS, Carroll BJ, Gresshoff PM (1986) Regulation of the soybean-*Rhizobium* nodule symbiosis by shoot and root factors. Plant Physiol 82:588–590

Dubrovsky JG, Napsucialy-Mendivil S, Duclercq J, Cheng Y, Shishkova S, Ivanchenko MG, Friml J, Murphy AS, Benkovà E (2011) Auxin minimum defines a developmental window for lateral root initiation. New Phytol 191:970–983

Forde BG (2002) Local and long-range signaling pathways regulating plant responses to nitrate. Annu Rev Plant Biol 53:203–224

Friml J (2003) Auxin transport – shaping the plant. Curr Opin Plant Biol 6:7–12

Fukaki H, Okushima Y, Tasaka M (2007) Auxin-mediated lateral root formation in higher plants. Int Rev Cytol 256:111–137

Garnett T, Conn V, Kaiser BN (2009) Root based approaches to improving nitrogen use efficiency in plants. Plant Cell Environ 32:1272–1283

Gifford ML, Dean A, Gutierrez RA, Coruzzi GM, Birnbaum KD (2008) Cell-specific nitrogen responses mediate developmental plasticity. Proc Natl Acad Sci USA 105:803–808

Guo YF, Chen FJ, Zhang FS, Mi GH (2005) Auxin transport from shoot to root is involved in the response of lateral root growth to localized supply of nitrate in maize. Plant Sci 169:894–900

Gutierrez RA, Lejay LV, Dean A, Chiaromonte F, Shasha DE, Coruzzi GM (2007) Qualitative network models and genome-wide expression data define carbon/nitrogen-responsive molecular machines in Arabidopsis. Genome Biol 8:R7

Hayashi S, et al 2013 Systemic signalling in legume nodulation: nodule formation and its regulation. In: Baluska F (ed) Long-distance systemic signaling and communication in plants. Springer, Heidelberg.

Hirsch AM, Bhuvaneswari TV, Torrey JG, Bisseling T (1989) Early nodulin genes are induced in alfalfa root outgrowths elicited by auxin transport inhibitors. Proc Natl Acad Sci USA 86:1244–1248

Jeudy C, Ruffel S, Freixes S, Tillard P, Santoni AL, Morel S, Journet E-P, Duc G, Gojon A, Lepetit M, Salon C (2010) Adaptation of *Medicago truncatula* to nitrogen limitation is modulated via local and systemic nodule developmental responses. New Phytol 185:817–828

Jin J, Watt M, Mathesius U (2012) The autoregulation gene *SUNN* mediates changes in root organ formation in response to nitrogen through alteration of shoot-to-root auxin transport. Plant Physiol 159:489–500

Kosslak RM, Bohlool BB (1984) Suppression of nodule development of one side of a split-root system of soybeans caused by prior inoculation of the other side. Plant Physiol 75:125–130

Krouk G, Crawford NM, Coruzzi GM, Tsay Y-F (2010a) Nitrate signaling: adaptation to fluctuating environments. Curr Opin Plant Biol 13:266–273

Krouk G, Lacombe B, Bielach A, Perrine-Walker F, Malinska K, Mounier E, Hoyerova K, Tillard P, Leon S, Ljung K, Zazimalova E, Benkova E, Nacry P, Gojon A (2010b) Nitrate-regulated auxin transport by NRT1.1 defines a mechanism for nutrient sensing in plants. Dev Cell 18:927–937

Krusell L, Madsen LH, Sato S, Aubert G, Genua A, Szczyglowski K, Duc G, Kaneko T, Tabata S, de Bruijn F, Pajuelo E, Sandal N, Stougaard J (2002) Shoot control of root development and nodulation is mediated by a receptor-like kinase. Nature 420:422–426

Kuppusamy KT, Ivashuta S, Bucciarelli B, Vance CP, Gantt JS, VandenBosch KA (2009) Knockdown of *CELL DIVISION CYCLE16* reveals an inverse relationship between lateral root and nodule numbers and a link to auxin in *Medicago truncatula*. Plant Physiol 151:1155–1166

Linkohr BI, Williamson LC, Fitter AH, Leyser HMO (2002) Nitrate and phosphate availability and distribution have different effects on root system architecture of Arabidopsis. Plant J 29:751–760

Liu J, An X, Cheng L, Chen F, Bao J, Yuan L, Zhang F, Mi G (2010) Auxin transport in maize roots in response to localized nitrate supply. Ann Bot 106:1019–1026

Malamy JE, Ryan KS (2001) Environmental regulation of lateral root initiation in Arabidopsis. Plant Physiol 127:899–909

Marchant A, Bhalerao R, Casimiro I, Eklof J, Casero PJ, Bennett M, Sandberg G (2002) AUX1 promotes lateral root formation by facilitating indole-3-acetic acid distribution between sink and source tissues in the Arabidopsis seedling. Plant Cell 14:589–597

Mathesius U (2008) Auxin: at the root of nodule development? Funct Plant Biol 35:651–668

Mathesius U, Schlaman HRM, Spaink HP, Sautter C, Rolfe BG, Djordjevic MA (1998) Auxin transport inhibition precedes root nodule formation in white clover roots and is regulated by flavonoids and derivatives of chitin oligosaccharides. Plant J 14:23–34

Mortier V, Den Herder G, Whitford R, Van de Velde W, Rombauts S, D'Haeseleer K, Holsters M, Goormachtig S (2010) CLE peptides control *Medicago truncatula* nodulation locally and systemically. Plant Physiol 153:222–237

Mortier V, De Wever E, Vuylsteke M, Holsters M, Goormachtig S (2012) Nodule numbers are governed by interaction between CLE peptides and cytokinin signaling. Plant J 70:367–376

Nishimura R, Hayashi M, Wu GJ, Kouchi H, Imaizumi-Anraku H, Murakami Y, Kawasaki S, Akao S, Ohmori M, Nagasawa M, Harada K, Kawaguchi M (2002) HAR1 mediates systemic regulation of symbiotic organ development. Nature 420:426–429

Novak K, Lisa L, Skrdleta V (2011) Pleiotropy of pea RisfixC supernodulation mutation is symbiosis-independent. Plant Soil 342:173–182

Okamoto S, Ohnishi E, Sato S, Takahashi H, Nakazono M, Tabata S, Kawaguchi M (2009) Nod factor/nitrate-induced CLE genes that drive HAR1-mediated systemic regulation of nodulation. Plant Cell Physiol 50:67–77

Pacios-Bras C, Schlaman HRM, Boot K, Admiraal P, Langerak JM, Stougaard J, Spaink HP (2003) Auxin distribution in *Lotus japonicus* during root nodule development. Plant Mol Biol 52:1169–1180

Penmetsa RV, Cook DR (1997) A legume ethylene-insensitive mutant hyperinfected by its rhizobial symbiont. Science 275:527–530

Penmetsa RV, Frugoli JA, Smith LS, Long SR, Cook DR (2003) Dual genetic pathways controlling nodule number in *Medicago truncatula*. Plant Physiol 131:998–1008

Perrine-Walker F, Doumas P, Lucas M, Vaissayre V, Beauchemin NJ, Band LR, Chopard J, Crabos A, Conejero G, Peret B, King JR, Verdeil JL, Hocher V, Franche C, Bennett MJ, Tisa LS, Laplaze L (2010) Auxin carriers localization drives auxin accumulation in plant cells infected by *Frankia* in *Casuarina glauca* actinorhizal nodules. Plant Physiol 154:1372–1380

Prayitno J, Rolfe BG, Mathesius U (2006) The ethylene-insensitive *sickle* mutant of *Medicago truncatula* shows altered auxin transport regulation during nodulation. Plant Physiol 142:168–180

Reed RC, Brady SR, Muday GK (1998) Inhibition of auxin movement from the shoot into the root inhibits lateral root development in Arabidopsis. Plant Physiol 118:1369–1378

Reid DE, Ferguson BJ, Gresshoff PM (2011a) Inoculation- and nitrate-induced CLE peptides of soybean control NARK-dependent nodule formation. Mol Plant-Microbe Interact 24:606–618

Reid DE, Ferguson BJ, Hayashi S, Lin Y-H, Gresshoff PM (2011b) Molecular mechanisms controlling legume autoregulation of nodulation. Ann Bot 108:789–795

Richardson AE, Barea J-M, McNeill AM, Prigent-Combaret C (2009) Acquisition of phosphorus and nitrogen in the rhizosphere and plant growth promotion by microorganisms. Plant Soil 321:305–339

Rightmyer AP, Long SR (2011) Pseudonodule formation by wild-type and symbiotic mutant Medicago truncatula in response to auxin transport inhibitors. Mol Plant-Microbe Interact 24:1372–1384

Ruffel S, Krouk G, Ristova D, Shashka D, Birnbaum KD, Coruzzi GM (2011) Nitrogen economics of root foraging: transitive closure of the nitrate-cytokinin relay and distinct systemic signaling for N supply vs. demand. Proc Natl Acad Sci USA 108:18514–18529

Sagan M, Morandi D, Tarenghi E, Duc G (1995) Selection of nodulation and mycorrhizal mutants in the model-plant *Medicago truncatula* (Gaertn) after gamma-ray mutagenesis. Plant Sci 111:63–71

Salon C, Lepetit M, Gamas P, Jeudy C, Moreau S, Moreau D, Voisin A-S, Duc G, Bourion V, Munier-Jolain N (2009) Analysis and modeling of the integrative response of Medicago truncatula to nitrogen constraints. C R Biol 332:1022–1033

Schnabel E, Journet EP, de Carvalho-Niebel F, Duc G, Frugoli J (2005) The *Medicago truncatula SUNN* gene encodes a CLV1-like leucine-rich repeat receptor kinase that regulates nodule number and root length. Plant Mol Biol 58:809–822

Schnabel E, Mukherjee A, Smith L, Kassaw T, Long S, Frugoli J (2010) The *lss* supernodulation mutant of *Medicago truncatula* reduces expression of the *SUNN* Gene. Plant Physiol 154:1390–1402

Searle IR, Men AE, Laniya TS, Buzas DM, Iturbe-Ormaetxe I, Carroll BJ, Gresshoff PM (2003) Long-distance signaling in nodulation directed by a CLAVATA1-like receptor kinase. Science 299:109–112

Streeter J (1988) Inhibition of legume nodule formation and N_2 fixation by nitrate. Crit Rev Plant Sci 7:1–23

Suzaki T, Yano K, Ito M, Umehara Y, Suganuma N, Kawaguchi M (2012) Positive and negative regulation of cortical cell division during root nodule development in *Lotus japonicus* is accompanied by auxin response. Development 139:3997–4006

Takanashi K, Sugiyama A, Yazaki K (2011) Involvement of auxin distribution in root nodule development of *Lotus japonicus*. Planta 234:73–81

Tamaki V, Mercier H (2007) Cytokinins and auxin communicate nitrogen availability as long-distance signal molecules in pineapple (*Ananas comosus*). J Plant Physiol 164:1543–1547

Tian Q, Chen F, Liu J, Zhang F, Mi G (2008) Inhibition of maize root growth by high nitrate supply is correlated with reduced IAA levels in roots. J Plant Physiol 165:942–951

van Noorden GE, Ross JJ, Reid JB, Rolfe BG, Mathesius U (2006) Defective long-distance auxin transport regulation in the *Medicago truncatula super numeric nodules* mutant. Plant Physiol 140:1494–1506

van Noorden GE, Kerim T, Goffard N, Wiblin R, Pellerone FI, Rolfe BG, Mathesius U (2007) Overlap of proteome changes in *Medicago truncatula* in response to auxin and *Sinorhizobium meliloti*. Plant Physiol 144:1115–1131

Vidal EA, Araus V, Lu C, Parry G, Green PJ, Coruzzi GM, Gutierrez RA (2010) Nitrate-responsive miRNA393/AFB3 regulatory module controls root system architecture in *Arabidopsis thaliana*. Proc Natl Acad Sci USA 107:4477–4482

Walch-Liu P, Filleur S, Gan YB, Forde BG (2005) Signaling mechanisms integrating root and shoot responses to changes in the nitrogen supply. Photosynth Res 83:239–250

Walch-Liu P, Ivanov II, Filleur S, Gan YB, Remans T, Forde BG (2006) Nitrogen regulation of root branching. Ann Bot 97:875–881

White J, Prell J, James EK, Poole P (2007) Nutrient sharing between symbionts. Plant Physiol 144:604–614

Wightman F, Schneider EA, Thimann KV (1980) Hormonal factors controlling the initiation and development of lateral roots. II. Effects of exogenous growth factors on lateral root formation in pea roots. Physiol Plant 49:304–314

Wopereis J, Pajuelo E, Dazzo FB, Jiang QY, Gresshoff PM, de Bruijn FJ, Stougaard J, Szczyglowski K (2000) Short root mutant of *Lotus japonicus* with a dramatically altered symbiotic phenotype. Plant J 23:97–114

Zhang H, Forde BG (1998) An *Arabidopsis* MADS box gene that controls nutrient-induced changes in root architecture. Science 279:407–409

Zhang H, Jennings A, Barlow PW, Forde BG (1999) Dual pathways for regulation of root branching by nitrate. Proc Natl Acad Sci USA 96:6529–6534

Systemic Signalling in Legume Nodulation: Nodule Formation and Its Regulation

Satomi Hayashi, Peter M. Gresshoff, and Brett J. Ferguson

Abstract Legume plants are able to enter into a symbiotic relationship with rhizobia bacteria. This results in the formation of a novel organ on the root called the nodule, where the rhizobia are housed. The rhizobia provide the host plant with nitrogen in exchange for carbohydrates. Successful nodule formation and sustainable nodulation involve complex signalling events. This includes systemic signalling between the symbiotic partners, and also signalling between the root and shoot of the plant. Factors such as plant hormone levels and environmental conditions for growth influence these systemic signalling pathways. This chapter investigates the different types of long-distance signalling events that are necessary for the development and regulation of legume nodulation.

Keywords Auto Regulation of Nodulation (AON) • Shoot Derived Inhibitor (SDI) • CLE peptide • Rhizobia • Flavonoid • Hormone • Acid

1 Introduction

Legume plants are characterised by their ability to enter into a symbiotic relationship with N-fixing bacteria collectively called rhizobia. This partnership requires a complex exchange involving a number of systemic signals. An effective relationship between the two symbionts results in the plant host forming a novel root organ called the nodule (Ferguson et al. 2010; Ferguson 2013). The nodule not only houses the rhizobia and provides them with resource, but also creates a suitable environment for the bacteria to fix atmospheric di-nitrogen into forms of nitrogen that can be used by the plant.

S. Hayashi • P.M. Gresshoff (✉) • B.J. Ferguson
Australian Research Council Centre of Excellence for Integrative Legume Research, The University of Queensland, St. Lucia, Brisbane, QLD 4072, Australia
e-mail: s.hayashi@uq.edu.au; p.greasshoff@uq.edu.au; b.ferguson@uq.edu.au

F. Baluška (ed.), *Long-Distance Systemic Signaling and Communication in Plants*, Signaling and Communication in Plants 19, DOI 10.1007/978-3-642-36470-9_11,

The process of nodule formation and development is termed "nodulation". Two main systemic signalling events are known to be critical for nodulation. The first occurs in the rhizosphere, and involves the recognition of the two symbiotic partners and the initiation of nodulation events. The second occurs within the plant to regulate nodule numbers as a means of balancing the resource cost and nitrogen benefit associated with nodulation.

2 Systemic Signalling in the Rhizosphere: Partner Recognition and the Initiation of Nodulation

Nodulation is a highly specific process in which only specific strains of rhizobia form a symbiotic relationship with a limited range of host legume species (Dénarié et al. 1996; Pueppke and Broughton 1999). This specificity is largely defined by signals exchanged between the symbiotic partners in the rhizosphere.

Flavonoids are one of the largest classes of phenylpropanoid-derived metabolites produced by the plant. They contain approximately 10,000 identified compounds and have diverse roles in plants. Flavonoids include compounds involved in antimicrobial and anti-herbivorous processes, plant defence, regulation of auxin transport, pigmentation and protection against UV rays (Dixon and Pasinetti 2010). In legume species, flavonoids also play a critical role in host–rhizobia signalling to initiate nodulation. Specific flavonoid molecules are exuded to the rhizosphere by the host root, acting as a chemoattractant for the bacteria. To a large extent, the specificity of the host–rhizobia symbiosis is defined by the structure of the exuded flavonoid.

Recognition of a specific flavonoid by the rhizobia induces the expression of nodulation (nod) genes that are required for the production and secretion of lipo-chito-oligosaccharides called the Nod factors (Peters et al. 1986; Redmond et al. 1986; Caetano-Anollés and Gresshoff 1991; Dénarié et al. 1996; Spaink 2000). All Nod factors have a backbone made of four to five β-1-4-linked *N*-acetylglucosamine residues with a fatty acyl chain attached to the non-reducing sugar. Each strain of rhizobia produces a Nod factor that is structurally unique to them, by means of length of the backbone, length and saturation of the acyl chain, and type and number of substituents attached to the basic backbone. It is these structural differences amongst the Nod factors that help to define the host–rhizobia specificity (Dénarié et al. 1996; Lerouge et al. 1990).

The Nod factor signal is recognised by receptors on the host root (Indrasumunar et al. 2010, 2011) and is one of the key steps for initiating nodule formation, inducing both biochemical and morphological changes in the root (Dénarié et al. 1996; Spaink 2000). In fact, Nod factor alone is able to induce root hair deformation, root hair curling and the induction of several early nodulation genes. Transcriptional changes induced in the epidermis lead to the production of secondary signals. It is currently postulated that at least some of these signals move into the root interior to promote cortical cell division and nodule primordia formation (Ferguson et al. 2010; Rival et al. 2012). In addition, bacteria exopolysacharides (EPS) and lipopolysachaides (LPS) are thought to play an important role in rhizobia infection events (Giraud et al. 2007; Jones et al. 2008; Kelly et al. 2012; Leigh et al. 1985; Mathesius et al. 2003).

3 Systemic Signalling in the Host Plant: The Autoregulation of Nodulation

The process of nodulation is costly to the plant in terms of the energy and resources required to form and maintain the nodule and its bacterial symbionts. To balance these costs with the benefit of acquiring fixed nitrogen, the plant limits its number of nodules via a systemic feedback inhibition signalling network called the autoregulation of nodulation (AON; Delves et al. 1986; Reid et al. 2011a).

On legume roots where AON is functional, most nodules form close to the crown of the root. Plants defective in AON show a hyper- or super-nodulating phenotype which is characterised by an extended region of nodulation as well as an increased nodule density (Carroll et al. 1985a, b). Cloning of genes responsible for the super-nodulating phenotype in several species has revealed that a transmembrane Leucine Rich Repeat (LRR) receptor-like kinase (GmNARK, LjHAR1, PsSYM29 and MtSUNN) is one of the key regulators in the AON pathway (Krusell et al. 2002; Nishimura et al. 2002; Searle et al. 2003; Schnabel et al. 2005). Grafting experiments using plants mutated in this gene have shown that it acts in the shoot (e.g. Delves et al. 1986), and hence demonstrating that AON involves long-distance signalling between the root and the shoot. These findings lead to the classical model of the AON signalling pathway, where a root-derived signal (Q) is transported to the shoot in response to the early nodulation events (Fig. 1; Delves et al. 1986; Reid et al. 2011a, b). Perception of Q by the LRR receptor-like kinase initiates the production of a shoot-derived inhibitor (SDI) signal that is transported back to the root to inhibit further nodule development.

4 Rhizobia-Induced CLE Peptides as a Candidate for Q

The LRR receptor-like kinases acting in AON are genetically and structurally similar to CLAVATA1 (AtCLV1) of *Arabidopsis* (Clark et al. 1997). The known ligand for AtCLV1 is AtCLV3, a CLAVATA/ESR-related (CLE) peptide. CLE peptides are characterised by having an N-terminal signal peptide and a conserved CLE domain made up of 12–13 amino acids that is thought to be the active signal. AtCLV3 acts in regulating the stem cell population of the shoot apical meristem. High resemblance of the AON LRR receptor-like kinase to AtCLV1 suggested that

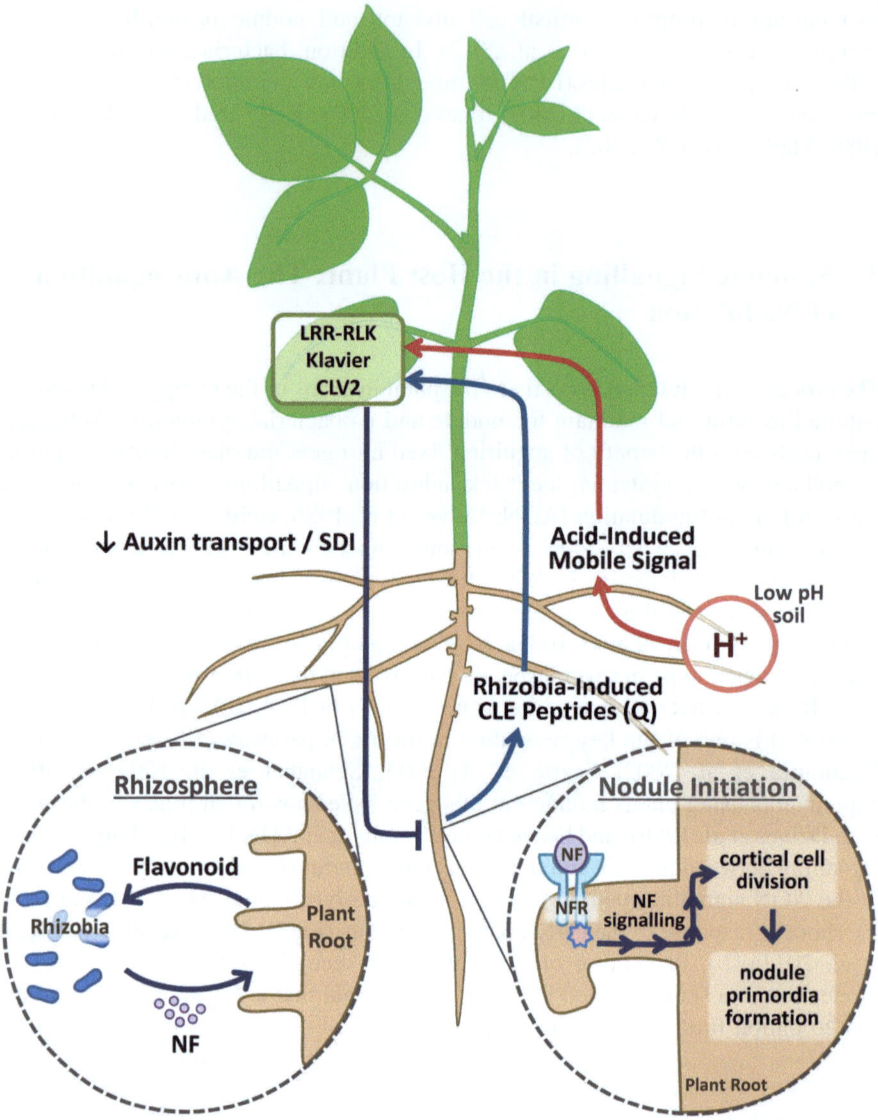

Fig. 1 Model of systemic siganlling events in legume nodulation. The roots of legume plants release flavonoid molecules into the rhizosphere. This triggers the production of Nod factor (NF) signals by compatible rhiobia species. NF is percieved by the Nod factor receptors (NFR) on the host plant and activates a series of signalling cascades (i.e. NF signalling) to initiate cortical devision that leads to the formation of nodule primordia. Nodule initiation triggers the production and transport of rhizobia-induced CLE peptides to the shoot, where they are thought to act as a ligand for a LRR receptor-like kinase (LRR-RLK; e.g. GmNARK, MtSUNN, LjHAR1 and PsSYM29) that is required to control nodule numbers. Klavier and CLAVATA2 (CLV2) receptors are also thought to be important for the perception of the root-derived signal and downstream signalling events. A shoot-derived inhibitor (SDI) molecule is produced following the percption of the root-derived signal. SDI is transported from the shoot to root where it inhibits further noudule

it might also act to perceive a CLE peptide signal similar to AtCLV3. To date, CLE peptides from several legume species, soybean (GmRIC1/2; Lim et al. 2011; Reid et al. 2011b), *Lotus japonicus* (LjCLE-RS1; Okamoto et al. 2009), and *Medicago truncatula* (MtCLE12/13; Mortier et al. 2010), have been identified as candidates as the ligand for the LRR receptor-like kinase acting in AON (i.e. Q). Expression of all of the abovementioned genes encoding these CLE peptides is strongly induced upon rhizobia inoculation (Hayashi et al. 2012; Lim et al. 2011; Mortier et al. 2010; Okamoto et al. 2009; Reid et al. 2011b). However, the timing in which these CLE genes are induced can be categorised into to two distinct stages of nodule development. GmRIC1 and LjCLE-RS1 are induced at very early stages of nodulation, as early as 12 h post-inoculation with rhizobia, whereas the induction of GmRIC2 and MtCLE12 and MtCLE13 occurs at much later stages of nodulation when nodules are maturing or have matured (Lim et al. 2011; Mortier et al. 2010; Okamoto et al. 2009; Reid et al. 2011b). The different timing of CLE induction post-rhizobia inoculation is supported by the other reports describing that the onset of AON occurs in multiple stages during nodule development for activation and maintenance of AON (Li et al. 2009). Recent work in soybean using GmRIC1 has also identified critical domains and amino acid residues required for optimal function of the CLE signal in the AON process (Reid et al. 2013). A predicted model of the unmodified GmRIC1 backbone is shown in Fig. 2.

The signal Q is predicted to travel from root to shoot, most likely via the xylem (Reid et al. 2011a, 2012). Although experiments have been carried out to identify the signal *in planta*, it has not been discovered to date. Overexpression of the rhizobia-induced CLE peptide-encoding genes can reduce nodulation, or even completely inhibit it, in a NARK-dependent manner (Lim et al. 2011; Mortier et al. 2010; Okamoto et al. 2009; Reid et al. 2011b). This further supports the idea that rhizobia-induced CLE peptides act in the systemic AON pathway.

5 Perception of Q and the Production of a Shoot-Derived Inhibitor

As mentioned earlier, Q is predicted to be perceived by an LRR receptor-like kinase in the shoot as part of the AON pathway. AON is also dependent on the function of the KLAVIER (Oka-Kira et al. 2005; Miyazawa et al. 2010) and CLAVATA2 (CLV2; Krusell et al. 2011) receptors functioning in the shoot, suggesting that a

Fig. 1 (continued) development. Hormones such as jasmonic acid (JA) and auxin also appear to be involved in this downstream signalling; however, their precise role in regulating of nodule numbers is yet to be fully established. In acidic growth conditions, plants send an acid-induced mobile signal to the shoot, which is percived by the LRR-RLK of the AON pathway in the shoot to trigger the inhibition of nodulation. The factors involved in this process are yet to be identified

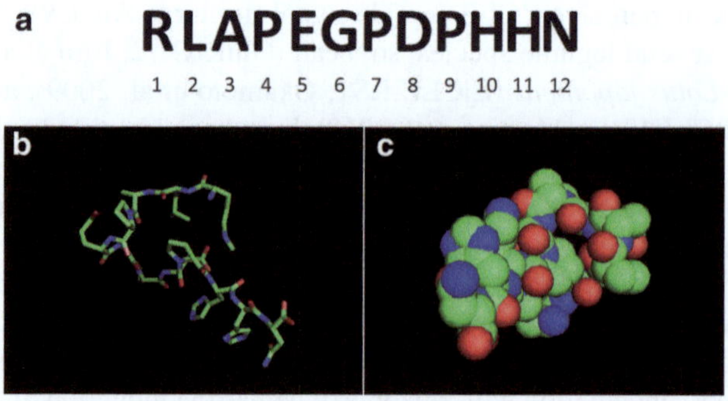

Fig. 2 Predicted model of the soybean nodulation CLE peptide, GmRIC1. (**a**) Amino acid composition of the GmRIC1 CLE domain thought to represent the final signal ligand; (**b**) molecular model of the GmRIC1 peptide backbone illustrating a pronounced boomerang shape. The central Gly residue seems critical to allow curvature; (**c**) filled in space model of the unmodified GmRIC1 peptide ligand. Molecular modelling of GmRIC1 was carried out using amino acid residues 1 to 12 the GmRIC1 CLE domain (Reid et al. 2011b) and the Phyre web server (Kelley and Sternberg 2009), before visualization in the PyMOL molecular graphics system (DeLano 2002)

receptor complex may be required to perceive Q. Interestingly, the homolog of CLV2 in *Arabidopsis*, AtCLV2, is also involved in the CLAVATA pathway with AtCLV1 and AtCLV3 (Jeong et al. 1999), demonstrating the high degree of similarity between the components of CLAVATA and AON pathways.

Physiological evidence indicates that following the perception of Q in the shoot by the LRR receptor-like kinase of AON, a novel SDI is produced. SDI is predicted to travel via the phloem to the root, where it inhibits further nodulation (Fig. 1). In the search to discover SDI, Lin et al. (2010, 2011) established a bioassay involving the feeding of plant extracts to recipient plants to observe the suppression of nodulation. Feeding leaf extracts from wild-type plants inhibited nodulation in *nark* mutant plants, whereas extracts from *nark* mutant plants could not suppress nodule numbers, confirming that the production of SDI is indeed NARK dependent. Using this bioassay, the authors further characterised SDI and demonstrated it is a small, heat-stable molecule that is not protein or RNA.

6 Inhibition of Nodulation by Acid Growth Conditions

Environmental stresses, such as acid soil, not only affect the health of the plant, but also negatively affect nodule development (Ferguson et al. 2013). Recent work by Lin et al. (2012) has shown that acid-treated (pH 4.0) soybean plants not only form a reduced number of nodules, but also exhibit a significantly reduced level of nod factor signalling at the early stages of nodulation (12–96 h post-inoculation). Using split root and grafting experiments, the authors demonstrated that the inhibition acts systemically via NARK in the shoot, similar to that of rhizobia-induced AON signalling pathway. This suggests that there is likely a mobile, acid-induced signal produced in the root, possibly a CLE peptide, similar to the rhizobia-induced CLE peptides that act in AON.

7 Hormones in Systemic Nodulation Signalling

Plant hormones play key roles during all plant developmental processes. To date, many types of hormone have been reported to modulate nodulation in legume roots, both positively and negatively (Ferguson and Mathesius 2003). In particular, jasmonic acid (JA), brassinosteroids and auxin have been proposed to act in systemic pathways during nodulation.

Comparative transcript analysis of soybean indicated that both JA biosynthesis and response genes were systemically down-regulated in the shoot of rhizobia-inoculated plants (Kinkema and Gresshoff 2008). The expression of JA-related genes and the level of JA itself were found to be regulated in a NARK-dependent manner, being more prominent in *nark* mutants (Kinkema and Gresshoff 2008; Seo et al. 2007). Foliar application of a JA biosynthesis inhibitor leads to a significant reduction in nodule number in *nark* mutants, further supporting the idea that reduced level of JA in the leaf during the early stages of nodulation is key to the regulation of nodule number (Kinkema and Gresshoff 2008). In contrast, foliar application of methyl-jasmonate has been shown to hinder nodulation in both *L. japonicus* and soybean, regardless of LjHAR1/GmNARK activity (Nakagawa and Kawaguchi 2006; Seo et al. 2007). These results suggest that JA acts as a systemic signal to negatively regulate nodulation, contradicting the above-mentioned studies. However, foliar application studies can be temperamental, depending on the level of uptake by the leaf, and it is likely that precise level and location of the hormone are required to regulate nodulation.

Other hormones such as brassinosteroids (BR) and auxin have also been reported to systemically modulate nodule numbers (Ferguson et al. 2005; Terakado et al. 2005; van Noorden et al. 2007). Grafting studies using a BR biosynthesis mutant of pea showed that the genotype of the shoot was responsible for the nodule number of the root (Ferguson et al. 2005). A study involving foliar application of BR and BR inhibitor in soybean has also suggested the involvement of BR in nodulation (Terakado et al. 2005). However, as described earlier, foliar application studies are associated with indirect issue related to uptake and concentration of the hormone, which makes the results difficult to interpret. Nonetheless, many active BR compounds are not thought to be transported in the plant and thus these findings could be the result of the indirect systemic mechanism influenced by the BR levels in the shoot (Symons and Reid 2004).

In nodulating *M. truncatula*, shoot-to-root transport of auxin is reduced at the timing corresponding to the onset of AON (van Noorden et al. 2007). Such a reduction in auxin transport is absent in super-nodulating *Mtsunn* mutant plants, suggesting that this decline in auxin-transport is MtSUNN dependent, and may possibly be a part of a long-distance signalling mechanism involved in the control of nodule numbers (van Noorden et al. 2007). An extended analysis of the role of auxin transport in nodulation control is further discussed in Mathesius (2013).

8 Conclusion and Future Perspectives

The establishment and regulation of effective nodulation involve complex systemic signalling events (Fig. 1). These can be categorised into two main types (1) rhizospheric signalling between the plant root and rhizobia for the initiation of nodulation, and (2) signal exchange between the root and shoot for nodulation control. The latter involves changes in the amount and transport of signalling molecules within the plant. However, there is still a great deal to be known about the specific mechanism in which these molecules act to regulate legume nodulation.

References

Caetano-Anollés G, Gresshoff PM (1991) Plant genetic control of nodulation. Annu Rev Microbiol 45:345–382

Carroll BJ, McNeil DL, Gresshoff PM (1985a) A supernodulation and nitrate-tolerant symbiotic (*nts*) soybean mutant. Plant Physiol 78:34–40

Carroll BJ, McNeil DL, Gresshoff PM (1985b) Isolation and properties of soybean [*Glycine max* (L.) Merr.] mutants that nodulate in the presence of high nitrate concentrations. Proc Natl Acad Sci USA 82:4162–4166

Clark SE, Williams RW, Meyerowitz EM (1997) The *CLAVATA1* gene encodes a putative receptor kinase that controls shoot and floral meristem size in *Arabidopsis*. Cell 89:575–585

DeLano WL (2002) The PyMOL molecular graphics system. DeLano Scientific LLC, San Carlos, CA [http://www.pymol.org]

Delves AC, Mathews A, Day DA, Carter AS, Carroll BJ, Gresshoff PM (1986) Regulation of the soybean-*Rhizobium* nodule symbiosis by shoot and root factors. Plant Physiol 82:588–590

Dénarié J, Debellé F, Promé JC (1996) *Rhizobium* lipo-chitooligosaccharide nodulation factors: signaling molecules mediating recognition and morphogenesis. Annu Rev Biochem 65:503–535

Dixon RA, Pasinetti GM (2010) Flavonoids and Isoflavonoids: from plant biology to agriculture and neuroscience. Plant Physiol 154:453–457

Ferguson BJ, Mathesius U (2003) Signaling interactions during nodule development. J Plant Growth Regul 22:47–72

Ferguson BJ, Ross JJ, Reid JB (2005) Nodulation phenotypes of gibberellin and brassinosteroid mutants of pea. Plant Physiol 138:2396–2405

Ferguson BJ, Indrasumunar A, Hayashi S, Lin M-H, Lin Y-H, Reid DE, Gresshoff PM (2010) Molecular analysis of legume nodule development and autoregulation. J Integr Plant Biol 52:61–76

Ferguson BJ (2013) The development and regulation of soybean nodules. In: Board JE (ed) A comprehensive survey of international soybean research - genetics, physiology, agronomy, and nitrogen relationships. InTech—Open Access, http://dx.doi.org/105772/52573, pp 31–47

Ferguson BJ, Lin MH, Gresshoff PM (2013) Regulation of legume nodulation by acidic growth conditions. Plant Signal Behav 8:e23426

Giraud E, Moulin L, Vallenet D, Barbe V, Cytryn E, Avarre JC, Jaubert M, Simon D, Cartieaux F, Prin Y, Bena G, Hannibal L, Fardoux J, Kojadinovic M, Vuillet L, Lajus A, Cruveiller S, Rouy Z, Mangenot S, Segurens B, Dossat C, Franck WL, Chang WS, Saunders E, Bruce D, Richardson P, Normand P, Dreyfus B, Pignol D, Stacey G, Emerich D, Verméglio A, Médigue C, Sadowsky M (2007) Legumes symbioses: absence of *Nod* genes in photosynthetic bradyrhizobia. Science 316:1307–1312

Hayashi S, Reid DE, Lorenc MT, Stiller J, Edwards D, Gresshoff PM, Ferguson BJ (2012) Transient Nod factor-dependent gene expression in the nodulation-competent zone of soybean (*Glycine max* [L.] Merr.) roots. Plant Biotechnol J 10:995–1010

Indrasumunar A, Kereszt A, Searle I, Miyagi M, Li D, Nguyen CDT, Men A, Carroll BJ, Gresshoff PM (2010) Inactivation of duplicated Nod factor receptor 5 (NFR5) genes in recessive loss-of-function nonnodulation mutants of allotetraploid soybean (*Glycine max* L. Merr.). Plant Cell Physiol 51:201–214

Indrasumunar A, Searle I, Lin M-H, Kereszt A, Men A, Carroll BJ, Gresshoff PM (2011) Nodulation factor receptor kinase 1α controls nodule organ number in soybean (*Glycine max* L. Merr.). Plant J 65:39–50

Jeong S, Trotochaud AE, Clark SE (1999) The Arabidopsis *CLAVATA2* gene encodes a receptor-like protein required for the stability of the CLAVATA1 receptor-like kinase. Plant Cell 11:1925–1933

Jones KM, Sharopova N, Lohar DP, Zhang JQ, VandenBosch KA, Walker GC (2008) Differential response of the plant *Medicago truncatula* to its symbiont *Sinorhizobium meliloti* or an exopolysaccharide-deficient mutant. Proc Natl Acad Sci USA 105:704–709

Kelley LA, Sternberg MJE (2009) Protein structure prediction on the Web: a case study using the phyre server. Nat Protocols 4:363–371

Kelly SJ, Muszyński A, Kawaharada Y, Hubber AM, Sullivan JT, Sandal N, Carlson RW, Stougaard J, Ronson CW (2012) Conditional requirement for exopolysaccharide in the *Mesorhizobium-Lotus* symbiosis. Mol Plant-Microbe Interact 26:319–329

Kinkema M, Gresshoff PM (2008) Investigation of downstream signals of the soybean autoregulation of nodulation receptor kinase GmNARK. Mol Plant-Microbe Interact 21:1337–1348

Krusell L, Madsen LH, Sato S, Aubert G, Genua A, Szczyglowski K, Duc G, Kaneko T, Tabata S, de Bruijn F, Pajuelo E, Sandal N, Stougaard J (2002) Shoot control of root development and nodulation is mediated by a receptor-like kinase. Nature 420:422–426

Krusell L, Sato N, Fukuhara I, Koch BEV, Grossmann C, Okamoto S, Oka-Kira E, Otsubo Y, Aubert G, Nakagawa T, Sato S, Tabata S, Duc G, Parniske M, Wang TL, Kawaguchi M, Stougaard J (2011) The *Clavata2* genes of pea and *Lotus japonicus* affect autoregulation of nodulation. Plant J 65:861–871

Leigh JA, Signer ER, Walker GC (1985) Exopolysaccharide-deficient mutants of *Rhizobium meliloti* that form ineffective nodules. Proc Natl Acad Sci USA 82:6231–6235

Lerouge P, Roche P, Faucher C, Maillet F, Truchet G, Promé JC, Dénarié J (1990) Symbiotic host-specificity of *Rhizobium meliloti* is determined by a sulphated and acylated glucosamine oligosaccharide signal. Nature 344:781–784

Li D, Kinkema M, Gresshoff PM (2009) Autoregulation of nodulation (AON) in *Pisum sativum* (pea) involves signalling events associated with both nodule primordia development and nitrogen fixation. J Plant Physiol 166:955–967

Lim CW, Lee YW, Hwang CH (2011) Soybean nodule-enhanced CLE peptides in roots act as signals in GmNARK-mediated nodulation suppression. Plant Cell Physiol 52:1613–1627

Lin Y-H, Ferguson BJ, Kereszt A, Gresshoff PM (2010) Suppression of hypernodulation in soybean by a leaf-extracted, NARK- and Nod factor-dependent small molecular fraction. New Phytol 185:1074–1086

Lin Y-H, Lin M-H, Gresshoff PM, Ferguson BJ (2011) An efficient petiole-feeding bioassay for introducing aqueous solutions into dicotyledonous plants. Nat Protoc 6:36–45

Lin M-H, Gresshoff PM, Ferguson BJ (2012) Systemic regulation of soybean nodulation by acid growth condition. Plant Physiol 160(4):2028–2039

Mathesius U, Mulders S, Gao M, Teplitski M, Caetano-Anollés G, Rolfe BG, Bauer WD (2003) Extensive and specific responses of a eukaryote to bacterial quorum-sensing signals. Proc Natl Acad Sci USA 100:1444–1449

Mathesius U (2013) SUNN mediates long-distance signalling in nodulation. In: Baluska F (ed) Long-distance systemic signaling and communication in plants, signaling and communication in plants. Springer, Heidelberg

Miyazawa H, Oka-Kira E, Sato N, Takahashi H, Wu G-J, Sato S, Hayashi M, Betsuyaku S, Nakazono M, Tabata S, Harada K, Sawa S, Fukuda H, Kawaguchi M (2010) The receptor-like kinase KLAVIER mediates systemic regulation of nodulation and non-symbiotic shoot development in *Lotus japonicus*. Development 137:4317–4325

Mortier V, Den Herder G, Whitford R, Van de Velde W, Rombauts S, D'haeseleer K, Holsters M, Goormachtig S (2010) CLE peptides control *Medicago truncatula* nodulation locally and systemically. Plant Physiol 153:222–237

Nakagawa T, Kawaguchi M (2006) Shoot-applied MeJA suppresses root nodulation in *Lotus japonicus*. Plant Cell Physiol 47:176–180

Nishimura R, Hayashi M, Wu G-J, Kouchi H, Imaizumi-Anraku H, Murakami Y, Kawasaki S, Akao S, Ohmori M, Nagasawa M, Harada K, Kawaguchi M (2002) HAR1 mediates systemic regulation of symbiotic organ development. Nature 420:426–429

Oka-Kira E, Tateno K, Miura K, Miura K, Haga T, Hayashi M, Harada K, Sato S, Tabata S, Shikazono N, Tanaka A, Watanabe Y, Fukuhara I, Nagata T, Kawaguchi M (2005) *klavier* (*klv*), a novel hypernodulation mutant of *Lotus japonicus* affected in vascular tissue organization and floral induction. Plant J 44:505–515

Okamoto S, Ohnishi E, Sato S, Takahashi H, Nakazono M, Tabata S, Kawaguchi M (2009) Nod factor/nitrate-induced *CLE* genes that drive HAR1-mediated systemic regulation of nodulation. Plant Cell Physiol 50:67–77

Peters NK, Frost JW, Long SR (1986) A plant flavone, luteolin, induces expression of Rhizobium meliloti nodulation genes. Science 233:977–980

Pueppke SG, Broughton WJ (1999) *Rhizobium sp.* strain NGR234 and *R. fredii* USDA257 share exceptionally broad, nested host ranges. Mol Plant-Microbe Interact 12:293–318

Redmond JW, Batley M, Djordjevic MA, Innes RW, Kuempel PL, Rolfe BG (1986) Flavones induce expression of nodulation genes in *Rhizobium*. Nature 323:632–635

Reid DE, Ferguson BJ, Hayashi S, Lin Y-H, Gresshoff PM (2011a) Molecular mechanisms controlling legume autoregulation of nodulation. Ann Bot 108:789–795

Reid DE, Ferguson BJ, Gresshoff PM (2011b) NARK-dependent nodule regulation is activated in the shoot and root in response to inoculation and nitrate induced CLE peptides of soybean. Mol Plant-Microbe Interact 24:606–618

Reid DE, Hayashi S, Lorenc M, Stiller J, Edwards D, Gresshoff PM, Ferguson BJ (2012) Identification of systemic responses in soybean nodulation by xylem sap feeding and complete transcriptome sequencing reveal a novel component of the autoregulation pathway. Plant Biotechnol J 10:680–689

Reid D, Li D, Ferguson BJ, Gresshoff PM (2013) Structure-function analysis of the GmRIC1 signal peptide and CLE domain required for nodulation control in soybean. J Exp Bot. doi: 10.1093/jxb/ert008

Rival P, de Billy F, Bono JJ, Gough C, Rosenberg C, Bensmihen S (2012) Epidermal and cortical roles of *NFP* and *DMI3* in coordinating early steps of nodulation in *Medicago truncatula*. Development 139:3383–3391

Schnabel E, Journet EP, de Carvalho-Niebel F, Duc G, Frugoli J (2005) The *Medicago truncatula SUNN* gene encodes a *CLV1*-like leucine-rich repeat receptor kinase that regulates nodule number and root length. Plant Mol Biol 58:809–822

Searle IR, Men AE, Laniya TS, Buzas DM, Iturbe-Ormaetxe I, Carroll BJ, Gresshoff PM (2003) Long-distance signaling in nodulation directed by a CLAVATA1-like receptor kinase. Science 299:109–112

Seo HS, Li J, Lee SY, Yu JW, Kim KH, Lee SH, Lee IJ, Paek NC (2007) The hypernodulating *nts* mutation induces jasmonate synthetic pathway in soybean leaves. Mol Cell 24:185–193

Spaink HP (2000) Root nodulation and infection factors produced by rhizobial bacteria. Annu Rev Microbiol 54:257–288

Symons GM, Reid JB (2004) Brassinosteroids do not undergo long-distance transport in pea. Implications for the regulation of endogenous brassinosteroid levels. Plant Physiol 135:2196–2206

Terakado J, Fujihara S, Goto S, Kuratani R, Suzuki Y, Yoshida S, Yoneyama T (2005) Systemic effect of a brassinosteroid on root nodule formation in soybean as revealed by the application of brassinolide and brassinazaole. Soil Sci Plant Nutr 51:389–395

van Noorden GE, Kerim T, Goffard N, Wiblin R, Pellerone FI, Rolfe BG, Mathesius U (2007) Overlap of proteome changes in *Medicago truncatula* in response to auxin and *Sinorhizobium meliloti*. Plant Physiol 144:1115–1131

Systemic Signaling in Light Acclimation of Leaves

Grzegorz Konert, Moona Rahikainen, Andrea Trotta, and Saijaliisa Kangasjärvi

Abstract Plants have evolved a multitude of mechanisms that adjust photosynthetic functions in the constantly fluctuating light environment. Perception of light stress in chloroplasts initiates local and systemic acclimation processes that involve complex interactions among apoplastic, chloroplastic, and mitochondrial pathways of cellular signaling. Moreover, distinct cell types seem to comprise cell-specific metabolic programs and signaling components, which elicit strictly coordinated changes in gene expression, optimization of photosynthetic machineries, and reprogramming of metabolic pathways and developmental cascades. In this chapter, we discuss the current understanding of systemic signaling in light acclimation in plants.

Keywords Light acclimation • Reactive oxygen species (ROS) • Excess excitation energy • Systemic acquired acclimation (SAA) • Signaling • High light

1 Introduction

Light is an important environmental factor that provides an energy source for photosynthesis, guides the activity of various metabolic programs, and modulates the growth and development during the entire life cycle of plants. On the other hand, like all living organisms, plants may also suffer detrimental effects if the incoming light energy is not handled with care. In the nature, light conditions undergo constant fluctuations due to climatic variations and seasonal alterations. Changes in cloudiness, wind, or the position of the sun may thus expose plants to light energy in amounts that exceed the need to fuel the ongoing metabolic reactions

G. Konert • M. Rahikainen • A. Trotta • S. Kangasjärvi (✉)
Biochemistry and Food Chemistry, Molecular Plant Biology, University of Turku, Turku 20014, Finland
e-mail: grzkon@utu.fi; mmrahi@utu.fi; andtro@utu.fi; saijaliisa.kangasjarvi@utu.fi

F. Baluška (ed.), *Long-Distance Systemic Signaling and Communication in Plants*, Signaling and Communication in Plants 19, DOI 10.1007/978-3-642-36470-9_12,

in chloroplasts. Under such conditions, the prevalence of excess light energy may imbalance the function of the photosynthetic machinery, resulting in decreased photosynthetic activity or even oxidative damage in plant cells. On the other hand, as this chapter emphasizes, perturbations in cellular metabolism elicit signals that activate various acclimation strategies in plant cells. The extent to which a plant is vulnerable to light-induced stress can vary considerably depending on the life history and pre-acclimation status of the plant. Thus, to warn shaded leaf tissues about the possibility of forthcoming scorching by sunshine, light-exposed parts of the plant elicit signals that allow shaded leaves to prepare themselves for potentially deleterious exposure to sunlight. As a whole, light acclimation is an outcome of complex interactions among developmental, environmental, and metabolic signals, which originate from different cellular compartments and converge to ensure growth and survival in changing environmental cues.

2 Local and Systemic Responses in Light Acclimation

2.1 The Concept of Light Stress

Growth and well-being of plants necessitate highly balanced light-harvesting systems that shuttle light energy for a range of light-driven redox reactions, which are tightly coupled with a multitude of pathways inside and outside the chloroplasts. Besides fueling the photo-assimilation of carbohydrates, photosynthetic redox chemistry provides NADPH and ATP for energy-demanding steps in the biosynthesis of amino acids, lipids, hormones, vitamins, and secondary metabolites, which are not only vital for basic metabolism but also have a great impact on the acclimation status of plant cells (Kirk and Leech 1972).

In linear photosynthetic electron flow, electrons are extracted from water at photosystem II (PSII) and shuttled through plastoquinone (PQ), cytochrome b_6f complex (Cytb_6f), plastocyanin (PC), and photosystem I (PSI) to ferredoxin–NADPH oxidoreductase (FNR) to generate NADPH. Electron transfer through Cytb_6f is coupled with proton pumping to thylakoid lumen, generating a *trans*-thylakoid proton motive force that drives the ATP synthase. The formation of the proton gradient and the production of ATP may become enhanced through cyclic electron transfer reactions, whereby electrons become redirected from PSI to Cytb_6f (Arnon 1959; Johnson 2011; Shikanai 2007). A fraction of reductive power is also targeted via PSI and the ferredoxin-thioredoxin reductase (FTR) to thioredoxins, which are small redox-active proteins that modulate the activation state of various target proteins through dithiol/disulfide exchange reactions (Keryer et al. 2004).

Absorption of excess excitation energy may cause over-reduction of the photosynthetic electron transfer chain, thus promoting light stress, which, depending on the severity of stress, may manifest itself as photoinhibition of PSII (Aro et al. 1993;

Kok 1956) or enhanced accumulation of reactive oxygen species (ROS). Upon saturation of photosynthetic carbon metabolism and a consequent limitation in the availability of PSI electron acceptors, electron transfer to molecular oxygen (O_2) leads to the formation of superoxide (O_2^-) and hydrogen peroxide (H_2O_2) in chloroplast stroma (Liu et al. 2001, 2004). In PSII, imbalanced electron transfer reactions may additionally lead to the formation of singlet oxygen ($^1O_2$) (Hideg et al. 1998; Macpherson et al. 1993). Accumulation of various types of ROS is an intrinsic property of the photosynthetic electron transfer chain, and is increasingly recognized among the key mechanisms that relay stress signals in photosynthetic tissues (Foyer and Noctor 2009; Neill et al. 2002).

Different biotic and abiotic stress conditions, which slow down photosynthesis and other enzymatic activities in chloroplasts, may also promote conditions where excitation energy is in excess. Changes in redox signaling pathways and ROS metabolism represent common nominators for different stress responses, and the mechanisms of signaling and resistance against stresses induced by light, salinity, drought, cold, osmotic stress, or plant pathogens are partially overlapping (Li et al. 2009; Navrot et al. 2007). It is therefore not surprising that light acclimation and plant immunity share a level of commonalities in plants (Kangasjärvi et al. 2012; Mühlenbock et al. 2008; Trotta et al. 2011).

2.2 *Biochemical and Physiological Adjustments in Local and Systemic Tissues*

Plants have evolved a multitude of strategies to cope with short-term and long-term variations in light intensity. Physiological and cellular responses include leaf bending and light-induced chloroplast movements, both of which attempt to decrease the input of excess excitation energy by the photosynthetic pigment–protein complexes. In addition, light acclimation promotes distinct morphological adjustments, including more compact rosettes, thickening of the leaves, and a smaller overall leaf area than the leaves of low-light-grown plants (Kagawa et al. 2001; Park et al. 1996; Walters 2005). On biochemical level, plant acclimation to high light commonly involves enhancement of photosynthetic capacity and upregulation of antioxidant enzymes (Karpinski et al. 1999). In short term, excess excitation energy may become dissipated through the PSII subunit S (PsbS) and xanthophyll-dependent non-photochemical quenching of excitation energy (NPQ) in PSII, redox shuttling to mitochondria and photorespiration (Nunes-Nesi et al. 2011; Peterson and Havir 2001). In addition, plants have evolved tightly redox-controlled mechanisms to regulate the efficiency of the light-harvesting antenna systems to maintain redox balance in chloroplasts. A major contributor in this system is a strictly redox-regulated phosphorylation of light-harvesting (LHCII) antenna proteins by the STN7 kinase, which becomes activated upon reduction of the plastoquinone pool, and inhibited via thioredoxin activity upon reduction of chloroplast stroma (Bellafiore et al. 2005; Rintamäki et al. 2000). By these means, STN7 regulates the association of the mobile LHCII antenna complexes between PSII and PSI, a phenomenon that is a major determinant of the redox status of the thylakoid electron transfer chain (Tikkanen et al. 2006). Currently, these short-term mechanisms are understood as balancing agents that prevent the formation of unnecessary acclimation signals from chloroplasts (Tikkanen et al. 2010).

Extended periods of high light will ultimately lead to a condition where the capacity of the above-mentioned short-term regulation mechanisms become exceeded, resulting in the onset of chloroplast signals that mediate local and systemic acclimation processes in leaves (Karpinski et al. 1999). These responses involve coordinated changes in the expression of chloroplast and nuclear encoded genes, which direct optimization of photosynthetic membrane protein complexes, metabolic pathways, and reprogramming of developmental cascades to best meet the needs of the prevailing environmental cues. Thus, long-term growth under high light intensities results in optimization of photosynthetic reactions through structural modulations in the composition of the photosynthetic protein complexes and their light-harvesting antennae (Anderson et al. 1995; Walters 2005). Such adjustments necessitate tight regulation of genes related to photosynthesis and other metabolic and regulatory pathways (Sheen 1990).

During the recent years, it has become clear that perception of light stress in chloroplasts initiates acclimation signals, which are not only sensed locally in the adjacent leaf cells but also in distal tissues to initiate a systemic acquired acclimation (SAA) in nonexposed parts of the plant (Mühlenbock et al. 2008; Pogson et al. 2008; Rossel et al. 2007). Such signaling mechanisms may be highly relevant in dense populations and canopies, where sun flecks appear during the day when the position of the sun is changing. Whereas the mechanisms of SAA are starting to emerge, reciprocal signaling from shaded leaves to sun-exposed parts has not yet been recognized. Here, we discuss the current understanding of systemic signaling in light acclimation in plants.

3 Protein Components and Metabolic Interactions in Chloroplast Retrograde Signaling

During evolution, a significant proportion of genes encoding the structural and regulatory components of the photosynthetic machinery have been transferred from the chloroplast genome to the nucleus through a process denoted as horizontal gene transfer (Martin et al. 1998). Consequently, photosynthetic pigment–protein complexes are formed by chloroplast and nuclear encoded subunits, and their expression must be tightly coordinated to ensure that the photosynthetic complexes are developed and adjusted in a stoichiometric fashion under constantly changing environmental cues. Although it is clear that the chloroplast and nuclear genomes must be coordinated, the mechanisms of signaling have only started to emerge

(Leister 2012). By now, a number of initiating sources for chloroplast signals as well as their consequences on the highly dynamic regulation of nuclear gene expression level have become well established, whereas the mechanisms of signal transduction are not so well understood.

Light receptors, including phytochromes and the blue light receptors CRYPTOCHROME (CRY) 1 and 2, have crucial signaling functions during the greening process, albeit CRY1 is known to mediate chloroplast signals in mature leaves as well (Kleine et al. 2007). In green leaves with fully developed chloroplasts, light-induced acclimation processes essentially involve self-regulatory feedback loops from photosynthetic components (Anderson et al. 1995; Huner et al. 1998; Walters and Horton 1995). Such self-regulation involves chloroplast retrograde signaling whereby organellar signals adjust nuclear gene expression to allow acclimation in changing environmental cues. According to Pogson et al. (2008), these signals can be divided into two classes (1) biogenic control, which ensures stoichiometric synthesis and assembly of the photosynthetic machinery, and (2) operational control, which aims at adjusting and optimizing the function of preexisting photosynthetic complexes under environmental challenges, such as light stress.

Natural environments are characterized by constant changes in temperature, water availability, and light conditions, and living organisms have to cope with these fluctuations by adjusting their metabolism accordingly. Photosynthetic organisms have evolved versatile and partially overlapping systems for sensing, signaling, and responding to a wide range of imbalances in chloroplasts. Plants monitor changes in light conditions by sensing modulations in the redox state of chloroplastic redox components, the PQ-pool PSI electron acceptors, and thiol redox components as well as formation of ROS. Depending on the eliciting signal/environmental condition, these redox components or their combinations mediate key early signaling functions to allow specific adjustment in cell functions (Tikkanen et al. 2010).

Signaling components that participate in the relay of information from chloroplasts to the nucleus have mainly been identified through genetic approaches and include the FLUORESCENT (FLU) protein, which controls the accumulation of $^1O_2$ in chloroplasts (Baruah et al. 2009). The signaling cascade triggered by $^1O_2$ in turn includes the chloroplastic EXECUTER (EX) proteins 1 and 2 as well as the blue light receptor CRYPTOCHROME1 (CRY1), which has also been shown to modulate the high light response in Arabidopsis (Kleine et al. 2007; Lee et al. 2007). The series of *genomes uncoupled* (*gun*) mutants revealed a set of chlorophyll biosynthesis enzymes, designated GUN2-5, which contribute to chloroplast signaling cascades through yet undefined mechanisms, albeit the role of ROS or redox-dependent signaling cascades has been implicated (Mochizuki et al. 2008; Moulin et al. 2008). GUN1 has a different function and signals through a chloroplast-localized transcription factor and connects the status of chloroplast gene expression with the regulation of photosynthesis-associated nuclear genes (Kindgren et al. 2012; Koussevitzky et al. 2007). The GUN1 pathway likely operates through the abscisic acid (ABA)-responsive transcription factor ABI4 (ABSCISIC ACID-

INSENSITIVE 4), which mediates the repression of nuclear photosynthesis-associated genes, and provides a point of cross-talk with light, ABA, and sugar signaling. In fact, organellar redox signals from both chloroplasts and mitochondria are known to converge at ABI4 in the nucleus (Giraud et al. 2009; Kerchev et al. 2011).

Besides redox-based chloroplast retrograde signals, key roles for chloroplast-derived metabolites in cellular communication systems have also become evident. These include 3′-phosphoadenosine 5′-phosphate (PAP), which translocates from chloroplasts to the nucleus and modulates the expression of specific genes through inhibition of exonuclease activity (Estavillo et al. 2011). Another example of a chloroplast-derived metabolite with a role in retrograde signaling is methylerythritol cyclodiphosphate (MEcPP), an intermediate in the chloroplastic isoprenoid biosynthesis pathway (Xiao et al. 2012). Upon exposure to abiotic stresses (i.e., wounding or high light), MEcPP is accumulated and moves to the nucleus where it induces the expression of the stress marker gene *HPL* (*Hydroperoxide lyase*), whereas the expression of photosynthesis-related genes is not changed. This observation suggests that MEcPP operates on a signaling pathway that is distinct from the one elicited in the *gun* mutants. Altogether, these findings highlight the complexity of mechanisms that allow chloroplasts to respond to an environmental change in a highly specific manner.

4 The Role of Mitochondrial Pathways in Light Acclimation

Metabolic intermediates and reducing equivalents undergo extensive exchange between chloroplasts, cytoplasm, mitochondria, and peroxisomes (Padmasree et al. 2002). These mechanisms are particularly important under excess light, when over-reduction of electron transfer components starts to limit photosynthetic reactions and may lead to photodamage of core enzymes. Plants have evolved processes in which mitochondrial pathways have a particularly important role in the alleviation of cellular stress. In chloroplasts, the level of NADPH can be controlled through oxidation of NADPH to $NADP^+$ by NADP-dependent malate dehydrogenase (NADP-MDH) with simultaneous reduction of oxaloacetate (OAA) to malate (Nunes-Nesi et al. 2011; Taniguchi et al. 2002). NADP-MDH in turn becomes activated upon reduction of the chloroplast Fd-thioredoxin system, and is therefore strictly regulated by light (Schepens et al. 2000). Malate is exported from chloroplasts by a process known as the malate–OAA shuttle, the translocation occurring in exchange for OAA into chloroplast stroma. Malate can be directed into the mitochondrial tricarboxylic acid cycle (TCA) (Yoshida et al. 2007). Alternatively, malate can be oxidized to OAA in the cytoplasm by a simultaneous reduction of NAD^+ to NADH, and the resulting OAA and NADH can be consumed in TCA and the mitochondrial electron transport chain (mitETC), respectively. Mitochondrial pathways also participate in quenching of excess reductants through photorespiration, which involves oxidation of glycine by mitochondrial enzymes

(Berry et al. 1978). This step results in the formation of NADH, which is used to reduce OAA to malate or directly incorporated into mitETC.

MitETC resides in the inner mitochondrial membrane and involves transport of electrons from NAD(P)H through NADH dehydrogenase (complex I), succinate dehydrogenase (complex II), and to ubiqinone, localized in innermembrane space, from which it is transported to cytochrome bc1 (complex III). Cytochrome c as a free protein transports electrons from complex III to cytochrome c oxidase (complex IV), and finally to oxygen with the formation of water. During these processes, complexes I, III, and IV pump H^+ from the mitochondrial matrix to the intermembrane space, creating a proton gradient that drives the ATP synthase (complex V, ATPase).

Plant mitochondria also possess an alternative pathway of mitETC through the alternative oxidase (AOX), which is not present in animal cells. AOX accepts electrons directly from ubiquinone pool, thus limiting electron flow to cytochrome c oxidase pathway. The AOX-dependent pathway does not generate proton gradient and thus does not participate in ATP synthesis through mitochondrial ATPase (Wilson 1980). In addition, plant mitochondria contain non-proton-pumping NAD(P)H dehydrogenases, which allow bypass of complex I as well (Rasmusson et al. 1998). All mitochondria possess also the uncoupling protein (UCP) which serves as yet another non-phosphorylating bypass of mitochondrial electron transfer chain by allowing relaxation of proton gradient between mitochondrial intermembrane space and matrix (Sweetlove et al. 2006).

Under long-term high light conditions, plants upregulate the level of cytochrome c oxidase rather than AOX (Yoshida et al. 2007), presumably to enhance the generation of ATP. The opposite is observed when plants are exposed to a short high light treatment when respiratory chain becomes over-reduced in part due to the high flow of reductants from chloroplasts to mitochondria and AOX becomes highly upregulated. Indeed AOX activity is critical for the maintenance of redox balance in mitochondria, and can be considered as an emergency mechanism activated upon abrupt exposure to stressful light conditions (Yoshida et al. 2011). Tight regulation of redox shuttles lowers the risk for generation of ROS (Dutilleul et al. 2003; Yoshida et al. 2007). AOX1a was found to be induced within 10–60 min after the onset of high light stress. AOX1a promoter activity and transcription abundance were shown to increase in response to treatment with ABA (Giraud et al. 2009). This regulation is likely to be mediated by the transcription factor ABI4, which was shown to interact with a promoter element of *AOX1a* in biochemical binding assays; moreover, *aox1a* promoter activity was found to be fully derepressed in *abi4* mutant leaves (Giraud et al. 2009). Regulation of the expression of *AOX1a* via ABI4 therefore provides a direct link between chloroplastic and mitochondrial retrograde signaling (Kerchev et al. 2011; Koussevitzky et al. 2007; Pesaresi et al. 2007). Altogether, leaf mitochondria play a crucial role in light acclimation. The capacity to dissipate excess reductants through photorespiration and the malate–OAA shuttle and consequent alleviation of light stress represent a major subject in the future investigation of the topic of light tolerance in plants.

5 The Role of Bundle Sheath Cells in Systemic Signaling

The process of systemic light acclimation involves a complicated network of signals that spread through the vasculature and induce changes in gene expression and establishment of stress tolerance in the entire plant. Considering the different physiological roles of distinct leaf cell types, it is conceivable that mesophyll cells, vascular tissues, and stomatal guard cells carry out specialized roles in light acclimation. Leaf veins transport water, micronutrients, and metabolic end products between different plant organs, and have also been demonstrated to carry signaling molecules from one part to another (Robert and Friml 2009). In C3 plants, leaf veins are flanked by a layer of elongated chloroplast-containing bundle sheath cells, which due to their central position between the vein and the surrounding mesophyll tissue carry out a specific role in the relay of light-dependent acclimation signals (Leegood 2008). Vascular bundle sheath cells therefore possess far-reaching and thus far largely unidentified regulatory roles in light acclimation.

Arabidopsis mutants with reticulated leaf phenotypes have provided evidence suggesting that bundle sheath-specific metabolic processes specifically modulate leaf development and acclimation in a light-dependent manner (Barth and Conklin 2003; Kinsman and Pyke 1998; Knappe et al. 2003; Lepistö et al. 2009; Overmyer et al. 2008). These mutants typically exhibit a conditional phenotype with dark green paraveinal and light green interveinal regions caused by metabolic perturbations in at least some experimental conditions or during a specific developmental stage. Classical examples include *cue1* (*chlorophyll a/b binding protein underexpressed 1*; Li et al. 1995), *reticulata* (*re*; González-Bayón et al. 2006), and the more recently identified *ntrc* deficient in a chloroplastic NADPH-dependent thioredoxin reductase (NTRC; Lepistö et al. 2009).

CUE1 encodes the phosphoenolpyruvate/phosphate translocator of the chloroplast envelope inner membrane (Knappe et al. 2003; Streatfield et al. 1999). Deficiencies in the function of CUE1 therefore deteriorate several chloroplastic metabolic routes and drastically affect the biosynthesis of aromatic amino acids (Knappe et al. 2003; Voll et al. 2003). The *reticulata* (*re*) mutant visually resembles *cue1* but exhibits the reticulate phenotype conditionally when grown under long day conditions (González-Bayón et al. 2006). *RE* encodes a putative transmembrane protein likely to be localized in the chloroplast envelope (Zybailov et al. 2008). The same mutant was also identified as *lower cell density1/radical-induced cell death2* (*lcd1/rcd2*) in two separate screens for ozone sensitivity (Barth and Conklin 2003; Overmyer et al. 2008). It is notable that *re* plants were also found to be sensitive to infection by the bacterial plant pathogen *Pseudomonas syringae*, suggesting that RE forms a component common for light acclimation and immunity in plants (Barth and Conklin 2003; Overmyer et al. 2008). Indeed, *re* plants accumulate H_2O_2 in the vascular leaf tissues, a phenomenon commonly observed when leaves are exposed to light stress (Karpinski et al. 1999; Overmyer et al. 2008). These findings on *cue* and *re* highlight the importance of transporters and signaling components of the chloroplast envelope membrane in plant development and acclimation.

Apart from *cue1* and *re,* young *ntrc* leaves exhibit a uniform pale green phenotype, but undergo a transient phase of reticulation during greening that occurs in aging *ntrc* leaves (Lepistö et al. 2009). Upon aging, the veins of *ntrc* plants turn green, and gradually the entire leaf acquires wild-type levels of chlorophyll (Lepistö et al. 2009). NTRC is likely to regulate a multitude of thiol-redox regulated processes in chloroplasts, such as starch metabolism (Michalska et al. 2009), chlorophyll biosynthesis, amino acid metabolism, and the biosynthesis of auxin (Lepistö et al. 2009). Moreover, NTRC has been shown to act as a reducing agent for 2-cysteine peroxiredoxin, and may thus be crucial for ROS-based signaling in chloroplasts (Pérez-Ruiz et al. 2006).

Altogether, although the molecular mechanisms underlying the reticulated pigmentation in *cue1*, *re*, and *ntrc* remain unresolved, these mutant characteristics have unequivocally demonstrated the existence of specified metabolic properties and unique characteristics of ROS tolerance in bundle sheath cells. Because of different physiological properties of bundle sheath cells, it can be assumed that the signaling and gene expression regulated by these signals vary as well (Fryer et al. 2003; Kangasjärvi et al. 2008, 2009).

6 Interplay Among Chloroplastic and Apoplastic Pathways in Systemic Signaling

The pioneering observation of systemic light acclimation suggested that light-induced accumulation of H_2O_2 in vascular leaf areas initiated systemic signaling, which induced stress resistance in high-light-illuminated Arabidopsis plants (Fryer et al. 2003; Karpinski et al. 1999). In further studies on the mechanisms of systemic signaling, ROS production in the apoplast of bundle sheath cells was associated with NADPH activity and found to become enhanced upon biosynthesis of ABA in adjacent cells of vascular parenchyma (Galvez-Valdivieso et al. 2009). In addition, Szechyńska-Hebda et al. (2010) proposed the involvement of photoelectron-physiological signaling (PEPS) in long-distance mechanisms of light acclimation.

These findings suggested the involvement of plasma membrane electrical potential, NADPH oxidases, and chloroplasts as sources of ROS in systemic signaling, and this signaling interaction most likely involves also the heterotrimeric G-proteins (Joo et al 2005). Moreover, besides H_2O_2, singlet oxygen also seems to accumulate predominantly along leaf vasculature (Suorsa et al. 2012), but the significance of this observation with respect to light acclimation remains obscure. NADPH oxidases are integral plasma membrane proteins composed of six trans-membrane domains supporting two heme groups. The two independent hemes and FAD are required for transfer of electrons from NADPH, which serves as cytoplasmic electron donor across the membrane to oxygen, the extracellular electron acceptor, to generate the superoxide radical (Sagi and Fluhr 2006). In a secondary reaction, O_2^- is dismutated to hydrogen peroxide.

Whereas the involvement of the above-mentioned components in systemic light acclimation is clear, the sequence of events remains poorly understood. Evidence has accumulated suggesting that NADPH oxidase-dependent apoplastic ROS production is followed by ROS production in the chloroplast, and that there is regulatory hierarchy in signal transfer from the apoplast to the chloroplast (Joo et al. 2005; Nomura et al. 2012). Such close signaling interaction is facilitated by the small physical distance between the chloroplast and the apoplast. In plant cells, the chloroplast envelope is often observed in a very close proximity with the plasma membrane, which should allow rapid and efficient communication between the two cellular compartments.

The finding that NADPH oxidases contribute to systemic signaling over long distances led to the formation of the concept termed the "ROS wave" (Mittler et al. 2011). Besides high light, this mechanism was shown to be induced by wounding, cold, heat, and salinity (Miller et al. 2009). Intriguingly, in all cases, formation of the ROS signal was found to depend on the specific NADPH oxidase isoform AtRbohD. Analysis of an *Arabidopsis* line expressing a luciferase reporter gene under the control of a ROS-responsive *ZAT12* promoter revealed that application of a local stress treatment elicited luciferase activity both at the site of treatment and in systemic untreated tissues. The signal propagation was rapid, traveling at a rate of 8.4 cm/min along the stem axis, and coincided with apoplastic ROS accumulation in wild-type plants but was suppressed in *atrbohD* mutant plants (Miller et al. 2009). These findings led to a conclusion that AtRbohD generates a ROS signal capable of triggering and maintaining an auto-propagating ROS wave that travels from cell to cell across long distances. This also clarifies the early observations where H_2O_2 was shown to have a key role in systemic acclimation responses under high light stress (Fryer et al. 2003; Rossel et al. 2007). How the ROS signal is perceived and how the ROS burst becomes triggered in the neighboring cells, however, will be matters of future breakthrough.

With respect to bundle sheath-specific pathways of ROS signaling, it is intriguing that chloroplasts located along the veins in Arabidopsis leaves were found to be less sensitive to light stress than the mesophyll cell chloroplasts. This became evident upon phenotypic characterization of *tapx sapx* double mutants deficient in the chloroplastic stromal and thylakoid-bound ascorbate peroxidases, or *tapx sapx vtc2* triple mutants deficient in both chloroplast ascorbate peroxidase and ascorbate. Both mutants showed bleaching of interveinal leaf tissues upon methyl-viologen treatment or illumination under high light (Giacomelli et al. 2007; Kangasjärvi et al 2008). Studies have also revealed that genes related to plant antioxidant network may be preferentially expressed in the paraveinal regions of leaves. These gene products include the cytoplasmic ascorbate peroxidase 2 (APX2), a microRNA (miR398) that targets the Cu/Zn superoxide dismutase (Sunkar et al. 2006) and a chloroplastic glutaredoxin (Cheng et al. 2006). The physiological significance of *APX2* induction or the extent to which APX2 is required to quench the ROS in the cytoplasm of bundle sheath cells, however, is not clear. As a whole, the physiological significance of cell-type-specific expression of antioxidant components in systemic light signaling remains to be elucidated.

7 Implications for Hormonal Signaling in Systemic Light Acclimation

Whereas the role for ROS as a messenger in systemic signaling seems clear, it is likely that the SAA response additionally involves other thus far unknown components, which promote gene expression changes in shaded leaves. SAA might be caused by a mobile signaling molecule eliciting the response, a system analogous to the regulation of flowering through FLORIGEN, which migrates from the shoot apex to initiate flowering in shoot apical meristem of buds in response to photoperiodism (Turck et al. 2008) (Figs. 1 and 2).

The mechanism of SAA seems to share a level of commonalities with those mediating the systemic acquired resistance (SAR), which becomes induced in uninfected parts of a pathogen-exposed plant (Mullineaux et al. 2000). Upon systemic induction of immune reactions, a mobile signaling molecule is transported to distal uninfected parts of the plant where it induces formation of ROS and the expression of pathogenesis-related genes.

In SAR, multiple phytohormones and other small signaling compounds including H_2O_2, jasmonic acid, salicylic acid, and azelaic acid have been suggested to function as a mobile signal from the infection site to distal leaves. However, no single substance has been found to mediate all features of SAR. Signals are transduced through the vascular tissue as well as air by volatile compound (Shah 2009). Heil and Ton (2008) proposed a model where distal defense responses are self-primed by volatile signals, while vascular signaling is needed for their full elicitation. Upon systemic signaling taking place after wounding, it has been proposed that the signal to distal leaves is transmitted through the cells of vascular parenchyma as a signal where jasmonate induces its own biosynthesis in the neighboring cells, thus moving the signal along the vascular tissue (Heil and Ton 2008). A similar kind of model might explain also the systemic acclimation to abiotic stresses.

Recent findings suggest that light acclimation involves cross-talk with components related to salicylic acid, jasmonic acid, and ethylene dependent defense signaling pathways (Mühlenbock et al. 2008). SAA, however, does not seem to depend on the perception or signaling through these stress hormones, and in this respect seems to be mediated by mechanisms different from SAR. Rossel et al. (2007) analyzed the jasmonic acid- or salicylic acid-related signaling mutants *jar1-1*, *jin1*, *npr1*, and *NahG*; all exhibited the systemic light acclimation response, albeit to a lesser extent than wild-type plants. Indeed, as yet there is no evidence that any of these hormones would directly act as a mobile signal for triggering systemic light acclimation responses in the nonexposed tissues.

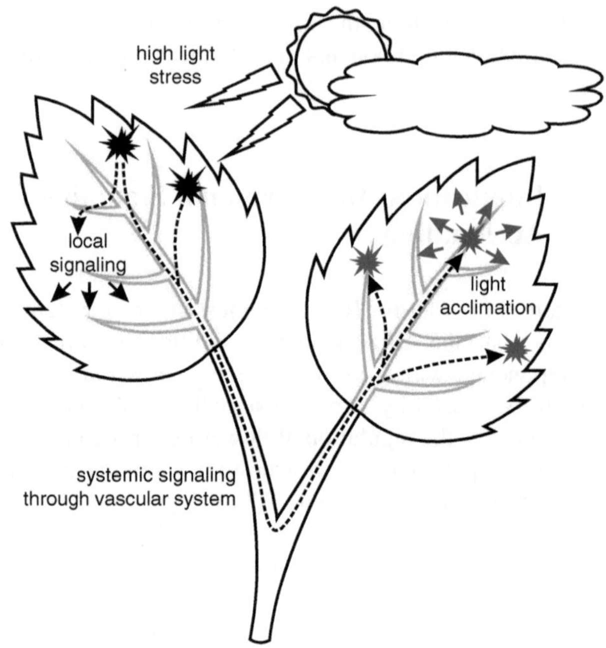

Fig. 1 Systemic light acclimation. Light stress signals are perceived in exposed leaves and transduced through the apoplast and vascular system to induce pre-acclimation in shaded leaves

A zinc finger transcription factor ZAT10 has been shown to regulate the expression of a set of same genes that are activated in distal and high light exposed leaves during SAA. ZAT10 becomes rapidly induced in leaf vasculature in both the exposed and the shaded tissues, with the exception of roots, after the onset of light stress, pointing to the role of ZAT10 in only green parts of plant. The induction of ZAT10 correlates with enhanced expression of antioxidant genes and increased tolerance against light-induced damage of PSII in the light-exposed leaves as well as tolerance to exogenous H_2O_2 (Rossel et al. 2007). Overexpression of ZAT10 results in a stunted mutant phenotype and promotes enhanced tolerance to various abiotic stresses including drought and high light. Thus ZAT10 may regulate the genes activated both in high light and drought (Mittler et al. 2006). The MAP kinases MAPK3 and MAPK6 have been shown to directly interact and phosphorylate ZAT10 (Nguyen et al 2012). Moreover, MAPK3 and MAPK6 mediate the onset of defense signaling in response to pathogens (Asai et al. 2002). This interaction with ZAT10 offers a route through which the responses to various stresses can be regulated and explains the elevated resistance to multiple stress conditions.

ZAT10 and another similar transcription factor ZAT12 are induced under a range of stress conditions, including high light, drought, salt, and cold (Lee et al. 2002: Sakamoto et al. 2000). Intriguingly, the induction of ZAT10 and ZAT12 was delayed upon high light illumination of the chloroplast signaling mutants *gun1* and *abi4* compared to wild-type plants (Koussevitzky et al. 2007), which raises the question whether the GUN1/ABI4 pathway also mediates systemic signals. A

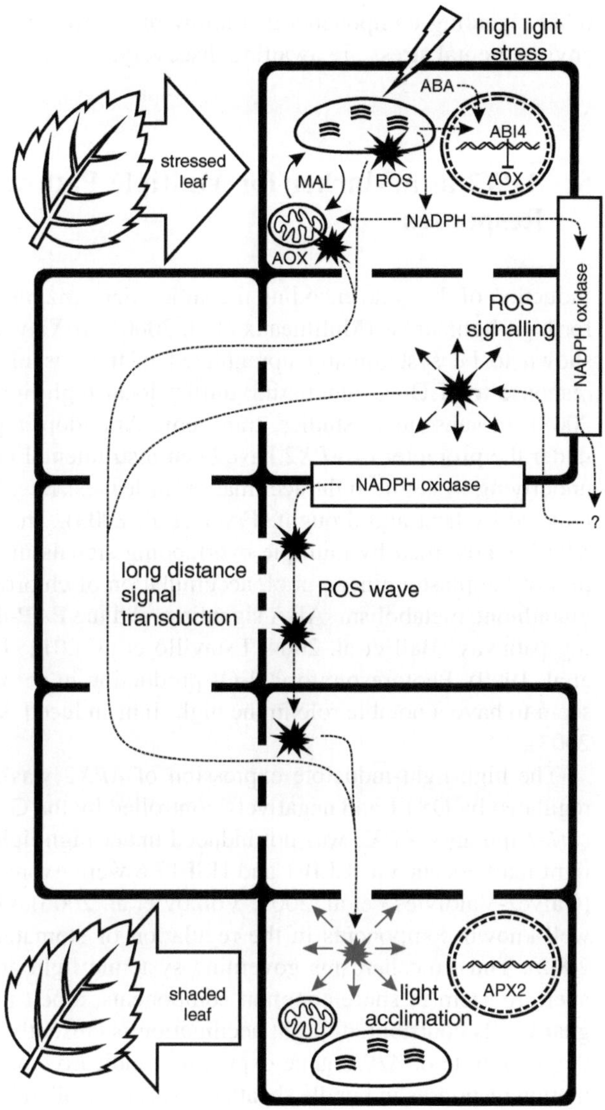

Fig. 2 Emerging mechanisms of systemic light signaling. Under excess light energy, chloroplasts initiate signals that trigger acclimation processes by adjusting gene expression in the nucleus. Over-reduction of photosynthetic electron acceptors also activates metabolic shuttles, whereby reducing equivalents are translocated to mitochondria, where the AOX pathway plays a major role in dissipating excess reducing energy under abrupt light stress. In the cytoplasm, NADPH serves as a substrate for plasmamembrane NADPH oxidases, the activity of which leads to generation of ROS in the apoplast. Through yet unknown mechanisms, successive bursts in ROS allow spreading of the stress signals to neighboring cells through apoplastic "ROS waves." Long-distance signal transduction occurs through vascular tissues. Spreading of high light signals from exposed leaves triggers gene expression changes, classically exemplified by induction of *APX2*, also in the shaded leaves allowing the entire plant to prepare itself for potential future light stress

notable overlap exists in the gene expression changes taking place after shift to high light and the treatment with exogenous ABA (Kimura et al. 2003). Some of the overlap can be explained with the sudden decrease in the cell water potential after the shift to high light. What is more, functional ABA biosynthesis is required for the high light gene expression. The regulation may take place through a pathway where the protein kinase OPEN STOMATA 1 (OST1) positively regulates the accumulation of H_2O_2, which in turn induces the high light genes (Galvez-Valdivieso et al. 2009). One can assume that a number of yet unidentified transcription factors and other signaling components that allow plants to record and signal the severity of environmental stress are awaiting discovery.

8 *APX2* as a Marker for Multiple Pathways in Light Stress Responses

Induction of the gene encoding the antioxidant enzyme APX2 is a classical marker for high light stress (Mullineaux et al. 2000). *APX2* was also one of the first genes shown to be systemically upregulated in the vascular tissue of distal leaves in response to H_2O_2 accumulating during local high light exposure (Bechtold et al. 2008). In subsequent studies, transgenic Arabidopsis plants expressing luciferase under the promoter of *APX2* have been instrumental in dissecting the components underlying systemic light acclimation in leaves. Besides high light, *APX2* can be induced by heat and drought (Fryer et al. 2003). The transcriptional induction of *APX2* is governed by multiple overlapping signals including light-induced reduction of the plastoquinone pool, accumulation of chloroplastic and apoplastic ROS, glutathione metabolism, ABA signaling, and the PAP-dependent retrograde signaling pathway (Ball et al. 2004; Estavillo et al. 2011; Fryer et al. 2003; Karpinski et al. 1999). Photorespiratory H_2O_2 production in peroxisomes, however, does not seem to have a notable role in the high-light-induced APX2 expression (Fryer et al 2003).

The high-light-inducible expression of *APX2* was also found to be positively regulated by OST1 and negatively controlled by the G-protein coupled receptor. In *ost1-1* mutants, *APX2* was not induced under high light, whereas some other high light markers such as ELIP1 and HSP17.6 were expressed like in wild-type plants (Galvez-Valdivieso et al. 2009; Volkov et al. 2006). OST1 and the G-proteins are well-known components in the regulation of stomatal movements (Merlot et al. 2002). Thus, mechanisms governing systemic light acclimation and the stomatal aperture seem to share common components, albeit the extent to which stomatal guard cells contribute to light acclimation is currently not understood. Altogether, the amplitude of *APX2* gene expression, indicative of the onset of systemic acclimation responses in bundle sheath cells, is controlled by overlapping pathways that respond to the water balance, metabolic homeostasis, and cellular ROS/redox state in Arabidopsis leaves.

9 Conclusions

Maintenance of cellular homeostasis plays a crucial role in plant "survival of the fittest." During evolution, plants have evolved a catalogue of mechanisms to cope with harsh environmental conditions that can be expected to change in time. A key phenomenon in plant acclimation is known as systemic light signaling, whereby shaded parts of the plant receive signals from light-stressed leaf tissues and develop adjustments for potentially upcoming stress conditions. Tight regulation of redox balance and metabolite shuttling between organelles represent vital mechanisms in the maintenance of metabolic homeostasis, and allow dissipation of excess reductants that otherwise would block and damage the photosynthethic apparatus.

Systemic signaling upon perception of high light stress is of basic importance in the evolutionary calendar of plants as sessile organisms. The emerging picture on systemic light acclimation indicates that chloroplast retrograde signals cross-communicate with apoplastic pathways through yet undefined molecular interactions. Considerable research efforts are therefore still needed to identify the signaling components and receptors that mediate and determine the acclimation strategies in local and systemic leaf tissues under light stress.

Acknowledgments This work was supported by Academy of Finland projects 263772, 218157, and 130595 for SK and Turku University Finnish Doctoral Program in Plant Science for GK.

References

Anderson JM, Chow WS, Park YI (1995) The grand design of photosynthesis: acclimation of the photosynthetic apparatus to environmental cues. Photosynth Res 46:129–139

Arnon DI (1959) Conversion of light into chemical energy in photosynthesis. Nature 184:10–21

Aro EM, Virgin I, Andersson B (1993) Photoinhibition of Photosystem II. Inactivation, protein damage and turnover. Biochim Biophys Acta 1143:113–134

Asai T, Tena G, Plotnikova J, Willmann MR, Chiu WL, Gomez-Gomez L, Boller T, Ausubel FM, Sheen J (2002) MAP kinase signalling cascade in Arabidopsis innate immunity. Nature 415:977–983

Ball L, Accotto GP, Bechtold U, Creissen G, Funck D, Jimenez A, Kular B, Leyland N, Mejia-Carranza J, Reynolds H, Karpinski S, Mullineaux PM (2004) Evidence for a direct link between glutathione biosynthesis and stress defense gene expression in Arabidopsis. Plant Cell 16:2448–2462

Barth C, Conklin PL (2003) The lower cell density of leaf parenchyma in the *Arabidopsis thaliana* mutant *lcd1-1* is associated with increased sensitivity to ozone and virulent *Pseudomonas syringae*. Plant J 35:206–218

Baruah A, Simková K, Apel K, Laloi C (2009) Arabidopsis mutants reveal multiple singlet oxygen signaling pathways involved in stress response and development. Plant Mol Biol 70:547–563

Bechtold U, Richard O, Zamboni A, Gapper C, Geisler M, Pogson B, Karpinski S, Mullineaux PM (2008) Impact of chloroplastic- and extracellular-sourced ROS on high light-responsive gene expression in Arabidopsis. J Exp Bot 59:121–133

Bellafiore S, Barneche F, Peltier G, Rochaix JD (2005) State transitions and light adaptation require chloroplast thylakoid protein kinase STN7. Nature 433:892–895

Berry JA, Osmond CB, Lorimer GH (1978) Fixation of $^{18}O_2$ during photorespiration. Plant Physiol 62:954–967

Cheng NH, Liu JZ, Brock A, Nelson RS, Hirschi KD (2006) AtGRXcp, an Arabidopsis chloroplastic glutaredoxin, is critical for protection against protein oxidative damage. J Biol Chem 281:26280–26288

Dutilleul C, Garmier M, Noctor G, Mathieu C, Chétrit P, Foyer CH, de Paepe R (2003) Leaf mitochondria modulate whole cell redox homeostasis, set antioxidant capacity, and determine stress resistance through altered signaling and diurnal regulation. Plant Cell 15:1212–1226

Estavillo GM, Crisp PA, Pornsiriwong W, Wirtz M, Collinge D, Carrie C, Giraud E, Whelan J, David P, Javot H, Brearley C, Hell R, Marin E, Pogson BJ (2011) Evidence for a SAL1-PAP chloroplast retrograde pathway that functions in drought and high light signaling in Arabidopsis. Plant Cell 23:3992–4012

Foyer CH, Noctor G (2009) Redox regulation in photosynthetic organisms: signaling, acclimation, and practical implications. Antioxid Redox Signal 11:861–905

Fryer MJ, Ball L, Oxborough K, Karpinski S, Mullineaux PM, Baker NR (2003) Control of *Ascorbate Peroxidase 2* expression by hydrogen peroxide and leaf water status during excess light stress reveals a functional organisation of Arabidopsis leaves. Plant J 33:691–705

Galvez-Valdivieso G, Fryer MJ, Lawson T, Slattery K, Truman W, Smirnoff N, Asami T, Davies WJ, Jones AM, Baker NR, Mullineaux PM (2009) The high light response in *Arabidopsis* involves ABA signaling between vascular and bundle sheath cells. Plant Cell 21:2143–2162

Giacomelli L, Masi A, Ripoll DR, Lee MJ, van Wijk KJ (2007) Ar*abidopsis thaliana* deficient in two chloroplast ascorbate peroxidases shows accelerated light-induced necrosis when levels of cellular ascorbate are low. Plant Mol Biol 65:627–644

Giraud E, Van Aken O, Ho LHM, Whelan J (2009) The transcription factor ABI4 is a regulator of mitochondrial retrograde expression of *ALTERNATIVE OXIDASE1a*. Plant Physiol 150:1286–1296

González-Bayón R, Kinsman EA, Quesada V, Vera A, Robles P, Ponce MR, Pyke KA, Micol JL (2006) Mutations in the *RETICULATA* gene dramatically alter internal architecture but have little effect on overall organ shape in Arabidopsis leaves. J Exp Bot 57:3019–3031

Heil M, Ton J (2008) Long-distance signalling in plant defence. Trends Plant Sci 13:264–272

Hideg E, Kálai T, Hideg K, Vass I (1998) Photoinhibition of photosynthesis in vivo results in singlet oxygen production detection via nitroxide-induced fluorescence quenching in broad bean leaves. Biochemistry 37:11405–11411

Huner NPA, Öquist G, Sarhan F (1998) Energy balance and acclimation to light and cold. Trends Plant Sci 6:224–230

Johnson GN (2011) Physiology of PSI cyclic electron transport in higher plants. Biochim Biophys Acta 1807:906–911

Joo JH, Wang S, Chen JG, Jones AM, Fedoroff NV (2005) Different signaling and cell death roles of heterotrimeric G protein alpha and beta subunits in the Arabidopsis oxidative stress response to ozone. Plant Cell 17:957–970

Kagawa T, Sakai T, Suetsugu N, Oikawa K, Ishiguro S, Kato T, Tabata S, Okada K, Wada M (2001) Arabidopsis NPL1: a phototropin homolog controlling the chloroplast high-light avoidance response. Science 291:2138–2141

Kangasjärvi S, Lepistö A, Hännikäinen K, Piippo M, Luomala EM, Aro EM, Rintamäki E (2008) Diverse roles for chloroplast stromal and thylakoid-bound ascorbate peroxidases in plant stress responses. Biochem J 412:275–285

Kangasjärvi S, Nurmi M, Tikkanen M, Aro EM (2009) Cell-specific mechanisms and systemic signalling as emerging themes in light acclimation of C3 plants. Plant Cell Environ 32:1230–1240

Kangasjärvi S, Neukermans J, Li S, Aro EM, Noctor G (2012) Photosynthesis, photorespiration, and light signalling in defence responses. J Exp Bot 63:1619–1636

Karpinski S, Reynolds H, Karpinska B, Wingsle G, Creissen G, Mullineaux P (1999) Systemic signaling and acclimation in response to excess excitation energy in Arabidopsis. Science 284:654–657

Kerchev PI, Pellny TK, Vivancos PD, Kiddle G, Hedden P, Driscoll S, Vanacker H, Verrier P, Hancock RD, Foyer CH (2011) The transcription factor ABI4 Is required for the ascorbic acid-dependent regulation of growth and regulation of jasmonate-dependent defense signaling pathways in Arabidopsis. Plant Cell 23:3319–3334

Keryer E, Collin V, Lavergne D, Lemaire S, Issakidis-Bourguet E (2004) Characterization of Arabidopsis mutants for the variable subunit of ferredoxin:thioredoxin reductase. Photosynth Res 79:265–274

Kimura M, Yamamoto YY, Seki M, Sakurai T, Sato M, Abe T, Yoshida S, Manabe K, Shinozaki K, Matsui M (2003) Identification of Arabidopsis genes regulated by high light-stress using cDNA microarray. Photochem Photobiol 77:226–233

Kindgren P, Kremnev D, Blanco NE, de Dios Barajas López J, Fernández AP, Tellgren-Roth C, Kleine T, Small I, Strand A (2012) The plastid redox insensitive 2 mutant of Arabidopsis is impaired in PEP activity and high light-dependent plastid redox signalling to the nucleus. Plant J 70:279–291

Kinsman EA, Pyke KA (1998) Bundle sheath cells and cell-specific plastid development in Arabidopsis leaves. Development 125:1815–1822

Kirk PR, Leech RM (1972) Amino acid biosynthesis by isolated chloroplasts during photosynthesis. Plant Physiol 50:228–234

Kleine T, Kindgren P, Benedict C, Hendrickson L, Strand A (2007) Genome-wide gene expression analysis reveals a critical role for CRYPTOCHROME1 in the response of Arabidopsis to high irradiance. Plant Physiol 144:1391–1406

Knappe S, Löttgert T, Schneider A, Voll L, Flügge UI, Fischer K (2003) Characterization of two functional phosphoenolpyruvate/phosphate translocator (PPT) genes in Arabidopsis–AtPPT1 may be involved in the provision of signals for correct mesophyll development. Plant J 36:411–420

Kok B (1956) On the inhibition of photosynthesis by intense light. Biochim Biophys Acta 21:234

Koussevitzky S, Nott A, Mockler TC, Hong F, Sachetto-Martins G, Surpin M, Lim J, Mittler R, Chory J (2007) Signals from chloroplasts converge to regulate nuclear gene expression. Science 316:715–719

Lee H, Guo Y, Ohta M, Xiong L, Stevenson B, Zhu JK (2002) *LOS2*, a genetic locus required for cold-responsive gene transcription encodes a bi-functional enolase. EMBO J 21:2692–2702

Lee KP, Kim C, Landgraf F, Apel K (2007) EXECUTER1- and EXECUTER2-dependent transfer of stress-related signals from the plastid to the nucleus of *Arabidopsis thaliana*. Proc Natl Acad Sci USA 104:10270–10275

Leegood RC (2008) Roles of the bundle sheath cells in leaves of C3 plants. J Exp Bot 59:1663–1673

Leister D (2012) Retrograde signaling in plants: from simple to complex scenarios. Front Plant Sci 3:135

Lepistö A, Kangasjärvi S, Luomala EM, Brader G, Sipari N, Keränen M, Keinänen M, Rintamäki E (2009) Chloroplast NADPH-thioredoxin reductase interacts with photoperiodic development in Arabidopsis. Plant Physiol 149:1261–1276

Li H, Culligan K, Dixon RA, Chory J (1995) CUE1: a mesophyll cell-specific positive regulator of light-controlled gene expression in Arabidopsis. Plant Cell 7:1599–1610

Li ZR, Wakao S, Fischer BB, Niyogi KK (2009) Sensing and responding to excess light. Annu Rev Plant Biol 60:239–260

Liu K, Sun J, Liu Y, Zhang QY, Kuang TY (2001) ESR study on superoxide radicals generated in photosystem II of higher plant. Prog Biochem Biophys 28:372–376

Liu K, Sun J, Song YG, Liu B, Xu YK, Zhang SX, Tian Q, Liu Y (2004) Superoxide, hydrogen peroxide and hydroxyl radical in D1/D2/cytochrome b-559 photosystem II reaction center complex. Photosynth Res 81:41–47

Macpherson AN, Telfer A, Truscott TG, Barber J (1993) Direct detection of singlet oxygen from isolated photosystem II reaction centres. Biochim Biophys Acta 1143:301–309

Martin W, Stoebe B, Goremykin V, Hapsmann S, Hasegawa M, Kowallik KV (1998) Gene transfer to the nucleus and the evolution of chloroplasts. Nature 393:162–165

Merlot S, Mustilli AC, Genty B, North H, Lefebvre V, Sotta B, Vavasseur A, Giraudat J (2002) Use of infrared thermal imaging to isolate Arabidopsis mutants defective in stomatal regulation. Plant J 30:601–609

Michalska J, Zauber H, Buchanan BB, Cejudo FJ, Geigenberger P (2009) NTRC links built-in thioredoxin to light and sucrose in regulating starch synthesis in chloroplasts and amyloplasts. Proc Natl Acad Sci USA 106:9908–9913

Miller G, Schlauch K, Tam R, Cortes D, Torres MA, Shulaev V, Dangl JL, Mittler R (2009) The plant NADPH oxidase RBOHD mediates rapid systemic signaling in response to diverse stimuli. Sci Signal 2:ra45

Mittler R, Kim Y, Song L, Coutu J, Coutu A, Ciftci-Yilmaz S, Lee H, Stevenson B, Zhu JK (2006) Gain- and loss-of-function mutations in *Zat10* enhance the tolerance of plants to abiotic stress. FEBS Lett 580:6537–6542

Mittler R, Vanderauwera S, Suzuki N, Miller G, Tognetti VB, Vandepoele K, Gollery M, Shulaev V, Van Breusegem F (2011) ROS signaling: the new wave? Trends Plant Sci 16:300–309

Mochizuki N, Tanaka R, Tanaka A, Masuda T, Nagatani A (2008) The steady-state level of mg-protoporphyrin IX is not a determinant of plastid-to-nucleus signalling in Arabidopsis. Proc Natl Acad Sci USA 105:15184–15189

Moulin M, McCormac AC, Terry MJ, Smith AG (2008) Tetrapyrrole profiling in Arabidopsis seedlings reveals that retrograde plastid nuclear signalling is not due to mg-protoporphyrin IX accumulation. Proc Natl Acad Sci USA 105:15178–15183

Mühlenbock P, Szechynska-Hebda M, Plaszczyca M, Baudo M, Mateo A, Mullineaux PM, Parker JE, Karpinska B, Karpinski S (2008) Chloroplast signaling and *LESION SIMULATING DISEASE1* regulate crosstalk between light acclimation and immunity in Arabidopsis. Plant Cell 20:2339–2356

Mullineaux P, Ball L, Escobar C, Karpinska B, Creissen G, Karpinski S (2000) Are diverse signalling pathways integrated in the regulation of arabidopsis antioxidant defence gene expression in response to excess excitation energy? Philos Trans R Soc Lond B Biol Sci 355:1531–1540

Navrot N, Rouhier N, Gelhaye E, Jacquot JP (2007) Reactive oxygen species generation and antioxidant systems in plant mitochondria. Physiol Plant 129:185–195

Neill S, Desikan R, Hancock J (2002) Hydrogen peroxide signalling. Curr Opin Plant Biol 5:388–395

Nguyen XC, Kim SH, Lee K, Kim KE, Liu XM, Han HJ, Hoang MH, Lee SW, Hong JC, Moon YH, Chung WS (2012) Identification of a C2H2-type zinc finger transcription factor (ZAT10) from Arabidopsis as a substrate of MAP kinase. Plant Cell Rep 31:737–745

Nomura H, Komori T, Uemura S, Kanda Y, Shimotani K, Nakai K, Furuichi T, Takebayashi K, Sugimoto T, Sano S, Suwastika IN, Fukusaki E, Yoshioka H, Nakahira Y, Shiina T (2012) Chloroplast-mediated activation of plant immune signalling in Arabidopsis. Nat Commun 3:926

Nunes-Nesi A, Araujo WL, Fernie AR (2011) Targeting mitochondrial metabolism and machinery as a means to enhance photosynthesis. Plant Physiol 155:101–107

Overmyer K, Kollist H, Tuominen H, Betz C, Langebartels C, Wingsle G, Kangasjärvi S, Brader G, Mullineaux P, Kangasjärvi J (2008) Complex phenotypic profiles leading to ozone sensitivity in *Arabidopsis thaliana* mutants. Plant Cell Environ 31:1237–1249

Padmasree K, Padmavathi L, Raghavendra AS (2002) Essentiality of mitochondrial oxidative metabolism for photosynthesis: optimization of carbon assimilation and protection against photoinhibition. Crit Rev Biochem Mol Biol 37:71–119

Park YI, Chow WS, Anderson JM (1996) Chloroplast movement in the shade plant *Tradescantia albiflora* helps protect photosystem II against light stress. Plant Physiol 111:867–875

Pérez-Ruiz JM, Spínola MC, Kirchsteiger K, Moreno J, Sahrawy M, Cejudo FJ (2006) Rice NTRC is a high-efficiency redox system for chloroplast protection against oxidative damage. Plant Cell 18:2356–2368

Pesaresi P, Schneider A, Kleine T, Leister D (2007) Interorganellar communication. Curr Opin Plant Biol 10:600–606

Peterson RB, Havir EA (2001) Photosynthetic properties of an *Arabidopsis thaliana* mutant possessing a defective PsbS gene. Planta 214:142–152

Pogson BJ, Woo NS, Förster B, Small ID (2008) Plastid signalling to the nucleus and beyond. Trends Plant Sci 13:602–609

Rasmusson AG, Heiser V, Irrgang KD, Brennicke A, Grohmann L (1998) Molecular characterisation of the 76 kDa iron-sulphur protein subunit of potato mitochondrial complex I. Plant Cell Physiol 39:373–381

Rintamäki E, Martinsuo P, Pursiheimo S, Aro EM (2000) Cooperative regulation of light-harvesting complex II phosphorylation via the plastoquinol and ferredoxin-thioredoxin system in chloroplasts. Proc Natl Acad Sci USA 97:11644–11649

Robert HS, Friml J (2009) Auxin and other signals on the move in plants. Nat Chem Biol 5:325–332

Rossel JB, Wilson PB, Hussain D, Woo NS, Gordon MJ, Mewett OP, Howell KA, Whelan J, Kazan K, Pogson BJ (2007) Systemic and intracellular responses to photooxidative stress in Arabidopsis. Plant Cell 19:4091–4110

Sagi M, Fluhr R (2006) Production of reactive oxygen species by plant NADPH oxidases. Plant Physiol 141:336–340

Sakamoto H, Araki T, Meshi T, Iwabuchi M (2000) Expression of a subset of the Arabidopsis Cys (2)/His(2)-type zinc-finger protein gene family under water stress. Gene 248:23–32

Schepens I, Johansson K, Decottignies P, Gillibert M, Hirasawa M, Knaff DB, Miginiac-Maslow M (2000) Inhibition of the thioredoxin-dependent activation of the NADP-malate dehydrogenase and cofactor specificity. J Biol Chem 275:20996–21001

Shah J (2009) Plants under attack: systemic signals in defence. Curr Opin Plant Biol 12:459–464

Sheen J (1990) Metabolic repression of transcription in higher plants. Plant Cell 2:1027–1038

Shikanai T (2007) Cyclic electron transport around photosystem I: genetic approaches. Annu Rev Plant Biol 58:199–217

Streatfield SJ, Weber A, Kinsman EA, Häusler RE, Li J, Post-Beittenmiller D, Kaiser WM, Pyke KA, Flugge UI, Chory J (1999) The phosphoenolopyruvate/phosphate translocator is required for phenolic metabolism, palisade cell development and plastid-dependent nuclear gene expression. Plant Cell 11:1609–1621

Sunkar R, Kapoor A, Zhu JK (2006) Posttranscriptional induction of two Cu/Zn superoxide dismutase genes in Arabidopsis is mediated by downregulation of miR398 and important for oxidative stress tolerance. Plant Cell 18:2051–2065

Suorsa M, Järvi S, Grieco M, Nurmi M, Pietrzykowska M, Rantala M, Kangasjärvi S, Paakkarinen V, Tikkanen M, Jansson S, Aro EM (2012) PROTON GRADIENT REGULATION5 is essential for proper acclimation of Arabidopsis photosystem I to naturally and artificially fluctuating light conditions. Plant Cell 24:2934–2948

Sweetlove LJ, Lytovchenko A, Morgan M, Nunes-Nesi A, Taylor NL, Baxter CJ, Eickmeier I, Fernie AR (2006) Mitochondrial uncoupling protein is required for efficient photosynthesis. Proc Natl Acad Sci USA 103:19587–19592

Szechyńska-Hebda M, Kruk J, Górecka M, Karpińska B, Karpiński S (2010) Evidence for light wavelength-specific photoelectrophysiological signaling and memory of excess light episodes in Arabidopsis. Plant Cell 22:2201–2218

Taniguchi M, Taniguchi Y, Kawasaki M, Takeda S, Kato T, Sato S, Tabata S, Miyake H, Sugiyama T (2002) Identifying and characterizing plastidic 2-oxoglutarate/malate and dicarboxylate transporters in *Arabidopsis thaliana*. Plant Cell Physiol 43:706–717

Tikkanen M, Piippo M, Suorsa M, Sirpiö S, Mulo P, Vainonen J, Vener AV, Allahverdiyeva Y, Aro EM (2006) State transitions revisited-a buffering system for dynamic low light acclimation of Arabidopsis. Plant Mol Biol 62:779–793

Tikkanen M, Grieco M, Kangasjärvi S, Aro EM (2010) Thylakoid protein phosphorylation in higher plant chloroplasts optimizes electron transfer under fluctuating light. Plant Physiol 152:723–735

Trotta A, Wrzaczek M, Scharte J, Tikkanen M, Konert G, Rahikainen M, Holmström M, Hiltunen HM, Rips S, Sipari N, Mulo P, Weis E, von Schaewen A, Aro EM, Kangasjärvi S (2011)

Regulatory subunit B′ gamma of protein phosphatase 2A prevents unnecessary defense reactions under low light in Arabidopsis. Plant Physiol 156:1464–1480

Turck F, Fornara F, Coupland G (2008) Regulation and identity of florigen: FLOWERING LOCUS T moves center stage. Annu Rev Plant Biol 59:573–594

Volkov RA, Panchuk II, Mullineaux PM, Schöffl F (2006) Heat stress-induced H2O2 is required for effective expression of heat shock genes in Arabidopsis. Plant Mol Biol 61:733–746

Voll L, Häusler RE, Hecker R, Weber A, Weissenböck G, Fiene G, Waffenschmidt S, Flügge UI (2003) The phenotype of the Arabidopsis *cue1* mutant is not simply caused by a general restriction of the shikimate pathway. Plant J 36:301–317

Walters RG (2005) Towards an understanding of photosynthetic acclimation. J Exp Bot 56:435–447

Walters RG, Horton P (1995) Acclimation of *Arabidopsis thaliana* to light environment: regulation of chloroplast composition. Planta 197:475–481

Wilson SB (1980) Energy conservation by the plant mitochondrial cyanide-insensitive oxidase. Some additional evidence. Biochem J 190:349–360

Xiao Y, Savchenko T, Baidoo EE, Chehab WE, Hayden DM, Tolstikov V, Corwin JA, Kliebenstein DJ, Keasling JD, Dehesh K (2012) Retrograde signaling by the plastidial metabolite MEcPP regulates expression of nuclear stress-response genes. Cell 149:1525–1535

Yoshida K, Terashima I, Noguchi K (2007) Up-regulation of mitochondrial alternative oxidase concomitant with chloroplast over-reduction by excess light. Plant Cell Physiol 48:606–614

Yoshida K, Watanabe CK, Hachiya T, Tholen D, Shibata M, Terashima I, Noguchi K (2011) Distinct responses of the mitochondrial respiratory chain to long- and short-term high-light Environments in *Arabidopsis thaliana*. Plant Cell Environ 34:618–628

Zybailov B, Rutschow H, Friso G, Rudella A, Emanuelsson O, Sun Q, van Wijk KJ (2008) Sorting signals, N-terminal modifications and abundance of the chloroplast proteome. PLoS One 3(4): e1994

Systemic Photooxidative Stress Signalling

Melanie Carmody and Barry Pogson

Abstract Systemic signalling of photooxidative stress from a high light (HL)-exposed leaf to a shaded leaf results in systemic acquired acclimation (SAA) in the distal tissue. As yet unanswered questions in systemic photooxidative stress signalling are in regard to what type of signal and what form of travel the signal takes from a small area of exposed tissue to as yet unstressed distal tissues. Issues such as the specificity of different stress responses, how different ROS signalling pathways converge, and antagonistically regulated systems are all currently being investigated. The majority of studies in this field, however, focus on the intercellular signalling aspects rather than leaf-to-leaf movement of the signal. Traditional studies of biotic long-distance signalling have not as yet been comprehensively applied to abiotic stress signalling research, particularly in regard to whether an abiotic signal is able to rapidly travel through the vasculature from leaf to leaf. This review covers literature relating to the effects that HL intensity and the production of ROS have on the stress signalling processes of light perception, retrograde and intercellular signalling, as well as leaf-to-leaf systemic signalling in the model organism *Arabidopsis thaliana*.

1 Introduction

One of the most crucial classes of signals plants produced in response to their environment are arguably those induced by light, as photosynthesis is the driving force behind energy production and therefore plant growth and survival. While plants have developed a considerable number of different light responses and

M. Carmody • B. Pogson (✉)
School of Biochemistry and Molecular Biology, Australian Research Council Centre of Excellence in Plant Energy Biology, Australian National University, Canberra, ACT, Australia
e-mail: barry.pogson@anu.edu.au

F. Baluška (ed.), *Long-Distance Systemic Signaling and Communication in Plants*, Signaling and Communication in Plants 19, DOI 10.1007/978-3-642-36470-9_13,

Acclimation

Short term	Long term
Chloroplast movement	Increased chloroplast numbers
Redox adjustment	Developmental changes in newly formed tissue
Antioxidant production	Pigment accumulation

Fig. 1 Summary of oxidative stress and acclimation processes in Arabidopsis. (**a**) The effect of oxidative stress, (**b**) acclimation processes. *PET* photosynthetic electron transport, *EEE* excess excitation energy, *ROS* reactive oxygen species

adaptive mechanisms that allow them to acclimate to different ambient light environments, the dynamic nature of a plant's daily environment can result in fleeting yet damaging exposure to stressful intensities of light resulting in cell death. Light quantity and quality are perceived by the plant, initiating adjustment of photosystems and photoprotective systems in accordance with subtle and prolonged changes in light availability during the day. Photooxidative stress induced by rapid increases in light intensity results in an imbalance in the photosystems found within chloroplasts, increasing reactive oxygen species (ROS) production and cellular damage that can consequently lead to cell death (Kato et al. 2007). Symptoms of photooxidative stress manifest themselves as photobleaching and necrosis of leaf tissue as shown in Fig. 1.

A number of short- and long-term acclimation processes exist in plants to reduce the amount of damage caused by photooxidative stress. These include the short-term changes in chloroplast movement, redox adjustment, and antioxidant production for chemical quenching of ROS (Tikkanen et al. 2010) and the long-term changes in photoprotective pigment accumulation, developmental changes to chloroplast numbers, and leaf thickness (Iida et al. 2001; Lake et al. 2002; Terashima and Hikosaka 1995; Thomas et al. 2003). These mechanisms are in reaction to either direct damage by ROS or the reduction in the efficiency of photosynthesis. While photooxidative stress can be induced by multiple stresses at the same time, an aspect that has been reviewed extensively (Karpinski 2006; Karpinski et al. 2003; Karpinski and Muhlenbock 2007; Mittler et al. 2004; Mullineaux and Baker 2010; Van Breusegem et al. 2008), this chapter focuses specifically on the systemic signalling of rapid high light (HL) stress and its impacts on photosynthetic electron transport, photooxidative stress, as well as ROS minimisation and detoxification.

2 Systemic Acquired Acclimation

The term systemic acquired acclimation (SAA) was originally coined by Karpinski et al. (1999) to describe a mechanism by which leaves exposed to fluctuating environmental conditions signal to distal, unexposed leaves as a way of pre-acclimating them to future stress events. This seminal study demonstrated that a cytosolic scavenger of H_2O_2, *ASCORBATE PEROXIDASE 2* (*APX2*), is induced in distal leaves of a partially HL-treated plant and that there were changes in distal photochemistry that were marginally higher than in the LL-untreated controls. The involvement of H_2O_2 in SAA was subsequently hypothesised (Karpinski et al. 1999). A second study of SAA was published by (Rossel et al. 2007) a decade later, demonstrating the extent of transcriptional responses to HL in distal tissues, as well as the rapid nature of the signal's induction. Since then a series of models have been developed highlighting a universal systemic stress signalling network in plants where processes related to programmed cell death (PCD) replace those involved in acclimation once the effects of oxidative stress reach a certain threshold (Karpinski and Szechynska-Hebda 2010; Miller et al. 2009; Mittler et al. 2011; Mühlenbock et al. 2008; Mullineaux and Baker 2010; Szechynska-Hebda et al. 2010).

While these studies include SAA as part of a more general stress signalling network, there exist discrepancies in our understanding. The experimental variations used in different studies make it hard for comparisons to be made and include differences in light intensity, duration of treatments, as well as tissue sampling from different organs and at different developmental stages. The optimal growth of Arabidopsis in the laboratory is at 130–150 μmol s^{-1} m^{-2} (Hensel et al. 1993). The original SAA studies both use HL intensities of around 1,000–3,000 μmol s^{-1} m^{-2} (Karpinski et al. 1999; Rossel et al. 2007), whereas other studies have reported systemic signals induced by moderate light (ML) intensities of only 300–400 μmol s^{-1} m^{-2} (Araya et al. 2008; Yano and Terashima 2001). Treatments at lower, shade response-inducing light intensities of 50–100 μmol s^{-1} m^{-2} have also induced excess energy flow through the photosystems by reducing CO_2 availability (Coupe et al. 2006; Lake et al. 2002). Other, end-point HL-acclimation studies observe whole-plant tissues 'primed' with ML intensities from 200 to 800 μmol s^{-1} m^{-2} for 24 h as well as different light qualities (white, blue, and red) before subjecting plants to a HL intensity of 1,800 μmol s^{-1} m^{-2} for 24 h (Iida et al. 2001). While this last study does not look at the distal responses to the priming or stress treatments, it does indicate that long-term acclimation is specifically more responsive to blue light.

Another point of difference between systemic studies is the varying definitions of what constitutes 'distal' tissue. The original SAA studies mostly treated 1/3 of a rosette with HL and collected random samples of distal, untreated, leaf tissue (Karpinski et al. 1999; Rossel et al. 2007). More recently, as studies have focused on the molecular mechanisms of the HL response, 'distal tissue' has been used to describe intercellular aspects of how signals get out of the cytosol and into neighbouring cells, along the floral bolt (Miller et al. 2009) as well as within the same leaf (Mühlenbock et al. 2008; Szechynska-Hebda et al. 2010). Long-term developmental studies into long-distance signalling of the ambient light environment categorise developing, newly formed leaves as 'distal' tissue (Coupe et al. 2006; Lake et al. 2002). Due to these inter-study variables, the possibility that these different systemic signals are independent of one another or whether they are the same or related cannot be ruled out. Also, aside from long-term developmental studies between mature and newly developing tissues, the immediate leaf-to-leaf signalling aspect of SAA has mostly been overlooked.

A distinguishing feature of the SAA signal that places it apart from other systemic plant signals described in the literature is that it is very rapid. SAA has been shown to be induced in as little as 15 min (Rossel et al. 2007) and possibly travels at a rate of ~8.4 cm min^{-1} up and down the floral bolt (Miller et al. 2009). Transcriptional studies of other types of stress such as ROS treatments on seedlings (Davletova et al. 2005; Vanderauwera et al. 2005), drought (Kimura et al. 2003), and herbivory (Kerchev et al. 2011) indicate that their immediate transcriptional response is not as large as changes occurring several hours, or even days, after the stress treatment. There is also little overlap between these and the systemic 1 h HL-response microarrays of Rossel et al. (2007) that removed UV and some of the IR through the use of a water bath. Differences in HL treatments between studies include heat and HL combination (Rossel et al. 2002) and very low light shade-grown plants subjected to 150 μmol s^{-1} m^{-2} 'HL' for 3 h (Kimura et al. 2001, 2003), though these avoided some of the effects of UV and heat with glass filters that removed wavelengths shorter than 400 nm and longer than 700 nm.

The Rossel et al. (2007) study demonstrated that gene expression changes between HL and distal tissues were very similar: 86 % of genes upregulated in HL-exposed leaves were also induced in distal leaves, and 71 % of genes downregulated in HL were likewise suppressed in distal leaves. Other rapidly induced stresses include wounding and heat shock responses in Arabidopsis (Desikan et al. 2001) and cyanobacteria (Hihara et al. 2001). Comparison of HL microarray results (Davletova et al. 2005) with wounding responses from the GENEVESTIGATOR database (Miller et al. 2009) demonstrates a large crossover between these two types of stress. Nonetheless, very few (Rossel et al. 2007) genes were also upregulated in this comparison (Miller et al. 2009), again highlighting the difficulties in comparing results from different HL or stress studies.

SAA marker or reporter genes that have been assessed in detail are also general stress response genes. These include *ZAT12* (Kimura et al. 2003; Miller et al. 2009), *APX1* (Miller et al. 2009), *APX2* (Karpinski et al. 1999), and *ZAT10* (Rossel et al. 2007). The Karpinski et al. (1999) and Rossel et al. (2007) HL studies use reporter gene transgenics to observe luciferin luminescence driven by the promoters of *APX2* and *ZAT10* in distal tissue, respectively. Both studies demonstrate systemic induction of these genes throughout the rosette, in the case of the latter the floral bolt, after a 1/3 rosette HL treatment. Other studies have analysed the distal responses of constructs to multiple stresses with *ZAT12* (Miller et al. 2009) and *APX1* (Szechynska-Hebda et al. 2010) promoters to the same effect. Caution should be taken when interpreting these results, however, as the nature of *promoter::luciferin* systems may not accurately reflect the speed of the signal due to variations in substrate tissue penetration and hydraulic transport through the vasculature. While immediate responses such as SAA, wounding, or heat shock might be more accurately measured with qRT-PCR, this is not as readily amenable to temporal and spatial studies.

The rapid nature of the SAA signal, as well as the large number of transcript changes in distal tissue with relatively minor photosynthetic and photoprotective changes within Arabidopsis, indicates that the initial SAA response should be considered as a 'priming response', rather than acclimation per se. As yet unanswered questions in this field are in regard to what type of signal and what form of travel the signal takes from a small area of exposed tissue to as yet unstressed distal tissues in order to acclimate them against future stress events. The majority of studies in this field, however, focus on the intercellular signalling aspects rather than leaf-to-leaf movement of the signal. This review covers literature relating to the effect that HL intensity has on the stress signalling processes of light perception, retrograde and intercellular signalling, as well as leaf-to-leaf systemic signalling in the model organism *Arabidopsis thaliana*.

3 Intra- and Intercellular ROS Signalling

Many of the proteins involved in the photooxidative stress response are nuclear encoded so signals must be directed out of the chloroplast through the cytosol via retrograde signalling. After induction of large transcriptional changes in the nucleus, signals can either be directed back towards the chloroplast (anterograde signalling) or directed out into the apoplast via the plasma membrane to neighbouring cells. This section provides an overview of ROS as signalling molecules, ROS-induced retrograde signalling pathways, and intercellular communication. Current issues in regard to specific ROS signalling relate to the specificity of these different signals (D'Autréaux and Toledano 2007; Gadjev et al. 2006; Møller and Sweetlove 2010), how different ROS signalling pathways converge (crosstalk) (Koussevitzky et al. 2008), and antagonistically regulated systems (Mullineaux and Baker 2010).

3.1 ROS as Signalling Molecules

One hypothesis for how stress signals are relayed from the chloroplast to the nucleus is that ROS themselves are the signalling molecules, though it is unlikely that a large amount would be able to dissipate into the cytosol due to the highly reactive nature and subsequently limited lifetimes of most ROS in the chloroplast (Table 1). $OH^{\bullet}$ radicals are thought to survive for <1 μs and so would not be able to diffuse more than 1 nm, if at all (Puppo 1992). $^{1}O_2$ is also highly reactive, with a

Table 1 Known half-lives of active oxygen species and the distance they could travel under oxidative stress

Species	Half-life	Distance of diffusion	Compartment of origin	References
$OH^{\bullet}$	1 ns	–	chl., apo., cyt.	Mubarakshina et al. (2010)
$^1O_2$	200 ns–4 μs	10–600 nm	chl.	Triantaphylidès et al. (2008)
O_2^-	2–4 μs	30 nm	chl., apo., cyt.	Vranova et al. (2002)
H_2O_2	1 ms	1 μm +	chl., apo., cyt.	Bhattacharjee (2005)

chl chloroplast, *apo* apoplast, *cyt* cytosol

half-life of up to 1 μs and a 30 nm diffusion distance (Gorman and Rodgers 1992; Hatz et al. 2007; Sies and Menck 1992). Recently, the potential for $^1O_2$ to diffuse out of the chloroplast and travel distances of up to 600 nm under HL exposure was demonstrated in *Chlamydomonas*; however, this was elicited under unrealistic single cell conditions (Breitenbach et al. 2009; Hatz et al. 2008; Kuimova et al. 2009). It is thought that the lifetime of $^1O_2$ is shortened within the cell under normal in vivo conditions by reactions with cellular molecules, though this is still unclear (Fischer et al. 2007; Redmond and Kochevar 2006). Similar to $^1O_2$, O_2^- is extremely reactive and has a relatively short half-life of 2–4 μs (Polle 2001; Vranova et al. 2002).

In contrast to the other ROS, H_2O_2 has relatively weak reactivity, a substantially longer half-life of 1 ms, and a further diffusion range of 1 μm, making it the ROS with the longest survival and diffusion distance (Bhattacharjee 2005; Polle 2001). Originally, as cytosolic APXs (cAPXs) are induced under HL (Chang et al. 2004; Karpinski et al. 1999), it was suggested that chloroplastic H_2O_2 might be a redox signal that activates their expression (Karpinski et al. 1999). As with $^1O_2$, there have since been contradictory reports as to whether H_2O_2 is able to diffuse out of chloroplasts. The original studies combined HL and methyl viologen (MV) treatments in order to induce nonphysiological levels of chloroplast-derived H_2O_2 to study whether this molecule could trigger the same transcriptional changes as under HL alone. These studies resulted in inactivation of *tAPX*, *sAPX*, and *cAPXs* during this treatment instead of the usual HL activation of APXs. This, along with the idea that H_2O_2 would be completely quenched by powerful antioxidant systems exclusive to chloroplasts to protect the photosystems, leads to the notion that H_2O_2 could not be a primary retrograde signal (Kitajima et al. 2007; Mano et al. 2001; Miyake et al. 2006; Pnueli et al. 2003; Polle 2001).

Recent studies demonstrating that H_2O_2 is still part of the retrograde signal network show that the inactivation of APXs is due to the MV treatment rather than reflecting the HL part of the signal. *cAPX* expression was much lower in a chloroplast-specific catalase mutant that also had lower chloroplastic levels of H_2O_2 than in wild-type plants (Yabuta et al. 2004). Another study, using intact isolated chloroplasts under physiological conditions with short durations of illumination and in the absence of MV application, demonstrated that while most chloroplast-produced H_2O_2 was detoxified inside chloroplasts, chemical inhibition of APXs resulted in much higher concentrations of H_2O_2 outside of chloroplasts (Mubarakshina et al. 2010). Under severe stress conditions, when APXs might be completely inhibited, the appearance of H_2O_2 in the cytoplasm could be very significant and could achieve up to 75 % of the H_2O_2 produced via photosynthetic electron transport (Mubarakshina et al. 2010). The rate of H_2O_2 appearance outside the chloroplasts increased with an increase in light intensity and time of illumination, though the authors concluded that it is unlikely that H_2O_2 can pass through the chloroplast envelope membrane simply by diffusion and it seems more likely that the molecule would require something akin to the transport action of aquaporins (Bienert et al. 2007; Henzler and Steudle 2000).

3.2 Retrograde Signalling

ROS may also take part in retrograde signal transduction pathways that are induced at various points of photosynthetic electron transport (PET), depicted in Fig. 2. Under normal conditions, electron flow is directed towards PSI where any excess is either managed through existing energy dissipation mechanisms or via quenching of H_2O_2/O_2^-. Under stress conditions, energy flow can be allowed to pass through PSII and induce photo-inhibition where high levels of $^1O_2$ are produced. Because conditions leading to higher levels of H_2O_2/O_2^- differ from those required to generate $^1O_2$, the perception of these two different types of ROS is thought to induce different responses, and therefore different chloroplast-to-nucleus signalling pathways (Kim et al. 2008; Leisinger et al. 2001; Maruta et al. 2012; Mullineaux 2009). While many of the photosynthetic and transcriptional components of retrograde signalling have been characterised, intermediate steps of separate retrograde signals are mostly unknown. Additional complexity can be seen where signals generated at different points in PET regulate similar and different subsets of genes in the nucleus, all of which need to be taken into account when assessing HL signalling.

When excess energy is directed towards PSII and photochemistry, tocopherol and carotenoid photoprotection is insufficient leading to accumulation of $^1O_2$ and the activation of cell death pathways. Genetic control of $^1O_2$ signalling is known through study of the conditional *flu* mutant (Meskauskiene et al. 2001; Wagner et al. 2004). Protochlorophyllide accumulates in this mutant in the dark, which then produces $^1O_2$ when it is reilluminated (Przybyla et al. 2008). While $^1O_2$ is not associated with PSII in this mutant, the activation of cell death is nonetheless controlled by two chloroplast-localised proteins: *EXECUTER1* (*EX1*) and *EXECUTER2* (*EX2*) (Lee et al. 2007; Przybyla et al. 2008; Wagner et al. 2004). Inactivation of the *EX1* gene is sufficient to induce $^1O_2$-related growth inhibition and cell death responses (Wagner et al. 2004), and inactivation of both genes results in the loss of nearly all $^1O_2$-specific transcriptional changes in the nucleus. Interestingly, a recent study has additionally shown how an oxidative cleavage product of β-carotene, β-cyclocitral, can also specifically induce singlet oxygen responsive

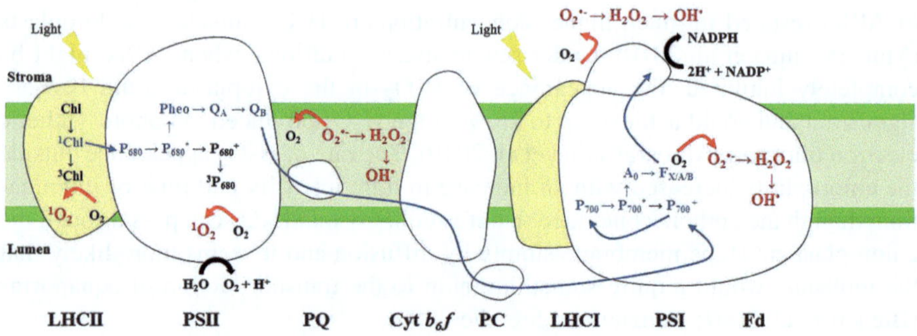

Fig. 2 Photosynthetic electron transport and ROS production. Linear electron flow between the lumen and stroma of the chloroplast occurs within the thylakoid membrane from PSII to PSI. A natural by-product of photosynthesis is the generation of ROS within the LHC, the thylakoid membrane, and into the stroma. *LHCII* light-harvesting complex II, *PSII* photosystem II, *PQ* plastoquinone, *PC* plastocyanin, *LHCI* light-harvesting complex I, *PSI* photosystem I, *Fd* ferredoxin

genes (Ramel et al. 2012). While this may act as an intermediate in the $^1O_2$ signalling pathway, it could also be acting in an EXECUTER-independent manner as there is still a HL-induced transcriptional response in *ex1* at 6 h and 12 h time points (Oelze et al. 2012).

Most other studies into ROS-specific retrograde signalling have focused on the H_2O_2/O_2^- retrograde pathway(s). Gene induction by these pathways often involves the induction of general stress response genes such as *APX2* and *APX1*, as well as transcription factors such as *ZAT10* and *ZAT12*, that regulate different subsets of HL-inducible genes (Karpinski et al. 1999; Miller et al. 2009; Pogson et al. 2008). Transcription of specific light-inducible *ELIP* genes increases in response to H_2O_2 as well as to HL (Kimura et al. 2001). Microarrays of H_2O_2 treatments have identified a large number of H_2O_2-specific transcription factors and there are 400 known H_2O_2-responsive protein families in existence (Desikan et al. 2001; Vandenbroucke et al. 2008). Many of the genes mentioned here are widely known as general stress response genes, although other subsets of genes have been found that respond specifically to H_2O_2/O_2^- such as *FER1* (Gadjev et al. 2006; Miller et al. 2009). Apart from the possibility of the H_2O_2 molecule acting as a retrograde signal from its PSI source, H_2O_2/O_2^- can also occur near PSII when electron flow becomes too much for nonphotochemical quenching (Fryer et al. 2003).

Apart from the direct ROS retrograde pathways, there are also redox-regulated signals based on the plastoquinone redox state. The oxidation of plastiquinol (PQH_2) is the rate-limiting step in PET between PSII and PSI (Li et al. 2009). Under HL conditions, the rate of electron flow from PSII exceeds the rate of oxidation of PQH_2 resulting in overreduction of the PQ pool. The chemical DCMU blocks PQ reduction and has been used to demonstrate the importance of PQ redox regulation in HL signalling as a small subset of DCMU-induced genes, including *ELIPs*, were also induced under HL (Kimura et al. 2003; Yang et al. 2001). The induction of *ELIP1* and *ELIP2* is also known to be mediated specifically by CRY1, a blue light photoreceptor. Further downstream, several phenylpropanoid metabolism genes, including the anthocyanin transcriptional regulators *PAP1* and *PAP2*, are also thought to be involved as their induction was also inhibited in *cry1*. Interestingly, anthocyanin accumulation was also inhibited in *cry1* under HL conditions (Kleine et al. 2007). *APX1* and *APX2*, known HL and SAA-responsive genes, are also thought to be regulated by PQ reduction (Karpinski et al. 1997; Rossel et al. 2007). Transcription of these genes was either unaffected in *cry1* or basally induced in LL *phyA*/*phyB*, *phot1*/*phot2*, and *cry1*/*cry2* plants but again having comparable HL responses to wild-type plants across all genotypes (Kleine et al. 2007; Ruckle et al. 2007; Szechynska-Hebda et al. 2010). This suggests that *APX1* and *APX2* are regulated by a redox signalling pathway other than those initiated by photoreceptors as is the case with the *ELIP* and *PAP* genes.

The redox state of the PQ pool can also alter protein phosphorylation by protein kinases involved in phosphorylation of PSII antenna and reaction centre proteins (Li et al. 2009; Liu et al. 2007). Phosphorylation of LHC proteins is required for the energy dissipation mechanism of state-transition quenching (qT). This mechanism is involved in redistribution of light-harvesting complexes between PSII and PSI to maintain PQ redox homeostasis during short-term light quality acclimation (Rochaix 2007; Tikkanen et al. 2010). The STN7 protein is required for the phosphorylation of some LHC proteins and the initiation of state transitions (Bellafiore et al. 2005; Ihnatowicz et al. 2008; Tikkanen et al. 2006, 2010). Both long- and short-term acclimation pathways involve STN7 (Pesaresi et al. 2009). In terms of long-term acclimation, no differences in RNA levels were detected in microarrays of *stn7* plants grown under HL, and so STN7 may be involved in PQ-regulated acclimation at the level of protein abundance (Bonardi et al. 2005). The effect of short-term HL exposure of *stn7* mutants was higher induction of stress responsive genes and therefore more tolerance to HL than in wild type. While the exact nature of how STN7 is involved in short-term acclimation is still unclear, it may require the involvement of a PSII phosphatase (Oelze et al. 2012; Samol et al. 2012; Sztatelman et al. 2010).

While extensive work has been done to pinpoint the initiation of retrograde signals in the chloroplast and their transcriptional responses in the nucleus, exact mechanisms for how retrograde signals are transmitted across the cytosol are relatively unknown. A recent study, however, demonstrates how the plastidial methylerythritol phosphate (MEP) pathway may be involved in biotic stress signalling in addition to the production of isoprenoids via an isoprenoid precursor, MEcPP (Xiao et al. 2012). Another possible retrograde signalling molecule is a phosphonucleotide (3′-phosphoadenosine 5′-phosphate [PAP]), which accumulates in response to drought and HL stress (Estavillo et al. 2011). In this study, PAP levels were shown to be regulated by the chloroplastic SAL1 enzyme through dephosphorylation of PAP to AMP. PAP can be transported out of the chloroplast (Gigolashvilia et al. 2012) and into the cytosol and nucleus thereby inhibiting exoribonucleases resulting in changes to RNA metabolism and gene expression

(Estavillo et al. 2011). This pathway could account for up to 35 % of the HL-inducible genes. There are, however, likely to be additional unknown pathways present, further complicating the signalling processes.

Additional complications from the crosstalk that occurs between many of the pathways under HL as well as between signals from different abiotic stresses make it difficult to isolate individual pathways (Laloi et al. 2007; Li et al. 2009). One type of antagonistic crosstalk particularly relevant to HL signalling is between the two distinct ROS signalling pathways. The presence of an antagonistic 'anti-cell death' system involving H_2O_2/O_2^- signalling has been demonstrated (Mullineaux and Baker 2010) that counteracts the EX1/EX2 'pro-cell death' pathway (Laloi et al. 2007; op den Camp et al. 2003; Wagner et al. 2004). A downstream regulatory protein that is involved in processing both H_2O_2/O_2^- -derived signals (Mateo et al. 2004) and in mediating EX1/EX2 $^1O_2$-regulated cell death (Ochsenbein et al. 2006) is ENHANCED DISEASE SUSCEPTIBILITY1, EDS1, and is required of intercellular stress signalling.

3.3 *Intercellular Signalling*

While there is debate as to whether ROS act as signalling molecules in retrograde signalling, intercellular signalling requires the accumulation of NADPH oxidase-derived O_2^-, and therefore H_2O_2, within the apoplast (Mullineaux and Baker 2010), depicted in Fig. 3. NADPH oxidases are situated in the plasma membrane and are encoded by respiratory burst oxidase genes (*RBOHs*) (Sagi and Fluhr 2006). In Arabidopsis, these include *RBOHD* and *RBOHF* known mostly through studies done on ABA signalling in guard cells (Kwak et al. 2003; Miao et al. 2006; Miyao 1994). In guard cells, the OPEN STOMATA1 (OST1) protein kinase, involved in ABA-dependent signalling, phosphorylates RBOHF (Sirichandra et al. 2009) and RBOHD needs to be phosphorylated in order for ROS production to occur under biotic stress (Nühse et al. 2007). Apoplastic ROS generation also requires the presence of EDS1 which mediates signals derived from both H_2O_2 and $^1O_2$ signalling pathways. The basal presence of EDS1 favours the default response of cell death (Straus et al. 2010). EDS1 was originally associated with the development of the hypersensitive response (HR) to pathogen attack and innate immunity (Keen 1990; Shinozaki and Yamaguchi-Shinozaki 2000). The major characteristic of HR is the formation of a zone of dead cells around the infected area similar to the development of patches of bleached tissue within leaves undergoing oxidative stress. It is proposed that the EDS1 regulatory switch controlling accumulation of ROS in the apoplast is part of a universal stress signalling network and part of cross-tolerance between biotic and abiotic stresses (Miller et al. 2009; Mullineaux and Baker 2010).

Downstream and interacting components of the EDS1 'pro-/anti-death' regulatory pathways include stress threshold-dependent NUDIX HYDROLASE7 (NUDT7), which induces moderate defences within stressed tissue (Miller et al. 2009; Wiermer et al. 2005), and LESION SIMULATING DISEASE1 (LSD1),

a

Label	State	Phenotype	Genes
(c)	Moderate defense	Green	EDS1 NUDT7
(d)	Cell death	Bleached	EDS1
(e)	High defense	Green, (adjacent to bleached tissue)	EDS1 NUDT7 LSD1

b

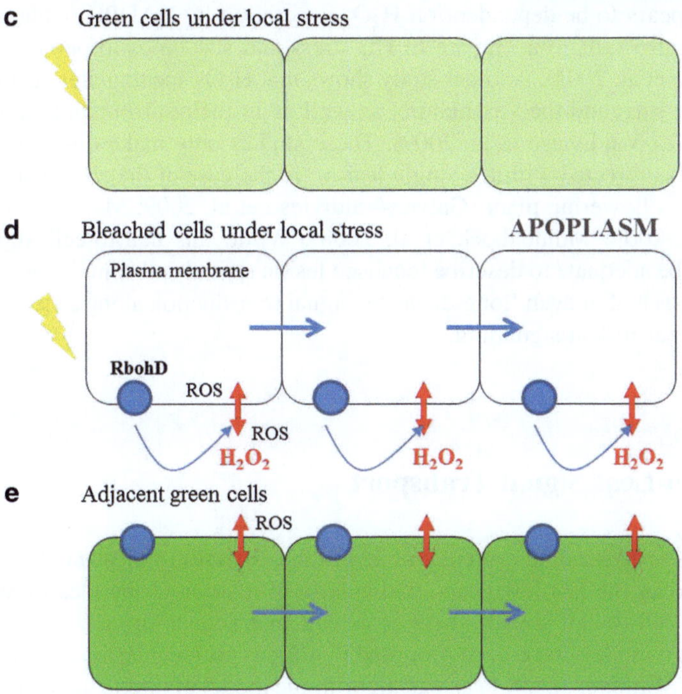

Fig. 3 The auto-propagating signal model. (**a**) Summary of cell phenotype, state, and gene induction between diagrams (**c–e**); (**b**) photobleaching pattern in a HL-treated Arabidopsis leaf; (**c**) green cells under local HL stress, moderate defence; (**d**) bleached cells under local HL stress, cell death; (**e**) green cells adjacent to bleached cells, high defence

responsible for increasing production of SODs and CATs on top of defences controlled by NUDT7 exclusively in cells adjacent to those succumbing to oxidative stress (Straus et al. 2010). NUDT7 acts by controlling redox homeostasis and raises the threshold at which EDS1 can trigger cell death (Ishikawa et al. 2010; Jambunathan et al. 2010). Lesion containment within the leaf involves the activation of LSD1 and SA signalling pathways that are independent of the EX1/EX2 cell death pathway (Mullineaux and Baker 2010; Rusterucci 2001; Torres et al. 2005). How signal transduction is mediated from the burst of RBOHD-catalysed ROS to different signalling pathways is not known, but it could involve enhanced antioxidant capacity through SA-induced GPXs that is known to occur through abscisic acid (ABA)-mediated signalling (Mateo et al. 2006; Mou et al. 2003). Current understanding does not explain how a cell finds itself switching between NUDT7-EDS1 and LSD1-EDS1 control to, for example, form patches of bleaching even though all tissues were exposed to HL.

Intercellular signalling has been a major focus of more recent SAA studies. Miller et al. (2009) demonstrate that signals from multiple abiotic stimuli are impaired in *rbohD* as well as by the NADPH oxidase inhibitor, DPI, and catalase, a scavenger of H_2O_2, along the floral bolt. The intercellular part of this signal therefore appears to be dependent on H_2O_2 generation by NADPH oxidases. The *lsd1* mutant also fails to acclimate to HL stress and succumbs to photooxidative stress (Mateo et al. 2004). Another study shows that H_2O_2 accumulates specifically in BSCs that surround the vasculature, as well as in their surrounding apoplastic spaces (Galvez-Valdivieso et al. 2009). These studies only make observations on how a signal is directed within a single leaf or, in the case of the *rbohD* study, the floral bolt of a flowering plant (Galvez-Valdivieso et al. 2009; Mateo et al. 2004; Miller et al. 2009; Mühlenbock et al. 2008). While this cell-to-cell signalling system may be adequate to describe localised lesion spread and containment within an Arabidopsis leaf or even 'long-distance' signal transduction along a stem, it does not explain leaf-to-leaf signalling.

4 Leaf-to-Leaf Signal Transport

The discourse surrounding systemic abiotic stress signalling in plants has gained momentum over the last few years, particularly in relation to the idea of an auto-propagating cell-to-cell signal where local cell interactions drive a larger global response. Two models have been proposed that focus on local lesion spread within the one leaf (Mittler et al. 2011) and an algorithmic model of intercellular signalling that works as a universal whole-plant stress signalling network in Arabidopsis (Karpinski and Szechynska-Hebda 2010). While both explain the local response of a signal between different cell types such as the mesophyll and bundle sheath cells of a single leaf, it is still not understood how the signal can then travel through the vasculature to other leaves within the Arabidopsis rosette, or how a plant produces different responses to specific types of stress. Further application of techniques used to study biotic long-distance signalling needs to be applied to abiotic stress signalling research, particularly in regard to how an abiotic signal is able to rapidly travel through the vasculature from leaf to leaf.

4.1 The Importance of the Vasculature

The different cell types (e.g. mesophyll, epidermal, guard cells, bundle sheath) within a leaf are spatially arranged around the vascular tissue that contains xylem and phloem (Esau 1953; Evert and Esau 2006; Nelson and Dengler 1997). An Arabidopsis leaf is characterised by a midvein which is continuous with the stem vascular bundles. A layer of bundle sheath cells (BSCs) entirely surrounds the vascular tissues and acts as an interface between these and the mesophyll tissues (Kinsman and Pyke 1998; Leegood 2008).

While BSCs are able to sense stress signals within xylem sap and can control the flux of solutes and water from the xylem to the mesophyll via the downregulation of plant aquaporin (AQP) activity (Shatil-Cohen et al. 2011), very little is known about how solutes or water enter the phloem stream from leaf tissue. Possible mechanisms involve either free movement through the apoplast (Evert et al. 1985; Harris and Chaffey 1985), the osmotic permeability of BSC membranes (Shatil-Cohen et al. 2011), or that BSCs act as a barrier and active solute transport is required to cross the BSC plasma membrane (Heinen et al. 2009). Early studies of xylem-mesophyll flow speculated that the osmotic water permeability of the BSC membrane was too low to support the transpiration stream (Boyer 1974). The apoplastic route was therefore initially considered to be the major route for xylem-sap transport; however, the discovery of suberin deposits in leaf BSCs has since indicated otherwise (Sack and Holbrook 2006). Mutant analysis of *gdu1*, with a mutation in a vascular glutamine transporter, and *ost1-2*, defective in ABA signalling and stomata closure, also demonstrates the existence of a hydraulic and solute transport barrier within BSCs (Ache et al. 2010; Pilot et al. 2004). A more recent study additionally demonstrates that BSCs are not only able to block small solutes but also water (Shatil-Cohen et al. 2011). Along with the discovery of plant aquaporins (AQPs), these findings indicate that the BSC plasma membrane is able to support active water and solute transport.

Interestingly, both mesophyll and BSCs contain chloroplasts, and an as yet unanswered question remains as to whether BSC chloroplasts have a specialised purpose in sensing rapid stress induction and relaying these signals directly into the phloem vascular stream. Fluxes of water are controlled by the hydraulic conductivity of the leaf (K_{leaf}) (Martre et al. 2002), which is known to decrease dramatically in response to many environmental and abiotic stresses (Sack and Holbrook 2006). Also, stress-induced H_2O_2 was shown to accumulate locally in the surrounding intercellular spaces of BSCs, but not in the mesophyll (Fryer et al. 2003; Galvez-Valdivieso et al. 2009). Given that BSCs also have a higher abundance of AQPs compared with their adjacent mesophyll (Frangne et al. 2001), they could be as important for signals entering the phloem stream as they are for those entering the leaf via the xylem tissue.

At the whole-plant level, plasmodesmata (Pd) have important roles in photoassimilate translocation from source to sink leaves through the phloem. In apoplastic-loading plants, such as tobacco and Arabidopsis, sucrose from bundle sheath and phloem parenchyma cells is first exported into the apoplast and is then actively loaded into the sieve element companion cell (CC) complex (Lalonde et al. 2004) where Pd are subsequently involved in the transport from the companion cell to the sieve element (Zavaliev et al. 2010). The signal must be loaded into the phloem stream via CC then move from this site to the shoot apical meristem (SAM). Arrival at the SAM requires post-phloem transport involving local cell-to-cell transport (Turnbull 2011). For a rapid signal, questions remain regarding whether a signal could bypass the SAM and travel directly to distal leaves.

4.2 Types of Leaf-to-Leaf Signals

There are many known possibilities for how a signal can travel systemically through a plant; these can fall into three categories: internal translocation of signals through the vasculature, external movement of a signal through production of airborne volatiles, and electrochemical movement through plant membranes. A widely accepted hypothesis is that environmental cues are sensed by mature organs and this information is transported to meristematic regions where newly formed tissues adopt a new developmental fate (Lough and Lucas 2006). Two outcomes of long-distance pathogen defence signalling are to either pre-emptively activate defence mechanisms in distal parts of the plant or to prime those tissues as a way of preparing them for an augmented response to future attack. Priming is less costly than activating defence mechanisms (Heil and Ton 2008), although whether abiotic SAA can be classified as a priming response is as yet unclear.

Three strategies to prove signal transmission according to Turnbull (2011) is to demonstrate the movement of the putative signal molecule, a phenotypic change, and the altered expression of a molecular target. While transcriptional and phenotypic changes have been predominantly studied, proof of signal movement is more difficult. Traditional methods for studying long-distance signals have involved two different approaches: biochemical approaches that follow the physical movement of the signal and grafting experiments that include mutants defective in either the production or the perception of the signal (Heil and Ton 2008). The vascular system of a plant acts as a transport pathway for water and nutrients as well as a long-distance communication network, specifically interorgan communication. Plants with large vasculature such as tomato and tobacco have been the model organisms of choice for the study of long-distance signalling in plants. However, adapting these techniques for use in Arabidopsis where most molecular and physiological studies have been performed would be of more use (Bainbridge et al. 2006; Turnbull 2011).

4.2.1 Phloem-Mobile Signals

Traditional studies of long-distance signalling due to plant defence have focused mostly on phloem-mobile elements such as plant hormones; however, very few of these have been attributed to SAA. Mutant analysis looking specifically at the SAA induction of *ZAT10* in distal tissue suggests that many of the major phytohormones of ABA, JA, and SA (*aba2-3*, *abi1-1*, *abi2-1*, *rax1-1*, *jar1-1*, *jin1*, *aos*) are not involved in SAA as the signal was not impaired in these mutants. It was also recently shown that microarrays of 10- and 100-fold HL treatments of 6 h and 12 h duration showed that the ABA response is not associated with the HL response at these time points (Oelze et al. 2012). There is, however, strong evidence supporting the involvement of ABA signalling, if not the hormone itself, in certain cell types such as guard cells and BSCs that could then initiate the unknown long-distance signal (Mullineaux and Baker 2010).

ABA is thought to travel from the vasculature into BSCs where a rapid accumulation of extracellular H_2O_2 occurs in order to regulate HL-responsive genes (Galvez-Valdivieso et al. 2009). Although some ABA mutants were shown not to disrupt SAA, it is interesting to note that ROS accumulation and expression of APX2, *ZAT10::LUC,* and the retrograde regulator, SAL1, are all localised in the vasculature (Estavillo et al. 2011; Fryer et al. 2003; Karpinski et al. 1999; Rossel et al. 2007). This suggests the involvement of ABA in retrograde signalling or indeed a novel plant hormone.

While chemical signals have been increasingly demonstrated to be responsible for root-shoot signalling, particularly in response to drought stress, there is also evidence supporting hydraulic signals, usually in a source-to-sink direction (Jia and Zhang 2008). Hydraulic root-derived signals have been shown to be major regulators of stomatal behaviour potentially through modification of chemical concentration in the vasculature as a result of changes in water flux (Fuchs and Livingston 1996). Hydraulics have the potential to transport complex signals as even though very low water potential does not appear to have a direct effect on stomatal aperture, it has still been shown to promote ABA-induced stomatal closure (Tardieu and Davies 1993). Whether changes in plant hydraulics could result in a HL-derived signal being transported between photosynthesising tissues in the aerial part of the plant is unknown; however, signal movement from the vasculature to BSCs may very well be functioning in a similar way to ABA-transported vascular signals into stomatal guard cells.

There is also evidence of a rapid pH-based systemic drought signalling mechanism where foliar and xylem sap pH changes have been linked with ABA concentration within leaves (Wilkinson and Davies 2008). Alterations in pH can be one of the first chemical changes measurable in xylem sap from plants exposed to drying soil (Sobeih et al. 2004), soil flooding (Jackson et al. 2003), and salt stress (Gao et al. 2004). More recently, changes in shoot or leaf xylem/apoplastic sap pH have also been detected in response to natural or imposed changes in the light from the aerial environment (Muhling and Lauchli 2000; Wilkinson and Davies 2002).

In terms of the root-shoot systemic regulation of stomatal aperture, ABA promotes ROS production that results in increases in cytosolic calcium leading to stomatal closure (Kwak et al. 2006). The symplast of an adjacent cell is thought to remove ABA from the xylem or a leaf apoplast and is pH dependent. Generally, a more acidic xylem/apoplast pH exists in the sap of unstressed plants and allows the greatest removal of ABA from the xylem and leaf apoplast. More alkaline sap pH values have been detected under stress conditions which reduced ABA-removing pH gradient across the membrane. In the case of guard cells, ABA accumulates to high enough levels to cause stomatal closure (Jia and Zhang 2008; Wilkinson and Davies 2008).

In addition to water and nutrients, recent studies of phloem sap have revealed the presence of numerous RNA transcripts and proteins (Buhtz et al. 2010; Lough and Lucas 2006). Small RNAs (sRNAs) would be ideal candidates for a long-distance signal as they are able to spread from cell to cell as well as through the vasculature in a non-cell autonomous manner (Chitwood and Timmermans 2010). sRNAs have gained a lot of recent attention as systemic signalling molecules in all fields of study, as reviewed in Kehr and Buhtz (2008), and phloem sap specifically contains distinct sets of sRNAs between leaves and roots in plants undergoing nutrient deficiency (Buhtz et al. 2010) and biotic stress (Katiyar-Agarwal and Jin 2010). They also move in a predominantly source-to-sink direction leading to the hypothesis that sRNAs, along with photoassimilates, are transported into growing meristems (Palauqui et al. 1997; Schwach et al. 2005). Their function as developmental systemic signals is evident; however, their role in responding rapidly to sudden stress events has not been studied extensively.

4.2.2 Airborne Signals

In addition to internal long-distance signal transport involving the vasculature, volatile organic compounds (VOCs) exist that can signal externally to other leaves in the rosette as well as to other plants (Heil and Ton 2008; Orians 2005). Airborne signals reach distal tissues at much lower concentrations and so are thought to cause priming rather than a full induction of defence responses; however, both airborne and vascular long-distance signals may also act together (Heil and Ton 2008). Ethylene is the most commonly known VOC triggered as part of the wounding response; however, other hormones such as JA and SA can become volatile through methylation (MeJA and MeSA) during biotic stress (Frost et al. 2007). Within-plant signalling by volatiles overcomes the spatial and temporal restrictions of the vascular system. While these compounds are known to be a beneficial defence mechanism against the random nature of herbivory or pathogen attack (Arimura et al. 2011; Heil et al. 2007; Heil and Ton 2008), it is unknown whether they are involved during abiotic stress. Interestingly, the HL-induced oxidative product of β-carotene, β-cyclocitral, is also a known volatile compound and would be worth further investigation (Ramel et al. 2012).

4.2.3 Electrochemical Signals

The third type of signalling, other than internally through the vasculature and externally in the form of VOCs, is electrochemically along plant membranes. Early studies of how electrical signals are able to regulate plant responses (Darwin 1875) have been revisited recently as a possible SAA and abiotic stress signal (Szechynska-Hebda et al. 2010). The electrical potential of BSC cells in the midvein of exposed leaves and in distal leaves undergoing SAA was measured simultaneously. Several seconds of excess light illumination was sufficient to induce an electrochemical signal, and the propagation speed between two different leaves was estimated to be around 0.3 cm s^{-1} after induction by light. The speed of the electrochemical signal between different leaves was also shown to be impacted by reduced induction of *APX2*, expressed exclusively in BSCs (Fryer et al. 2003), as the systemic propagation of this electrical signal between different leaves is more than twice as fast in the null *apx2-1* mutant than in wild-type plants. This suggests that *APX2* is involved in the regulation of the electrochemical signal. It is important to note, however, that these results were not correlated with the known later induction of SAA marker genes or changes in photosynthetic parameters (Karpinski et al. 1999; Rossel et al. 2007).

5 Conclusions

The precise nature of a systemic signal generated specifically by oxidative stress is as yet unknown. Questions remain in regard to whether, or in what combination, the leaf-to-leaf signal travels through the vasculature, as a volatile, or even electrochemically. Interestingly, the recent discovery of the HL-induced oxidative product of β-carotene, β-cyclocitral, links chloroplast-generated $^1O_2$ with a molecule that is capable of both volatile and vascular movement. This discovery sheds light on a plant's ability to respond specifically to different types of stress. Future work in this area should focus on determining which oxidative by-products are generated by individual stresses, as well as how these products contribute to distinct and shared signalling pathways.

References

Ache P, Bauer H, Kollist H, Al-Rasheid KAS, Lautner S et al (2010) Stomatal action directly feeds back on leaf turgor: new insights into the regulation of the plant water status from non-invasive pressure probe measurements. Plant J 62:1072–1082

Araya T, Noguchi K, Terashima I (2008) Manipulation of light and CO2 environments of the primary leaves of bean (*Phaseolus vulgaris* L.) affects photosynthesis in both the primary and the first trifoliate leaves: involvement of systemic regulation. Plant Cell Environ 31:50–61

Arimura G-I, Ozawa R, Maffei ME (2011) Recent advances in plant early signaling in response to herbivory. Int J Mol Sci 12:3723–3739

Bainbridge K, Bennett T, Turnbull C, Leyser O (2006) Grafting. Methods Mol Biol 323:39–44

Bellafiore S, Barneche F, Peltier G, Rochaix J-D (2005) State transitions and light adaptation require chloroplast thylakoid protein kinase STN7. Nature 433:892–895

Bhattacharjee S (2005) Reactive oxygen species and oxidative burst: roles in stress, senescence and signal transduction in plants. Curr Sci 89:1113–1121

Bienert GP, Moller ALB, Kristiansen KA, Schulz A, Moller IM et al (2007) Specific aquaporins facilitate the diffusion of hydrogen peroxide across membranes. J Biol Chem 282:1183–1192

Bonardi V, Pesaresi P, Becker T, Schleiff E, Wagner R et al (2005) Photosystem II core phosphorylation and photosynthetic acclimation require two different protein kinases. Nature 437:1179–1182

Boyer JS (1974) Water transport in plants: mechanism of apparent changes in resistance during absorption. Planta 117:187–207

Breitenbach T, Kuimova MK, Gbur P, Hatz S, Schack NB et al (2009) Photosensitized production of singlet oxygen: spatially-resolved optical studies in single cells. Photochem Photobiol 8:442–452

Buhtz A, Pieritz J, Springer F, Kehr J (2010) Phloem small RNAs, nutrient stress responses, and systemic mobility. BMC Plant Biol 10:64

Chang CC-C, Ball L, Fryer MJ, Baker NR, Karpinski S, Mullineaux PM (2004) Induction of ASCORBATE PEROXIDASE 2 expression in wounded Arabidopsis leaves does not involve known wound-signalling pathways but is associated with changes in photosynthesis. Plant J 38:499–511

Chitwood DH, Timmermans MCP (2010) Small RNAs are on the move. Nature 467:415–419

Coupe SA, Palmer BG, Lake JA, Overy SA, Oxborough K et al (2006) Systemic signalling of environmental cues in Arabidopsis leaves. J Exp Bot 57:329–341

D'Autréaux B, Toledano MB (2007) ROS as signalling molecules: mechanisms that generate specificity in ROS homeostasis. Nat Rev Mol Cell Biol 8:813–824

Darwin C (1875) Insectivorous plants. John Murray, London

Davletova S, Schlauch K, Coutu J, Mittler R (2005) The zinc-finger protein Zat12 plays a central role in reactive oxygen and abiotic stress signaling in Arabidopsis. Plant Physiol 139:847–856

Desikan R, Mackerness SAH, Hancock JT, Neill SJ (2001) Regulation of the Arabidopsis transcriptome by oxidative stress. Plant Physiol 127:159–172

Esau K (1953) Plant anatomy, vol xii. Wiley, New York, 735 p

Estavillo GM, Crisp PA, Pornsiriwong W, Wirtz M, Collinge D et al (2011) Evidence for a SAL1-PAP chloroplast retrograde pathway that functions in drought and high light signaling in Arabidopsis. Plant Cell 23:3992–4012

Evert RF, Esau K (2006) Esau's plant anatomy: meristems, cells, and tissues of the plant body. Wiley, New York

Evert RF, Botha CEJ, Mierzwa RJ (1985) Free-space marker studies on the leaf of *Zea mays* L. Protoplasma 126:62–73

Fischer BB, Krieger-Liszkay A, Hideg E, Snyrychová I, Wiesendanger M, Eggen RIL (2007) Role of singlet oxygen in chloroplast to nucleus retrograde signaling in *Chlamydomonas reinhardtii*. FEBS Lett 581:5555–5560

Frangne N, Maeshima M, Schäffner AR, Mandel T, Martinoia E, Bonnemain JL (2001) Expression and distribution of a vaculoar aquaporin in young and mature leaf tissues of Brassica napus in relation to water fluxes. Planta 212:270–278

Frost CJ, Appel HM, Carlson JE, De Moraes CM, Mescher MC, Schultz JC (2007) Within-plant signalling via volatiles overcomes vascular constraints on systemic signalling and primes responses against herbivores. Ecol Lett 10:490–498

Fryer MJ, Ball L, Oxborough K, Karpinski S, Mullineaux PM, Baker NR (2003) Control of Ascorbate Peroxidase 2 expression by hydrogen peroxide and leaf water status during excess light stress reveals a functional organisation of Arabidopsis leaves. Plant J 33:691–705

Fuchs EE, Livingston NJ (1996) Hydraulic control of stomatal conductance in Douglas fir [*Pseudotsuga menziesii* (Mirb.) Franco] and alder [*Alnus rubra* (Bong)] seedlings. Plant Cell Environ 19:1091–1098

Gadjev I, Vanderauwera S, Gechev TS, Laloi C, Minkov IN et al (2006) Transcriptomic footprints disclose specificity of reactive oxygen species signaling in Arabidopsis. Plant Physiol 141:436–445

Galvez-Valdivieso G, Fryer MJ, Lawson T, Slattery K, Truman W et al (2009) The high light response in Arabidopsis involves ABA signaling between vascular and bundle sheath cells. Plant Cell 21:2143–2162

Gao DJ, Knight MR, Trewavas AJ, Sattelmacher B, Plieth C (2004) Self-reporting Arabidopsis expressing pH and [Ca^{2+}] indicators unveil ion dynamics in the cytoplasm and in the apoplast under abiotic stress. Plant Physiol 134:898–908

Gigolashvilia T, Geierb M, Ashykhmina N, Frerigmann H, Wulfert S et al (2012) The Arabidopsis thylakoid ADP/ATP carrier TAAC has an additional role in supplying plastidic phosphoadenosine 5′-phosphosulfate to the cytosol. Plant Cell 24(10):4187–4204

Gorman AA, Rodgers MAJ (1992) New trends in photobiology. J Photochem Photobiol B 14:159–176

Harris N, Chaffey NJ (1985) Plasmatubules in transfer cells of pea (*Pisum sativum* L.). Planta 165:191–196

Hatz S, Lambert JDC, Ogilby PR (2007) Measuring the lifetime of singlet oxygen in a single cell: addressing the issue of cell viability. Photochem Photobiol Sci 6:1106–1116

Hatz S, Poulsen L, Ogilby PR (2008) Time-resolved singlet oxygen phosphorescence measurements from photosensitized experiments in single cells: effects of oxygen diffusion and oxygen concentration. Photochem Photobiol 84:1284–1290

Heil M, Ton J (2008) Long-distance signalling in plant defence. Trends Plant Sci 13:264–272

Heil M, Bueno S, Carlos J (2007) Within-plant signaling by volatiles leads to induction and priming of an indirect plant defense in nature. Proc Natl Acad Sci USA 104:5467–5472

Heinen RB, Ye Q, Chaumont F (2009) Role of aquaporins in leaf physiology. J Exp Bot 60:2971–2985

Hensel LL, Grbić V, Baumgarten DA, Bleecker AB (1993) Developmental and age-related processes that influence the longevity and senescence of photosynthetic tissues in Arabidopsis. Plant Cell 5:553–564

Henzler T, Steudle E (2000) Transport and metabolic degradation of hydrogen peroxide in *Chara corallina*: model calculations and measurements with the pressure probe suggest transport of H_2O_2 across water channels. J Exp Bot 51:2053–2066

Hihara Y, Kamei A, Kanehisa M, Kaplan A, Ikeuchi M (2001) DNA microarray analysis of cyanobacterial gene expression during acclimation to high light. Plant Cell 13:793–806

Ihnatowicz A, Pesaresi P, Lohrig K, Wolters D, Müller B, Leister D (2008) Impaired photosystem I oxidation induces STN7-dependent phosphorylation of the light-harvesting complex I protein Lhca4 in *Arabidopsis thaliana*. Planta 227:717–722

Iida A, Kazuoka T, Torikai S, Kikuchi H, Oeda K (2001) A zinc finger protein RHL41 mediates the light acclimatization response in Arabidopsis. Plant J 24:191–203

Ishikawa K, Yoshimura K, Ogawa T, Shigeoka S (2010) Distinct regulation of Arabidopsis ADP-ribose/NADH pyrophosphohydrolases, AtNUDX6 and 7, in biotic and abiotic stress responses. Plant Signal Behav 5:839–841

Jackson MB, Saker LR, Crisp CM, Else MA, Janowiak F (2003) Ionic and pH signalling from roots to shoots of flooded tomato plants in relation to stomatal closure. Plant Soil 253:103–113

Jambunathan N, Penaganti A, Tang Y, Mahalingam R (2010) Modulation of redox homeostasis under suboptimal conditions by Arabidopsis nudix hydrolase 7. BMC Plant Biol 10:173

Jia W, Zhang J (2008) Stomatal movements and long-distance signaling in plants. Plant Signal Behav 3:772–777

Karpinski S (2006) Holistic analysis of stress signalling network; new reactive oxygen species- and sugar-specific DNA cis-regulatory elements and functional interdependence between salicylic acid and glutathione signalling. Free Radic Res 40:S43–S43

Karpinski S, Muhlenbock P (2007) Genetic, molecular and physiological mechanisms controlling cell death, defenses, and antioxidant network in response to abiotic and biotic stresses in plants. Comp Biochem Physiol 146:S60

Karpinski S, Szechynska-Hebda M (2010) Secret life of plants: from memory to intelligence. Plant Signal Behav 5:1391–1394

Karpinski S, Escobar C, Karpinska B, Creissen G, Mullineaux PM (1997) Photosynthetic electron transport regulates the expression of cytosolic ascorbate peroxidase genes in Arabidopsis during excess light stress. Plant Cell 9:627–640

Karpinski S, Reynolds H, Karpinska B, Wingsle G, Creissen G, Mullineaux P (1999) Systemic signaling and acclimation in response to excess excitation energy in Arabidopsis. Science 284:654–657

Karpinski S, Gabrys H, Mateo A, Karpinska B, Mullineaux PM (2003) Light perception in plant disease defence signalling. Curr Opin Plant Biol 6:390–396

Katiyar-Agarwal S, Jin H (2010) Role of small RNAs in host-microbe interactions. Annu Rev Phytopathol 48:225–246

Kato Y, Miura E, Matsushima R, Sakamoto W (2007) White leaf sectors in yellow variegated2 are formed by viable cells with undifferentiated plastids. Plant Physiol 144:952–960

Keen NT (1990) Gene-for-gene complementarity in plant-pathogen interactions. Annu Rev Genet 24:447–463

Kehr J, Buhtz A (2008) Long distance transport and movement of RNA through the phloem. J Exp Bot 59:85–92

Kerchev PI, Pellny TK, Vivancos PD, Kiddle G, Hedden P et al (2011) The transcription factor ABI4 Is required for the ascorbic acid-dependent regulation of growth and regulation of jasmonate-dependent defense signaling pathways in Arabidopsis. Plant Cell 23:3319–3334

Kim C, Meskauskiene R, Apel K, Laloi C (2008) No single way to understand singlet oxygen signalling in plants. EMBO Rep 9:435–439

Kimura M, Yoshizumi T, Manabe K, Yamamoto YY, Matsui M (2001) Arabidopsis transcriptional regulation by light stress via hydrogen peroxide-dependent and -independent pathways. Genes Cells 6:607–617

Kimura M, Yamamoto YY, Seki M, Sakurai T, Sato M et al (2003) Identification of Arabidopsis genes regulated by high light-stress using cDNA microarray. Photochem Photobiol 77:226–233

Kinsman EA, Pyke KA (1998) Bundle sheath cells and cell-specific plastid development in Arabidopsis leaves. Development 125:1815–1822

Kitajima S, Shimaoka T, Kurioka M, Yokota A (2007) Irreversible cross-linking of heme to the distal tryptophan of stromal ascorbate peroxidase in response to rapid inactivation by H_2O_2. FEBS J 274:3013–3020

Kleine T, Kindgren P, Benedict C, Hendrickson L, Strand A (2007) Genome-wide gene expression analysis reveals a critical role for CRYPTOCHROME1 in the response of Arabidopsis to high irradiance. Plant Physiol 144:1391–1406

Koussevitzky S, Suzuki N, Huntington S, Armijo L, Sha W et al (2008) Ascorbate peroxidase 1 plays a key role in the response of *Arabidopsis thaliana* to stress combination. J Biol Chem 283:34197–34203

Kuimova MK, Yahioglu G, Ogilby PR (2009) Singlet oxygen in a cell: spatially dependent lifetimes and quenching rate constants. J Am Chem Soc 131:332–340

Kwak JM, Mori IC, Pei Z-M, Leonhardt N, Torres MA et al (2003) NADPH oxidase AtrbohD and AtrbohF genes function in ROS-dependent ABA signaling in Arabidopsis. EMBO J 22:2623–2633

Kwak JM, Nguyen V, Schroeder JI (2006) The role of reactive oxygen species in hormonal responses. Plant Physiol 141:323–329

Lake JA, Woodward FI, Quick WP (2002) Long-distance CO2 signalling in plants. J Exp Bot 53:183–193

Laloi C, Stachowiak M, Pers-Kamczyc E, Warzych E, Murgia I, Apel K (2007) Cross-talk between singlet oxygen- and hydrogen peroxide-dependent signaling of stress responses in *Arabidopsis thaliana*. Proc Natl Acad Sci USA 104:672–677

Lalonde S, Wipf D, Frommer WB (2004) Transport mechanisms for organic forms of carbon and nitrogen between source and sink. Annu Rev Plant Biol 55:341–372

Lee KP, Kim C, Landgraf F, Apel K (2007) EXECUTER1- and EXECUTER2-dependent transfer of stress-related signals from the plastid to the nucleus of *Arabidopsis thaliana*. Proc Natl Acad Sci USA 104:10270–10275

Leegood RC (2008) Roles of the bundle sheath cells in leaves of C3 plants. J Exp Bot 59:1663–1673

Leisinger U, Rüfenacht K, Fischer B, Pesaro M, Spengler A et al (2001) The glutathione peroxidase homologous gene from *Chlamydomonas reinhardtii* is transcriptionally up-regulated by singlet oxygen. Plant Mol Biol 46:395–408

Li Z, Wakao S, Fischer BB, Niyogi KK (2009) Sensing and responding to excess light. Annu Rev Plant Biol 60:239–260

Liu GS, Greenshields DL, Sammynaiken R, Hirji RN, Selvaraj G, Wei YD (2007) Targeted alterations in iron homeostasis underlie plant defense responses. J Cell Sci 120:596–605

Lough TJ, Lucas WJ (2006) Integrative plant biology: role of phloem long-distance macromolecular trafficking. Annu Rev Plant Biol 57:203–232

Mano J, Ohno C, Domae Y, Asada K (2001) Chloroplastic ascorbate peroxidase is the primary target of methylviologen-induced photooxidative stress in spinach leaves: its relevance to monodehydroascorbate radical detected with in vivo ESR. Biochim Biophys Acta 1504:275–287

Martre P, Morillon R, Barrieu F, North GB, Nobel PS, Chrispeels MJ (2002) Plasma membrane aquaporins play a significant role during recovery from water deficit. Plant Physiol 130:2101–2110

Maruta T, Noshi M, Tanouchi A, Tamoi M, Yabuta Y et al (2012) H_2O_2-triggered retrograde signaling from chloroplasts to nucleus plays specific role in response to stress. J Biol Chem 287:11717–11729

Mateo A, Mühlenbock P, Rustérucci C, Chang CC-C, Miszalski Z et al (2004) LESION SIMULATING DISEASE 1 is required for acclimation to conditions that promote excess excitation energy. Plant Physiol 136:2818–2830

Mateo A, Funck D, Mühlenbock P, Kular B, Mullineaux PM, Karpinski S (2006) Controlled levels of salicylic acid are required for optimal photosynthesis and redox homeostasis. J Exp Bot 57:1795–1807

Meskauskiene R, Nater M, Goslings D, Kessler F, op den Camp R, Apel K (2001) FLU: a negative regulator of chlorophyll biosynthesis in *Arabidopsis thaliana*. Proc Natl Acad Sci USA 98:12826–12831

Miao Y, Lv D, Wang P, Wang X-C, Chen J et al (2006) An Arabidopsis glutathione peroxidase functions as both a redox transducer and a scavenger in abscisic acid and drought stress responses. Plant Cell 18:2749–2766

Miller G, Schlauch K, Tam R, Cortes D, Torres MA et al (2009) The plant NADPH oxidase RBOHD mediates rapid systemic signaling in response to diverse stimuli. Sci Signal 2:ra45

Mittler R, Vanderauwera S, Gollery M, Van Breusegem F (2004) Reactive oxygen gene network of plants. Trends Plant Sci 9:490–498

Mittler R, Vanderauwera S, Suzuki N, Miller G, Tognetti VB et al (2011) ROS signaling: the new wave? Trends Plant Sci 16:300–309

Miyake C, Shinzaki Y, Nishioka M, Horiguchi S, Tomizawa K-I (2006) Photoinactivation of ascorbate peroxidase in isolated tobacco chloroplasts: *Galdieria partita* APX maintains the electron flux through the water-water cycle in transplastomic tobacco plants. Plant Cell Physiol 47:200–210

Miyao M (1994) Involvement of active oxygen species in degradation of the D1 protein under strong illumination in isolated subcomplexes of photosystem II. Biochemistry 33:9722–9730
Møller IM, Sweetlove LJ (2010) ROS signalling—specificity is required. Trends Plant Sci 15:370–374
Mou Z, Fan W, Dong X (2003) Inducers of plant systemic acquired resistance regulate NPR1 function through redox changes. Cell 113:935–944
Mubarakshina MM, Ivanov BN, Naydov IA, Hillier W, Badger MR, Krieger-Liszkay A (2010) Production and diffusion of chloroplastic H_2O_2 and its implication to signalling. J Exp Bot 61:3577–3587
Mühlenbock P, Szechynska-Hebda M, Plaszczyca M, Baudo M, Mateo A et al (2008) Chloroplast signaling and LESION SIMULATING DISEASE1 regulate crosstalk between light acclimation and immunity in Arabidopsis. Plant Cell 20:2339–2356
Muhling KH, Lauchli A (2000) Light-induced pH and K^+ changes in the apoplast of intact leaves. Planta 212:9–15
Mullineaux P (2009) ROS in retrograde signalling from the chloroplast to the nucleus. In: Rio LA, Puppo A (eds) Reactive oxygen species in plant signaling. Springer, Berlin, pp 221–240
Mullineaux PM, Baker NR (2010) Oxidative stress: antagonistic signaling for acclimation or cell death? Plant Physiol 154:521–525
Nelson T, Dengler N (1997) Leaf vascular pattern formation. Plant Cell 9:1121–1135
Nühse TS, Bottrill AR, Jones AME, Peck SC (2007) Quantitative phosphoproteomic analysis of plasma membrane proteins reveals regulatory mechanisms of plant innate immune responses. Plant J 51:931–940
Ochsenbein C, Przybyla D, Danon A, Landgraf F, Göbel C et al (2006) The role of EDS1 (enhanced disease susceptibility) during singlet oxygen-mediated stress responses of Arabidopsis. Plant J 47:445–456
Oelze M-L, Vogel MO, Alsharafa K, Kahmann U, Viehhauser A et al (2012) Efficient acclimation of the chloroplast antioxidant defence of *Arabidopsis thaliana* leaves in response to a 10- or 100-fold light increment and the possible involvement of retrograde signals. J Exp Bot 63:1297–1313
op den Camp RGL, Przybyla D, Ochsenbein C, Laloi C, Kim C et al (2003) Rapid induction of distinct stress responses after the release of singlet oxygen in Arabidopsis. Plant Cell 15:2320–2332
Orians C (2005) Herbivores, vascular pathways, and systemic induction: facts and artifacts. J Chem Ecol 31:2231–2242
Palauqui JC, Elmayan T, Pollien JM, Vaucheret H (1997) Systemic acquired silencing: transgene-specific post-transcriptional silencing is transmitted by grafting from silenced stocks to non-silenced scions. EMBO J 16:4738–4745
Pesaresi P, Hertle A, Pribil M, Kleine T, Wagner R et al (2009) Arabidopsis STN7 kinase provides a link between short- and long-term photosynthetic acclimation. Plant Cell 21:2402–2423
Pilot G, Stransky H, Bushey DF, Pratelli R, Ludewig U et al (2004) Overexpression of GLUTAMINE DUMPER1 leads to hypersecretion of glutamine from hydathodes of Arabidopsis leaves. Plant Cell 16:1827–1840
Pnueli L, Liang H, Rozenberg M, Mittler R (2003) Growth suppression, altered stomatal responses, and augmented induction of heat shock proteins in cytosolic ascorbate peroxidase (Apx1)-deficient Arabidopsis plants. Plant J 34:187–203
Pogson BJ, Woo NS, Förster B, Small ID (2008) Plastid signalling to the nucleus and beyond. Trends Plant Sci 13:602–609
Polle A (2001) Dissecting the superoxide dismutase-ascorbate-glutathione-pathway in chloroplasts by metabolic modeling. Computer simulations as a step towards flux analysis. Plant Physiol 126:445–462
Przybyla D, Göbel C, Imboden A, Hamberg M, Feussner I, Apel K (2008) Enzymatic, but not non-enzymatic, $^1O_2$-mediated peroxidation of polyunsaturated fatty acids forms part of the

EXECUTER1-dependent stress response program in the flu mutant of *Arabidopsis thaliana*. Plant J 54:236–248
Puppo A (1992) Effect of flavonoids on hydroxyl radical formation by fenton-type reactions; influence of the iron chelator. Phytochemistry 31:85–88
Ramel F, Birtic S, Ginies C, Soubigou-Taconnat L, Triantaphylides C, Havaux M (2012) Carotenoid oxidation products are stress signals that mediate gene responses to singlet oxygen in plants. Proc Natl Acad Sci USA 109:5535–5540
Redmond RW, Kochevar IE (2006) Spatially resolved cellular responses to singlet oxygen. Photochem Photobiol 82:1178–1186
Rochaix J-D (2007) Role of thylakoid protein kinases in photosynthetic acclimation. FEBS Lett 581:2768–2775
Rossel JB, Wilson IW, Pogson BJ (2002) Global changes in gene expression in response to high light in Arabidopsis. Plant Physiol 130:1109–1120
Rossel JB, Wilson PB, Hussain D, Woo NS, Gordon MJ et al (2007) Systemic and intracellular responses to photooxidative stress in Arabidopsis. Plant Cell 19:4091–4110
Ruckle ME, DeMarco SM, Larkin RM (2007) Plastid signals remodel light signaling networks and are essential for efficient chloroplast biogenesis in Arabidopsis. Plant Cell 19:3944–3960
Rusterucci C (2001) The disease resistance signaling components EDS1 and PAD4 are essential regulators of the cell death pathway controlled by LSD1 in Arabidopsis. Plant Cell 13:2211–2224
Sack L, Holbrook NM (2006) Leaf hydraulics. Annu Rev Plant Biol 57:361–381
Sagi M, Fluhr R (2006) Production of reactive oxygen species by plant NADPH oxidases. Plant Physiol 141:336–340
Samol I, Shapiguzov A, Ingelsson B, Fucile G, Crèvecoeura M et al (2012) Identification of a photosystem II phosphatase involved in light acclimation in Arabidopsis. Plant Cell 24:2596–2609
Schwach F, Vaistij FE, Jones L, Baulcombe DC (2005) An RNA-dependent RNA polymerase prevents meristem invasion by potato virus X and is required for the activity but not the production of a systemic silencing signal. Plant Physiol 138:1842–1852
Shatil-Cohen A, Attia Z, Moshelion M (2011) Bundle-sheath cell regulation of xylem-mesophyll water transport via aquaporins under drought stress: a target of xylem-borne ABA? Plant J 67:72–80
Shinozaki K, Yamaguchi-Shinozaki K (2000) Molecular responses to dehydration and low temperature: differences and cross-talk between two stress signaling pathways. Curr Opin Plant Biol 3:217–223
Sies H, Menck CFM (1992) Singlet oxygen induced DNA damage. Mutat Res 275:367–375
Sirichandra C, Gu D, Hu H-C, Davanture M, Lee S et al (2009) Phosphorylation of the Arabidopsis AtrbohF NADPH oxidase by OST1 protein kinase. FEBS Lett 583:2982–2986
Sobeih WY, Dodd IC, Bacon MA, Grierson D, Davies WJ (2004) Long-distance signals regulating stomatal conductance and leaf growth in tomato (*Lycopersicon esculentum*) plants subjected to partial root-zone drying. J Exp Bot 55:2353–2363
Straus MR, Rietz S, Loren V, van Themaat E, Bartsch M, Parker JE (2010) Salicylic acid antagonism of EDS1-driven cell death is important for immune and oxidative stress responses in Arabidopsis. Plant J 62:628–640
Szechynska-Hebda M, Kruk J, Gorecka M, Karpinska B, Karpinski S (2010) Evidence for light wavelength-specific photoelectrophysiological signaling and memory of excess light episodes in Arabidopsis. Plant Cell 22:2201–2218
Sztatelman O, Waloszek A, Banaś AK, Gabryś H (2010) Photoprotective function of chloroplast avoidance movement: in vivo chlorophyll fluorescence study. J Plant Physiol 167:709–716
Tardieu F, Davies WJ (1993) Integration of hydraulic and chemical signalling in the control of stomatal conductance and water status of droughted plants. Plant Cell Environ 16:341–349
Terashima I, Hikosaka K (1995) Comparative ecophysiology of leaf and canopy photosynthesis. Plant Cell Environ 18:1111–1128

Thomas PW, Woodward FI, Quick WP (2003) Systemic irradiance signalling in tobacco. New Phytol 161:193–198
Tikkanen M, Piippo M, Suorsa M, Sirpiö S, Mulo P et al (2006) State transitions revisited—a buffering system for dynamic low light acclimation of Arabidopsis. Plant Mol Biol 62:779–793
Tikkanen M, Grieco M, Kangasjärvi S, Aro E-M (2010) Thylakoid protein phosphorylation in higher plant chloroplasts optimizes electron transfer under fluctuating light. Plant Physiol 152:723–735
Torres MA, Jones JDG, Dangl JL (2005) Pathogen-induced, NADPH oxidase-derived reactive oxygen intermediates suppress spread of cell death in *Arabidopsis thaliana*. Nat Genet 37:1130–1134
Triantaphylidès C, Krischke M, Hoeberichts FA, Ksas B, Gresser G et al (2008) Singlet oxygen is the major reactive oxygen species involved in photooxidative damage to plants. Plant Physiol 148:960–968
Turnbull C (2011) Long-distance regulation of flowering time. J Exp Bot 62:4399–4413
Van Breusegem F, Bailey-Serres J, Mittler R (2008) Unraveling the tapestry of networks involving reactive oxygen species in plants. Plant Physiol 147:978–984
Vandenbroucke K, Robbens S, Vandepoele K, Inze D, de Peer YV, Van Breusegem F (2008) Hydrogen peroxide-induced gene expression across kingdoms: a comparative analysis. Mol Biol Evol 25:507–516
Vanderauwera S, Zimmermann P, Rombauts S, Vandenabeele S, Langebartels C et al (2005) Genome-wide analysis of hydrogen peroxide-regulated gene expression in Arabidopsis reveals a high light-induced transcriptional cluster involved in anthocyanin biosynthesis. Plant Physiol 139:806–821
Vranova E, Inze D, Van Breusegem F (2002) Signal transduction during oxidative stress. J Exp Bot 53:1227–1236
Wagner D, Przybyla D, Op den Camp R, Kim C, Landgraf F et al (2004) The genetic basis of singlet oxygen-induced stress responses of *Arabidopsis thaliana*. Science 306:1183–1185
Wiermer M, Feys BJ, Parker JE (2005) Plant immunity: the EDS1 regulatory node. Curr Opin Plant Biol 8:383–389
Wilkinson S, Davies WJ (2002) ABA-based chemical signalling: the co-ordination of responses to stress in plants. Plant Cell Environ 25:195–210
Wilkinson S, Davies WJ (2008) Manipulation of the apoplastic pH of intact plants mimics stomatal and growth responses to water availability and microclimatic variation. J Exp Bot 59:619–631
Xiao Y, Savchenko T, Baidoo EEK, Chehab WE, Hayden DM et al (2012) Retrograde signaling by the plastidial metabolite MEcPP regulates expression of nuclear stress-response genes. Cell 149:1525–1535
Yabuta Y, Maruta T, Yoshimura K, Ishikawa T, Shigeoka S (2004) Two distinct redox signaling pathways for cytosolic APX induction under photooxidative stress. Plant Cell Physiol 45:1586–1594
Yang D, Andersson B, Aro E, Ohad I (2001) The redox state of the plastoquinone pool controls the level of the light-harvesting chlorophyll a/b binding protein complex II (LHC II) during photoacclimation. Photosynth Res 68:163–174
Yano S, Terashima I (2001) Separate localization of light signal perception for sun or shade type chloroplast and palisade tissue differentiation in *Chenopodium album*. Plant Cell Physiol 42:1303–1310
Zavaliev R, Sagi G, Gera A, Epel BL (2010) The constitutive expression of Arabidopsis plasmodesmal-associated class 1 reversibly glycosylated polypeptide impairs plant development and virus spread. J Exp Bot 61:131–142

Macromolecules Trafficking in the Phloem and Interorgan Communication

Ziv Spiegelman, Guy Golan, and Shmuel Wolf

Abstract The phloem is a major component of the vascular tissue responsible for the delivery of photoassimilates and nutrients from source (photosynthetically active) tissues to sink organs. The presence of active plasmodesmata between the companion cells–sieve element complex and adjacent cells creates a symplastic continuum connecting almost all cells, even at distant tissues. That phloem sap contains a wide repertoire of proteins has long been established. It is, however, only been recently established that thousands of RNA molecules are also present within the sieve tube. While a large number of these macromolecules were identified through the use of modern analytical tools combined with bioinformatics methods, a biological role explaining their presence in the sieve tube is assigned to only a few. Insights provided by long-distance movement of viral particles conjoined with the characterization of several phloem sap proteins and RNA molecules form the foundation of the hypothesis that macromolecules play a role in the plant's long-distance communication signaling system. A future challenge is to dissect the mechanism by which plants control trafficking of these macromolecules from their site of synthesis through the sieve tube and all the way to their target cells.

Keywords Cell-to-cell movement • Grafting • Phloem sap • Plasmodesmata • Proteins • RNA • viroids

Z. Spiegelman • G. Golan • S. Wolf (✉)
The Robert H. Smith Faculty of Agriculture, Food and Environment, The Otto Warburg Minerva Centre for Agricultural Biotechnology, The Robert H. Smith Institute of Plant Sciences and Genetics in Agriculture, The Hebrew University of Jerusalem, Rehovot 76100, Israel
e-mail: swolf@agri.huji.ac.il

F. Baluška (ed.), *Long-Distance Systemic Signaling and Communication in Plants*, Signaling and Communication in Plants 19, DOI 10.1007/978-3-642-36470-9_14,

1 Introduction

Embarking on a journey from source to sink tissues, we begin by describing the sieve element–companion cell complex and the role of interconnecting plasmodesmata in the trafficking of macromolecules into this complex. Before detailing any major pertinent findings, however, phloem sampling and its limitations are described. This is followed by reviewing the translocation of RNA molecules through cell-to-cell and long-distance movement of viroids as well as the presence of endogenous RNA in the phloem. We then delved deeper into specific proteins found in the sieve tube, their trafficking mechanism, and their potential role in long-distance interorgan movement. All evidence is thus combined to show that in higher plants, the phloem serves as a superhighway for the delivery of macromolecules to communicate between distant organs.

2 The Sieve Element–Companion Cell Complex

The evolution of land plants required the development of vascular systems to enable efficient response to altering environmental conditions. Plant vascular systems comprise two major components: the xylem whose function is to transport water and soluble mineral from the roots to the shoots and the phloem that is traditionally considered responsible for the distribution of photoassimilates and nutrients among various plant tissues. Xylem development is thought to have occurred in order to ensure tissue hydration of land plants by transporting water from the plant–soil interface to all aerial organs (Raven 1993; Scarpella and Helariutta 2010). Phloem evolution, on the other hand, is probably a consequence of the increase in plant size and development of distinct organs (Graham 1993; van Bel 1999) under the assumption that the transition from multicellular to supracellular organism demanded long-distance transport of photoassimilates and signal molecules. In creating an efficient long-distance delivering tube, a phloem "mother cell" is divided asymmetrically and the two emergent cells undergo unique differentiation to form the companion cell–sieve element (CC–SE) complex. CCs typically contain rather large nuclei, numerous elongated mitochondria and plastids, and an abundance of free ribosomes (Behnke 1989). The resulting dense cytoplasm indicates high metabolic activity that also supports the neighboring SEs (Schultz 1998; van Bel 1999). Differentiation of the SE is characterized by loss of organelles, including nuclei and ribosomes, changes in endoplasmic reticulum (ER) formation, and widening of symplastic connections between SEs to form the sieve-plate pores (Esau 1969; Schultz 1998). Longitudinal files of SEs that are symplastically connected via sieve-plate pores form a long-distance delivering tube, called sieve tube. An interesting characteristic feature of mature SEs is the dense sieve element reticulum (SER) (Sjolund and Shin 1983). While its role is still unclear, it is likely that the SER interferes with, or selectively anchors, macromolecules that are present in the translocation stream. The two cell types, CCs and SEs, function interdependently through the abundant interconnecting plasmodesmata. These are modified plasmodesmata known as plasmodesmata pore units (PPUs), which usually have a single branch on the SE side and several branches on the CC side of the common wall (Esau 1969).

Plants can be divided into two major groups in terms of symplastic conductivity between the CC–SE complex and the neighboring phloem parenchyma or bundle sheath cells. Numerous plant species that are characterized by a high frequency of plasmodesmata interconnecting the entire pathway from the mesophyll to the CC–SE complex (e.g., Cucurbitaceae) are defined as symplastic phloem loaders (Turgeon and Beebe 1991). Minor vein companion cells in these plants are termed intermediary cells (IC) and are located at the periphery of the vein, in contact with bundle sheath cells (Turgeon et al. 1975). However, in many other plants the number of plasmodesmata interconnecting the CC–SE complex and the neighboring phloem parenchyma or bundle sheath cells is significantly smaller (Gamalei 1989). These plants (e.g., Solanaceae) are defined as apoplastic phloem loaders (van Bel 1993). Regardless of frequency, it is important to note that ultrastructural studies show that plasmodesmata interconnecting the CC–SE complex with neighboring cells exist in all studied plant species, including apoplastic loaders (Robards and Lucas 1990). Microinjection experiments (Madore et al. 1986) as well as electrical coupling studies (van Bel and van Rijen 1994) indicate that these plasmodesmata are functional.

3 Trafficking of Macromolecules into the CC–SE Complex

That plasmodesmata interconnecting the CC–SE complex and their neighboring cells are able to deliver macromolecules across this boundary was first evidenced by systemic movement of viral particles (Esau 1968; Esau et al. 1967). Similar to photoassimilates, long-distance transport of viruses comprises three stages: loading of virus particles unto the CC–SE complex, long-distance trafficking, and unloading in sink organs. Indeed, the interface between the bundle sheath/phloem parenchyma cells and the CC–SE complex was found to be a boundary for the systemic movement of *cowpea chlorotic mottle virus* infecting soybean plants (Goodrick et al. 1991) and TMV mutant infecting tobacco plants (Ding et al. 1996), both of which are apoplastic-loading plants. But all other viruses that infect this type of plants could efficiently overcome this “barrier” to complete systemic infection. Interestingly, the bundle sheath (BS)-phloem interface in cucumber (symplastic loader) plants, at which the frequency of plasmodesmata is high, was also found to impose a boundary to viral movement (Thompson and García-Arenal 1998). These results further support the notion that even in symplastic loaders, special plasmodesmata interconnect the intermediary and BS cells to maintain the IC–SE complex as a distinct domain, even though trafficking of small molecules (sucrose) across this boundary is probably not inhibited.

Expression of green fluorescent protein (GFP) under a tissue-specific promoter in transgenic tobacco plants provided the first experimental evidence for the ability of plant-expressed proteins to enter the vasculature and move long distance (Peleg et al. 2007). The *gfp* encoded gene was introduced into tobacco plants under the fructose-1,6-bisphosphatase (FBPase) promoter, resulting in expression of the heterologous protein in mesophyll and bundle sheath cells only. Infection of these plants with either cucumber mosaic virus (CMV), tobacco mosaic virus (TMV), or potato virus Y (PVY) enabled the movement of GFP into the CC–SE complex, while long-distance movement was evidenced to the shoot apex. In this respect, it is important to note that the above-described experiments failed to demonstrate long-distance trafficking of the *gfp* mRNA (Peleg et al. 2007). These studies show that plasmodesmata interconnecting the CC–SE complex with neighboring cells act as a selective entrance for proteins and RNA molecules into the sieve tube.

4 Sampling the Phloem

Development of modern analytical tools along with expanded genomic databases allows for the detailed analyses of molecule profiles within vascular tissues (Haebel and Kehr 2001; Nakazono et al. 2003; Vilaine et al. 2003; Giavalisco et al. 2006; Kehr and Buhtz 2008). These studies established that in addition to small molecules such as sugars, minerals, amino acids, and growth substances, the sieve tube contains a large repertoire of macromolecules such as proteins and RNAs. One can assume that all these macromolecules are destined for degradation in sink organs where they will provide building blocks and substrates for the synthesis of protein and nucleic acids. However, accumulated evidence provides the foundation for the hypothesis that at least some of these macromolecules act as signaling agents involved in the plant's endogenous network that is required for the communication between distant organs (Ruiz-Medrano et al. 1999; Lough and Lucas 2006).

A prerequisite for our understanding of the biological role of macromolecules present within the sieve tube is their identification and full characterization. A major obstacle in identifying the whole cassette of molecules present in the phloem is the sampling procedure. One of the first attempts to perform high-throughput analyses of the vascular tissue transcriptome was performed on vascular bundles of common plantain leaves (Pommerrenig et al. 2006). Almost 6,000 expressed sequence tags (ESTs) were sequenced, indicating a specific vascular profile. However, this profile represented all phloem (and xylem) cells and could not be limited to those cells only that are present within the sieve tube. Further attempts to analyze the content of specific phloem cells made use of a laser capture microdissection (LCM) system. Marked differences were observed between transcripts extracted by the LCM technique from epidermis cells and vascular tissue cells (Asano et al. 2002; Nakazono et al. 2003). As yet, however, even this precise technology is unable to isolate clean, specific SEs. One direct elegant means to collect phloem sap is provided by phloem-feeding insects (Fukumorita and Chino 1982; Fisher and

Frame 1984). Using this technique, a few microliters of phloem sap may be collected following microcauterization of the feeding insect's stylet. While this volume is sufficient for transcriptome analyses, it is a limiting factor in proteomic studies (Gaupels et al. 2008). In addition, aphids exude saliva that alters the chemistry of sap properties (Moran et al. 2002). Plant species that bleed copious amounts of phloem sap from cut stems or petioles are major candidates for the identification and characterization of macromolecules present within the sieve tube (Ruiz-Medrano et al. 1999; Barnes et al. 2004; Giavalisco et al. 2006; Omid et al. 2007). However, collection of phloem exudates from cut sections is invasive and, therefore, raises serious questions regarding potential contamination by surrounding cells. The sudden pressure release within the high SE hydrostatic pressure probability also affects neighboring cells which in turn may alter the profile of molecules in the collected sap. Nevertheless, special caution such as sharp cuts and the discard of the first few bleeding drops facilitates the extraction of relatively clean sieve tube sap. In contrast to all other phloem cells, SEs contain disaccharides (sucrose) and polysaccharides and do not contain reducing sugars such as glucose or fructose. Lack of reducing sugars in collected exudates thus provides sound confirmation for its purity, attesting that contamination is minimal from neighboring CC or phloem parenchyma cells.

5 Long-Distance Movement of RNA Molecules

Endogenous RNA molecules were identified in phloem sap over 40 years ago (Kollmann et al. 1970; Ziegler 1975). At first, these "traces" of nucleic acids were considered to be a contamination originating in neighboring cells rather than an integral constituent of the sieve elements. This view was based on the assumption that RNA molecules must be cell autonomous, acting within their transcription site. As RNA molecules are vulnerable to degradation, it was difficult to explain the ability of mRNA to be active following movement in the phloem. Moreover, it is still difficult to comprehend the complex mechanism of the loading and unloading of relatively large nucleic acid into and out of the sieve tube and to explain the biological rationale for long-distance trafficking of these molecules in plants.

5.1 Cell-to-Cell and Long-Distance Movement of Viroids

First insight into the mechanism of long-distance trafficking of RNA molecules was provided through the examination of the systemic spread of viroids. Viroids are small (275–400 nucleotides) single-stranded circular RNAs that infect plants (Diener 1979). They do not encode proteins and are not encapsidated, rendering them a unique system for the study of cell-to-cell and long-distance movement of

RNA molecules. As viroids do not encode any proteins, they have evolved structural motifs that interact with host cell factors to mediate their replication and trafficking throughout the plant. Exploration of the mechanism underlying viroid trafficking may therefore shed some light on the manner by which endogenous RNA molecules traffic long distance in plants (Wang and Ding 2010). Indication for the trafficking of viroids via the phloem bulk flow, from source to sink organs, was provided following the inoculation of tomato source leaves with potato spindle tuber viroid (PSTVd). Accumulation of the viroid was evident in shoot tips, young leaves, and roots, but not in leaves positioned below the infected source leaf (Palukaitis 1987). To date, there are no indications for the transport of nucleic acids across cell membranes. It thus stands to reason that the entry of viroids into the sieve tube must follow a symplastic route via plasmodesmata interconnecting the companion cells and the sieve elements. Microinjection experiments indeed provide direct evidence for cell-to-cell viroid movement via plasmodesmata. Rapid movement of fluorescently labeled PSTVd was observed when the viroid was injected into a single mesophyll cell, while the fluorescent signal was restricted to one cell following the injection into a symplastically isolated guard cell (Ding et al. 2005). Moreover, the fusion of a 1,400 nucleotide sequence to PSTVd enabled its cell-to-cell movement via plasmodesmata, while no movement of this sequence could be detected following its injection as a naked sequence. These findings indicate that movement of PSTVd via plasmodesmata is mediated by a specific sequence or a structural motif.

A molecular genetic approach was applied to further identify the specific sequence motifs regulating the entry of viroids into the phloem. Single point mutations (substitution of C318 to A or U43 to G) were sufficient to abolish long-distance movement of the mutated PSTVd in *Nicotiana benthamiana* plants, while replication at the cellular level was not affected (Zhong et al. 2007). Interestingly, both wild-type and mutant viroids were capable of trafficking between mesophyll and bundle sheath cells while the mutants failed to enter the phloem, indicating that those motifs are crucial for movement across the boundary between bundle sheath and phloem cells to enable systemic spread. These results also suggest that molecular constituents of plasmodesmata interconnecting mesophyll cells differ from those of plasmodesmata interconnecting bundle sheath and phloem cells and that specific viroid motifs are required for the interaction with host factors to enable systemic movement across diverse cellular boundaries. Further analysis establishes that an essential motif for the trafficking of the viroid from the bundle sheath to the phloem is the formation of a water-inserted *cis* Watson–Crick/Watson–Crick base pair flanked by short helices comprising canonical Watson–Crick/Watson–Crick base pairs (Zhong et al. 2007). This seminal work demonstrated, for the first time, that a specific tertiary structural RNA motif is required for molecules to enter the CC–SE complex in source leaves.

The complexity of viroid systemic movement was further demonstrated by two PSTVd strains differing in five spontaneous nucleotide substitutions (Qi et al. 2004). While both PSTVd^{NB} and PSTVd^{NT} replicated (and to a similar level) in inoculated source leaves, only PSTVd^{NB} was capable of establishing infection in systemic young sink leaves. In situ hybridization experiments establish the presence of $PSTVd^{NB}$ in all cells of the systemic leaf, while $PSTVd^{NT}$ was detected in vascular and bundle sheath cells alone. These findings indicate that $PSTVd^{NT}$ was not able to move from bundle sheath to mesophyll cells in those young leaves despite its ability to move in the other direction, from the mesophyll via the bundle sheath and into the sieve tube, in infected source leaves (Qi et al. 2004).

The accumulated data thus indicate that cell-to-cell and long-distance movement of RNA molecules is regulated by the function of plasmodesmata at specific boundaries. In other words, plasmodesmata interconnecting bundle sheath and phloem cells act as checkpoints and control the direction for the delivery of viroids into and out of the sieve tubes to control long-distance trafficking and systemic infection.

5.2 Endogenous RNA in the Phloem

A pioneering study aimed at high-throughput analyses of the sieve tube transcription profile made use of a cDNA library constructed from melon phloem sap mRNA (Omid et al. 2007). Sequencing of about 2,000 randomly selected clones resulted in almost 1,000 unique transcripts. Annotation of these transcripts revealed that 40 % of them were related to stress and defense response while 15 % were related to signal transduction and transcriptional control. Significant differences found between this profile and the profile of transcripts identified in Arabidopsis CCs (Ivashikina et al. 2003) indicate that the transcript profile of phloem sap does not reflect the profile of the neighboring CCs, suggesting tight regulation on cell-to-cell movement of mRNA molecules from CCs to SEs. Compared to functional transcripts identified in CCs, only about a tenth of the transcripts identified in the phloem sap were associated with metabolism (2.3 % vs. 16 % in CCs) and redox regulation (3.5 % vs. 29 % in CCs).

It is now assumed that the population of mRNA molecules within the sieve tube is of some several thousands. However, questions are still open regarding the ability of all these molecules to move long distance and the target cells of the trafficking transcripts. Melon–pumpkin heterografting experiments demonstrated that only 6 of 43 examined transcripts were capable of long-distance movement. Annotation of the six transcripts revealed that two of them were identified as *Aux/IAA* and one as *small auxin-up RNA* (SAUR), while the other three were encoded for "hypothetical proteins" (Omid et al. 2007). The fact that the three identified long-distance trafficking transcripts are related to auxin signal transduction raises the possibility that auxin may exert control over developmental processes at the level of the whole plant both by polar cell-to-cell transport down the plant axis as well as through its influence on phloem mobility of specific auxin-sensitive transcripts. We have further confirmed the ability of one melon *Aux/IAA* transcript to move long distance in *N. benthamiana* and have succeeded to alter sensitivity to auxin in transgenic tomato shoots and roots overexpressing this gene in the phloem only (Golan et al. unpublished data).

Data published over the past decade establishes that in addition to mRNA, phloem sap contains a variety of small RNA (sRNA) molecules (Buhtz et al. 2008, 2010; Kehr and Buhtz 2008; Pant et al. 2008). Over 1,000 sRNA and numerous microRNA molecules were identified in phloem sap of *Cucurbita maxima* (Yoo et al. 2004). A more recent study characterized small (30–90 bases) noncoding RNAs in pumpkin phloem sap (Zhang et al. 2009). In the course of these studies, a battery of tRNAs, ribosomal RNAs, and spliceosomal RNAs were identified whose biological role within the sieve tube is yet perplexing. Readers are directed to recent reviews (Thompson and Schulz 1999; Vilaine et al. 2003; Lough and Lucas 2006; Kehr and Buhtz 2008) and other chapters in this book covering the current knowledge on the potential long-distance signaling role of several mobile mRNA in controlling developmental processes and miRNA with respect to nutrient stress response.

6 Proteins in the Sieve Tube

Various sets of proteins were identified in phloem sap collected from several plant species (Fisher et al. 1992; Ishiwatari et al. 1995, 1998; Golecki et al. 1999; Thompson and Schulz 1999; Schobert et al. 2000). The development of mass spectrometry-based analytical tools over the last decade together with increasingly expanding databases has conjointly enabled high-throughput analyses of phloem sap protein profiles. These analyses were performed mainly on bleeding plant species such as castor bean (Barnes et al. 2004), cucurbits (Haebel and Kehr 2001; Lin et al. 2009), and *Brassica napus* (Giavalisco et al. 2006). Subsequently, it is generally accepted that around one thousand proteins are present in phloem sap with a size distribution in the range of several kDa up to over 100 kDa. The current assumption is that mature SEs do not possess the machinery required for the completion of the genetic translation process. Therefore, it is likely that most of the phloem sap proteins enter the sieve tube via the unique branched, delta shape, structure of PPUs. The fact that molecular masses of several phloem sap proteins are higher than 100 kDa suggests that the size exclusion limits (SELs) of the PPUs are significantly higher than that determined for plasmodesmata interconnecting other plant cell types (Fisher et al. 1992). More recent studies established that GFP fusion proteins as large as 67 kDa traffic from CCs to SEs (Stadler et al. 2005), providing direct experimental evidence for the ability of PPUs to deliver macromolecules into the sieve tube. The fact that phloem sap contains numerous proteins originating in neighboring cells raised questions regarding the ability of these proteins to move long distance and the biological significance of such movement.

The ability of proteins to move long distance was demonstrated in Arabidopsis and tobacco plants expressing GFP under the companion cell-specific *AtSuc2* promoter (Imlau et al. 1999). The heterologous protein was accumulated in sink organs indicating its long-distance trafficking and unloading from sieve tubes in flowers, roots, and young leaves. That GFP is capable of long-distance movement was further demonstrated by plant–parasite interaction (Haupt et al. 2001; Nadler-Hassar et al. 2004). GFP that was expressed in tobacco plants under the *AtSuc2* promoter trafficked to *Cuscuta reflexa* phloem via absorbing hyphae and was unloaded in meristematic sink organs. This apparent exchange of proteins between host plants and parasites suggests an exciting mode of interspecies communication. Such experiments provide the impression that trafficking of proteins via the phloem is passive and directed from source to sink along the bulk flow. While this is the general notion, an elegant study performed by Aoki et al. (2005) provides experimental evidence for the ability of phloem proteins to also traffic against the bulk flow. Using a unique aphid-assisted method, pumpkin (*Cucurbita maxima*) phloem proteins were directly introduced into rice (*Oryza sativa*) sieve tubes. Tagged proteins were then detected and profiled in different sink tissues (roots and shoot meristems). Surprisingly, *C. maxima* PP1 (CmPP16-1) was found to move rootward, by a mechanism(s) distinct from bulk flow transport, dependent on its interaction with specific rice phloem proteins. This study established, for the first time, that control over protein long-distance trafficking via the phloem is selectively regulated by protein–protein interaction. It may, however, be significant that to date this is the only study showing directionality trafficking of phloem proteins in a mechanism dissimilar to the bulk flow.

Numerous studies published over the last 20 years characterized phloem sap trafficking proteins, their function, and the mechanism that allows for their mobility. In a preliminary attempt to explore these questions, Balachandran et al. (1997), using microinjection experiments, explored cell-to-cell transport of some phloem sap proteins between mesophyll cells. Injection of *C. maxima* phloem lectin (PP2), Knotted1, and *R. communis* cystatin and glutaredoxin along with size-fractionated protein samples was found to promote both cell-to-cell trafficking of large fluorescently labeled dextran molecules and long-distance transport of the injected proteins themselves. This phenomenon was attributed to an increase in plasmodesmatal size exclusion limit that subsequently allowed for the transport of large molecules, suggesting some interaction between a possible array of mobile long-distance proteins within the phloem sap and the plasmodesmata. Other phloem proteins that were found to interact with plasmodesmata and facilitate their own cell-to-cell transport are rice (*O. sativa*) thioredoxin h (Ishiwatari et al. 1998) and *R. communis* cyclophilin (Gottschalk et al. 2008).

Heterografting experiments provide a common means by which long-distance trafficking of endogenous phloem sap proteins may be demonstrated (Golecki et al. 1999). Correlating with the formation of a graft union, some 9 or 10 days after grafting, proteins originating from *C. maxima* could be isolated from the grafted *C. sativus* phloem sap. Similarly, using genus-specific antibodies against CmPP1 and CmPP2, it was shown that these proteins can traffic to grafted *C. sativus* via fascicular phloem. In contrast, *CmPP1* and *CmPP2* mRNAs were not found to be phloem-mobile, suggesting that the protein is the trafficking unit rather than a mobile mRNA that is further translated. Another protein found to be phloem-mobile is the serine protease inhibitor (Serpin) CmPS-1 (Petersen et al. 2005). This protein was found to be exclusively localized in sieve elements, accumulated in phloem exudates, and graft transmissible. The authors proposed that Serpin is essential for the protection of phloem from feeding insects that use serine proteases to degrade and digest plant proteins.

Perhaps the most characterized phloem-mobile protein is FLOWERING LOCUS T (FT), which is involved in flowering induction. It has long been established that plants respond to day length perceived by the leaves within which an unknown signal molecule (known as "florigen") is formed, trafficking to the shoot apical meristem where it induces the transition from vegetative to reproductive development (King and Zeevaart 1973). FT was found to be restricted to vascular tissue (Takada and Goto 2003) and, similar to florigen, its function was mainly assigned to the shoot apical meristem (Abe et al. 2005; Wigge et al. 2005). Direct correlation between FT expression and flowering induction was observed in Arabidopsis (Corbesier et al. 2007), rice (Tamaki et al. 2007), and pumpkin (Lin et al. 2007). Heterografting experiments established that the FT protein is capable of long-distance movement, and its accumulation in the shoot apex was associated with the transition to flowering. Interestingly, inhibition of long-distance trafficking of FT by nuclear signal peptide or by fusion to 3XYFP abolishes its influence over flower production in Arabidopsis (Jaeger and Wigge 2007; Mathieu et al. 2007) providing support for the notion that FT is a long-distance signaling molecule. Collectively, these studies show that long-distance FT trafficking via the phloem plays a significant role in communication between source leaves and developmental processes occurring in the shoot apex meristem. Notwithstanding, the complete signal transduction pathway starting in one tissue and ending in a distant meristem has yet to be explored. It seems reasonable to assume that more phloem sap proteins will be identified in the near future and characterized as long-distance signaling molecules that are involved in orchestrating developmental processes at the level of the whole plant.

An interesting group of proteins identified in phloem sap is that of RNA-binding proteins (RBPs) (Giavalisco et al. 2006; Lin et al. 2009). It is now generally assumed that RNA molecules do not move naked in the phloem but rather as part of an efficiently packed RNA protein (RNP) complex. The interaction of viroids with phloem proteins to enable their long-distance movement laid down the notion that viruses and viroids hitchhike on the plant's endogenous mechanism which is responsible for the delivery of endogenous RNA molecules (Lucas and Wolf 1993; Lough and Lucas 2006). Supporting this notion is the interaction of selected RBPs with viral RNA that was demonstrated in vitro. Hop stunt viroid (HSVd) was shown to bind to CsPP2 and CsLec17, phloem-mobile proteins that are presumably involved in the intraplant viroid transport (Gomez and Pallas 2004; Gomez et al. 2005). The first phloem sap protein that was found to interact with endogenous RNA molecules was the *Cucurbita maxima* PP16 (Xoconostle-Cazares et al. 1999). This protein was identified as a plant paralog of viral movement proteins, capable of binding its own mRNA resulting in movement of the RNA complex from rootstock to scion in a heterografted plant system. Another phloem sap RNA-binding protein able to translocate from source to sink tissues is the CmRBP50 (Ham et al. 2009), which is evolutionarily related to animal polypyrimidine tract-binding proteins. In addition to its RNA-binding properties, this protein also binds additional phloem sap proteins, providing the basis for a model in which phloem delivery of RNA molecule occurs via multiple interactions with an RBP-50-based RNP complex (Ham et al. 2009). The role of phloem sap proteins in the long-distance delivery of RNA molecules was further supported by the identification of the *Cucurbita maxima* SMALL RNA-BINDING PROTEIN1 (CmPSRP1). Interaction between CmPSRP1 and a 25-nucleotide ssRNA to enable cell-to-cell transport of this RNP complex suggests that long-distance transmission of silencing signals is controlled by specific interaction between unique phloem proteins and their small RNA partners (Yoo et al. 2004). The accumulated data support the notion that phloem sap RBPs are required for long-distance delivery of RNA molecules as part the plant's overall communication network. Future studies should clarify the exact mode by which these proteins act. For example, they may possibly be required for the delivery of special RNA molecules into the sieve tube via PPUs, targeting of RNPs into specific cellular cites, or the protection of RNA molecules from degradation.

Protein-mediated signal transduction commonly involves protein modifications such as phosphorylations and glycosylations. Therefore, it is logical to assume that as a long-distance signaling network, the phloem sap contains an array of both modified proteins and their modifiers. Indeed, a battery of phosphorylated proteins were identified in the phloem sap of rice (*O. sativa*) by Nakamura et al. (1993). Moreover, changes in the proportion of phosphorylated proteins were observed in response to alteration in light intensity. In another study, these authors demonstrated the activation of phloem sap protein phosphorylation by Ca^{2+} and Mg^{2+} and the inhibition of phosphorylation by K^{+} ions. Initial characterization of phosphorylated proteins revealed a 17-kDa protein as the most highly phosphorylated protein in rice phloem sap alongside a 65-kDa autophosphorylated protein (Nakamura et al. 1995). A recent molecular study establishes that in order for the interaction between *Nicotiana tabacum* NON-CELL-AUTONOMOUS PATHWAY PROTEIN1 (Nt-NCAPP1) and pumpkin phloem sap non-cell-autonomous proteins (NCAPs) to take place, phosphorylation and glycosylation of both protein partners is required (Taoka et al. 2007). A 36-amino acid residue motif was identified as the site on which these processes must occur. More importantly, microinjection into *N. benthamiana* mesophyll cells establishes that this 36-amino acid peptide is sufficient to enable cell-to-cell movement of proteins that cannot normally cross the interconnecting plasmodesmata between those cells. It should be noted that CmRBP50, the phloem sap RNA-binding protein mentioned earlier as involved in long-distance RNA trafficking, performs its role through the formation of a phloem protein complex in a phosphorylation-dependent manner (Li et al. 2011). A phosphorylation–glycosylation recognition motif thus functions to control the interaction between targeted phloem sap proteins and their specific chaperones; an interaction that, presumably, facilitates movement of the protein complex via PPUs into (and out of) the CC–SE complex for long-distance trafficking.

7 Concluding Remarks

In higher plants, plasmodesmata provide a major route for cell-to-cell communication. As early as 1930, Munch advanced the hypothesis that plasmodesmata establish the continuity of the cytosolic compartments of neighboring cells. This symplastic continuity together with long-distance transport tissue located in the vasculature conjointly provides a pathway for solute movement between distant plant organs. The fact that this symplastic continuum enables movement of macromolecules such as protein and RNA molecules sets the stage for the notion that certain macromolecules act within a communication system operating at whole-plant level. While the accumulated data establishes that the phloem stream contains thousands of macromolecules, it is still unclear which of these are translocated and, more importantly, whose function is dependent on their long-distance transport. To date only certain molecules were found to play a role in the control over developmental processes responding to environmental inputs such as day length or nutrient status. Future studies should aim at identifying additional macromolecules that act as signaling agents in the communication process between distant plant organs. A major challenge is expected in the identification of the cellular site in which these molecules are transcribed or translated, the molecular mechanism responsible for their delivery into and out of the sieve tube, their precise target cells, and their mode of action once reaching their target tissues.

Acknowledgments Research in the authors' laboratory was supported by the United States–Israel Binational Science Foundation (BSF 2007052) and by the Israel Science Foundation (ISF 380/06).

References

Abe M, Kobayashi Y, Yamamoto S, Daimon Y, Yamaguchi A, Ikeda Y, Ichinoki H, Notaguchi M, Goto K, Araki T (2005) FD, a bZIP protein mediating signals from the floral pathway integrator FT at the shoot apex. Science 309:1052–1056

Aoki K, Suzui N, Fujimaki S, Dohmae N, Yonekura-Sakakibara K, Fujiwara T, Hayashi H, Yamaya T, Sakakibara H (2005) Destination-selective long-distance movement of phloem proteins. Plant Cell 17:1801–1814

Asano T, Masumura T, Kusano H, Kikuchi S, Kurita A, Shimada H, Kadowaki K (2002) Construction of a specialized cDNA library from plant cells isolated by laser capture micro-dissection: toward comprehensive analysis of the genes expressed in the rice phloem. Plant J 32:401–408

Balachandran S, Xiang Y, Schobert C, Thompson GA, Lucas WJ (1997) Phloem sap proteins from *Cucurbita maxima* and *Ricinus communis* have the capacity to traffic cell to cell through plasmodesmata. Proc Natl Acad Sci USA 94:14150–14155

Barnes A, Bale J, Constantinidou C, Ashton P, Jones A, Pritchard J (2004) Determining protein identity from sieve element sap in *Ricinus communis* L. by quadrupole time of flight (Q-TOF) mass spectrometry. J Exp Bot 55:1473–1481

Behnke H-D (1989) Structure of the phloem. In: Baker DA, Milburn JA (eds) Transport of photoassimilates. Longman Scientific & Technical, Harlow, pp 79–137

Buhtz A, Springer F, Chappell L, Baulcombe DC, Kehr J (2008) Identification and characterization of small RNAs from the phloem of *Brassica napus*. Plant J 53:739–749

Buhtz A, Pieritz J, Springer F, Springer F, Kehr J (2010) Phloem small RNAs, nutrient stress responses, and systemic mobility. BMC Plant Biol 10:64

Corbesier L, Vincent C, Jang S, Fornara F, Fan Q, Searle I, Giakountis A, Farrona S, Gissot L, Turnbull C, Coupland G (2007) FT protein movement contributes to long-distance signaling in floral induction of Arabidopsis. Science 316:1030–1033

Diener TO (1979) Viroids: structure and function. Science 205:859–866

Ding X, Shintaku MH, Carter SA, Nelson RS (1996) Invasion of minor veins of tobacco leaves inoculated with tobacco mosaic virus mutants defective in phloem-dependent movement. Proc Natl Acad Sci USA 93:11155–11160

Ding B, Itaya A, Zhong X (2005) Viroid trafficking: a small RNA makes a big move. Curr Opin Plant Biol 8:606–612

Esau K (1968) Viruses in plant hosts: form, distribution, and pathogenic effects. The University of Wisconsin Press, Madison, WI

Esau K (1969) Encyclopedia of plant anatomy. Gebruder Borntraeger, Berlin

Esau K, Cronshaw J, Hoefert LL (1967) Relation of beet yellows virus to the phloem and to movement in the sieve tube. J Cell Biol 32:71–87

Fisher DB, Frame JM (1984) A guide to the use of the exuding-stylet technique in phloem physiology. Planta 161:385–393

Fisher DB, Wu Y, Ku MS (1992) Turnover of soluble proteins in the wheat sieve tube. Plant Physiol 100:1433–1441

Fukumorita T, Chino M (1982) Sugar, amino acid and inorganic contents in rice phloem sap. Plant Cell Physiol 23:273–283

Gamalei YV (1989) Structure and function of leaf minor veins in trees and herbs. A taxonomic review. Trees 3:96–110

Gaupels F, Buhtz A, Knauer T, Deshmukh S, Waller F, van Bel AJE, Kogel K-H, Kehr J (2008) Adaptation of aphid stylectomy for analyses of proteins and mRNA in barley phloem sap. J Exp Bot 59:3297–3306

Giavalisco P, Kapitza K, Kolasa A, Buhtz A, Kehr J (2006) Towards the proteome of *Brassica napus* phloem sap. Proteomics 6:896–909

Golecki B, Schulz A, Thompson GA (1999) Translocation of structural P proteins in the phloem. Plant Cell 11:127–140

Gomez G, Pallas V (2004) A long-distance translocatable phloem protein from cucumber forms a ribonucleoprotein complex in vivo with Hop stunt viroid RNA. J Virol 78:10104–10110

Gomez G, Torres H, Pallas V (2005) Identification of translocatable RNA-binding phloem proteins from melon, potential components of the long-distance RNA transport system. Plant J 41:107–116

Goodrick BJ, Kuhn CW, Hussey RS (1991) Restricted systemic movement of cowpea chlorotic mottle virus in soybean with nonnecrotic resistance. Phytopathology 81:1426–1431

Gottschalk M, Dolgener E, Xoconostle-Cazares B, Lucas WJ, Komor E, Schobert C (2008) *Ricinus communis* cyclophilin: functional characterisation of a sieve tube protein involved in protein folding. Planta 228:687–700

Graham LE (1993) Origin of land plants. Wiley, New York

Haebel S, Kehr J (2001) Matrix-assisted laser desorption/ionization time of flight mass spectrometry peptide mass fingerprints and post source decay: a tool for the identification and analysis of phloem proteins from *Cucurbita maxima* Duch. separated by two-dimensional polyacrylamide gel electrophoresis. Planta 213:586–593

Ham BK, Brandom JL, Xoconostle-Cazares B, Ringgold V, Lough TJ, Lucas WJ (2009) A polypyrimidine tract binding protein, pumpkin RBP50, forms the basis of a phloem-mobile ribonucleoprotein complex. Plant Cell 21:197–215

Haupt S, Oparka KJ, Sauer N, Neumann S (2001) Macromolecular trafficking between *Nicotiana tabacum* and the holoparasite *Cuscuta reflexa*. J Exp Bot 52:173–177

Imlau A, Truernit E, Sauer N (1999) Cell-to-cell and long-distance trafficking of the green fluorescent protein in the phloem and symplastic unloading of the protein into sink tissues. Plant Cell 11:309–322

Ishiwatari Y, Honda C, Kawashima I, Nakamura S, Hirano H, Mori S, Fujiwara T, Hayashi H, Chino M (1995) Thioredoxin h is one of the major proteins in rice phloem sap. Planta 195:456–463

Ishiwatari Y, Fujiwara T, McFarland KC, Nemoto K, Hayashi H, Chino M, Lucas WJ (1998) Rice phloem thioredoxin h has the capacity to mediate its own cell-to-cell transport through plasmodesmata. Planta 205:12–22

Ivashikina N, Deeken R, Ache P, Kranz E, Pommerrenig B, Sauer N, Hedrich R (2003) Isolation of AtSUC2 promoter-GFP-marked companion cells for patch-clamp studies and expression profiling. Plant J 36:931–945

Jaeger KE, Wigge PA (2007) FT protein acts as a long-range signal in Arabidopsis. Curr Biol 17:1050–1054

Kehr J, Buhtz A (2008) Long distance transport and movement of RNA through the phloem. J Exp Bot 59:85–92

King RW, Zeevaart JA (1973) Floral stimulus movement in perilla and flower inhibition caused by noninduced leaves. Plant Physiol 51:727–738

Kollmann R, Dörr I, Kleinig H (1970) Protein filaments-structural components of the phloem exudate. Planta 95:86–94

Li P, Ham BK, Lucas WJ (2011) CmRBP50 protein phosphorylation is essential for assembly of a stable phloem-mobile high-affinity ribonucleoprotein complex. J Biol Chem 286:23142–23149

Lin MK, Belanger H, Lee YJ, Varkonyi-Gasic E, Taoka K, Miura E, Xoconostle-Cazares B, Gendler K, Jorgensen RA, Phinney B, Lough TJ, Lucas WJ (2007) FLOWERING LOCUS T protein may act as the long-distance florigenic signal in the cucurbits. Plant Cell 19:1488–1506

Lin M-K, Lee Y-J, Lough TJ, Phinney BS, Lucas WJ (2009) Analysis of the pumpkin phloem proteome provides functional insights into angiosperm sieve tube function. Mol Cell Proteomics 8:343–356

Lough TJ, Lucas WJ (2006) Integrative plant biology: role of phloem long-distance macromolecular trafficking. Annu Rev Plant Biol 5:203–232

Lucas WJ, Wolf S (1993) Plasmodesmata: the intercellular organelle of green plants. Trends Cell Biol 3:308–315

Madore MA, Oross JW, Lucas WJ (1986) Symplastic transport in Ipomoea tricolor source leaves: demonstration of functional symplastic connections from mesophyll to minor veins by a novel dye-tracer method. Plant Physiol 82:432–442

Mathieu J, Warthmann N, Kuttner F, Schmid M (2007) Export of FT protein from phloem companion cells is sufficient for floral induction in Arabidopsis. Curr Biol 17:1055–1060

Moran PJ, Cheng Y, Cassell JL, Thompson GA (2002) Gene expression profiling of *Arabidopsis thaliana* in compatible plant-aphid interactions. Arch Insect Biochem Physiol 51:182–203

Munch E (1930) Die stoffbewegung in der pflanze. Gustav Fischer, Jena

Nadler-Hassar T, Goldshmidt A, Rubin B, Wolf S (2004) Glyphosate inhibits the translocation of green fluorescent protein and sucrose from a transgenic tobacco host to *Cuscuta campestris* Yunk. Planta 219:790–796

Nakamura S, Hayashi H, Mori S, Chino M (1993) Protein phosphorylation in the sieve tubes of rice plants. Plant Cell Physiol 34:927–933

Nakamura S, Hayashi H, Mori S, Chino M (1995) Detection and characterization of protein kinases in rice phloem sap. Plant Cell Physiol 36:19–27

Nakazono M, Qiu F, Borsuk LA, Schnable PS (2003) Laser capture microdissection, a tool for the global analysis of gene expression in specific cell types: identification of genes expressed differentially in epidermal cells or vascular tissues of maize. Plant Cell 15:583–596

Omid A, Keilin T, Glass D, Leshkowitz D, Wolf S (2007) Characterization of phloem-sap transcription profile in melon plants. J Exp Bot 58:3645–3656

Palukaitis P (1987) Potato spindle tuber viroid. Investigation of the long-distance intra-plant transport route. Virology 158:239–241

Pant BD, Buhtz A, Kehr J, Scheible WR (2008) MicroRNA399 is a long-distance signal for the regulation of plant phosphate homeostasis. Plant J 53:731–738

Peleg G, Malter D, Wolf S (2007) Viral infection enables phloem loading of GFP and long-distance movement of the protein. Plant J 51:165–172

Petersen MC, Hejgaard J, Thompson GA, Schulz A (2005) Cucurbit phloem serpins are graft-transmissible and appear to be resistant to turnover in the sieve element–companion cell complex. J Exp Bot 56:3111–3120

Pommerrenig B, Barth I, Niedermeier M, Kopp S, Schmid J, Dwyer RA, McNair RJ, Klebl F, Sauer N (2006) Common plantain. A collection of expressed sequence tags from vascular tissue and simple and efficient transformation method. Plant Physiol 142:1427–1441

Qi Y, Pelissier T, Itaya A, Hunt E, Wassenegger M, Ding B (2004) Direct role of a viroid RNA motif in mediating directional RNA trafficking across a specific cellular boundary. Plant Cell 16:1741–1752

Raven JA (1993) The evolution of vascular plants in relation to quantitative functioning of water-conducting cells and stomata. Biol Dev 68:337–363

Robards AW, Lucas WJ (1990) Plasmodesmata. Annu Rev Plant Physiol Plant Mol Biol 41:369–419

Ruiz-Medrano R, Xoconostle-Cazares B, Lucas WJ (1999) Phloem long-distance transport of CmNACP mRNA: implications for supracellular regulation in plants. Development 126:4405–4419

Scarpella E, Helariutta Y (2010) Vascular pattern formation in plants. Curr Top Dev Biol 91:221–265

Schobert C, Gottschalk M, Kovar DR, Staiger CJ, Yoo BC, Lucas WJ (2000) Characterization of *Ricinus communis* phloem profilin, RcPRO1. Plant Mol Biol 45:719–730

Schultz A (1998) Phloem: structure related to function. Cell Biol Physiol 59:429–475

Sjolund RD, Shin CY (1983) Freeze-fracture analysis of phloem structure in plant tissues culture. I. The sieve element reticulum. J Ultrastruct Res 82:111–121

Stadler R, Wright KM, Lauterbach C, Amon G, Gahrtz M, Feuerstein A, Oparka KJ, Sauer N (2005) Expression of GFP-fusions in *Arabidopsis* companion cells reveals nonspecific protein trafficking into sieve elements and identifies a novel postphloem domain in roots. Plant J 41:319–331

Takada S, Goto K (2003) Terminal flower2, an Arabidopsis homolog of heterochromatin protein1, counteracts the activation of flowering locus T by constans in the vascular tissues of leaves to regulate flowering time. Plant Cell 15:2856–2865

Tamaki S, Matsuo S, Wong HL, Yokoi S, Shimamoto K (2007) Hd3a protein is a mobile flowering signal in rice. Science 316:1033–1036

Taoka K, Ham BK, Xoconostle-Cazares B, Rojas MR, Lucas WJ (2007) Reciprocal phosphorylation and glycosylation recognition motifs control NCAPP1 interaction with pumpkin phloem proteins and their cell-to-cell movement. Plant Cell 19:1866–1884

Thompson JR, García-Arenal F (1998) The bundle sheath-phloem interface of *Cucumis sativus* is a boundary to systemic infection by tomato aspermy virus. Mol Plant Microbe Interact 11:109–114

Thompson GA, Schulz A (1999) Macromolecular trafficking in the phloem. Trends Plant Sci 4:354–360

Turgeon R, Beebe DU (1991) The evidence for symplastic phloem loading. Plant Physiol 96:349–354

Turgeon R, Webb JA, Evert RF (1975) Ultrastructure of minor veins in *Cucurbita pepo* leaves. Protoplasma 83:217–232

van Bel AJE (1993) Strategies of phloem loading. Annu Rev Plant Physiol Plant Mol Biol 44:253–281

van Bel AJE (1999) Evolution, polymorphology and multifunctionality of the phloem system. Perspect Plant Ecol Evol Syst 2:163–184

van Bel AJE, van Rijen HVM (1994) Microelectrode-recorded development of the symplasmic autonomy of the sieve element/companion cell complex in the stem phloem of *Lupinus luteus* L. Planta 192:165–175

Vilaine F, Palauqui J-C, Amselem J, Kusiak C, Lemoine R, Dinant S (2003) Towards deciphering phloem: a transcriptome analysis of the phloem of *Apium graveolens*. Plant J 36:67–81

Wang Y, Ding B (2010) Viroids: small probes for exploring the vast universe of RNA trafficking in plants. J Integr Plant Biol 52:28–39

Wigge PA, Kim MC, Jaeger KE, Busch W, Schmid M, Lohmann JU, Weigel D (2005) Integration of spatial and temporal information during floral induction in Arabidopsis. Science 309:1056–1059

Xoconostle-Cazares B, Xiang Y, Ruiz-Medrano R, Wang HL, Monzer J, Yoo BK, MacFarland KC, Franceschi VR, Lucas WJ (1999) Plant paralog to viral movement protein that potentiates transport of mRNA into the phloem. Science 283:94–98

Yoo BC, Kragler F, Varkonyi-Gasic E, Haywood V, Archer-Evans S, Lee YM, Lough TJ, Lucas WJ (2004) A systemic small RNA signaling system in plants. Plant Cell 16:1979–2000

Zhang S, Sun L, Kragler F (2009) The phloem-delivered RNA pool contains small noncoding RNAs and interferes with translation. Plant Physiol 150:378–387

Zhong X, Tao X, Stombaugh J, Leontis N, Ding B (2007) Tertiary structure and function of an RNA motif required for plant vascular entry to initiate systemic trafficking. EMBO J 26:3836–3846

Ziegler H (1975) Nature of transported substances in the phloem. In: Zimmermann MH, Milburn JA (eds) Encyclopedia of plant physiology, vol 1. Springer, Berlin, pp 59–100

Electrical Long-Distance Signaling in Plants

Matthias R. Zimmermann and Axel Mithöfer

Abstract In higher plants at least three different types of electrical long-distance signaling exist: action potential (AP), variation potential (VP), and system potential (SP), all of which have their own characteristics concerning their generation, duration, amplitude, velocity, and propagation. Whereas both AP and VP are due to a transient depolarization of the plasma membrane, the SP is based on hyperpolarization. For more than 100 years the AP is known and described for some specialized plants such as the Venus flytrap. Meanwhile, all three types of electrical signaling have been shown for many common plants, monocots as well as dicots, indicating that the capability to generate long-distance electrical signals is not the exception but a general physiological feature of plants. In spite of this, positive proofs for the involvement of these kinds of electrical signaling in the induction of many different plant responses to (a)biotic stresses or in developmental processes still wait to be established.

Keywords Action potential • Variation potential • System potential • Downstream signaling • DIR1

1 Introduction

1.1 Electrical Long-Distance Signals: Brief History and Presence

The unequal separation of charged molecules and ions across the plasma membrane to generate a membrane potential is usually due to the activity of plasma

M.R. Zimmermann
Botanisches Institut I, Justus-Liebig-Universität, Gießen, Germany

A. Mithöfer (✉)
Bioorganische Chemie, Max-Planck-Institut für Chemische Ökologie, Hans-Knöll-Str 8, Jena, Germany
e-mail: amithoefer@ice.mpg.de

F. Baluška (ed.), *Long-Distance Systemic Signaling and Communication in Plants*, Signaling and Communication in Plants 19, DOI 10.1007/978-3-642-36470-9_15,

membrane-located channels and pumps. The presence of a membrane potential is an essential feature for all kinds of living cells and necessary for the uptake of nutrients and minerals as well as to establish and regulate intracellular physiological processes. In addition, the interaction between different cells or different tissue and organs via an electrical signaling for communication over a long distance is a well-established phenomenon for animals and known for more than 200 years. Even for plants fast electrical signals were described about 140 years ago: In 1873, Burdon-Sanderson studied the leaf closing mechanism in the carnivorous Venus flytrap, *Dionaea muscipula*. In addition, Darwin (1875) investigated tentacle movements of sundew plants, *Drosera spp*. Some years later, Kunkel (1898) and Bose (1907) showed that in *Mimosa pudica* the fast leaf(let) movements depended on electrical signals. All those plants represent remarkable and very specialized species, which, thus, were highly attractive to the scientists. General studies on long-distance electrical signaling in other common plants were not very popular among physiologists for many decades although some studies suggested that upon wounding electrical signals may travel through phloem and/or xylem elements (Pickard 1973; Davies 1987; Rhodes et al. 1996). Fortunately, in recent years an increasing number of studies are published dealing with the phenomenon of electrical long-distance signaling in higher plants.

2 Electrical Signals in Plants

In animals only one genuine electrical signal is known, the action potential. In contrast, three different signal types have been identified in plants—(1) action potential, (2) variation potential, and (3) system potential; all of which will be described in the following.

2.1 Action Potential

The action potential (AP) is a transient depolarization of the plasma membrane with a very typical voltage signature that is interestingly found in animals as well as in plants. Plant- and animal-originated APs share fundamental characteristics: (1) the distinctive voltage signature, (2) the all-or-nothing law, (3) the ability of self-propagation, and (4) a refractory period, but they differ in (1) time frame and (2) molecular compounds involved (Pickard 1973; Beilby 2007; Pyatygin et al. 2008).

Most details about the complex and fine-adjusted molecular mechanism of APs were found for lower plants, e.g., *Acetabularia spec.* (Gradmann 1976), *Chara spec.* (Hope and Findlay 1964; Beilby 2007), *Nitella spec.* (Blatt 1974; Kikuyama 1987), and *Conocephalum conicum* (Dziubinska et al. 1983; Trebacz et al. 1989, 1994, 1997; Trebacz 1992). In higher plants, APs have been mainly examined in plants with motor activity, i.e., *Mimosa pudica, Drosera spp.*, or *Dionaea*

Table 1 Action potential characteristics and various examples for higher plants

Species	Stimulus	Amplitude (mV)	Time (s)	Velocity (cm min^{-1})	Setup	Location	Reference
Zea mays	Electrical/cold-shock	+50–80	30–120	180–300	Intra	Sieve tubes/leaf	Fromm and Bauer (1994)
Z. mays	Electrical/cold-shock			12–120	Intra	Mesophyll cells/leaf	Fromm and Bauer (1994)
Salix viminalis	Electrical	+30–50	10	120	Intra	Stem, cortex	Fromm and Spanswick (1993)
Vicia faba	Heat	+100–110			Intra	Sieve tubes/leaf	Furch et al. (2007)
Lycopersicon esculentum	Heat	+79			Intra		Rhodes et al. (1996)
Cucurbita pepo	Sucrose (100 mM)	+90	60		Intra	Sieve tubes/petiole	Eschrich et al. (1988)
C. pepo	Heat	+60	20–30		Intra	Sieve tubes/fruit	Eschrich et al. (1988)
L. esculentum	Heat	+50	120		Intra	Main vein	Herde et al. (1998)
Lupinus angustifolius	Electrical	−60	180	3.6–5.6	Extra	Stem	Paszewski and Zawadzki (1976)
Mimosa pudica	Heat	−74–60			Extra	Stem	Roblin (1985)
Salix viminalis	Electrical	−50		120	Extra	Stem, cortex	Fromm and Spanswick (1993)
Helianthus annuus	Electrical	−50	20–30	11	Extra	Stem, leaf	Dziubinska et al. (2001)
Arabidopsis thaliana	KCl (1 M) + prick	−43–8	60	0.78–9	Extra	Midrib and petiole/leaf	Favre et al. (2001)
Vicia faba	Heat	−76–56		6.3–12.1	Extra	Stem, leaf	Dziubinska et al. (2003)
C. pepo	Cooling	−120	120–600		Extra	Hypocotyl	Opritov et al. (2005)
Hordeum vulgare	KCl, $CaCl_2$, glutamic acid	−80–70	900	20–30	Extra	Substomatal cavity/leaf	Felle and Zimmermann (2007)

Note. Voltage changes of intracellular recordings are positive going (=depolarization) and all kinds of extracellular measurements show a negative shift of voltage (= "hyperpolarization"). Prefix indicates direction of voltage change. Extracellular (extra) recordings were executed with surface, blindly pierced, or substomatal-placed electrodes. Intracellular (intra) measurements were achieved with severed stylets of aphids or impaled microelectrodes. Velocities are converted to cm min^{-1}. Data are arranged to the category setup

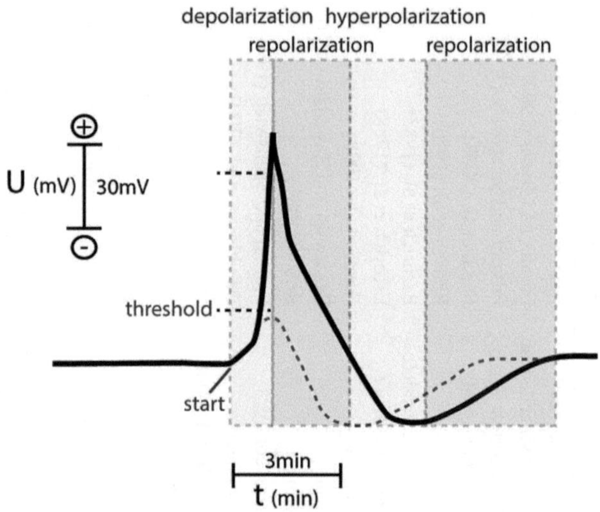

Fig. 1 Illustration of an intracellular recorded action potential (AP) in higher plants. An AP is separated into three steps: depolarization, repolarization, and hyperpolarization. The depolarization is again separated in three different steps. First, a minor slope is observed within the initiate phase of the depolarization. Passing the threshold, the typical rapid and steep depolarization is observed and persists merely a few seconds (app. 10 s). At the end of the depolarization, the kinetics is getting slower again (*dashed line*). The subsequent repolarization is divided into two steps separated by a distinct edge and fading to hyperpolarization. U = voltage

muscipula (Houwink 1935; Sibaoka 1969; Hodick and Sievers 1988; Volkov et al. 2010a, b). In addition, numerous studies with higher plant species without motor activities confirmed the general occurrence and characterized APs in plants (Table 1).

Up to now, diverse stimuli (touch, heat, cold, KCl, $CaCl_2$, glutamic acid, electrical current) have been identified to initiate the orchestrated molecular cascade of an AP (Fig. 1) via three different mechanisms: (1) a depolarization (KCl, electrical current), (2) an increase of apoplastic Ca^{2+} ($CaCl_2$), or (3) a ligand-binding receptor (glutamic acid; Felle and Zimmermann 2007). In each case, the consequence is a Ca^{2+} influx from the apoplast into the cytoplasm. This increase of cytosolic Ca^{2+} causes the opening of anion channels in the plasma membrane and, subsequently, an efflux of anions (mainly Cl^-) along the electrochemical gradient from the cytoplasm into the apoplast (Bradley and Williams 1966; Tarr et al. 1970; Lunevsky et al. 1983; Tsutsui et al. 1986; Felle and Zimmermann 2007). Interestingly, Mg^{2+} is not able to replace Ca^{2+} confirming the specific role of Ca^{2+} during an AP (Tarr et al. 1970). Only if a certain threshold is passed all anion channels will open and the characteristic rapid and strong depolarization ("break-through") can be observed (Fig. 1). The massive loss of negative charges affects the membrane potential and the Nernstian potential of potassium. This results in a passively driven K^+ efflux from the cytoplasm into the apoplast (Tsutsui et al. 1986; Felle and Zimmermann 2007). At the end of the depolarization phase when anion- and K^+-

efflux is balanced and the H^+-ATPases start working, the peak of the depolarization is reached (Fig. 1). To some extent, the single phases are reflected in the voltage pattern of the depolarization phase with its typical kinetics (Fig. 1). The repolarization is based upon an active symport of H^+/Cl^- and a passively driven K^+-influx from the apoplast into the cytoplasm (Felle and Zimmermann 2007).

The reported rate of transmission (velocity) ranges from 0.78 to 300 cm min^{-1} (Table 1) suggesting differences in the electrical coupling within diverse plant species. In general, the phloem is supposed to be the main responsible cellular translocation pathway for an AP within a plant (Eschrich et al. 1988; Fromm and Bauer 1994). Sieve elements and companion cells exhibit a relative high electrical coupling via abundant cell-to-cell contacts (plasmodesmata; van Bel and Ehlers 2005). Direct local and systemic measurements of APs within sieve elements supported the thesis (Eschrich et al. 1988). But also the apoplast was suggested to have a crucial role in long-distance transmission of APs (Herde et al. 1998; Felle and Zimmermann 2007). Without doubt, plants have to cope with the problem of a missing electrical shield around the phloem (vascular system). The consequence is a leakage current and, thus, a longitudinal loss of voltage (see also cable theory; Adam et al. 2009). In mammalian, a myelin sheath acts as an electrical shield. That handicap might be compensated in plants—at least in part—by an intensive interplay of the phloem and the corresponding apoplast.

Finally, any AP-originated disturbance of the membrane potential has to be reset to the original conditions. During this time frame, called (absolute/relative) refractory period, cells are restricted for a next excitation. The refractory period is characteristic of APs. In the literature, a time range from 50 s (*Mimosa pudica*) to 2 h is reported indicating the refractory period to be plant specific (Paszewski and Zawadzki 1976; Fromm and Spanswick 1993; Fromm and Bauer 1994).

2.2 Variation Potential

The variation potentials (VPs), also known as slow wave potentials (SWPs), are transient depolarizations of the plasma membrane with variable shape, amplitude, and time fame (Fig. 2; Table 2) (Houwink 1935; Sibaoka 1953; Roblin and Bonnemain 1985; Stahlberg and Cosgrove 1996). VPs have been reported for a wide range of herbaceous plants (Table 2) and have been detected solely or in combination with APs (Roblin 1985; Furch et al. 2007). Diverse mechanical/physical stimuli such as heat, wounding (cutting, excision), and pressure as well as chemical factors (Ricca's factor, cyanide, aziden 2,4-dinitrophenol, sodium cholate) have been identified as external stimuli (Table 2) (Ricca 1916; van Sambeek and Pickard 1976b; Stahlberg and Cosgrove 1997a; Stanković et al. 1997). The observed inconsistent but characteristic voltage pattern is a result of the underlying molecular mechanism.

The depolarization of VPs results primarily from an inhibition of the plasma membrane H^+-ATPases. It was shown that during a VP, no changes in cell input

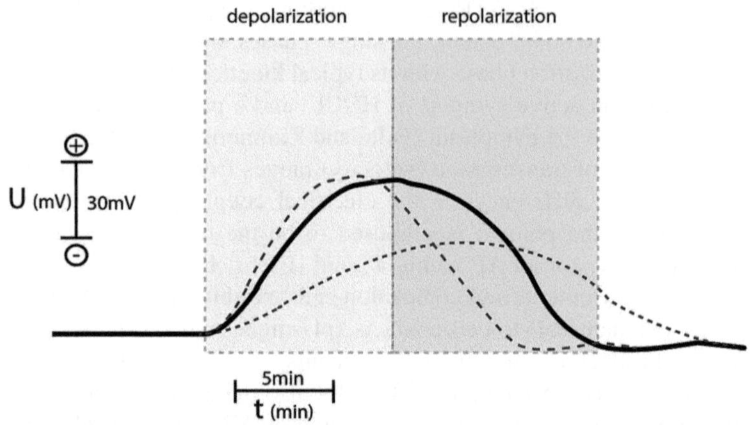

Fig. 2 Illustration of an intracellular recorded variation potential (VP) in higher plants. VPs reflect a wave-like, transient depolarization with variable shape and duration displayed here with three different voltage patterns (*continuous, dashed, and dotted lines*). Amplitude and duration depend on the distance between stimulus and recording site. A subsequent hyperpolarization can be observed to some extent. U = voltage

resistance occurred, which actually is typical for APs, and a decreased activity of proton pumps was concluded (Julien et al. 1991; Stahlberg and Cosgrove 1992). Congruously, a study with diverse proton pump inhibitors showed voltage patterns (Stahlberg and Cosgrove 1996) that are comparable with those evoked by mechanical stimuli. In addition, heat, cutting, or exogenous pressure application induces an alteration of the hydraulic pressure conditions within the xylem/apoplast via an import of energy (heat, pressure) or an opening of the vascular system (cutting). Indeed, changes of turgor pressure could be measured locally and systemically following application of heat or exogenous pressure (Malone and Stanković 1991; Malone 1992). Moreover, a combined determination of stem length and leaf thickness via transducers as a sign of hydraulic pressure alteration and electrical recordings showed a temporal correlation (Malone 1992; Stanković et al. 1997). The change of hydraulic pressure propagates in terms of a diminished wave along the xylem vessels and affects the surrounding cells by an inhibition of the proton pumps in the plasma membrane causing a transient depolarization. An essential turgor sensitivity of proton pumps was reported earlier (ref. see Stahlberg and Cosgrove 1992, 1996). The decline of proton pump activity results in reduced H^+-extrusion from the cytoplasm into the xylem/ apoplast. There, the consequent alkalization could be determined with a pH-sensitive fluorescent dye, for example, in *Pisum sativum* from pH 4.5 to 6.0 (Stahlberg and Cosgrove 1996); this is additionally supported by the observed pH dependence of VPs (Julien et al. 1991). So VPs reflect the local electrical consequence of a hydraulic pressure wave running along xylem/apoplast (Malone 1992; Stahlberg et al. 2005).

The variable strength (amplitude and duration, time frame) of a VP is positively correlated with the force of the locally affecting hydraulic pressure wave. It could

Table 2 Variation potential characteristics and various examples for higher plants

Species	Stimulus	Amplitude (mV)	Time (min)	Velocity (cm min^{-1})	Setup	Location	Reference
Pisum sativum	Wound	+40–45	5–8	6	Intra	Epidermal cells	Stahlberg and Cosgrove (1992)
Cucumis sativus	Pressure	−100–40	20–30	0.5–3	Intra	Epidermal cells	Stahlberg and Cosgrove (1997b)
Vicia faba	Heat	+90	20		Intra	Sieve tube/leaf	Furch et al. (2007)
Cucurbita maxima	Heat	+50–75	20		Intra	Sieve Tube/leaf	Furch et al. (2010)
Lycopersicon esculentum	Heat	−40–30	3–8		Extra	Stem, leaflet	van Sambeek and Pickard (1976a)
L. esculentum	Crude extract + wound	−35			Extra	Rhachis	van Sambeek and Pickard (1976b)
L. esculentum	Mannitol (1 M) + wound	−12			Extra		van Sambeek and Pickard (1976b)
L. esculentum	KCl (1 M) + wound	−22			Extra		van Sambeek and Pickard (1976b)
L. esculentum	KCl (0.1 M) + wound	−7			Extra		van Sambeek and Pickard (1976b)
Mimosa pudica	Heat	−82–52			Extra	Stem	Roblin (1985)
Vicia faba	Heat	−59–17		9	Extra	Stem	Roblin (1985)
L. esculentum	Heat	−54–33			Extra	Stem	Roblin (1985)
Triticum durum	Heat	−20–10	10–15		Extra	Leaf	Malone and Stanković (1991)
P. sativum	Wound	−35	5–8	6	Extra	Epidermal cells	Stahlberg and Cosgrove (1992)
Triticum aestivum	Heat	20		600	Extra	Leaf	Malone (1992)
P. sativum	Pressure/heat	−80–60	5–10		Extra	Epicotyl	Stahlberg and Cosgrove 1996
L. esculentum	Heat	−23			Extra		Rhodes et al. (1996)
Helianthus annuus	Pressure + wound	−50–10		5–100	Extra	Stem	Stanković et al. (1997)
P. sativum	Pressure	−80–40	5	2–3	Extra	Stem, hypocotyl	Stahlberg and Cosgrove (1997b)
Helianthus annuus	Heat	−100	8–12	5–25	Extra	Stem, leaf	Dziubinska et al. (2001)
Arabidopsis thaliana	KCl (1 M) + pricks			1.8	Extra	Midrib and petiole/leaf	Favre et al. (2001)
Vicia faba	Heat	−100–20		5.1	Extra	Stem, leaf	Dziubinska et al. (2003)

(continued)

Table 2 (continued)

Species	Stimulus	Amplitude (mV)	Time (min)	Velocity (cm min^{-1})	Setup	Location	Reference
Nicotiana tabacum	Heat	−25–10	30–60	90–120	Extra	Leaf	Hlavácková et al. (2006)
C. maxima	Heat	−115–66	60	0.6	Extra	Vein, leaf	Furch et al. (2010)
C. maxima	Wound	−52–38	60–180	240–480	Extra	Vein/petiolus	Zimmermann et al. (2013)
C. maxima	Wound	−28–16	60–240	240–480	Extra	Vein/petiolus	Zimmermann et al. (2013)

Note. Voltage changes of intracellular recordings are positive going (=depolarization) and all kinds of extracellular measurements show a negative shift of voltage (= "hyperpolarization"). Prefix indicates direction of voltage change. Extracellular (extra) recordings were executed with surface, blindly pierced, or substomatal-placed electrodes. Intracellular (intra) measurements were achieved with severed stylets of aphids or impaled microelectrodes. Velocities are converted to cm min^{-1}. Wound includes damage, excision and cutting. Data are arranged to the category setup

be observed that an increasing exogenous pressure application (30–100 kPa) at the root of *Pisum sativum* provoked stronger shapes of VPs, and simultaneously, a negative regression analysis displayed a decline of VPs with an increase of distance to the given continuous stimulus (50 kPa). The related radial leak of xylem pressure was calculated to be approximately 4 kPa cm^{-1} and is in direct context to the strength of a VP (Stahlberg and Cosgrove 1997a).

Between the occurrence of a hydraulic pressure wave and the VP, a lag time was observed depending on the strength of hydraulic pressure (Malone and Stanković 1991; Malone 1992; Stanković et al. 1997; Stahlberg et al. 2005). A faster stimulation of a VP was measured with stronger hydraulic pressure explaining why a wide range of measured velocities can be found in the literature (Table 2) due to the fact that the rate of propagation depends on the distance between stimulus and recording site and declined with an increasing distance (Stanković et al. 1997; Hlavácková et al. 2006). Here, it has to be noticed that in contrast to APs, VPs are not able to self-propagate excluding them to be a genuine long-distance signal (Stahlberg and Cosgrove 1997a). That aspect is additionally supported by the results that chemically induced VPs via proton pump inhibitors could be merely measured at the site of application with the exception of Ricca's factor and sodium cholate (Ricca 1916; Stahlberg and Cosgrove 1997a).

Concerning physical stimuli, it is likely that at least mechanosensitive channels are also influenced during a VP. Depolarization with various proton pump inhibitors did not evoke APs, thus indicating a participation of mechanosensitive Ca^{2+} channels during VPs only (Stahlberg and Cosgrove 1997b). The observed stronger Ca^{2+} release was arrestingly demonstrated by the Ca^{2+}-dependent forisome dispersion within sieve elements of *Vicia faba* plants after a heat stimulus (Furch et al. 2007, 2009).

The transmission of VPs was demonstrated for both directions—acropetal and basipetal (van Sambeek and Pickard 1976a)—and is even able to pass dead tissue being in good accordance with the concept of a running hydraulic wave along xylem vessels (Roblin 1985; Roblin and Bonnemain 1985). However, a recent study observed discrimination between the acro- and basipetal pathway. Following a cut at a petiole of *Cucurbita maxima* plants, the recorded VPs were stronger at the basal side (basipetal direction) (Zimmermann et al. 2013). In addition, it could be shown that the electrical reaction after application of heat was strongest in phloem cells (sieve elements and companion cells) in comparison to other cell types, i.e., epidermal, cortical, and vascular parenchyma cells. This may be founded by the intensive interaction of xylem and phloem (Eschrich et al. 1988; Rhodes et al. 1996).

Heat is a favored stimulus to provoke electrical signals (see Tables 1 and 2). After a heat stimulus, a combined spatiotemporal appearance of APs and VPs has been reported (Roblin 1985; Furch et al. 2007, 2009). Typically, electrical recordings close to the stimulus site are characterized by an overlap of AP and VP. This phenomenon complicates an exact determination of the particular signaling type and is referred to as electropotential wave (EPW; Fig. 3) (Furch et al. 2007). Any increase of the distance between stimulus and recording site thus

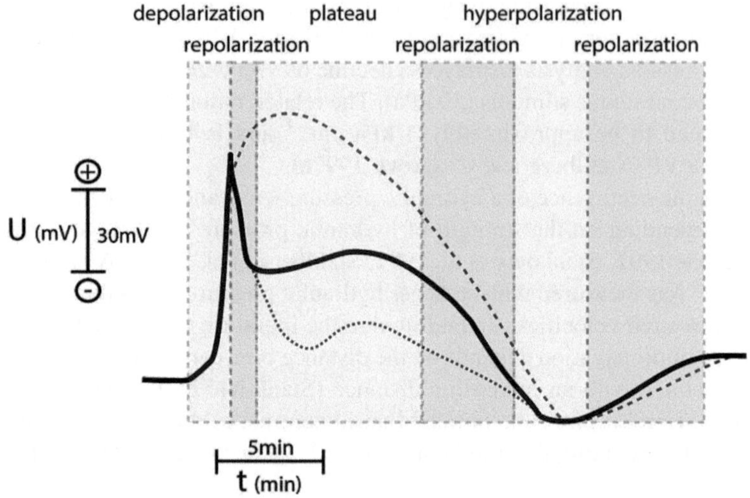

Fig. 3 Illustration of an intracellular recorded electropotential wave (EPW) in higher plants. An EPW reflects a mixed potential of AP and VP, generally observed after a heat stimulus close to the application site. Voltage pattern of EPWs is also variable and depends on the distance to the stimulus site, indicated here with different lines (*continuous, dashed, and dotted lines*). With increasing distance to the stimulus site, VPs are diminished and the voltage pattern of APs is solidified. U = voltage

strongly decreases the part of VP and in many cases the AP remains. Consequently, systemic measurements in a distant leaf only showed an AP (Zimmermann and Felle 2009).

2.3 System Potential

System potentials (SPs), in contrast to APs and VPs, reflect a systemic self-propagating hyperpolarization of the plasma membrane or depolarization of apoplastic voltage, respectively (Fig. 4) (Zimmermann et al. 2009). The term SP considers the striking fact this electrical signals can be reliably recorded on systemically, in contrast to APs and VPs. An occurrence of SPs was demonstrated for monocots as well as for dicots indicating a general ability of higher plants to transmit hyperpolarizations systemically (Zimmermann et al. 2009; Zimmermann 2010).

The combined application of wounding (cutting) and different chemical substances (e.g., glutamic acid, $LiCl_2$, $CuCl_2$, aspartic acid, glutamate) as well as stimulation with heat (scorching) evoked SPs (Zimmermann 2010). However, closer analysis was executed with diverse salt solutions (KCl, NaCl, $MgCl_2$, and $CaCl_2$). SPs were systemically recorded with a noninvasive approach of substomatal-placed microelectrodes (Felle et al. 2000; Zimmermann et al. 2009)

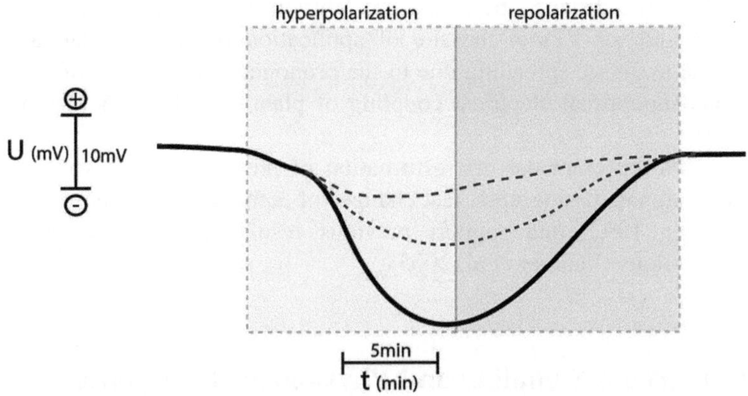

Fig. 4 Illustration of an intracellular recorded system potential (SP) in higher plants. SPs are self-propagating hyperpolarizations with variable shapes displayed here by different line types (*continuous, dashed, and dotted*). The slope of the hyperpolarization is mainly steeper than the repolarization. U = voltage

by giving a stimulus at one leaf and recording of the electrical reaction at a distant leaf. A constant propagation rate of SPs was detected with 5–6 cm min^{-1} for both directions acropetal and basipetal. The determined range of velocity indicates rather an electrotonic propagation of SPs than a possible induction of wound-allocated electrogenic substances, which has been mentioned for VPs and was determined to be approximately 0.8–1.7 cm min^{-1} (Canny 1975).

Studies with various chemical stimuli (KCl, NaCl, $MgCl_2$, and $CaCl_2$) at different concentrations (10–100 mM) showed that (1) cations rather than anion trigger SPs, (2) shapes of SPs are adapted to the mode of the stimulus, and (3) strength (in terms of concentration) of the stimulus; all of which indicate the possibility of encoding more sophisticated information (Zimmermann et al. 2009).

An analysis of diverse ion activities (Ca^{2+}, K^+, H^+, Cl^-) within the apoplast of substomatal cavity during a SP again indicated the participation of proton pumps. This hypothesis was proven with the application of fusicoccin, a toxin of the phytopathogenic fungus *Fusicoccum amygdali,* causing an activation of proton pumps (Marrè 1979; Hager 2003). Besides the induction of SPs by fusicoccin, the transmission of the electrical signal could be prevented with the application of the proton pump inhibitor vanadate (Zimmermann et al. 2009). Hence, these results suggest that the activated status of proton pumps may be relayed from cell to cell reasoning the self-propagation of SPs. Although the same applied stimuli are well known to provoke APs (i.e., glutamic acid, KCl, $CaCl_2$) or VPs (heat), SPs were systemically recorded in the most cases, indicating a higher probability of long-distance transmission for SPs in comparison to APs and VPs (Zimmermann 2010). However, it seems contradictory that the same stimuli evoke different electrical signaling types; but the depolarizations of APs and VPs can be seen as "regulated" disturbances of the membrane potential that has to be recovered with the activation of proton pumps; the latter, on the other hand, possess a sufficient electrical coupling for long-distance transmission. Without a doubt, the diverse stimuli trigger APs and/ or VPs at the site of application but the depolarization gets "lost" on the systemic spreading due to the pronounced decrement of VPs and/or the obvious suboptimal electrical coupling of plant cells for APs (Zimmermann et al. 2009).

Hence, SPs represent the first affirmation of previously proposed concept of propagating signals basing upon fast changes of active pumps (refs. see Stahlberg and Cosgrove 1992) and support previous results of systemically recorded hyperpolarizations (Lautner et al. 2005).

3 Downstream Signaling and Physiological Responses

The evidence of electrical signals in plant tissues per se, generated upon artificial treatments such as heat or high salt concentrations, does not necessarily mean that those signals are of physiological relevance. However, up to now various studies demonstrated an involvement of electrical signals in various physiological reactions in higher plants (Table 3). Numerous indications were given for a proper role of electrical signals within intra- and distant intercellular communication and for the regulation of physiological processes at the molecular, cellular, and the organism level (Davies 1987; Fromm and Bauer 1994; Fromm and Lautner 2007). The most obvious results were obtained for plants with motor activity, i.e., tentacle bending in *Drosera spp.*, or leaf movements in *Mimosa pudica* or *Dionaea muscipula*. The necessity of mechanically triggered action potentials for the induction of rapid, thigmonastic leaf movement has been well established (Braam 2004; Volkov et al. 2007, 2009, 2010b). All these motor activities depend on the generation of an AP (Sibaoka 1969; Williams and Pickard 1972; Fromm and Eschrich 1988).

Heat induces both AP and VP that accompany the enhancement of systemic ethylene emission, for example, in leaves of *Vicia faba* seedlings (Dziubinska et al. 2003). Other studies focused on general metabolic processes (photosynthesis respiration, gas exchange, stomata movements) following diverse abiotic stimuli (Table 3).

Encouraging results have been published in the field of plant defense supporting the idea of electrical signaling as an integral signaling event in the initiation of defensive reactions (Stahlberg and Cosgrove 1994; Favre et al. 2001). Electrical signals have been shown to cause effects in systemic leaves, for example, the regulation of various genes (Graham et al. 1986; Wildon et al. 1992; Stanković and Davies 1997; Herde et al. 1998). In tomato (*Lycopersicon esculentum)*, the first results described that proteinase inhibitor (*pin*) as well as calmodulin genes were upregulated due to wounding and heat stimuli. Consequentially, plants that generated no electrical signal could not accumulate pin mRNA (Stanković and Davies 1997). Moreover, a heat-induced and VP-mediated accumulation of proteinase inhibitor genes (*pin II*) and jasmonic acid was described in potato (*Solanum*

Table 3 Electrical signal types and proposed mediated physiological impacts

Species	Stimulus	Signal type	Physiological reaction	Reference
Dionaea muscipula	Mechanical	AP	Trap closure, delivery of enzymes for digestion	Sibaoka (1969)
Drosera spec.	Mechanical	AP	Movement of tentacles to catch insects	Williams and Pickard (1972)
Mimosa pudica	Cold shock, mechanical	AP	Regulation of leaf movement	Fromm and Eschrich (1988), Sibaoka (1969)
Salix viminalis	Wounding	AP	Transpiration and photosynthesis	Fromm and Eschrich (1993)
Zea mays	Irrigation	AP	Increase of gas exchange	Fromm and Fei (1998)
Z. mays	Cold shock	AP	Reduction of phloem transport	Fromm and Bauer (1994)
L. esculentum	Electrical	AP	Induction of *pin2* gene expression	Stanković and Davies (1996)
Cucumis sativus	Wounding	VP	Growth inhibition	Stahlberg and Cosgrove (1994)
Pisum sativum	Wounding	VP	Drop of growth rate	Stahlberg and Cosgrove (1992)
Solanum tuberosum	Heat	VP	Induction of jasmonic acid biosynthesis and *pin2* gene expression	Fisahn et al. (2004)
Populus nigra	Heat	VP	Transient reduction of photosynthesis	Lautner et al. (2005)
Nicotiana tabacum	Heat	VP	Systemic induced stomatal closure and increase of ABA and JA	Hlavácková et al. (2006)
Bidens pilosa	Heat	AP/VP	Calmodulin mRNA accumulation	Vian et al. (1996)
Vicia faba	Heat	AP/VP	Induce systemic enhancement of ethylene emission	Dziubinska et al. (2003)
Vicia faba	Heat	EPW	Occlusion of sieve elements via forisome dispersion	Furch et al. (2007, 2009)
Cucurbita maxima	Heat	EPW	Occlusion of sieve elements via proteins	Furch et al. (2010)
L. esculentum	Mechanical wounding	?	Accumulation of mRNA encoding proteinase inhibitor	Wildon et al. (1992)
L. esculentum	Heat	?	Accumulation of PIN2, JA; involvement of ABA	Herde et al. (1998)

Wounding includes damage, excision, and cutting. Data are arranged to the category signal type. *AP* action potential, *VP* variation potential, *?* electrical signal is not specified

tuberosum) plants (Fisahn et al. 2004). Proteinase inhibitors negatively affect the digestion of herbivores and jasmonic acid is a well-established regulator within herbivory-induced plant defense responses (Mithöfer and Boland 2012). A previous report of a general decline of phloem content translocation (Fromm and Bauer

1994) has been specified in recent studies where a distinct relationship of transmitted electrical signals and the release of Ca^{2+} into sieve elements were shown. The consequence was the occlusion of sieve elements via Ca^{2+}-dependent forisome dispersion in *Vicia faba* plants or protein plugs in *Cucurbita maxima* (Furch et al. 2007, 2009, 2010). Sealing of sieve elements can be seen as an early response to pathogen and/or herbivore attack to prevent transmission of attacker-released toxins and loss of valuable phloem sap.

The biological significance of SPs as a common signal asks for a natural trigger and an involvement within distinct physiological processes. First indications were found following feeding of herbivorous lepidopteran caterpillars (*Spodoptera littoralis*). Various hyperpolarizations could be recorded in distant leaves indicating SPs. Moreover, preliminary results suggest a systemic decrease of jasmonic acid accumulation after induction of SPs (Zimmermann 2010).

4 Conclusions and Outlook

For higher plants, the existence of electrical long-distance signals is well established and accepted. The three types of electrical signaling, AP, VP, and SP, have been demonstrated for various plants, monocots as well as dicots. Regardless of how the different types of electrical signals are generated, realized, and propagated, this indicates that this kind of signaling is widespread and a general feature in plants, though the best-noticed studies have been done with specialized plants such as the carnivorous *Dionaea muscipula* or touch-sensitive *Mimosa pudica*.

Although there is a growing body of evidence strongly suggesting that electrical signals in plants are indeed constituents of abiotic and biotic stress-induced signaling cascades, a final proof is still missing. Unfortunately, still very often the applied stimuli used to induce electrical signals are artificial and far from being physiological. There is no question that such experiments clearly showed the ability of plants to generate electrical signals, but it is difficult to generally conclude that these signals are naturally occurring and involved in plant's physiology, particularly in the communication between distant cells or organs. Thus, in order to understand electrical long-distance signaling in plants, the elucidation of a signaling cascade that includes or depends on an electrical signal and leads to the initiation of a defined physiological response is still a task that urgently needs to be addressed. For sure, this will remain a major scientific challenge for plant physiologists in the future.

References

Adam G, Läuger P, Stark G (2009) Physikalische Chemie und Biophysik. Springer, Heidelberg

Beilby MJ (2007) Action potential in charophytes. Int Rev Cytol 257:43–82

Blatt FJ (1974) Temperature dependence of the action potential in *Nitella flexilis*. Biochim Biophys Acta 339:382–389

Bose JCH (1907) Plant response as a means of physiological investigation. Longman, Green & Co, London

Braam J (2004) In touch: plant responses to mechanical stimuli. New Phytol 165:373–389

Bradley J, Williams EJ (1966) Chloride electrochemical potentials and membrane resistances in *Nitella translucens*. J Exp Bot 18:241–253

Burdon-Sanderson J (1873) Note on the electrical phenomena which accompany irritation of the leaf of *Dionaea muscipula*. Proc R Soc Lond 21:495–496

Canny MJP (1975) Mass transfer. In: Zimmermann HM, Milburn JA (eds) Encyclopedia of plant physiology. Springer, Berlin, pp 139–153

Darwin CR (1875) Insectivorous plants. John Murray, London

Davies E (1987) Action potentials as multifunctional signals in plants: a unifying hypothesis to explain apparently disparate wound responses. Plant Cell Environ 10:623–631

Dziubinska H, Paszewski A, Trebacz K, Zawadzki T (1983) Electrical activity of the liverwort *Conocephalum conicum*: The all-or-nothing law, strength-duration relation, refractory periods and intracellular potentials. Physiol Plantarum 57:279–284

Dziubinska H, Trebacz K, Zawadzki T (2001) Transmission route for action potentials and variation potentials in *Helianthus annuus* L. J Plant Physiol 158:1167–1172

Dziubinska H, Filek M, Koscielniak J, Trebacz K (2003) Variation and action potentials evoked by thermal stimuli accompany enhancement of ethylene emission in distant non-stimulated leaves of *Vicia faba* minor seedlings. J Plant Physiol 160:1203–1210

Eschrich W, Fromm J, Evert RF (1988) Transmission of electric signals in sieve tubes of zucchini plants. Bot Acta 101:327–331

Favre P, Greppin H, Agosti RD (2001) Repetitive action potentials induced in *Arabidopsis thaliana* leaves by wounding and potassium chloride application. Plant Physiol 39:961–969

Felle HH, Zimmermann MR (2007) Systemic signalling in barley through action potentials. Planta 226:203–214

Felle HH, Hanstein S, Steinmeyer R, Hedrich R (2000) Dynamics of ionic activities in the apoplast of the sub-stomatal cavity of intact *Vicia faba* leaves during stomatal closure evoked by ABA and darkness. Plant J 24:297–304

Fisahn J, Herde O, Willmitzer L, Peña-Cortés H (2004) Analysis of the transient increase in cytosolic Ca^{2+} during the action potential of higher plants with high temporal resolution: requirement of Ca^{2+} transients for induction of jasmonic acid biosynthesis and PINII gene expression. Plant Cell Physiol 45:456–459

Fromm J, Bauer T (1994) Action potentials in maize sieve tubes change phloem translocation. J Exp Bot 45:463–469

Fromm J, Eschrich W (1988) Transport processes in stimulated and non-stimulated leaves of *Mimosa pudica*. Trees 2:18–24

Fromm J, Eschrich W (1993) Electric signals released from roots of willow *Salix viminalis* L. change transpiration and photosynthesis. J Plant Physiol 141:673–680

Fromm J, Fei HM (1998) Electrical signaling and gas exchange in maize plants of drying soil. Plant Sci 132:203–213

Fromm J, Lautner S (2007) Electrical signals and their physiological significance in plants. Plant Cell Environ 30:249–257

Fromm J, Spanswick R (1993) Characteristics of action potentials in willow (*Salix viminalis* L.). J Exp Bot 44:1119–1125

Furch ACU, Hafke JB, Schulz A, van Bel AJE (2007) Ca^{2+}-mediated remote control of reversible sieve tube occlusion in *Vicia faba*. J Exp Bot 61:3697–3708

Furch ACU, van Bel AJE, Fricker MD, Felle HH, Fuchs M, Hafke JB (2009) Sieve element Ca^{2+} channels as relay stations between remote stimuli and sieve tube occlusion in *Vicia faba*. Plant Cell 21:2118–2132

Furch ACU, Zimmermann MR, Will T, Hafke JB, van Bel AJE (2010) Remote-controlled stop of phloem mass flow by biphasic occlusion in *Cucurbita maxima*. J Exp Bot 61:3697–3708

Gradmann D (1976) "Metabolic" action potentials in *Acetabularia*. J Membr Biol 29:23–45

Graham JS, Hall G, Pearce G, Ryan CA (1986) Regulation of proteinase inhibitors I and II mRNAs in leaves of wounded tomato plants. Planta 169:399–405
Hager A (2003) Role of the plasma membrane H^+-ATPase in auxin-induced elongation growth: historical and new aspects. J Plant Res 116:483–505
Herde O, Peña-Cortés H, Willmitzer L, Fisahn J (1998) Remote stimulation by heat induces characteristic membrane-potential responses in the veins of wild-type and abscisic acid-deficient tomato plants. Planta 206:146–153
Hlavácková V, Krchnák P, Naus J, Novák O, Spundová M, Strnad M (2006) Electrical and chemical signals involved in short-term systemic photosynthetic responses of tobacco plants to local burning. Planta 225:235–244
Hodick D, Sievers A (1988) The action potential of *Dionaea muscipula* Ellis. Planta 174:8–18
Hope AB, Findlay GP (1964) The action potential in *Chara*. Plant Cell Physiol 5:377–379
Houwink AL (1935) The conduction of excitation in *Mimosa pudica*. Recueil des Travaux Botaniques Neerlandais 32:51–91
Julien JL, Desbiez MO, de Jeagher G, Frachisse JM (1991) Characteristics of the wave of depolarization induced by wounding in *Bidens pilosa*. J Exp Bot 42:131–137
Kikuyama M (1987) Ion efflux during a single action potential of *Nitella axilliformis* in medium lacking Ca^{2+}. Plant Cell Physiol 28:179–186
Kunkel KAJ (1898) Über elektromotorische Wirkungen an unverletzten lebenden Pflanzenteilen. Arbeiten des Botanischen Instituts Würzburg 2:1–17
Lautner S, Grams EET, Matyssek R, Fromm J (2005) Characteristics of electrical signals in poplar and responses in photosynthesis. Plant Physiol 139:2200–2209
Lunevsky VZ, Zherelova OM, Yostrikov IY, Berestovsky GN (1983) Excitation of *Characeae* cell membranes as a result of activation of calcium and chloride channels. J Membr Biol 72:43–58
Malone M (1992) Kinetics of wound-induced hydraulic signals and variation potentials in wheat seedlings. Planta 187:505–510
Malone M, Stanković B (1991) Surface potentials and hydraulic signals in wheat leaves following localized wounding by heat. Plant Cell Environ 14:431–436
Marrè E (1979) Fusicoccin: a tool in plant physiology. Ann Rev Plant Physiol 30:273–288
Mithöfer A, Boland W (2012) Plant defense against herbivores: chemical aspects. Ann Rev Plant Biol 63:431–450
Opritov VA, Lobov SA, Pyatgin SS, Mysyagin SA (2005) Analysis of possible involvement of local bioelectric responses in chilling perception by higher plants exemplified by *Cucurbita pepo*. Russ J Plant Physiol 52:801–808
Paszewski A, Zawadzki T (1976) Action potentials in *Lupinus angustifolius* L. shoots. III. Determination of the refractory periods. J Exp Bot 27:369–374
Pickard BG (1973) Action potentials in higher plants. Bot Rev 39:172–201
Pyatygin SS, Opritov VA, Vodeneev VA (2008) Signaling role of action potentials in higher plants. Russ J Plant Physiol 55:285–291
Rhodes JD, Thain JF, Wildon DC (1996) The pathway for systemic electrical signal conduction in the wounded tomato plant. Planta 200:50–57
Ricca U (1916) Soluzione d´un problema di fisiologia: la propagazione di stimulo nella *Mimosa*. Nuovo Giornale Botanico Italiano 23:51–170
Roblin G (1985) Analysis of the variation potential induced by wounding in plants. Plant Cell Physiol 26:455–461
Roblin G, Bonnemain JL (1985) Propagation in *Vicia faba* stem of a potential variation induced by wounding. Plant Cell Physiol 26:1273–1283
Sibaoka T (1953) Some aspects on the slow conduction of stimuli in the leaf of *Mimosa pudica*. Sci Rep Tohoku Univ Biol 20:72–88
Sibaoka T (1969) Physiology of rapid movements in higher plants. Ann Rev Plant Physiol 20:165–184
Stahlberg R, Cosgrove DJ (1992) Rapid alterations in growth rate and electrical potentials upon stem excision in pea seedlings. Planta 187:523–531

Stahlberg R, Cosgrove DJ (1994) Comparison of electric and growth responses to excision in cucumber and pea seedlings. I. Short-distance effects are a result of wounding. Plant Cell Environ 17:1143–1151
Stahlberg R, Cosgrove DJ (1996) Induction and ionic basis of slow wave potentials in seedlings of *Pisum sativum* L. Planta 200:416–425
Stahlberg R, Cosgrove DJ (1997a) The propagation of slow wave potentials in pea epicotyls. Plant Physiol 113:209–217
Stahlberg R, Cosgrove DJ (1997b) Slow wave potentials in cucumber differ in form and growth effect from those in pea seedlings. Physiol Plantarum 101:379–388
Stahlberg R, Cleland RE, van Volkenburgh E (2005) Decrement and amplification of slow wave potentials during their propagation in *Helianthus annuus* L. shoots. Planta 220:550–558
Stanković B, Davies E (1996) Both action potentials and variation potentials induce proteinase inhibitor gene expression in tomato. FEBS Lett 390:275–279
Stanković B, Davies E (1997) Intercellular communication in plants: electrical stimulation of proteinase inhibitor gene expression in tomato. Planta 202:402–406
Stanković B, Zawadzki T, Davies E (1997) Characterization of the variation potential in sunflower. Plant Physiol 115:1083–1088
Tarr SE, Lannoye RJ, Dainty J (1970) Changes of ionic and electrical properties of *Chara australis* when the external Ca is replaced by Mg. J Exp Bot 21:552–557
Trebacz K (1992) Measurements of intra- and extracellular pH in the liverwort *Conocephalum conicum* during action potentials. Physiol Plantarum 84:448–452
Trebacz K, Tarnecki R, Zawadzki T (1989) The effects of ionic channel inhibitors and factors modifying metabolism on the excitability of the liverwort *Conocephalum conicum*. Physiol Plantarum 75:24–30
Trebacz K, Simonis W, Schönknecht G (1994) Cytoplasmatic Ca^{2+}, K^{+}, Cl^{-}, and NO_3^{-} activities in the liverwort *Conocephalum conicum* L. at rest and during action potentials. Plant Physiol 106:1073–1084
Trebacz K, Simonis W, Schönknecht G (1997) Effects of anion channel inhibitors on light-induced potential changes in the liverwort *Conocephalum conicum*. Plant Cell Physiol 38:550–557
Tsutsui I, Taka-aki O, Kishimoto U (1986) Inhibition of Cl^{-} channel activation in *Chara corallina* membrane by lanthanum ion. Plant Cell Physiol 27:1197–1200
van Bel AJE, Ehlers K (2005) Electrical signalling via plasmodesmata. In: Oparka KJ (ed) Plasmodesmata. Blackwell, Oxford, pp 263–278
van Sambeek JW, Pickard BG (1976a) Mediation of rapid electrical, metabolic transpirational, and photosynthetic changes by factors released from wounds. I. Variation potentials and putative action potentials in intact plants. Can J Bot 54:2642–2650
van Sambeek JW, Pickard BG (1976b) Mediation of rapid electrical, metabolic transpirational, and photosynthetic changes by factors released from wounds. II. Mediation of the variation potential by Ricca's factor. Can J Bot 54:2651–2661
Vian A, Vian-Henry C, Schantz R, Ledoigt G, Frachisse JM, Desbiez MO, Julien JL (1996) Is membrane potential involved in calmodulin gene expression after external stimulation in plants? FEBS Lett 380:93–96
Volkov AG, Adesina T, Markin VS, Jovanov E (2007) Kinetics and mechanism of *Dionaea muscipula* trap closing. Plant Physiol 146:694–702
Volkov AG, Carrell H, Markin VS (2009) Biologically closed electrical circuits in venus flytrap. Plant Physiol 149:1661–1667
Volkov AG, Foster JC, Ashby TA, Walker RK, Johnson JA, Markin VS (2010a) *Mimosa pudica*: Electrical and mechanical stimulation of plant movement. Plant Cell Environ 33:163–173
Volkov AG, Foster JC, Markin VS (2010b) Signal transduction in *Mimosa pudica*: biologically closed electrical circuits. Plant Cell Environ 33:816–827
Wildon DC, Thain JF, Minchin PEH, Gubb IR, Reilly AJ, Skipper YD, Doherty HM, O'Donnell PJ, Bowles DJ (1992) Electrical signalling and systemic proteinase inhibitor induction in the wounded plant. Nature 360:62–65

Williams SE, Pickard BG (1972) Properties of action potentials in *Drosera* tentacles. Planta 103:222–240

Zimmermann MR (2010) Analyse der (elektrophysiologisch vermittelten) systemischen Signalübertragung bei *Hordeum vulgare* L. und *Vicia faba* L. Doctoral thesis, Justus-Liebig-University Gießen, Germany

Zimmermann MR, Felle HH (2009) Dissection of heat-induced systemic signals: Superiority of ion fluxes to voltage changes in substomatal cavities. Planta 229:539–547

Zimmermann MR, Maischak H, Mithöfer A, Boland W, Felle HH (2009) System potentials, a novel electrical long-distance apoplastic signal in plants, induced by wounding. Plant Physiol 149:1593–1600

Zimmermann MR, Hafke JB, van Bel AJE, Furch ACU (2013) Interaction of xylem and phloem during exudation and wound occlusion in *Cucurbita maxima*. Plant Cell Environ 36:237–247. doi:10.1111/j.1365-3040.2012.02571.x

Intercellular Communication in Plants: Evidence for an EMF-Generated Signal that Evokes Local and Systemic Transcriptional Responses in Tomato

A. Vian, E. Davies, and P. Bonnet

Abstract Exposing the oldest leaf of a tomato plant to a short (10 min), low-amplitude (5 V.m^{-1}), high-frequency (900 MHz) electromagnetic field evoked a rapid (15 min) and systemic accumulation of the stress-related transcript LebZIP-1 in the exposed leaf and in the distant, terminal leaf that is protected from EMF radiation. The accumulation was prevented by calcium counteracting drugs both locally and systemically. It was also prevented, but only in the distant tissue, in the ABA tomato mutant *Sitiens* or in wild-type tomato grown in the presence of the ABA synthesis inhibitor naproxen.

Keywords Tomato • Electromagnetic field • Signal transmission • Mode stirred reverberation chamber • mRNA accumulation

1 Introduction

Nowadays, wireless devices such as cellular telephones, Wi-Fi accessories and telecommunication antennae are widespread in the environment causing growing concerns about possible interactions between the electromagnetic fields (EMF) that they generate and living organisms. Most of the reports address the problem in terms of public health (Schüz et al. 2009) or cognitive modifications (Ng et al.,

A. Vian (✉)
UMR1345 Institut de Recherche en Horticulture et Semences (Université d'Angers, INRA, Agrocampus-Ouest), Université d'Angers, SFR 4207 QUASAV, PRES L'UNAM, 49045 Angers, France
e-mail: alain.vian@univ-angers.fr

E. Davies
Department of Plant Biology, North Carolina State University, Raleigh, NC 27695-7612, USA

P. Bonnet
Institut Pascal, UMR 6602 CNRS-Université Blaise Pascal, BP 10448, 63000 Clermont-Ferrand, France

F. Baluška (ed.), *Long-Distance Systemic Signaling and Communication in Plants*, Signaling and Communication in Plants 19, DOI 10.1007/978-3-642-36470-9_16,

2012) both of which are highly integrated, terminal-level responses in which the direct and unequivocal contribution of an EMF is extremely difficult to assess (Vian et al. 2006). Several reports point out biological effects such as the possible impact on the nervous system in animals (D'Andrea et al. 2003; Zeni et al. 2012) or the modification of gene expression (Czyz et al. 2004; Pardo et al. 2012) or proteome structure (Fragopoulou et al. 2012). Changes in metabolism have also been reported in plants, especially after exposure to low-frequency EMF. Davies (1996) showed that plants exposed to EMF displayed an increase in height and mass compared to unexposed ones. Exposure of seeds to low-frequency EMF increases resistance to virus (Trebbi et al. 2007) and causes an increase in stress-related enzyme activity (ornithine decarboxylase and phenylalanine ammonia-lyase), a change in photosynthetic activity (Yano et al. 2004) and a reduction of root growth in pea (Robertson et al. 1981). Tafforeau et al. (2002, 2004) found that exposing flax plants to GSM (900 MHz) radiation or to 105 GHz using a Gunn oscillator in conditions of calcium deprivation increased the production of stem meristems. Tkalec et al. (2005) exposed *Lemna minor* (Lemnaceae) to EMF of different frequencies and noted that growth was reduced (particularly at 900 MHz) and may be related to changes observed in total peroxidase activity, suggesting a possible link with oxidative stress. Furthermore, Kouzmanova et al. (2009) demonstrated that *Plectranthus* (Lamiaceae) exposure to EMF caused very rapid alterations in the activity of several key enzymes in leaves (isocitrate dehydrogenase, malate dehydrogenase and glucose-6-phosphate dehydrogenase), which were registered immediately after the end of the exposure. Taken together, these reports strongly suggested that plants actually do perceive EMF exposures of different frequencies and respond to them in ways that mimic those evoked by other environmental cues. Based on these observations, we decided to use plants as models to establish a formal and unequivocal relationship between the application of a weak (nonthermal) high-frequency electromagnetic stimulation and a rapid biological response. We also wanted to determine if the response is displayed just locally (i.e. only in the stimulated area) or also systemically (occurring in distant regions after a local stimulus), as observed after many kinds of environmental stimulations including wounding (Davies and Schuster 1981; Vian et al. 1996; Stankovic and Davies 1996, 1997) and insect attacks (Heil and Bostock 2002). Plants are exceedingly well suited for such studies: they are immobile (thus keep a constant orientation to the EMF) and their high surface to volume ratio places a high proportion of cell in direct contact with the electromagnetic field (Vian et al. 2007).

In our first set of experiments, we used the immediate accumulation of stress-related transcripts as an internal marker of EMF perception, to demonstrate that low-amplitude (5 $V.m^{-1}$), high-frequency (900 MHz) electromagnetic fields evoke an almost immediate, strong yet transient accumulation of calmodulin and proteinase inhibitor mRNA (Roux et al. 2006; Vian et al. 2006). We therefore proposed that exposure to EMF constitutes a genuine environmental signal (Roux et al. 2008).

Plants subjected to a local traumatic stimulus almost always respond in a very similar way in distant, intact tissues, implying the transmission of information from the site of stimulation to the rest of the plant (Davies and Schuster 1981). Also, the work conducted by Peña-Cortés et al. (1991, 1995) and Herde et al. (1996, 1999) clearly showed that the transmission of the wound signal in tomato did not occur in abscisic acid mutants: the *Sitiens* mutant displayed the local but not the distant wound responses, suggesting that the transmission of the wound signal is impaired in ABA-deficient plants, even though Birkenmeieir and Ryan (1998) clearly showed that ABA is not the primary signal for the transmission of wound-evoked information throughout the plant.

Demonstrating such a systemic mechanism would constitute a strong argument to advocate EMF as a common environmental stimulus. To address this question, we needed to expose only a part of the plant to the EMF. Accordingly, we designed a culture chamber that allows exposure of an individual leaf, the rest of the plant being shielded (protected from EMF). Investigating the accumulation of stress-related transcript in both the exposed and protected tissues constitutes a powerful approach to determine if a brief exposure to EMF would evoke a systemic response. To our knowledge, we are the only group to investigate such signal transmission after a nonthermal electromagnetic stimulation.

2 Experimental Setup

We used a 3-week-old tomato (*Lycopersicon esculentum*, cv. VFN-8) grown in controlled environment (Vian et al. 2006). On the one hand, the complex shape of the plant makes it difficult to keep the orientation of the different parts constant from one sample to another toward electromagnetic waves of fixed polarization and angle of incidence. On the other hand, from an electromagnetic point of view, the physical parameters (frequency, amplitude, polarization) are fundamental in coupling phenomena. Studies for which these quantities are not precisely controlled may lead to ambiguous results. We used a mode-stirred reverberation chamber (MSRC, Fig. 1) that fulfils these requirements. It consists of a large, cuboid Faraday cage (8.4 × 6.2 × 3.5 m, about 200 m^3) that totally isolates the experiment from the external electromagnetic background. Inside this highly resonant enclosure, the MSRC acts as an overmoded cavity. Moreover, when the frequency is sufficiently high (i.e. greater than 300 MHz in this case), the multiple reflections enhanced by the rotation of a large metallic stirrer that continuously changes the geometric characteristics of the cavity induce the formation of an electromagnetic field that is statistically homogeneous and isotropic (Fig. 2). Within a specific volume (called the "working volume"), an experimental object is then irradiated from all directions with the same amplitude and polarization since the electromagnetic field characteristics fulfil the IEC 61000-4-21 specifications (2002). Thus, the dimensioning of this working volume is absolutely necessary in order to expose different plant samples in a comparable and controlled way.

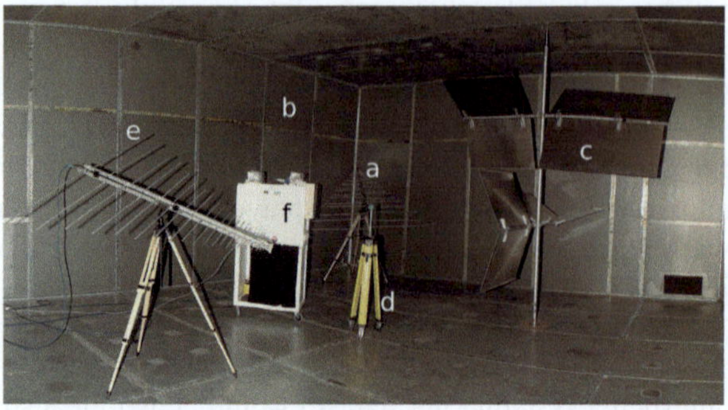

Fig. 1 *The mode-stirred reverberation chamber*. Internal view of the exposure facility, the mode-stirred reverberation chamber (MSRC). It is a vast metallic cavity (200 m^3) in which an electromagnetic signal is injected through a log-period antenna (*a*), this signal is reflected by the walls (*b*), the geometry of the cavity being continuously changed by the rotation of a large metallic stirrer (*c*). The field characteristics are measured through a triaxial probe (*d*) and a measuring antenna (*e*). This results in a homogeneous and isotropic electromagnetic field during one rotation of the stirrer that fulfils the IEC 61000-4-21 quality criteria (2002). The plants are grown in the culture chamber (*f*) that allows control over the culture conditions

We pioneered the use of the MSRC as a tool for exposure to EMF of both plant and human material (Vian et al. 2006; Lalléchère et al. 2010; Roux et al. 2011). We developed two different custom-made culture chambers in which both the temperature and the light (150–175 μmol.$m^{-2}s^{-2}$) are controlled. The first one is made of plywood and is transparent to EMF, while the second one is essentially identical but shielded with an 8-mm polymer mesh covered with a 0.1-mm-thick aluminium layer, which reduced the radiation to 10 % of normal. We made a small hole through which the oldest leaf protruded outside the shielded chamber so that it was the only part of the plant to be subjected to the full EMF irradiation (5 V.m^{-1}), while the remainder of the plant was subjected to the shield-reduced amplitude (0.5 V.m^{-1}). The mature (exposed) and terminal (protected) leaves were collected, frozen in liquid nitrogen and immediately subjected to RNA isolation using the Tri Reagent system (Sigma chemicals). Total RNA was tested for the abundance of a stress-related transcript, namely, LebZip1 (Stankovic et al. 2000), as previously described (Vian et al. 2006).

3 Field Characteristics

To define the working volume of the empty MSRC, we recorded the 3 axial components of the electric field (E_x, E_y and E_z) using an isotropic PMM-183 probe and demonstrated (Vian et al. 2006; Lalléchère et al. 2010) that the field is statistically homogeneous and isotropic within the neighbourhood of the culture chamber. First, following the IEC 61000-4-21 criteria (2002), we checked that the values of the standard deviation of the electric field Cartesian components, noted σ_x, σ_y, σ_z, and the standard deviation of the total field, noted σ_{xyz}, were below the

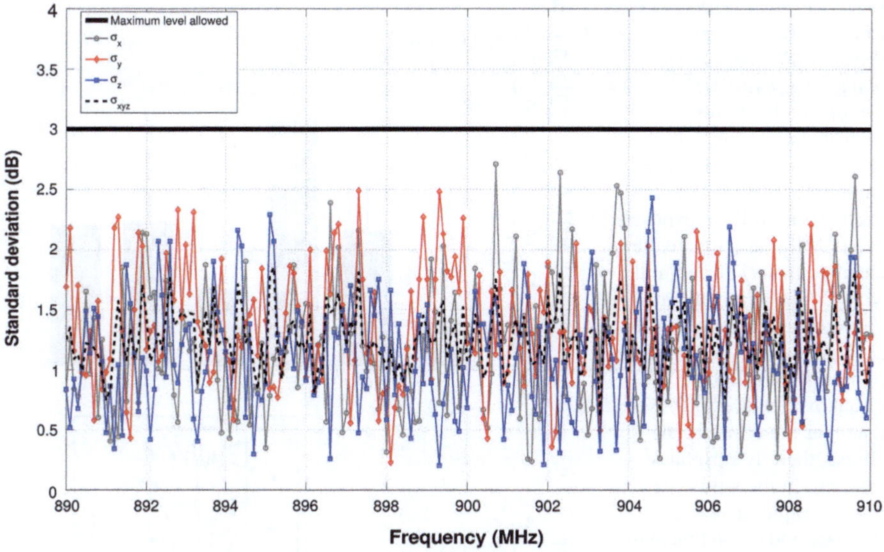

Fig. 2 *Characteristics of the electromagnetic field established inside the MSRC at 900 MHz*. In mode-stirred operation, the stirrer rotates continuously, at constant speed, and a set of electric field measurements are taken and averaged at each frequency, before moving on to a new frequency. We computed the standard deviations of probe data for each Cartesian field component and for the total electric field around the 900 MHz frequency. Experiments were carried out inside a volume where these standard deviations were consistently below the limit of 3 dB fixed by the IEC-61000-4-21 specifications (2002), verifying that the EMF was both homogeneous (the value of the total field was the same in all points of the test volume) and isotropic (the Cartesian components of the field were similar). The field uniformity is strongly driven (but not in a trivial way) by the stirrer geometry and the quality factor of the MSRC [Figure redrawn from Lalléchère et al. (2010), courtesy of The Electromagnetics Academy]

limit of 3 dB (Fig. 2) at each of eight space points defining a rectangular cuboid. Secondly, the loading factor (for the used frequency of 900 MHz) was very close to 1, indicating that the presence of the custom-made culture chambers did not affect these field characteristics.

4 Local Response to EMF Exposure

A short (10-min) exposure of the entire above-ground plant to a weak electromagnetic exposure (5 $V.m^{-1}$) leads to the very rapid accumulation and subsequent decrease of stress-related lebZIP-1 mRNA in both the wild-type (Fig. 3a) and the ABA-deficient *Sitiens* mutant (Fig. 3b). However, the response differs insofar as the wild-type exhibits a second, delayed transcript increase, whereas the mutant does not. An identical triphasic response in the wild-type and biphasic response in ABA mutant was also observed after wound (flame) stimulus (Vian et al. 1999).

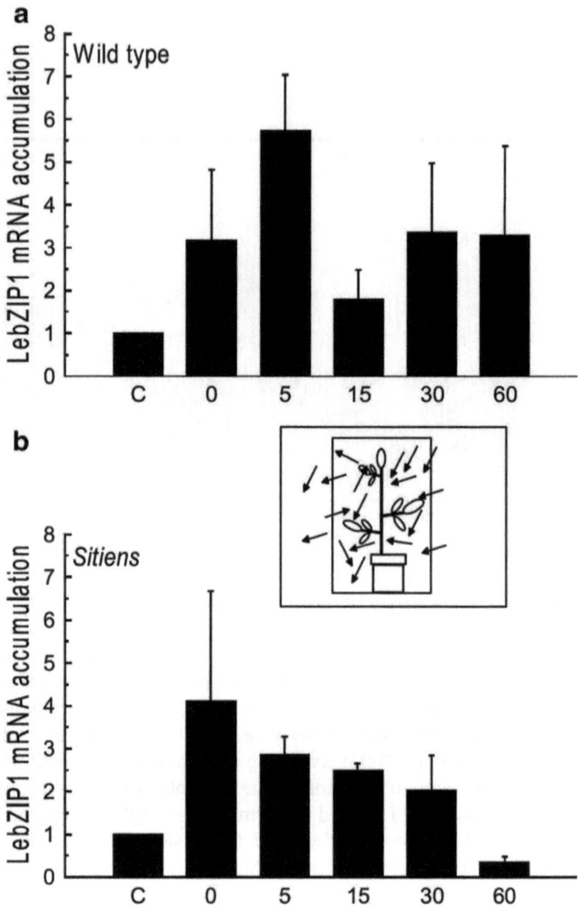

Fig. 3 *Typical responses of wild-type (a) and ABA-deficient (Sitiens) tomato plants to global EMF exposure*. Plants were subjected to a global, high-frequency (900 MHz), low-amplitude (5 $V.m^{-1}$) and short (10-min) EMF exposure in an EMF-transparent culture chamber (*inset*) placed in the MSRC. The terminal leaf was collected and assayed for the accumulation of the wound transcript LebZIP1 (Stankovic et al. 2000). Note that over the 60-min time course, there is a triphasic response in the wild-type mutant (rapid increase, decrease and second increase) but only a biphasic response (rapid increase and slow decrease) in the ABA-deficient mutant (*Sitiens*) [Figure redrawn from Beaubois et al. (2007)]

When wild-type plants were placed totally within the shielded chamber, no accumulation occurred in the aged (leaf 1) or terminal leaf (Fig. 4), demonstrating that it is, indeed, the EMF exposure that constitutes the activating stimulus. It also demonstrates that the shielding equipment of the culture chamber was efficient.

5 Signal Generation and Transmission

In a different experiment, we let just the oldest leaf protrude from the shielded culture chamber (thus being the only part exposed to EMF), while the rest of the plant was protected from EMF (Fig. 5, inset). Exposing the wild-type plant this way to EMF evoked not only the accumulation of stress-related mRNA in the exposed tissues (Fig. 5a, leaf 1) but also in distant regions that are protected from EMF by the shielding material of the culture chamber (Fig. 5a, terminal leaf). The

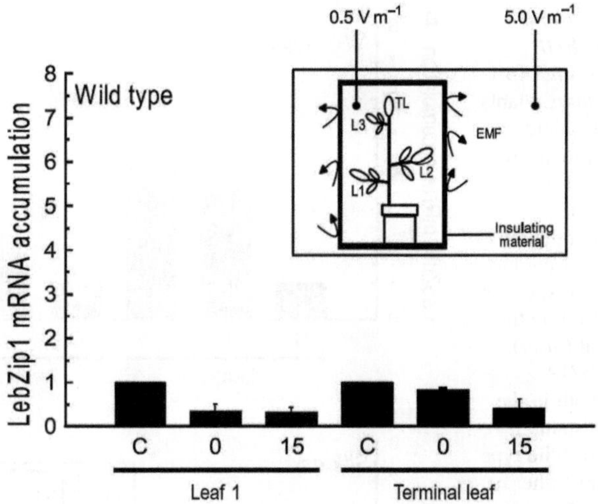

Fig. 4 *Accumulation of bZIP transcripts in plants placed in the shielded (EMF-proof) culture chamber*. Plants were cultivated in a culture chamber shielded with a polymer armoured with aluminium foil (*inset*). This insulating material caused the EMF amplitude to drop about tenfold from 5 $V.m^{-1}$ outside the chamber to 0.5 $V.m^{-1}$ inside the chamber. In this condition, no accumulation of LebZIP-1 occurred in leaf 1 or in the terminal leaf. *L1–L3,* leaf 1–leaf 3; *TL,* terminal leaf [Figure redrawn from Beaubois et al. (2007)]

accumulation there was essentially identical in amplitude but delayed by a few minutes compared to that observed in the exposed tissue (Fig. 5a, leaf 1). This implies the genesis and rapid transmission of a signal from the exposed tissue to the rest of the plant. Indeed, we interpreted the short delay observed in transcript accumulation in the distant tissue as the time needed for the signal to move from the stimulated to the terminal leaf.

The signal genesis appears to be strongly dependent upon the presence of calcium, since spraying the exposed leaf with 1 mM of the calcium chelator EGTA and 1 mM of the calcium channel blocker $LaCl_3$ totally prevented the response not only in the distant (protected) terminal leaf but also in the local (exposed) leaf 1 (Fig. 6a). Calcium is a major component of plant responses to environmental stresses (see Reddy et al. 2011 for a review). It is worth noting that both EGTA and $LaCl_3$ may act upstream of the activation of gene expression in the very first events of the stimulus-perception-transduction pathway (Luan et al. 2002) by perturbing normal calcium movements and metabolism. The initial steps of EMF perception have been proposed to take place within, or in the neighbourhood of, the plasma membrane (Gaber et al. 2005; Sun et al. 2012). Interestingly, Szczegielniak (2005) demonstrated in maize the presence of a protein kinase that is dependent upon both calcium and phospholipids, two major components of the early plant response pathway to injurious stimuli (Saltveit et al. 2005) that might therefore also be involved in EMF perception. Our results are consistent with the involvement of calcium movement in the very early steps of EMF stimulus perception and/or

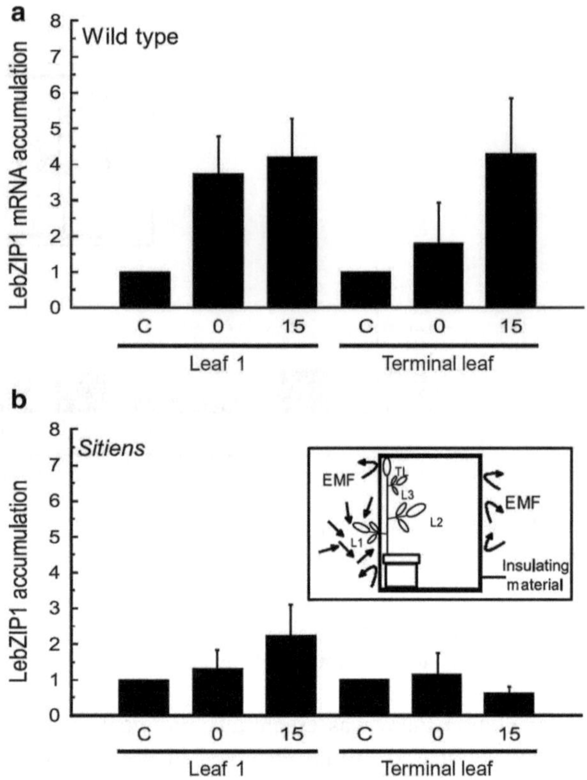

Fig. 5 *Local and distant accumulation of LebZIP1 mRNA in wild-type and ABA-deficient tomato plants.* Plants were placed in the shielded culture chamber in such a way that only the oldest leaf protruded outside the chamber, thus being the only part of the plant subjected to EMF exposure, the rest of the plant being protected by the armouring material (*inset*). Accumulation of bZIP mRNA occurred both in the exposed and the distant, protected leaf in the wild type (**a**), while in contrast, the *Sitiens* mutant (**b**) showed an accumulation only in the exposed tissues. Similar responses (Beaubois et al. 2007) were observed with the proteinase inhibitor (Pin2) mRNA. *EMF,* electromagnetic field; *L1–L3,* leaf 1–leaf 3; *TL,* terminal leaf [Figure redrawn from Beaubois et al. (2007)]

transduction insofar as a total inhibition of the plant response to EMF exposure occurred in the presence of calcium counteracting drugs. Additional evidence implicating biological membranes was furnished by Roux et al. (2008): tomato energy metabolism was transiently but strongly affected (30 % drop in ATP content and adenylate energy charge). Furthermore, transcript accumulation in response to EMF exposure was impaired when plants were cultivated in the presence of CCCP (a decoupling agent).

The perception and transduction of the EMF exposure occurred normally in the presence of the ABA synthesis inhibitor naproxen in the exposed leaf (Fig. 6b, leaf 1), suggesting that this phytohormone is not necessary for these events to take place. In contrast, the distant response was totally prevented (Fig. 6b, terminal leaf) showing that ABA is absolutely required for the transmission of the signal throughout the plant. We do not know if ABA is itself the intercellular signal or if it is needed to generate some other transmitted signal. Hlaváčková et al. (2006) investigated the events following heat stimulus in tobacco and proposed that the rapid propagation of an electrical wave of depolarization acted as an inducer for systemic synthesis of ABA, supporting an indirect role for ABA in the transmission of the actual signal. The study from Birkenmeieir and Ryan (1998), showing that ABA was not by itself the mobile signal but rather participated in the establishment

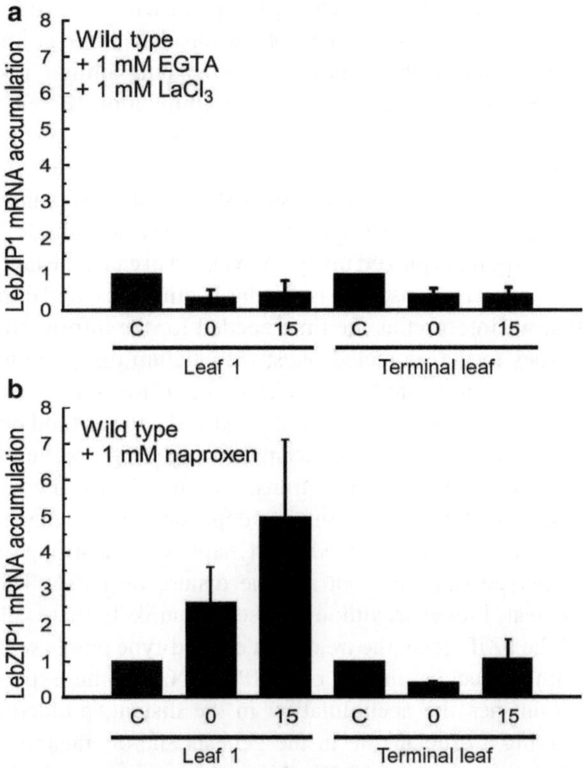

Fig. 6 *Effects of drugs on signal genesis and transmission.* Wild-type tomato plants treated with 1 mM of the ABA synthesis inhibitor naproxen (**a**) show a normal local response (leaf 1) in the exposed leaf but lack the distant response in the terminal, protected leaf. In contrast, wild-type tomato plants treated with 1 mM of the calcium chelator EGTA and 1 mM of the calcium channel blocker $LaCl_3$ lack responses in both the local, exposed leaf and in the distant, protected leaf [Figure redrawn from Beaubois et al. (2007)]

of a favourable metabolism for the genesis of the genuine signal, needs to be considered under the light of new data (Davies et al. 2005). Indeed, Sauter et al. (2001) pointed out that ABA can move through the phloem and the xylem. Recently, Goodger and Schachtman (2010) demonstrated in the context of water stress that the rapid increase of ABA concentration in the xylem originated from its biosynthesis in leaves and root suggested rapid ABA transport within the plant. The involvement of mechanosensitive channels, possibly activated after hydraulic (Malone 1994) or electrical/hydraulic signals (Stankovic and Davies 1996, 1997), may be related to the conclusions of Zhang et al. (2006), pointing out the fact that a plasma membrane sensor could, possibly through calcium signalling, turn on the genes of the ABA synthesis pathway. Again, a direct link is made between the plasma membrane and subsequent cellular mechanism (i.e. ABA biosynthesis).

6 Distant Response to EMF Exposure

Tomato plants impaired in the biosynthesis of ABA (the *Sitiens* tomato mutant) did not display the response in the distant, protected leaf (Fig. 5b, terminal leaf), while they did in the stimulated leaf (Fig. 5b, leaf 1). These results were very similar to those observed after treating the plant with the ABA synthesis inhibitor naproxen (Fig. 6b), supporting the conclusion that ABA is implicated in the transmission rather than in the genesis of the signal. Similar results were obtained for Pin-2 mRNA (proteinase inhibitor) accumulation (Beaubois et al. 2007) that has very different cellular functions from LebZIP-1, suggesting that these observations are potentially true for a wide variety of stress-related genes.

Taken together, these results show that (1) a plant organ (aged or terminal leaf) can perceive and respond to a short, low-amplitude EMF exposure; (2) once the EMF signal is picked up by an exposed organ, a distant tissue (e.g. the terminal leaf) displays a response that is similar in amplitude and duration, with only a short delay that we interpret as the time needed for the information to move from the exposed tissues to the protected ones; (3) calcium might be needed for the response itself (i.e. thus prevented in the exposed leaf treated with calcium counteracting drugs) but might as well be required for signal genesis (and possibly its transmission) since treatment with calcium counteracting drugs prevents both the local and distant responses; and (4) the transmission of this message—but not the molecular response itself—is somehow dependent upon the metabolism of abscisic acid. Wild-type plants treated with naproxen (a strong inhibitor of ABA synthesis) displayed the local but not the distant response. The exact role of ABA is still unclear; however, although it seems unlikely that it directly evokes the expression of the bZIP gene, the treatment of wild-type plants with naproxen does not interfere with the accumulation of bZIP mRNA in the exposed leaf, whereas it strongly diminishes the accumulation in the distant, protected leaf showing that ABA is playing a crucial role in the genesis and/or transmission of the signal evoked by plant exposure to EMF. As it does not perturb the response itself, transcript accumulation and signal transmission appear to be totally different events. Similarly, the distant accumulation of bZIP mRNA was impaired in the JL-5 tomato mutant deficient for the biosynthesis of jasmonic acid (JA), while it stayed unaffected in the exposed tissues (Beaubois et al. 2007). The very similar expression profiles obtained for both ABA and JA mutants suggest that both phytohormones are required for the efficient transmission of the signal from the exposed tissues to the shielded ones, JA and ABA being implicated in the same pathways (Anderson et al. 2004).

Based on our experiments, it is not possible to decipher if (1) the local and distant responses are both elicited by the transmitted signal (its propagation, but not its generation being perturbed by the lack of ABA or JA). In this case, the local accumulation of stress-related transcripts occurred indirectly, in response to the genesis of the signal or (2) the local response is elicited through a pathway that is separate from the genesis or transmission of the signal. Taken together, these results are summarized in a model of the plant response to EMF exposure (Beaubois et al. 2007). To our knowledge this aspect of plant responses to EMF has not been investigated by other teams, and the original work (Beaubois et al. 2007) remains the only dedicated work.

Based upon previous work done on tomato (Stankovic and Davies 1996, 1997) and *Bidens* (Vian et al. 1996; Vian and Davies 2006), we postulated that this information transmission may be mediated through the propagation of an electrical signal (action potential and/or variation potential). Indeed, the similarities of the responses (evocation of the accumulation of the same transcripts, local and distant responses, very little delay between the local and distant responses suggesting a very rapid transmission) all advocate for a similar transmission mechanism. Since the involvement of an electrical wave of depolarization was clearly demonstrated after environmental stimulation (Stankovic and Davies 1996, 1997; Zimmermann et al. 2009), we conducted experiments to decipher if such a mechanism is operating after EMF stimulation (Roux et al. unpublished data). Briefly, we connected inserted metallic electrodes into plants placed in the shielded culture chamber, with the oldest leaf protruding outside so that it would be exposed to EMF and thus generate the systemic signal. However, and despite intensive tests, no clear results have been found, mainly because of interactions between the electromagnetic field and the inserted electrodes that made the signal noisy and difficult to interpret.

Acknowledgements The authors wish to thank "l'Académie des Sciences de l'Institut de France" for the "Grand Prix de Physiologie Intégrative Pierre et Cyril Grivet 2008" awarded to A. Vian that partially supported this research.

References

Anderson JP, Badruzsaufari E, Schenk PM, Manners JM, Desmond OJ, Ehlert C, Maclean DJ, Ebert PR, Kazana K (2004) Antagonistic interaction between abscisic acid and jasmonate-ethylene signaling pathways modulates defense gene expression and disease resistance in Arabidopsis. Plant Cell 16:3460–3479

Beaubois É, Girard S, Lallechère S, Davies E, Paladian F, Bonnet P, Ledoigt G, Vian A (2007) Intercellular communication in plants: evidence for two rapidly-transmitted systemic signals generated in response to EMF stimulation in tomato. Plant Cell Environ 30:834–844

Birkenmeier GF, Ryan CA (1998) Wound signaling in tomato plants: evidence that ABA is not a primary signal for defense gene activation. Plant Physiol 117:687–693

Czyz J, Guan K, Zeng Q, Nikolova T, Meister A, Schonborn F, Schuderer J, Kuster N, Wobus AM (2004) High frequency electromagnetic fields (GSM signals) affect gene expression levels in tumor suppressor p53-deficient embryonic stem cells. Bioelectromagnetics 25:296–307

D'Andrea JA, Adair ER, de Lorge JO (2003) Behavioral and cognitive effects of microwave exposure. Bioelectromagnetics 6(Suppl):39–62

Davies MS (1996) Effects of 60 Hz electromagnetic fields on early growth in three plant species and a replication of previous results. Bioelectromagnetics 17:154–161

Davies E, Schuster A (1981) Intercellular communication in plants: evidence for a rapidly generated, bidirectionally transmitted wound signal. Proc Natl Acad Sci USA 78:2422–2426

Davies WJ, Kudoyarova G, Hartung W (2005) Long-distance ABA signaling and its relation to other signaling pathways in the detection of soil drying and the mediation of the plants response to drought. J Plant Growth Regul 24:285–295

Fragopoulou AM, Samara A, Antonelou MH, Xanthopoulou A, Papadopoulou A, Vougas K, Koutsogiannopoulou E, Anastasiadou E, Stravopodis DJ, Tsangaris GT, Margaritis LH (2012) Brain proteome response following whole body exposure of mice to mobile phone or wireless DECT base radiation. Electromagn Biol Med 31:250–274. doi:10.3109/15368378.2011.631068

Gaber MH, El Halim NA, Khalil WA (2005) Effect of microwave radiation on the biophysical properties of liposomes. Bioelectromagnetics 26:194–200

Goodger JQ, Schachtman DP (2010) Re-examining the role of ABA as the primary long-distance signal produced by water-stressed roots. Plant Signal Behav 5:1298–1301

Heil M, Bostock RM (2002) Induced systemic resistance (ISR) against pathogens in the context of induced plant defences. Ann Bot 89:503–512

Herde O, Fisanh J, Willmitzer L, Peña-Cortés H (1996) Localized wounding by heat initiates the accumulation of proteinase inhibitor II in ABA-deficient plants by triggering jasmonic acid biosynthesis. Plant Physiol 112:853–860

Herde O, Peña-Cortés H, Wasternack C, Willmitzer L, Fisahn J (1999) Electric signal and pin2 gene expression on different abiotic stimuli depend on a distinct threshold level of endogenous abscisic acid in several abscisic acid-deficient tomato plants. Plant Physyiol 119:213–218

Hlaváčková V, Krchňák P, Nauš J, Novák O, Špundová M, Strnad M (2006) Electrical and chemical signals involved in short-term systemic photosynthetic responses of tobacco plants to local burning. Planta 225:235–244

IEC-61000-4-21 (2002) Reverberation chamber test methods, International Electrotechnical Commission, Electromagnetic compatibility (EMC), part 4 section 21: testing and measurement techniques

Kouzmanova M, Dimitrova M, Dragolova D, Atanasova G, Atanasov N (2009) Alterations in enzyme activities in leaves after exposure of *Plectranthus sp.* plants to 900 MHz electromagnetic field. Biotechnol Equip 23:1310–1328

Lalléchère S, Girard S, Roux D, Bonnet P, Paladian F, Vian A (2010) Mode stirred reverberation chamber (MSRC): a large and efficient tool to lead high frequency bioelectromagnetic in vitro experimentation. Progr Electromagn Res B 26:257–290

Luan S, Kudlab J, Rodriguez-Concepcion M, Yalovskyd S, Gruissem W (2002) Calmodulins and calcineurin B–like proteins calcium sensors for specific signal response coupling in plants. Plant Cell 14:S389–S400

Malone M (1994) Wound-induced hydraulic signals and stimulus transmission in *Mimosa pudica* L. New Phytol 128:49–56

Ng TP, Lim ML, Niti M, Collinson S (2012) Long-term digital mobile phone use and cognitive decline in the elderly. Bioelectromagnetics 33:176–185

Pardo JCT, Grimaldi S, Taranta M, Naldi I, Cinti C (2012) Microwave electromagnetic field regulates gene expression in T-lymphoblastoid leukemia CCRF-CEM cell line exposed to 900 MHz. Electromagn Biol Med 31:1–18

Peña-Cortés H, Willmitzer L, Sanchez-Serrano JJ (1991) Abscisic acid mediates wound induction but not developmental-specific expression of the proteinase inhibitor II gene family. Plant Cell 3:963–972

Peña-Cortès H, Fisahn J, Willmitzer L (1995) Signals involved in wound-induced proteinase inhibitor II gene expression in tomato and potato plants. Proc Natl Acad Sci USA 92:4106–4113

Reddy ASN, Ali GS, Celesnik H, Day IS (2011) Coping with stresses: roles of calcium- and calcium/calmodulin-regulated gene expression. Plant Cell 23:2010–2032

Robertson D, Miller MW, Cox C, Davis HT (1981) Inhibition and recovery of growth processes in roots of *Pisum sativum* L. exposed to 60-Hz electric fields. Bioelectromagnetics 2:329–340

Roux D, Vian A, Girard S, Bonnet P, Paladian F, Davies E, Ledoigt G (2006) Electromagnetic fields (900 MHz) evoke consistent molecular responses in tomato plants. Physiol Plant 128:283–288

Roux D, Vian A, Girard S, Bonnet P, Paladian F, Davies E, Ledoigt G (2008) High frequency (900 MHz) low amplitude (5 V m^{-1}) electromagnetic field: a genuine environmental stimulus that affects transcription, translation, calcium and energy charge in tomato. Planta 227:883–891

Roux D, Girard S, Paladian F, Bonnet P, Lalléchère S, Gendraud M, Davies E, Vian A (2011) Human keratinocytes in culture exhibit no response when exposed to short duration, low amplitude, high frequency (900 MHz) electromagnetic fields in a reverberation chamber. Bioelectromagnetics 32:302–311

Saltveit ME, Choia YJ, Tomas-Barberan FA (2005) Involvement of components of the phospholipid-signaling pathway in wound-induced phenylpropanoid metabolism in lettuce (*Lactuca sativa*) leaf tissue. Physiol Plant 125:345–355

Sauter A, Davies WJ, Hartung W (2001) The long distance abscisic acid signal in the droughted plant: the fate of the hormone on its way from root to shoot. J Exp Bot 52:1991–1997

Schüz J, Lagorio S, Bersani F (2009) Electromagnetic fields and epidemiology: an overview inspired by the fourth course at the international school of bioelectromagnetics. Bioelectromagnetics 30:511–524

Stankovic B, Davies E (1996) Both action potentials and variation potentials induce proteinase inhibitor gene expression in tomato. FEBS Lett 390:275–279

Stankovic B, Davies E (1997) Intercellular communication in plants: electrical stimulation of proteinase inhibitor gene expression in tomato. Planta 202:402–406

Stankovic B, Vian A, Henry-Vian C, Davies E (2000) Molecular cloning and characterization of a tomato cDNA encoding a systemically wound-inducible bZIP DNA-binding protein. Planta 212:60–66

Sun WJ, Mogadam MK, Sommarin M, Nittby H, Salford LG, Persson BRR, Eberhardt JL (2012) Calcium efflux of plasma membrane vesicles exposed to ELF magnetic fields—test of a nuclear magnetic resonance interaction model. Bioelectromagnetics 33:535–542

Szczegielniak J, Klimecka M, Liwosz A, Ciesielski A, Kaczanowski S, Dobrowolska G, Harmon AC, Muszyńska GA (2005) Wound-responsive and phospholipid-regulated maize calcium-dependent protein kinase. Plant Physiol 139:1970–1983

Tafforeau M, Verdus MC, Norris V, White G, Demarty M, Thellier M, Ripoll C (2002) SIMS study of the calcium-deprivation step related to epidermal meristem production induced in flax by cold shock or radiation from a GSM telephone. J Trace Microprobe Techn 20:611–623

Tafforeau M, Verdus MC, Norris V, White GJ, Cole M, Demarty M, Thellier M, Ripoll C (2004) Plant sensitivity to low intensity 105 GHz electromagnetic radiation. Bioelectromagnetics 25:403–407

Tkalec M, Malarić K, Pevalek-Kozlina B (2005) Influence of 400, 900, and 1900 MHz electromagnetic fields on *Lemna minor* growth and peroxidase activity. Bioelectromagnetics 26:185–193

Trebbi G, Borghini F, Lazzarato L, Torrigiani P, Calzoni GL, Betti L (2007) Extremely low frequency weak magnetic fields enhance resistance of NN tobacco plants to tobacco mosaic virus and elicit stress-related biochemical activities. Bioelectromagnetics 28:214–223

Vian A, Davies E (2006) Two different wound signals evoke very rapid, systemic CMBP transcript accumulation in tomato. Plant Signal Behav 1:261–264

Vian A, Henry-Vian C, Schantz R, Ledoigt G, Frachisse JM, Desbiez MO, Julien JL (1996) Is membrane potential involved in calmodulin gene expression after external stimulation in plants? FEBS Lett 380:93–96

Vian A, Henry-Vian C, Davies E (1999) Rapid and systemic accumulation of chloroplast mRNA binding protein transcripts after flame stimulus in tomato. Plant Physiol 121:517–524

Vian A, Roux D, Girard S, Bonnet P, Paladian F, Davies E, Ledoigt G (2006) Microwave irradiation affects gene expression in plants. Plant Signal Behav 1:67–69

Vian A, Faure C, Girard S, Davies E, Hallé F, Bonnet P, Ledoigt G, Paladian F (2007) Plants respond to GSM like radiation. Plant Signal Behav 2:522–524

Yano A, Ohashi Y, Hirasaki T, Fujiwara K (2004) Effects of a 60 Hz magnetic field on photosynthetic CO_2 uptake and early growth of radish seedlings. Bioelectromagnetics 25:572–581

Zeni O, Sannino A, Sarti M, Romeo S, Massa R, Scarf MR (2012) Radiofrequency radiation at 1950 MHz (UMTS) does not affect key cellular endpoints in neuron-like PC12 cells. Bioelectromagnetics 33:497–507

Zhang J, Jia W, Yang J, Ismail AM (2006) Role of ABA in integrating plant responses to drought and salt stresses. Field Crop Res 97:111–119

Zimmermann MR, Maischak H, Mithöfer A, Boland W, Felle HH (2009) System potentials, a novel electrical long-distance apoplastic signal in plants, induced by wounding. Plant Physiol 149:1593–1600

Systemic Wound Signaling in Plants

John P. Delano-Frier, Gregory Pearce, Alisa Huffaker, and Johannes W. Stratmann

Abstract Ever since the seminal discovery of systemic wound signaling in tomato and potato plants by Green and Ryan (Science 1972), a number of candidate systemic wound signals have been proposed. These can be classified into three groups: (1) Chemical signals, including the alarm hormone systemin and other peptide hormones, jasmonic acid is a phytohormone, as well as reactive oxygen species (ROS); (2) physical signals, including electrical and hydraulic signals; and (3) herbivore-induced volatile compounds, including green leafy volatiles and terpenes. These signals are generated at or close to the site of herbivore-inflicted injury and systemically move to target tissues where they induce defense responses. Chemical and physical signals depend on the connectivity of the plant body, whereas volatile compounds are released into the airspace. Different plants with different morphologies and ecological niches employ different modes of systemic signaling.

J.P. Delano-Frier
Unit for Plant Biotechnology and Genetic Engineering, Center for Research and Advanced Studies (Cinvestav) at Irapuato, POB 36821, Irapuato, Gto., Mexico
e-mail: jdelano@ira.cinvestav.mx

G. Pearce (✉)
Division of Natural Sciences and Mathematics, Lewis-Clark State College, 500 8th Ave., Lewiston, ID 83501, USA
e-mail: pearce940@gmail.com

A. Huffaker
Chemistry Research Unit, USDA-ARS Center for Medical, Agricultural & Veterinary Entomology, 1700 SW 23rd Dr, Gainesville, FL 32608, USA
e-mail: alisa.huffaker@ars.usda.gov

J.W. Stratmann
Department of Biological Sciences, University of South Carolina, Coker Life Science Bldg 715 Sumter Street, Columbia, SC 29208, USA
e-mail: johstrat@mailbox.sc.edu

F. Baluška (ed.), *Long-Distance Systemic Signaling and Communication in Plants*, Signaling and Communication in Plants 19, DOI 10.1007/978-3-642-36470-9_17,

Keywords Systemic signaling • Systemins • Peps • Jasmonic acid • Phytohormones • ROS • Electrical signals • Hydraulic signals • Volatiles • Herbivory

1 Introduction

Plant life depends on accurate sensing of the environment. Growth, development, fitness, and survival are governed by abiotic factors such as light, temperature, and nutrients, as well as cues from pathogens, symbionts, competitors, and herbivores. Plant defenses against herbivory start when herbivore eggs are deposited onto a plant (Doss et al. 2000). However, gravid female insects may go through a meticulous host selection process to ensure that their offspring encounters optimal growth conditions (Thompson and Pellmyr 1991). Upon hatching, insect larvae immediately start feeding on plant tissue, which generally induces plant defense responses. Sensors and receptors at the wound site recognize mechanical signals generated by insect mouthparts or chemicals in the oral secretions of the larvae (Alborn et al. 1997). The localized wounding by small herbivore larvae can serve as a warning sign of impending large-scale herbivory and potential defoliation by larger and more mobile larvae. As a consequence of this selection pressure, plants evolved systemic defense responses that extend beyond the wound sites and can be quickly induced upon herbivory.

Defenses against insect pests include plant metabolites that reduce the fitness of the herbivore. It has been known for a long time that plants evolved an amazing capacity to synthesize toxic or repellent secondary metabolites for protection (Fraenkel 1959; Stahl 1888). Some of these metabolites are constitutively produced, while others are synthesized only in response to herbivory. In a groundbreaking study, C. A. Ryan and his coworker T. R. Green discovered that tomato plants respond to an attack by Colorado potato beetles (*Leptinotarsa decemlineata*) with the de novo synthesis of defensive proteinase inhibitors that interfere with digestive processes in the herbivore gut (Green and Ryan 1972). In the same study, they introduced a novel concept in plant–insect interactions, systemic signaling. After demonstrating that proteinase inhibitor synthesis was upregulated not only at wound sites but also throughout the plant, they postulated the presence of an "inducing factor" that moves through the vasculature. This initiated the search for systemic wound signals, leading to amazing novel insights into intraplant communication.

Cost–benefit tradeoffs in induced defense responses are highly complex (Steppuhn and Baldwin 2008), and additional factors must be considered for systemic induced defenses. A systemic defense response may be too costly and ineffective if the herbivores are relatively immobile, low in number, small in comparison to the plant, or evenly distributed on the plant. In these cases, localized defenses may be the preferred defense strategy (Zangerl 1999). In mature plants, induced leaf defenses may be reduced in favor of constitutive defenses for maturing seeds and fruits, or resource allocation may be shifted away from defense to seed and fruit development (Barton and Koricheva 2010; Boege and Marquis 2005; Van

Dam et al. 2001). On the other hand, systemic responses should be beneficial if a localized attack predicts imminent danger to other parts of the plant, e.g., in the presence of moving phytophagous insects. If the herbivore encounters fully activated defenses all over the plant, it has to cope with these defenses immediately. In contrast, in a scenario of multiple localized induced defenses, the herbivore may take advantage of the time period between wounding and mobilization of defenses to consume unspoiled food.

Another potential benefit of systemic defenses is that they may confer broad spectrum resistance against herbivores. For example, systemically upregulated anti-digestive proteins or nicotine affect many species of arthropod herbivores (Li et al. 2002a; Schmeltz 1971). Only specialists like the Colorado potato beetle or *Manduca sexta* have evolved mechanisms to cope with these defenses (Felton et al. 1992; Glendinning 2002).

A more subtle systemic response is "priming". During a priming response, certain aspects of defenses, often signaling components, are systemically upregulated without leading to synthesis of defense compounds. This heightened alert level avoids costly allocation of resources to defense in case the herbivore does not strike, but enables a swift and strong response in the case of attack (Beckers et al. 2009; Engelberth et al. 2004). In contrast, if herbivory is highly likely to occur and to afflict large parts of a plant, constitutive defenses are more effective than induced defenses, even though they entail costly resource allocation for defense (Wittstock and Gershenzon 2002; Zangerl and Rutledge 1996).

Adaptivity of systemic defenses depends on plant architecture and size. On a big plant, a few small herbivores may be tolerated without a significant investment in defenses. Trees can even tolerate and survive large herbivore outbreaks that result in complete defoliation, but lead to regrowth after the threat of herbivory is diminished (Schowalter et al. 1986). Even in herbal plants, a change in resource allocation from shoot to root allows the plant to tolerate severe leaf damage and enables regrowth (Schwachtje et al. 2006). In contrast, a systemic defense response throughout a large plant, such as a tree, would require a significant and possibly unnecessary investment in resources.

Another important factor that influences defense responses is the behavior of the herbivores. Herbivores in different feeding guilds, such as cell content feeders, phloem feeders, or chewing herbivores, inflict different types of damage with regard to size of affected area, affected tissue, severity of damage, frequency of occurrence, etc. As a consequence, plants mount different defenses against different herbivores (Kessler and Baldwin 2002; Thompson and Goggin 2006; Walling 2000). For systemic responses, plants must interpret a particular type of injury as a warning sign of impending damage elsewhere in the plant body. The mobility of the herbivore represents a predictive factor. For instance, if larvae move frequently around on a plant, possibly to avoid local defenses, it will be beneficial for the plant to induce defenses all over the plant. The benefit of systemic defenses against slow-moving herbivores may be reduced or less urgent. Herbivore size also matters. Mobilization of systemic defenses takes time during which large herbivores consume large amounts of tissue until the defenses are upregulated to levels that are harmful to the herbivore. In the case of a suddenly appearing, highly mobile, large herbivore such as a grass hopper, this may lead to significant tissue loss. If such herbivores are likely to be part of the habitat of a plant, constitutive defenses or tolerance mechanisms may be more efficient.

Systemic defenses may be direct or indirect. Direct defenses result in the systemic upregulation of secondary metabolites and anti-digestive proteins. Indirect defenses result in attraction of a third trophic level which reduces the fitness of the herbivore (De Moraes et al. 1998; Turlings et al. 1990). They are mediated via volatile compounds that can be released from the wounded and systemic unwounded leaves.

If systemic defenses are adaptive, the evolutionary origins may lay in the multicellular green algal ancestors of land plants, or in early land plants such as mosses and seedless vascular plants. Although it is known that multicellular green algae (Chlorophyta) exhibit induced defenses against herbivore grazers (Toth and Pavia 2007), to our knowledge, nothing is known about systemic defenses in these plants. The absence of vascular tissue would also require other means of systemic signaling, e.g., through diffusible factors (Stratmann et al. 1996). Interestingly, in the colonial green alga *Volvox carteri*, genes with a possible defensive function are upregulated in response to wounding, and a diffusible glycoprotein triggers transition to a sexual life cycle (Amon et al. 1998). In the mosses *Physcomitrella patens* and *Dicranum scoparium*, defenses and defense signals known from angiosperms are present (Ponce De León et al. 2012), and wounding resulted in the release of volatile oxylipins, which exhibited anti-feeding properties to a slug (Rempt and Pohnert 2010; Wichard et al. 2005). However, it remains to be determined whether these oxylipins play a role in systemic responses to herbivory. Also, not much is known about systemic defenses in seedless vascular plants. Studying defense against herbivory in green algae and primitive land plants should be fertile grounds to gain insights into the evolution of systemic induced defenses.

This chapter is focused on herbivory-induced systemic signals, which we broadly define as signals that move beyond the wound site inflicted by an herbivore. These may include short- and long-distance signals that can travel multidirectionally through the airspace or phloem, from cell to cell, or unidirectionally within the xylem. We define long- and short-distance signals as signals that do or do not extend beyond the wounded leaf or leaflet, respectively.

Primary signals directly generated by insects do not necessarily function as systemic signals. For example, it was demonstrated that the fatty acid–amino acid conjugate (FAC) volicitin, an elicitor present in oral secretions of lepidopteran larvae, does not move beyond the wound site (Truitt and Pare 2004). This also appears to be true for danger-associated molecular patterns (DAMPs) such as oligogalacturonides (Baydoun and Fry 1985). Herbivore-derived elicitors and DAMPs rather induce the synthesis or mobilization of plant-derived systemic signals. These signals move sufficiently fast to escape ingestion by large insects, such as 5th instar *Manduca sexta* larvae, which rapidly consume large areas of leaf tissue (Schittko et al. 2000).

Such rapid signals could be either volatiles or physical signals. Green leaf volatiles (GLVs) are generated immediately (within 20 s) upon tissue wounding and herbivory (Matsui et al. 2000), and are known to upregulate or prime defenses (Arimura et al. 2005; Engelberth et al. 2004; Farag and Paré 2002). Tissue damage by insect mandibles also generates electrical and hydraulic signals that are rapidly propagated via the phloem or xylem, respectively. Hydraulically dispersed chemical signals were estimated to be translocated at a speed of at least 450 mm/min (Malone 1993). Electrical signals depend on living phloem tissue and include wound-induced action potentials (60–270 mm/min) (Rhodes et al. 1996; Stankovic and Davies 1996; Wildon et al. 1992), slow wave or variation potentials caused by pressure-induced membrane depolarizations (6–250 mm/min) (Stahlberg and Cosgrove 1997; Stankovic and Davies 1996), and system potentials (50–100 mm/min) (Zimmermann et al. 2009). The wide ranges for the velocity of physical systemic signals may be due to decreases over time and distance, and species-specific factors. When voltage changes or hydraulic pressure were not determined directly, delaying factors like signal transduction may result in lower estimated velocities.

Chemical phloem-mobile signals are expected to move slower than physical signals at, approximately the velocity of sucrose, which was determined to be in the range of 10–20 mm/min, depending on the specifics of sink-source relationships (Aloni et al. 1986; Peuke et al. 2006). In Arabidopsis, systemic JA-inducing wound signals were estimated to move at least 20–30 mm/min, which was thought to exclude a slow-moving chemical signal (Glauser et al. 2009; Koo et al. 2009). Surprisingly, further analysis of the nature of the systemic wound signal in Arabidopsis revealed that it is an auto-propagated reactive oxygen species (ROS) wave reaching a speed of 84 mm/min signal (Miller et al. 2009).

Taken together, many different types of signals can move at a speed of faster than 1 mm/s, which may be fast enough to escape ingestion by the herbivore. In general, different systemic signals may have evolved as adaptations to specific selection pressures such as different feeding behaviors among herbivorous insects.

It is not clear whether the signal that is generated at the wound site is the same as the signal that moves to distant parts of the wounded leaf or plant. For example, in tomato, JA or a JA derivative is the long-distance signal that translocates from wounded to unwounded leaves. Systemic JA translocation requires the action of the alarm hormone systemin, which induces JA synthesis. Accordingly, systemic JA signaling is absent in the systemin-insensitive mutant *spr1* (Kandoth et al. 2007; Lee and Howe 2003; Li et al. 2002b). Systemin is thought to be active close to the wound site, although this has not been thoroughly investigated. Systemin, like JA, is capable of moving slowly in the phloem (Nárvaez-Vasquez et al. 1994; Pearce et al. 1991), and if it were only active at the wound site, it would likely be ingested by herbivores. On the other hand, systemin action requires an unknown mode of activation or release, such as processing from its preformed prohormone, prosystemin. A possible scenario would be that prosystemin processing is initiated in the vicinity of the wound site in response to a primary rapid herbivory-induced mechanical or volatile signal. Systemin then functions as a secondary local or short-distance signal which upregulates JA synthesis. The tertiary systemic signal JA then moves through the phloem to distant parts of the plant and activates expression of defense genes.

Vascular systemic signals, such as systemin and JA, are constrained by the orthostichy of the vasculature and can only reach target leaves that are connected to the wounded leaf (Jones et al. 1993; Orians et al. 2000; Schittko and Baldwin 2003). This limitation does not apply to volatile signals (Frost et al. 2007). On the other hand, volatile systemic signals function at the mercy of the weather. Optimal systemic defenses may therefore depend on a combination of vascular and volatile signals (Heil and Ton 2008). The size of a plant may pose another limitation for systemic signaling. In a big plant like a tree, relatively large quantities of long-distance signal must be produced to avoid dilution when moving long distances, regardless as to whether the signal is transported through the vasculature or airspace. In a large plant, it may be more adaptive to induce or prime systemic responses that gradually decrease with the distance from the wound site, where the imminent danger of attack is highest.

Additional factors potentially affecting systemic signaling are the herbivore feeding guild, the abiotic environment, and the developmental stage of a plant. Feeding guilds not only cause different types of damage and induce different defense responses, but they are also dissimilar with regard to the mechanics of wounding and ensuing mechanical signals. It has been shown that caterpillars like *Spodoptera littoralis* are capable of generating hydraulic (Alarcon and Malone 1994) and electrical (Maffei et al. 2004) long-distance signals. On the other hand, aphids and cell content feeders do not target the xylem, and the extent of damage over time is more confined compared to a chewing insect, so they are less likely to generate mechanical alarm signals. However, aphids may secrete or generate locally acting elicitors such as saliva factors which could trigger the production of physical or chemical systemic signals (Will and van Bel 2008).

The abiotic environment (e.g., drought, salt, heat, and humidity) affects transpiration and the water status of the plant which directly affects xylem-borne hydraulic signals (Malone and Alarcon 1995), and indirectly phloem-borne and airborne signals. Systemic signaling is also dependent on plant development (Van Dam et al. 2001). Flowers and fruits represent strong carbon sinks which may impede transport of a phloem-borne defense signal from a wounded carbon source leaf against the assimilate flow to another source leaf.

In the field, it is likely that a wide range of environmental and developmental stimuli affects systemic defense signaling and defense responses. This may be further complicated by root and shoot symbionts such as mycorrhizal fungi or fungal and bacterial endophytes, known to affect plant–insect interactions in various ways, which is are likely to involve shoot–root and root–shoot communication (Partida-Martinez and Heil 2011).

2 Peptide Hormones

2.1 Systemin

The pivotal discovery by Green and Ryan (1972) that mechanical wounding led to the accumulation of proteinase inhibitors (PIs) in wounded tomato plants provided the first hint of a mobile signal that was able to travel from the wound site to distal unwounded tissues to induce the bunkering of these defense-related proteins. It also provided a simple but powerful experimental tool with which to start the search for the elusive *Protease Inhibitor Inducing Factor*, or PIIF. After many false PIIF leads, mostly represented by fragments of polygalacturonic acid derived from cell wall extracts that looked promising until they proved to be rather immobile and bioactive at relatively high concentrations (Bishop et al. 1981, 1984; Baydoun and Fry 1985), the search finally yielded a most unexpected candidate: the bioactive 18 amino-acid peptide systemin, whose name was inspired by its presumed role as the systemic wound signal. When systemin was applied to wound sites, it was found to move in the phloem similarly to sucrose and to exert its bioactivity at the sub-nanomolar range (Pearce et al. 1991). These two characteristics, in addition to the wound inducibility of the gene coding for its precursor protein (see below), the effects of the antisense gene in blocking systemic wound signaling, the reestablishment of the systemic wound response in antisense plants by the exogenous addition of systemin, and the effectiveness of prosystemin overexpressing plants against plant-chewing insect pests, prompted the proposal that systemin was the long-sought mobile signal responsible for the induction of the systemic wound response in tomato (reviewed in Ryan 2000; Dombrowski 2003). However, elegant grafting experiments based on the insightful use of a number of jasmonic acid (JA) biosynthesis and perception mutants and systemin signaling mutants demonstrated that systemin is not the mobile signal in tomato. Instead, these series of experiments clearly showed that the production of JA in damaged local leaves and its perception in distal leaves are necessary for inducing systemic responses (Li et al. 2002a; Lee and Howe 2003; Howe 2004). However, wounded *spr1* (systemin-insensitive mutant) rootstock leaves were unable to send a systemic signal to wild-type scion leaves (Lee and Howe 2003), suggesting that systemin is required for mobilization of the long-distance signal. Thus, the currently accepted status of systemin is that of a bioactive peptide that functions locally by inducing and amplifying JA and its own production along the vascular tissues, a process which is believed to sustain the long-distant transport of the JA-derived systemic mobile wound signal (Ryan and Moura 2002; Schilmiller and Howe 2005). Recent studies also reported that systemin enhances the systemic production of bioactive volatile compounds capable of attracting beneficial parasitiod wasps by activating genes involved in their biosynthesis. This implied that systemin also plays a role in the systemic activation of indirect defenses in tomato (Corrado et al. 2007; Degenhardt et al. 2010).

Systemin is released from the C-terminal end of a 200-aa precursor protein, known as prosystemin (McGurl et al. 1992). A critical role for prosystemin/systemin in wound signaling was established by transforming tomato plants with the prosystemin cDNA in an antisense orientation, which proved unable to accumulate PIs in response to mechanical damage and were therefore compromised in their ability to resist damage by feeding *Manduca sexta* larvae (Orozco-Cárdenas et al. 1993). In contrast, prosystemin overexpressing tomato plants are known to accumulate high levels of PIs and other proteins, mostly believed to have a role in defense, and have been found to be more resistant to chewing insects and cell-content feeding herbivores than control plants (McGurl et al. 1994; Bergey et al. 1996; Li et al. 2002b; Chen et al. 2005). The role of the N-terminal domain of prosystemin is still the subject of speculation. However, work by Dombrowski et al. (1999) showed that the PI-inducing activity of prosystemin, although more active than systemin on a molar basis, had an absolute requirement for a functional C-terminal systemin domain.

Another unknown aspect of systemin's mode of action is the mechanism by which it is released from the prosystemin precursor protein. It has been suggested that the processing and subsequent release of systemin from its precursor result from the mixing of prosystemin with proteolytic enzymes from other compartments as a consequence of the disruption of cell integrity by insect feeding or mechanical wounding (Ryan 2000). An inkling of the possible existence of such processing enzymes was generated from transgenic tobacco plants that were engineered to produce a version of tomato prosystemin in which the systemin encoding region was replaced by a trypsin-modulating oostatic factor, a ten-amino-acid insect peptide hormone toxic to *Heliothis virescens* larvae. The plants were capable of correctly processing the cytoplasmic release of the peptide and, in consequence, exert a toxic activity against insect larvae feeding on them (Tortiglione et al. 2003).

Another important finding using tobacco as a model was derived from plants transformed with the C-terminal Sys coding sequence only. Interestingly, these plants showed a drastic phenotype, characterized by numerous morphological, physiological, and reproductive alterations, together with increased susceptibility to insect herbivory (Rocha-Granados 2004). These results strongly suggested that the N-terminal end of the prosystemin protein is important for the control of systemin activity, and may regulate functions other than defense in plants. This is supported by data that suggest that systemins are not exclusively involved in defensive functions. For instance, potato plants constitutively expressing the tomato prosystemin gene also accumulate major storage proteins in tubers, demonstrating that the systemin-processing machinery can operate in different tissues and trigger developmentally regulated genes (Narvaez-Vasquez and Ryan 2002).

Moreover, silencing of the endogenous prosystemin gene did not influence the synthesis of PIs in *Solanum nigrum* (Schmidt and Baldwin 2006), a finding that raised the possibility that systemin might help the plant to tolerate rather than resist folivory via an increase in the root allocation of photoassimilates. According to this proposal, such a mechanism may allow plants to compete more effectively with conspecifics and also to compensate for tissue losses during herbivory (Schmidt and Baldwin 2009).

Additional evidence for possible dual roles of systemin in defense and development came from a series of studies in mycorrhizal tomato that suggested that systemin plays an important role in the modulation of local and/or systemic resistance responses triggered in mycorrhizal plants infected with various fungal and oomycete pathogens and of the mycorrhizal colonization process itself. Thus, data by de la Noval et al. (2007) showed that changes in the accumulation of pathogenesis-related (PR) proteins such as chitinases (CHI), β-glucanases (BG), peroxidases, and phenylalanine ammonia lyase could be modulated by exogenous systemin in mycorrhizal tomato. This was in agreement with a previous report describing that application of exogenous systemin to mycorrhizal tomato plants in the early stages of colonization induced root and leaf accumulation of BG and CHI activity (Noval et al. 2004). These results indicated that the pattern of systemic disease resistance conferred by mycorrhizal colonization is influenced by the exogenous application of systemin, which operates by means of a still undefined mechanism. Conversely, evidence for an important role in the arbuscular mycorrhizal symbiosis process came from subsequent studies showing that all colonization parameters measured in mycorrhizal roots, including arbuscule abundance, were significantly higher in prosystemin-overexpressing transgenic tomato plants (Tejeda-Sartorius et al. 2008) and that increased colonization was positively correlated with the augmented expression of selective "early" and "late" prosystemin-dependent genes, such as polygalacturonases and polyphenol oxidases, respectively (Avilés-Arnaut H. unpublished data; Ryan 2000; Lee and Howe 2003; Li et al. 2004).

Finally, recently reported proteomic data demonstrated that the constitutive expression of the tomato prosystemin gene in tobacco plants highly affected host protein synthesis, which included not only proteins induced in response to oxidative stress or pathogen/herbivorous organisms, but also genes involved in energy production and carbon metabolism, such as mitochondrial ATP synthase, iron–sulfur subunit 1 precursor of cytochrome B6-F complex, oxygen evolving complex, chloroplast precursor of ribulose-phosphate-3-epimerase, and a BTF3b-like factor (Rocco et al. 2008). The contrasting results observed between systemin and prosystemin overexpressing tobacco plants strongly suggest that the release of systemin from its precursor protein must be a strictly regulated process to avoid the possible deleterious effects caused by a systemin "overdose."

2.2 Hydroxyproline-Rich Systemins

With the discovery of systemin in tomato, a search was initiated for systemins in other plant species. Systemin homologs were found to be specific to a single clade of the Solanaceae (Constabel et al. 1998), thus requiring a biochemical approach for purification of unrelated potential defense peptides. This was undertaken in tobacco (*Nicotiana tabacum*) which had previously been shown to have a wound-inducible long-distance defense response (Pearce et al. 1993). An "alkalinization assay" was developed from previous research demonstrating that systemin, added to tomato suspension cell cultures, caused a rapid increase in extracellular pH (Felix and Boller 1995; Schaller and Oecking 1999). Tobacco extracts were purified utilizing HPLC methodology and fractions were tested in the alkalinization assay utilizing tobacco suspension cell cultures. Two active fractions were identified and were also shown to induce tobacco trypsin inhibitor in a similar manner to systemin in tomato (Pearce et al. 2001). Peptides in the active fractions were sequenced revealing two 18 amino acid peptides containing multiple hydroxyprolines. Both peptides contained pentose sugar moieties attached to hydroxyprolines and were named *Nicotiana tabacum* hydroxyproline-rich systemin I and II (*Nt*HypSysI and II).

Both NtHypSys I and II are encoded within the same precursor, NtproHypSys, and like prosystemin, the gene is upregulated by mechanical wounding, MeJA and insect feeding (Pearce et al. 2001; Rocha-Granados et al. 2005). This, along with the ability of systemin and HypSys to induce protease inhibitors for defense against insect pests, has led to a classification of systemin and HypSys within the same group based on function rather than sequence homology (Ryan and Pearce 2003). HypSys precursors, along with many of their HypSys peptides, have now been isolated from tomato, petunia, and black nightshade. HysSys peptides probably evolved earlier than systemin peptides as they have now been identified outside the Solanaceae in sweet potato (*Ipomoea batatas*, Convolvulaceae) (Pearce 2011). All of the HypSys glycopeptides isolated to date are 18–20 amino acids in length and have a hydroxyproline-rich central core motif consisting of either –OOXO– or –OXOO– where –O– is hydroxyproline and –X– is hydroxyproline, serine, threonine, or alanine. The other characteristic of HypSys glycopeptides is conservation around the extreme amino and carboxyl termini, suggesting a conservation of the proteases responsible for processing the glycopeptides from their precursors (Pearce 2011).

The role of hydroxyproline-rich systemins in long-distance signaling is believed to be as an initiator of a receptor-mediated signaling cascade that leads to JA production for long-distance signaling, much the same as systemin. In tomato, where both systemin and HypSys are present, there is an extremely potent wound induction of protease inhibitors in the non-wounded distal leaves of wounded plants, and this has been attributed to a cooperative action of systemin and HypSys (Narvaez-Vasquez et al. 2007).

Localization studies have determined that tomato *preproHypSys* is synthesized in phloem parenchyma cells and is upregulated by wounding, systemin, and methyl jasmonate (Narvaez-Vasquez et al. 2005). The precursor is modified via the secretory pathway with hydroxylation of proline and the addition of pentose sugars before being sequestered in the cell wall matrix. ProHypSys localization within the phloem and in the cell wall matrix places it in an ideal location for release of the HypSys peptides for the production of JA for long-distance signaling. SlproHypSys processing most likely occurs in the apoplast after wounding, suggesting either that nonspecific cytoplasmic proteases are mixed with cell walls upon damage or that specific proteases are synthesized or activated in response to wounding.

In addition to the indirect evidence presented above for HypSys involvement in defense against herbivory, overexpression of the NtpreproHypSys gene in tobacco plants caused an increase in protease inhibitors and polyphenol oxidase, resulting in resistance to attack by *Helicoverpa armijera* larvae (Ren and Lu 2006). Also, *NtpreproHypSys* expression was induced in plants under attack by either *Bemisia tabaci* or *Manduca sexta* (Rocha-Granados et al. 2005). These studies indicate that HypSys is an important component for defense against insect attack.

HypSys peptides may have different functions in different species. In *Nicotiana tabacum*, it was shown that silencing the HypSys gene had very little effect on insect feeding or on levels of defense-related compounds (Berger and Baldwin 2007) and was instead involved in functions such as self-pollination and flower morphology (Berger and Baldwin 2009). These observations, coupled with the surprising finding that HypSys glycopeptides supplied to petunia plants elevated the levels of *defensin-1* (Pearce et al. 2007), a gene involved in pathogen defense, suggest that future studies on HypSys glycopeptides will reveal new, insightful discoveries.

2.3 Peps

The potency of systemin and HypSys peptides in activating systemic anti-herbivore defenses raises the intriguing question of whether peptide mediation of wound responses is a conserved plant regulatory motif. While systemin peptides are readily identifiable from solanales such as potato, pepper, and black nightshade, orthologues have not been identified in non-solanaceous species (Constabel et al. 1998). Whether the lack of candidate orthologues is because peptide regulation of wound responses evolved only in the Solanaceae, or because peptides of similar function in other species have unrecognizably divergent amino acid sequences has been unknown. Intriguingly, genes bearing sequence homology to HypSys precursors have been identified in nonsolanaceous species including poplar and coffee, but functional roles of the putative peptides have not been ascertained (Pearce 2011).

A recently discovered family of peptide signals, the plant elicitor peptides (Peps), appears to function similarly to systemin in a number of ways, and may provide a missing link to answer the question of whether peptide-mediated wound responses are ubiquitous in plants. Originally discovered using the alkalinization assay in *Arabidopsis thaliana*, Peps are 23 amino acids in length and, like systemin, are enriched in basic residues (Pearce et al. 1991; Huffaker et al. 2006). The active peptides are contained at the carboxyl terminus of larger precursor proteins (PROPEPs) that, like prosystemin, are enriched in Glutamic acid-lysine-glutamic acid (EKE) motifs and have no traditional signal sequence for translocation through the secretory pathway (McGurl et al. 1992; Huffaker et al. 2006). In contrast to systemin, candidate orthologues sharing sequence homology with AtPROPEPs occur in diverse plants, having been identified in over 50 species and functionally characterized in maize (Huffaker et al. 2011). PROPEPs constitute multigene families in most plants, and research with maize Peps indicates that individual peptides likely have specific functions (Huffaker et al. 2013).

Peps interact with PEP Receptor to stimulate the accumulation of transcripts and metabolites associated with plant defense (Yamaguchi et al. 2006, 2010; Huffaker et al. 2006, 2011, 2013). Signal transduction processes activated by Peps bear striking resemblance to those initiated by systemin (Ryan 2000). Both peptides result in production and accumulation of jasmonic acid, ethylene, and ROS, calcium flux, and inhibition of a plasma membrane proton-dependent ATP synthase (Howe et al. 1996; Meindl et al. 1998; Schaller and Oecking 1999; Ryan 2000;

Orozco-Cárdenas et al. 2001; Huffaker et al. 2006, 2011; Qi et al. 2010; Krol et al. 2010; Ma et al. 2012). Constitutive expression of AtPROPEP1 in Arabidopsis resulted in plants with enhanced resistance to a necrotrophic fungal pathogen (Huffaker et al. 2006). Similarly, application of AtPep1 peptide to Arabidopsis or ZmPep1 to maize plants prior to inoculation reduces disease symptoms caused by a variety of bacterial and fungal pathogens (Yamaguchi et al. 2010; Huffaker et al. 2011).

Expression of PROPEP genes is differentially activated in response to treatment with phytohormones such as JA, SA, and ethylene (Huffaker and Ryan 2007; Yamaguchi et al. 2010; Huffaker et al. 2011, 2013). Some PROPEP precursor genes are additionally transcribed in response to stimuli associated with pathogen attack, whereas others are specifically expressed in response to herbivory cues. For instance, in maize, expression of the ZmPROPEP1 gene is strongly induced by infection with fungal pathogens or fungus-derived elicitors (Huffaker et al. 2011). In contrast, the ZmPROPEP3 gene is transcribed in response to insect-derived elicitors and caterpillar herbivory (Huffaker et al. 2013). As indicated by the transcriptional specificity, the corresponding peptides play differing roles in regulating maize defense. ZmPep1 stimulates expression of genes encoding pathogen defense proteins such as chitinases and PR proteins and can confer enhanced resistance to fungal pathogens (Huffaker et al. 2011). ZmPep3 strongly induces genes encoding anti-herbivore defense proteins such as proteinase inhibitors, and causes emission of herbivory-associated volatiles (Huffaker et al. 2013). Treatment of maize leaves with ZmPep3 is sufficient to attract parasitoid *Cotesia marginiventris* wasps involved in indirect anti-herbivore defense and to reduce the growth of *Spodoptera exigua* larvae on leaves (Huffaker et al. 2013). Pep homologues from Fabaceous and Solanaceous plants similarly induce anti-herbivore defenses, indicating that plant elicitor peptides may function analogously to systemins in many plants (Huffaker et al. 2013).

As does systemin, Peps function in an amplification loop with second messenger signals such as jasmonic acid (Schilmiller and Howe 2005; Yamaguchi and Huffaker 2011). Precursor genes encoding the peptides are expressed upon treatment with jasmonic acid, and both peptides in turn activate synthesis of jasmonic acid. This relationship serves to propagate defense signaling both spatially and temporally (Schilmiller and Howe 2005; Ryan et al. 2007). Although not yet determined, it is likely that Peps are similar to systemins in their capability of inducing systemic wound responses while not being the mobile signals themselves (Stratmann 2003). Continuing research into Pep signaling as it relates to herbivory will reveal the extent of overlap between the two peptide families in this capacity.

2.4 Other Defense Peptides

Most of our knowledge on defense peptides is restricted to the Solanaceae family, where systemin and HypSys are involved in production/amplification of JA for long-distance signaling. An effort to discover new defense peptides via the alkalinization assay led to the discovery of the AtPep family (See 21.2.3 above). The AtPeps appear to play a more universal role within the Plant Kingdom and have now been shown to be involved in defense against pathogens and insect herbivores (Huffaker et al. 2013).

When soybean (*Glycine max*) was studied for the involvement of peptides in defense, two novel peptides were isolated that were not related to any of the known defense peptide. First, GmSubPep (*Glycine max* Subtilase Peptide), a 12-amino acid peptide, was isolated and found to be embedded in a putative extracellular protease (subtilase) (Pearce et al. 2010). GmSubPep appears to be proteolyzed out of the subtilase, possibly after attack by a pathogen, and initiates a receptor-mediated induction of pathogen defense genes. Second, GmPep914, an eight-amino acid peptide, was isolated from the same soybean leaf extract (Yamaguchi et al. 2011). Its origin is a putative 52-amino acid protein, GmproPep914, which, along with a closely related gene, GmproPep890, is induced by JA. Both GmPep914 and GmPep890 stimulated expression of pathogen defense genes in suspension cultures. At present, it is unknown whether these peptides activate long-distance signaling in whole plants. However, the discovery of two evolutionary unrelated soybean peptides, GmSubPep and GmPep914/890, that regulate the same set of defense genes in suspension cell cultures, raises the question as to whether these genes are induced by herbivores or pathogens in planta. In addition, like systemin and HypSys, both GmSubPep and GmPep914/890 appear to be restricted to a narrow range of the Kingdom Plantae: the nitrogen-fixing branch of the Rosids. This is perhaps indicative of the complexity and specificity of insect/pathogen–plant defense interactions.

3 Phytohormones

3.1 Jasmonic Acid

Jasmonic acid or jasmonates (JAs) are biochemically defined as cyclopente(a)none compounds derived from lipoxygenase-dependent oxidation of polyunsaturated fatty acids. Biosynthesis involves several steps and organelles, collectively known as the octadecanoid pathway, in order to yield the 3*R*, 7*S* stereoisomer of JA [also known as (+)-7-iso-JA]. The final steps involve transport of JA to the cytosol where it is conjugated specifically to isoleucine (Ile) by the enzyme JASMONIC ACID RESISTANT1 (JAR1) (Staswick and Tiryaki 2004; Suza and Staswick 2008). Following conjugation, JA–Ile presumably diffuses into the nucleus where it binds to CORONATINE-INSENSITIVE1 (COI1)–JASMONATE ZIM-domain (JAZ) receptor complexes to activate gene expression (Thines et al. 2007; see below).

Bioactive JAs perform a major role in regulating plant responses to both wound stress and arthropod herbivory, although the respective transcriptional changes induced by these two types of stress overlap only marginally (Reymond et al. 2000). The control of plant-arthropod herbivore and certain plant–pathogen interactions is also highly dependent on what appear to be two apparently antagonistic branches of JA signaling (Lorenzo and Solano 2005; Kazan and Manners 2008). One branch, believed to be based on the activation of the basic helix-loop-helix leucine zipper MYC transcription factor (TF) acting in concert with abscisic acid (ABA) signaling (Anderson et al. 2004; Lorenzo et al. 2004), controls the interaction with a whole gamut of arthropod herbivores (Walling 2008; Howe and Jander 2008; Browse and Howe 2008; Koo et al. 2009; Diezel et al. 2011; Avila et al. 2012; Kallenbach et al. 2012). The second branch, which controls growth and responses to necrotrophs and probably oomycetes, relies on several transcription factors (TFs) including ERF1 and ORA59, which belong to the APETALA2/ETHYLENE RESPONSE FACTOR (AP2/ERF) family of TFs, and EIN3 and EIL1 which are controlled by JA and ethylene (ET) (Verhage et al. 2011 and references therein).

Substantial knowledge regarding the function of JAs has been generated from the study of mutant plants that are blocked in the synthesis of linolenic acid (18:3), the fatty acid precursor of JA (McConn and Browse 1996; Li et al. 2003), or are deficient in enzymes of the JA synthesis pathway (Schaller et al. 2005). Other mutants defective in JA perception have also been instrumental for the study of induced resistance to herbivory and long-distance JA signaling (Browse 2009). An example of this was provided by grafting experiments based on tomato (*Solanum lycopersicum*) mutants defective in JA synthesis (*spr2*, *acx1a*) and perception (*jai1*), and a systemin insensitive mutant (*spr1*). These experiments demonstrated that not only a JA-derived compound, rather than systemin, is a component of the mobile signal for wound-induced systemic defense responses in tomato, but also that systemin is required for the synthesis of JA in the wounded leaf (Li et al. 2002, 2005). These elegant grafting experiments convincingly showed that long-distance systemic signaling requires both the biosynthesis of JA in rootstock tissues and the ability to perceive a jasmonate signal in distal leaves of the scion (Li et al. 2005, 2002). Accumulated evidence from these and other studies support the hypothesis that JA regulates the production of, or acts as, a mobile wound signal. A model proposed at the time suggested that the peptide signal systemin promotes long-distance defense responses by amplifying jasmonate production in vascular tissues, thereby reinforcing the concept that systemin and JA interact through a positive feedback loop to propagate the long-distance signal (Schilmiller and Howe 2005).

The model devised by Schilmiller and Howe (2005) to explain long-distance signaling in tomato plants proposed that binding of systemin to its receptor (SR160) activated the octadecanoid pathway for JA biosynthesis. It predicted that the still unresolved proteolytic processing of prosystemin to systemin, and its release from the phloem parenchyma led to the production of JA in the companion cell–sieve element complex followed by loading into sieve elements across plasmodesmatal connections which presumably allowed long-distance transportation in the phloem.

The model correctly foretold that modified forms of JA could play a role in systemic signaling and that the rate of movement of the signal in tomato, based on the capacity of the phloem to transport small molecules at rates up to 40 cm/h, oscillated between 1 and 5 cm/h. It has now been demonstrated that (+)-7-iso-jasmonoyl–isoleucine (JA–Ile) is the key molecule in JA-mediated signal transduction (see following paragraph), although structurally related JA-amino acid conjugates, such as JA-Val and JA-Leu, are also known to have weak biological activity (Katsir et al. 2008). Interestingly, JA, methyl jasmonate (MeJA), or 12-oxo-9(S),13(S)-phytodienoic acid (OPDA), the JA precursor, were found to have no biological activity (see below), acting only as substrates for the biosynthesis of JA–Ile conjugates catalyzed by JAR1, an ATP-dependent adenylate-forming enzyme (Staswick et al. 2005; Wu et al. 2008; Stitz et al. 2011). The use of deuterium-labeled analogs in tobacco and tomato confirmed the basic tenets of the mentioned model by finding that exogenously applied JA–Ile is transportable to distal sites to induce the plant defense system. Surprisingly, JA–Ile was found to move at a much faster rate than JA despite its lower polarity, and that a transient accumulation of JA and JA–Ile could be observed within 30 min of wounding in the damaged and distal leaves (Sato et al. 2009, 2011).

Exogenously applied JA–Ile was biologically active in the distal leaves of tobacco plants and the stem exudates from tomato plants showed the expected wound-induced accumulation of JA–Ile. The transport of the targeted compounds could not be observed without wounding, which suggested that the generation of a still undefined wound signal is essential for long-distance JA-based signaling. Parallel research on JA biosynthesis and signaling in Arabidopsis coincided with this observation by demonstrating that a mobile signal other than JA is produced in the damaged leaf and transported to the distal tissue. Oligosaccharides, ROS including S-nitrosothiols such as S-nitrosoglutathione (GSNO; Sect. 17.4), systemin (Sect. 17.2.1), volatile organic compounds (Sect. 17.6), other plant hormones (Sect. 17.3.2), and hydraulic (Sect. 17.5) and electrical signals (see Chap. 18) have all been proposed as wound-induced signals, able to operate at different temporal and spatial scales (Pearce et al. 1991; Leon et al. 2001; Wasternack 2007; Heil and Ton 2008; Koo et al. 2009; Zimmermann et al. 2009; Espunya et al. 2012).

In accordance with the signaling kinetics expected, JA accumulated rapidly in tissues proximal to injury sites within 30 s of injury in wounded Arabidopsis leaves. Moreover, a general JA increase distal to wounds was detected throughout the plant within 2–5 min of wounding. The accumulation was strongest in unwounded leaves with direct vascular connections to wounded leaves wherein JA levels increased significantly within 120 s of wounding. This implied that the long-distance wound signal leading to distal JA accumulation traveled to unwounded leaves at a velocity of at least 3 cm/min (Glauser et al. 2008, 2009; Koo et al. 2009). This was much faster than the reported transport of small molecules in the phloem of wheat plants (up to 0.7 cm/min) (Fisher 1990), thereby reinforcing the notion that the systemic signal was not a mobile jasmonate. Thus, the rapid JA accumulation in distal tissues in response to wounding was deemed to occur either from the metabolism of OPDA and dinor-OPDA pools (Koo et al. 2009; Schaller and Stintzi 2009) or from a de novo event beginning with fatty acid oxygenation (Koo et al. 2009; Glauser et al. 2008, 2009). The first scenario meant that Arabidopsides, which are esterified forms of OPDA and dinor-OPDA, could be potential precursors for JA synthesis, although the concept that unbound pools of OPDA could constitute the major precursors for rapid JA synthesis has also been considered (Chung et al. 2009). Other potential sources of nascent JA have been considered and include tri-unsaturated fatty acids, released by specific glycerolipases, (Kallenbach et al. 2010), other JA precursors such as MeJA (Stitz et al. 2011 and references therein), and probably other reversibly bound cyclopentenone jasmonates. The rapidity of the de novo synthesis of distal JA–Ile necessarily meant that a pathway enzyme preceding JAR1 was likely to be the regulated step in systemic JA–Ile production. The observed temporal correspondence observed between the rise in systemic JA–Ile and the decline in free OPDA suggested that OPDA metabolism could be a control point for JA/JA–Ile production in distal leaves (Koo et al. 2009).

It has been hypothesized that control of JA/JA–Ile biosynthesis might be achieved by the phosphorylation/dephosphorylation cycles of key enzymes regulated by mitogen-activated protein kinases (MAPKs), which are known to play a key role in the wound induction of JA synthesis. Support for a role of MAPKs in early wound signaling leading to JA accumulation comes from experimental evidence in tobacco and tomato showing that a rapid wound-generated xylem-borne signal is able to activate MAPKs (Kandoth et al. 2007; Seo et al. 1995, 1999; Stratmann and Ryan 1997; Wu et al. 2007). On the other hand, the peroxisomal flavoprotein enzyme OPR3 and downstream enzymes in the β-oxidation pathway (Schaller et al. 2005; Browse 2009) are also considered to be potential control points for the wound-induced posttranslational regulation of JA synthesis (Breithaupt et al. 2006). Also, transport of OPDA into peroxisomes via the ATP-binding cassette transporter COMATOSE (Theodoulou et al. 2005) could represent an additional potential point of posttranslational regulation of OPDA metabolism.

In addition, one or more lipoxygenases (LOX) contribute directly to rapid OPDA and JA synthesis upon wounding. This is in agreement with the finding that LOX2 contributes to approximately 75 % of JA measured shortly after wounding Arabidopsis leaves and with the observed suppression of 75 % of wound-induced JA synthesis in leaves by a dominant version of a microRNA (miR319a) that indirectly and negatively regulates *LOX2* gene expression (Schommer et al. 2008). LOX2 is also known to require divalent cations such as Ca^{2+} for activity. In fact, Ca^{2+} transients are believed to be important for initiating JA synthesis in potato, and *LOX2* transcripts and LOX enzyme activity are upregulated in a cation channel mutant known to alter K^+ and Ca^{2+} homeostasis in *Arabidopsis*. Despite its obvious importance in JA biogenesis, LOX2 is not considered essential for some of the most rapid events elicited by wounding (Glauser et al. 2009; Gfeller et al. 2010).

JA rapidly accumulates in wounded tomato leaves, increasing at least 40-fold within 90 min after a single wound (Weber et al. 1997). After 90 min, JA levels start to subside, reaching half-maximal values approximately 9 h after wounding (Reymond et al. 2000). In the past years, the mechanisms employed by the plant to reduce the jasmonate burst back to the resting levels found prior to wounding have been gradually elucidated. The detection of polar JA metabolites, predominantly 12-hydroxyjasmonoyl-isoleucine (12-HOJA-Ile), 12-hydroxyjasmonate (12-HOJA), and 11-hydroxyjasmonate (11-HOJA), that started to accumulate at ~45 min post-wounding in Arabidopsis signaled the start of an enzyme-mediated jasmonate clearance phase that clearly involved targeted hydroxylation steps (Glauser et al. 2009). This was confirmed by recent studies showing that JA–Ile turnover is mediated by a ω-oxidation pathway involving members of the CYP94 family of cytochromes P450 (Koo et al. 2011; Heitz et al. 2012). The ω-oxidation pathway, in which JA–Ile is converted to 12-hydroxy-JA-Ile (12-OH-JA-Ile) and then further oxidized to dicarboxy-JA-Ile (12- COOH-JA-Ile), is now recognized as a major route for catabolism of the hormone (Koo and Howe 2012). Oxidized JA–Ile derivatives are likely to exist but have yet to be reported. Two distinct possibilities are the formation of *O*-linked glucosyl derivatives of 12-OH-JA-Ile, which could be sequestered in the vacuole, and the esterification of the newly formed carboxyl group of 12-COOH-JA-Ile that could be targeted for subsequent degradation via the β-oxidization pathway in the peroxisome. Incomplete clearance of endogenous JA–Ile in mutants that are impaired in JA–Ile12-hydroxylation suggests that additional pathways may participate in JA–Ile metabolism. The alternative metabolic routes of hormone deactivation are formation of JA–Ile glucose esters and hydrolysis of JA–Ile to free JA and Ile. It has also been suggested that esterification of JA–Ile to JA–Ile–Me, as well as epimerization of (3*R*,7*S*)-JA–Ile to the less active (3*R*,7*R*) isomer, may be endogenous mechanisms to reduce the activity of the hormone, although recent in vivo studies indicate that epimerization of (3*R*,7*S*)-JA–Ile is unlikely to play a significant role in deactivation of JA–Ile (Suza et al. 2010; Koo and Howe 2012). Other major routes of JA metabolism are methylation and decarboxylation of C-1 to yield volatile MeJA and cis-jasmone, respectively, and reduction of C-6 to yield cucurbic acid, which may also be esterified to sugar residues.

3.2 *Other Phytohormones*

In addition to JA, other phytohormones have been implicated as contributing regulators of systemic wound responses, with the predominant interactors being salicylic acid (SA), ethylene (ET), and ABA (Bari and Jones 2009). Together, complex cross-talk between these signaling pathways fine tunes plant responses, and the relative contributions of each as well as antagonistic or synergistic effects on defense are highly situational. Experimental observations vary in relation to plant species, attacking organisms and environment. Herbivory and exposure to insect-derived elicitors induce production of JA, ET, and in some cases SA (Schmelz et al. 2009). ABA has not been observed to accumulate in response to herbivore elicitors, but is produced in local tissues after herbivory, potentially as a reaction to desiccation caused by tissue damage (Birkenmeier and Ryan 1998; Thaler and Bostock 2004).

An antagonistic relationship between SA and signaling mediated by JA has been widely observed (Dong 2004; Beckers and Spoel 2006; Thaler et al. 2010). SA was originally reported to inhibit wound or oligogalacturonic acid (OGA)-induced proteinase inhibitor (PIN) accumulation in tomato and subsequently demonstrated to also inhibit accumulation in response to systemin and JA (Doherty et al. 1988; Doares et al. 1995). A vast number of subsequent experiments have confirmed the antagonism between the two pathways, and it has been observed that in the context of herbivory, JA-mediated responses are induced by chewing insects, while SA-induced defenses are more strongly activated by piercing and sucking insect pests (Moran and Thompson 2001; de Vos et al. 2007). Elicitation of SA-associated signaling has been demonstrated to be directly detrimental to plant systemic defense against chewing herbivores as measured by relative plant damage and herbivore growth rates (Thaler et al. 2002, 2010; Mewis et al. 2006). However, experimental evidence for the role SA might play in systemic immunity against sucking insects such as aphids has been contradictory. Some studies report that increased SA accumulation enhances aphid resistance (Avila et al. 2012). However, screening of mutant plants compromised in SA signaling did not reveal any general trends relating to aphid resistance (Thompson and Goggin 2006; De Vos et al. 2007). Surprisingly, given that aphids stimulate SA production and associated defenses, mutant plants with impaired JA signal transduction were less resistant to aphid-induced damage and allowed increased proliferation (Mewis et al. 2006; Thompson and Goggin 2006).

Methylated salicylic acid (MeSA) is also produced in response to herbivory by both chewing and sucking/piercing insects, and has effects that are equally difficult to generalize (Chen et al. 2003). MeSA is a constituent of herbivory-induced volatile blends emitted by many plant species and is a predominant component for some (Michereff et al. 2011). Predatory mites that are natural enemies of spider mite herbivores are attracted to mite-infested tomato plants, but not to plants unable to convert SA to MeSA, implying a role for MeSA as an attractant (Ament et al. 2006). In contrast, MeSA was detrimental to attraction of parasitoid wasps by *Pieris rapae*-infested Arabidopsis plants (Snoeren et al. 2010). While MeSA has been demonstrated to act as a systemic signal with respect to responses against pathogens, regulation of systemic anti-herbivore defenses remains less clear (Shulaev et al. 1997).

Ethylene is intimately entangled with both SA and JA to modulate defense (O'Donnell et al. 1996; Adie et al. 2007; von Dahl and Baldwin 2007). While ethylene treatment is not sufficient to induce systemic anti-herbivore responses such as PIN accumulation in tomato, intact ethylene signaling is required for elicitation of systemic defense by signals such as JA, OGAs, systemin, and wounding (O'Donnell et al. 1996). Ethylene is concomitantly produced with JA by herbivory-associated stimuli and modulates the magnitude of JA-induced defenses (O'Donnell et al. 1996; Von Dahl and Baldwin 2007; Schmelz et al. 2009). Treatment of tomato plants with salicylic acid not only reduces PIN accumulation upon challenge with JA or systemin, but also diminishes ethylene emission in response to these stimuli (O'Donnell et al 1996). Similarly, in maize pretreatment of leaves with inhibitors of ethylene biosynthesis dramatically reduces accumulation of transcript encoding anti-herbivory defense genes and volatile metabolites upon caterpillar elicitation (Schmelz et al. 2003; Harfouche et al. 2006). Interestingly, the magnitude of JA accumulation was unchanged in these circumstances; indicating that ethylene modulation of responses is not dependent upon JA levels (Schmelz et al. 2003).

Although the mechanisms of cross-talk between the three hormone pathways are undoubtedly complex, NON-EXPRESSOR OF PATHOGENESIS-RELATED GENES1 (NPR1) protein has emerged as a node of great importance in signal convergence and modulation (Dong 2004). Although NPR1 is a modulator of transcription in the nucleus, it also mediates SA interference with JA signaling, with this function occurring in the cytosol (Dong 2004; Spoel and Loake 2011). This interaction is dependent upon cellular redox states and in particular the ratio of oxidized versus reduced glutathione; JA treatment results in reduced overall levels of glutathione and conversion of remaining pools to a predominantly oxidized state (Spoel and Loake 2011). SA induces increased levels of glutathione with ratios strongly favoring the reduced state, triggering an environment in which disulfide bridges of cytosolic NPR1 are reduced (Mou et al. 2003; Koornneef et al. 2008). This not only results in relocalization of some NPR1 to the nucleus, but also is necessary for SA antagonism of JA signaling (Koornneef et al. 2008). If SA-induced accumulation of glutathione is blocked, SA interference with JA responses is also eliminated (Koornneef et al. 2008).

Ethylene modulation of systemic defense signaling also involves NPR1. Ethylene enhances SA-mediated responses in an NPR1-dependent manner through potentiation of NPR1 activated transcription (Leon-Reyes et al. 2009). However, ethylene also prevents SA inhibition of JA responses through NPR1: for biotic interactions in which relatively great quantities of ethylene are produced simultaneously with JA, or in the case of pretreatment with ethylene, SA antagonism of JA is abolished (Leon-Reyes et al. 2009, 2010).

In addition to the complex web of JA, SA, and ethylene signaling, ABA has also been implicated as a signal mediating some wound responses. While ABA was once proposed to be a major contributor to wound-induced PIN accumulation in tomato, subsequent studies indicated that ABA is much less potent than signals such as systemin or JA in eliciting this response (Peña-Cortés et al. 1995; Birkenmeier and Ryan 1998). Plants deficient in ABA are less resistant to caterpillar herbivory, but the association of ABA with the wound response and the observed stimulatory effects have been attributed to ABA-mediated drought acclimation rather than direct synergism with JA-mediated responses (Birkenmeier and Ryan 1998; Thaler and Bostock 2004). Although these results indicate that ABA is not a direct regulator of systemic wound responses, they highlight the dependence of defense on abiotic stress conditions. Characterization of maize responses to herbivory at both roots and shoots indicated that ABA accumulated systemically, whereas JA was not observed to accumulate (Erb et al. 2009). Because root herbivory led to enhanced resistance to subsequent foliar herbivory, the possibility of ABA as a systemic regulator of wound responses was again raised (Erb et al. 2009). However, follow-up research demonstrated that ABA was not responsible for induced systemic resistance to herbivory, although it was not determined which molecules were alternatively acting as systemic signals (Erb et al. 2011).

Subsequent studies have introduced increased complexity to the interaction of ABA with JA and ethylene (Anderson et al. 2004; Dombrecht et al. 2007; Erb et al. 2009; Pauwels et al. 2010). ABA- and JA-induced signaling both activate the MYC2 transcription factor in Arabidopsis (Anderson et al. 2004; Lorenzo et al. 2004). This leads to differential transcriptional repression or stimulation of subsets of JA-responsive genes, with anti-herbivore defense genes among those that are positively regulated (Anderson et al. 2004; Lorenzo et al. 2004; Dombrecht et al. 2007). These mechanistic studies of ABA intersection with JA and wound signaling have been predominantly focused on Arabidopsis; future research into molecular mechanisms driving the interaction in other plants will establish whether such a link is common.

4 ROS Signals

4.1 The Central Role of ROS Signaling in Plant–Herbivore Interactions

There is a large body of evidence obtained by using diverse experimental approaches and arthropod herbivore–plant interactions that supports the view that ROS signals are integral to plant–insect interactions and the systemic wound response. A survey of wound-induced signaling in plant species demonstrated ROS production as a conserved response (Orozco-Cardenas and Ryan 1999). Many subsequent studies demonstrated local ROS accumulation in response to chewing insects, mites, and aphids (Leitner et al. 2005; Moloi and van derWesthuizen 2006; van Eck et al. 2010), and even oviposition (Little et al. 2007). NADPH oxidase and xanthine oxidase were involved in aphid-triggered ROS accumulation (Moloi and van der Westhuizen 2006, 2008; Berner and van der Westhuizen 2010). Several transcriptomic and proteomic studies have demonstrated the activation of ROS-responsive and ROS-metabolizing systems following insect infestation of plants (Arabidopsis, rice, Sitka spruce, poplar, *Brassica oleracea*, barley, and sorghum), particularly by phloem feeders (Park et al. 2006; Ralph et al. 2006, 2008; Couldridge et al. 2007; Kempema et al. 2007; Lippert et al. 2007; Broekgaarden et al. 2008; Kuĉśnierczyk et al. 2008; Gutsche et al. 2009; Wei et al. 2009; Philippe et al. 2010). ROS are critical second messenger contributors to systemic signaling, and when tomato H_2O_2 production was blocked via inhibition of NADPH oxidase, systemic activation of defense genes in response to systemin, OGAs, wounding, or MeJA was abolished

(Orozco-Cárdenas et al. 2001). Maffei et al. (2006) provided evidence for systemic ROS signaling by an insect-specific ROS signature in response to chewing and piercing-sucking insects and mechanical damage in lima bean. In an elegant study using a ROS-responsive promoter coupled to the luciferase reporter gene, Miller et al. (2009) uncovered the underlying mechanism for systemic ROS signaling in Arabidopsis. Mechanical wounding and extreme environmental conditions produced a rapid systemic ROS signal that traveled at a rate of 8.4 cm/ min and was dependent on the NADPH oxidase RbohD. ROS production by one cell triggered ROS production in an adjacent cell, thus leading to a systemic auto-propagating ROS wave. The signal could be inhibited by distal suppression of ROS accumulation. Additionally, it was proposed that different ROS wave signatures could be generated by the diffusion of ROS from different subcellular compartments and that such ROS waves could encode specific information (Mittler et al. 2011) relevant within the context of plant interactions with the environment (Kerchev et al. 2012).

4.2 Role of NO in Systemic Signaling

Nitric oxide (NO) is now considered a key signaling molecule in plants. NO is a lipophilic gas under atmospheric conditions that can rapidly diffuse through membranes (reviewed by Hong et al. 2008; Gupta 2011). The presence of an unpaired electron in NO is responsible for its high reactivity with oxygen, superoxide ion, transition metals, and thiols. These reactions are fundamental for the cellular functions of NO within the cell. Thus, NO can react with protein and nonprotein thiols to form nitrosothiols (SNOs) (Stamler 1995). A key representative of SNOs in the plants is GSNO, a bioactive, stable, and mobile reservoir of NO, that plays an important role in defense responses to herbivory and pathogen attack in plants (see below). Levels of GSNO are controlled mainly by the enzyme GSNO reductase (GSNOR) (Liu et al. 2001). Evidence supporting the phloem mobility of GSNO was provided by immunolocalization studies in Arabidopsis showing that GSNO labeling increased rapidly and uniformly in wounded leaves, whereas in systemic leaves, the GSNO signal was detected in vascular tissues prior to its spread to the parenchyma cells. Such movement pattern, together with the reported localization of GSNOR protein in the phloem (Rustérucci et al. 2007) and of its substrate, GSNO, in the collenchyma cells located adjacent to the vascular cambium (Barroso et al. 2006) strongly suggests that GSNO is involved in the transmission of the mobile wound signal through the vascular tissue (Espunya et al. 2012). Such results were also in agreement with previous findings in sunflower hypocotyls and Arabidopsis showing that wounding produces an accumulation of GSNO with a concomitant reduction of GSNOR content (Chaki et al. 2011), and a JA- and wound-dependent downregulation of GSNOR expression, respectively (Díaz et al. 2003). In turn, NO inhibits JA- and wound-induced defenses such as proteinase inhibitor accumulation, and when scavengers of NO were applied to tomato plants, this inhibition was ameliorated (Orozco-Cárdenas and Ryan 2002).

Although GSNOR appears to be a key regulator of systemic defense responses, in wounding (Espunya et al. 2012), insect herbivory (Wünsche et al. 2011), and pathogenesis (Feechan et al. 2005; Tada et al. 2008; Lindermayr et al. 2010), it remains to be determined whether GSNO itself, or a secondary signal generated at the wounded leaves, is the long-distance signal responsible for systemic accumulation of GSNO in distal uninjured tissues.

5 Physical Signals

Physical herbivory-induced systemic signals can be of either electrical or hydraulic nature. Electrical signals are covered by Zimmermann and Mithoefer in another chapter of this book (Chap. 18). Physical signals also include variation potentials (slow wave potentials), which are caused by self-propagated hydraulic signals (pressure waves), and result in local plasma membrane depolarizations, giving the appearance of a moving electrical signal.

The nature of hydraulic signaling has been reviewed comprehensively by M. Malone (1993, 1996). Briefly, there are two modes of hydraulic signal transmission, pressure waves and hydraulic dispersal of signaling compounds. Both are xylem borne and can thus be distinguished from electrical signals which depend on living tissue. Also, both involve mass flow of water. Hydraulic dispersal can be induced by mechanical wounding. Since the water column in the xylem is under negative tension, water from damaged cells is drawn into the xylem. This decreases the negative xylem tension at the wound site, which spreads as a pressure wave throughout the plant to establish hydraulic equilibrium. Along with water, chemicals from the wound site will be drawn into the xylem and dispersed away from the wound site, thus reversing the otherwise unidirectional root-to-shoot flow in the xylem (Malone 1993, 1996).

Here, we will focus on the role of hydraulic signaling in responses to herbivory and mechanical wounding. Unfortunately, not many experiments have been carried out using chewing insects as inducers of hydraulic signals. Frequently, stimuli such as extreme heat (flame torching) and pressure, which are not proper stimuli to mimic herbivory, were applied to induce hydraulic signals.

Malone et al. (1994) found that hydraulic pressure waves are not specific wound response-inducing signals since these signals could be generated without inducing PI activity. This prompted Malone and Alarcon (1995) to test three alternatives, phloem transport, electrical signaling, and hydraulic dispersal. Steam girdling, a procedure where living tissue on a stem or petiole is killed by heat within a narrow zone, did not prevent signal transmission through the steam girdle. Since the xylem remains intact in a steam girdle, it was concluded that only hydraulic dispersal can account for the systemic upregulation of PI activity through a steam girdle. This was confirmed by Rhodes and coworkers, even though they had previously favored electrical signals. Their results from steam girdling and dye movement experiments were consistent with a xylem-borne wound signal (Rhodes et al. 1996, 1999). It should be noted that hydraulic signals that pass a steam girdle will also activate slow wave potentials. In the aforementioned studies, hydraulic signals were triggered by severe mechanical wounding such as "flame wounding". Therefore, it was reassuring to learn that herbivorous *Spodoptera littoralis* larvae were capable of generating hydraulic signals in tomato plants, probably resulting in dispersal of chemical wound signals. These insect-induced hydraulic signals were estimated to have a theoretical range of up to 270 cm (Alarcon and Malone 1994).

One problem for most experiments investigating electrical and hydraulic signals in the wound response is a long lag period in response time. Often, a detectable molecular defense response can only be measured long after the signal has reached the target tissue. In most studies, PI activity was measured to link hydraulic or electrical signals to the wound response, but increases in PI activity can generally not be detected within an hour after the wound stimulus. A more rapid response to a systemic signal is the activation of MAPKs. MAPK activity can be measured conveniently by immunoblotting using antibodies against phosphorylated active MAPKs (Hind et al. 2010) or in-gel kinase assays (Stratmann and Ryan 1997), so it can serve as a convenient early response marker for herbivory-related systemic signals. A remarkably fast MAPK activation was found in systemic tillers of *Lolium temulentum* after wounding of an adjacent tiller. The signal migrated from the wounded tiller down to the root crown and back up to the adjacent tiller where it activated a MAPK within 5 min (Dombrowski et al. 2011). In two-leaf stage tomato seedlings, *Manduca sexta* larvae systemically activate the MAPKs MPK3 and MPK1/2 within 3 min after onset of feeding (Kandoth et al. 2007; Stratmann and Ryan 1997). These MAPKs were shown to be essential for tomato resistance against *M. sexta* and for synthesis of JA (Kandoth et al. 2007). The rapid MAPK activation is consistent with an insect-induced hydraulic or electrical signal. To address the nature of this signal, tomato seedlings were steam girdled on the stem and, after a 1-day recovery period, excised below the girdle and placed into water. The excision generated a signal that passed the steam girdle and transiently induced MAPK activity in leaves within less than 5 min. When the excised seedlings were placed into a solution containing the alarm hormone systemin, MAPK activity also increased within less than 5 min, but persisted for more than 30 min. This demonstrated that systemin rapidly passed the steam girdle. The results are consistent with a xylem-borne hydraulic MAPK-activating wound signal as well as with the hydraulic dispersal theory. It cannot entirely be excluded that volatiles represent the systemic signal. But it is neither known whether stem excision results in the rapid release of volatiles like "green leaf" volatiles, nor whether volatiles can induce MAPK activity.

These results give strong support for the role of hydraulic signaling in the systemic wound response in tomato, and MAPKs represent a link between the rapid hydraulic signal and the slower increase in transcripts or activity of defense proteins such as PIs. A possible scenario for MAPK function in systemic tissue would be the posttranslational modification of signaling proteins such as transcription factors (Popescu et al. 2009), or regulation of ethylene and JA biosynthesis (Kandoth et al. 2007; Kim et al. 2003; Liu and Zhang 2004; Wu et al. 2007).

Transcripts for JA biosynthetic enzymes also increase relatively fast, within less than 30 min after wounding (Heitz et al. 1997, Lee and Howe 2004, Li et al. 2004; Ryan 2000). However, their transcription is not regulated by MAPKs in tomato (Kandoth et al. 2007), and their localization in the chloroplast and peroxisomes make them inaccessible for direct modification by cytosolic MAPKs. Moreover, there is overwhelming evidence from grafting experiments that a slow-moving phloem-borne signal, most likely JA or a JA-related compound, functions as a long-distance signal in the systemic wound response in tomato (Lee and Howe 2003, Li et al. 2002, Stratmann 2003, Sect. 17.3.1).

Some of these grafting experiments are consistent with a hydraulic systemic signal that upregulates JA synthesis in systemic leaves via MAPKs. However, when a wild-type scion was grafted onto an *spr2* (JA biosynthesis mutant) rootstock, the scion did not show a wound response, even though the *spr2* rootstock could have generated a hydraulic signal. This indicates that hydraulic signals do not play a discernible role for the systemic wound response in tomato. Could the hydraulic signal at least result in rapid JA synthesis in systemic tissue? Again grafting experiments showed that this is unlikely since a wounded wild-type rootstock induced a full wound response in *spr2* scions, indicating that de novo JA synthesis is not required in systemic tissues (Li et al 2002). In contrast, de novo JA synthesis in systemic tissue is rapidly induced by another long-distance signal in Arabidopsis (Glauser et al. 2008; Koo et al. 2009), and in *Nicotiana attenuata*, application of *M. sexta* oral secretions generates different systemic responses compared to mechanical wounding (Wu et al. 2007).

So, what is the role of the hydraulic signal in tomato? A possible scenario would be that the systemically activated MAPK upregulates the synthesis of ethylene. Ethylene alone is not sufficient to activate the tomato wound response, but it is required for JA action (O'Donnell et al. 1996). Moreover, ethylene synthesis is directly activated by MAPKs via phosphorylation of the ethylene biosynthetic enzyme 1-aminocyclopropane-1-carboxylic acid synthase (Kim et al. 2003; Liu and Zhang 2004).

The function of the systemic MAPK activation could be further investigated by studying grafts between wild-type and MAPK-silenced plants to test whether MAPK activity is required for the systemic wound response in the scion. It should be mentioned that most studies that were discussed here were carried out with young tomato seedlings. In older tomato plants and in other plant species, systemic signaling may be different (see Introduction).

6 Volatiles

During herbivory, exposure to damage and to elicitors present in insect oral secretions triggers plants to emit metabolites ranging from terpenes to chemicals derived from fatty acid or shikimate precursors (Turlings et al. 1990; Alborn et al. 1997; Schmelz et al. 2006). Among these volatiles are phytohormone derivatives;

methylated forms of both jasmonate (MeJA) and salicylate (MeSA) are components of the chemical blends produced, and both have been demonstrated to promote systemic defense responses. Tomato plants accumulated large pools of foliar proteinase inhibitors when exposed to MeJA emanating from sagebrush (*Artemisia tridentata*) co-cultivated in close proximity (Farmer and Ryan 1994). Ectopic expression of the gene encoding jasmonate methyltransferase (JMT) in Arabidopsis resulted in a threefold higher level of MeJA emission and constitutive expression of jasmonate-regulated defense genes, and conferred enhanced resistance to the fungal pathogen *Botrytis cinerea* (Seo et al. 2001). Tobacco plants infected by tobacco mosaic virus (TMV) emitted MeSA which was perceived by neighboring "receiver" plants (Shulaev et al. 1997). Labeling studies demonstrated uptake of MeSA by receiver plants that were subsequently more resistant to TMV than were plants which had not been exposed to infection-induced MeSA emissions (Shulaev et al. 1997).

While volatile derivatives of phytohormones act as signals analogous to their nonvolatile forms, research into other emitted chemicals has primarily elucidated their role in defense rather than regulation thereof. Herbivore-induced volatile release is an elegant indirect strategy deployed by the plant to recruit natural enemies of herbivore pests, and has been demonstrated to occur both above- and belowground. In maize, for instance, *Cotesia marginiventris* wasps are attracted to volatiles emitted from leaves in response to *Spodoptera exigua* larval herbivory and parasitize the pest, reducing larval fitness (Turlings et al. 1990). Belowground, maize root damage caused by *Diabrotica virgifera* larvae also results in production of volatiles, among which the terpene *E*-beta-caryophyllene promotes attraction of entomopathogenic nematodes that prey upon the larvae (Rasmann et al. 2005; Köllner et al. 2008). This phenomenon of plant conscription of natural enemies through release of volatile attractants has been observed for many plant species, and these interactions aid in limiting plant damage.

Despite the well-documented contributions of volatiles to indirect anti-herbivore defense, attraction of natural enemy predator insects is unlikely to have driven evolution of this plant response (Peñuelas and Llusià 2004). Volatiles directly contribute to plant defense through repulsion of herbivores and ovipositioning females (Kessler and Baldwin 2001; de Moraes et al. 2001; Heil and Karban 2009). Furthermore, many of the chemicals emitted upon arthropod-induced damage have documented antimicrobial activity, indicating that they might have originated to discourage colonization and infection of wound sites. Plants compromised in the production of terpene or green leafy terpene volatiles (GLVs) are more susceptible to both herbivory and microbial infection (Huang et al. 2012; Shiojiri et al. 2006). While having a myriad of effects on both microbes and insects, volatile chemicals elicit plant responses as well. Beyond phytohormones such as MeJA and MeSA, in maize, GLV aldehydes and alcohols prime plant herbivore defenses, leading to more rapid responses of greater magnitude (Engelberth et al. 2004; Ruther and Kleier 2005). Similar effects have been demonstrated for Arabidopsis in response to monoterpene exposure, which stimulates expression of an extensive set of genes associated with response to stress or biotic attack (Godard et al. 2008). This transcriptional reprogramming was reduced in Arabidopsis mutants compromised in jasmonate or ethylene signal transduction. Activation of defense responses by emitted GLV and terpene volatiles indicates that these chemicals may have originated not only for defense but also as a means for signal transmission. However, the mechanisms by which plants perceive and initially transduce this signal are not yet understood (Dicke et al. 2003; Heil et al. 2008).

Despite the lack of information about perception of herbivore-induced volatile molecules, evidence for their role as mediators of defense responses has been accumulating. In addition to the aforementioned examples, exposure to chemicals emitted from neighboring plants subjected to herbivory has been implicated in promoting expression of defense genes and reducing subsequent arthropod damage in numerous laboratory and field settings (Arimura et al. 2000, 2010; Karban et al. 2000; Baldwin et al. 2002, 2006; Ramadan et al. 2011). "Eavesdropping" on neighboring plants to monitor the chances of imminent danger through perception of volatile emissions would be a very convenient tactical capability. However, chemical emissions disperse to concentrations below those required to elicit responses at distances more than a few dozen centimeters (Heil and Karban 2009). This indicates that neighbors must be very close to eavesdrop and also raises the possibility that volatile emissions might be most effectively detected by systemic tissues of the plant that is emitting them.

A number of recent studies have been designed to determine whether herbivore-induced volatiles emanating from the damaged part of a plant act as signals to induce systemic defense responses. The architecture of some plants is such that there is a great distance separating systemic tissues or vascular connections are limited. In these conditions, the ability to transmit signals through air rather than relying strictly on vasculature transmission would be a great advantage (Farmer 2001). For instance, mature lima bean plants grow as long branched vines which may share close contact spatially, but limited direct vascular connections. Lima bean tendrils exposed to volatiles through proximal placement to distal parts of the plant subjected to beetle herbivory had increased extra-floral nectar secretion, displayed reduced damage to subsequent herbivory, and attracted more protective ants (Heil and Silva Bueno 2007). When volatiles were prevented from escaping through isolation of the herbivore damaged leaves inside plastic bags, the systemically induced defenses were eliminated.

Experiments with blueberry bushes yielded similar results. Blueberry bushes have constrained anatomical vascular connections; connectivity of vasculature is limited between branches of the same shoot and absent between different shoots of the same plant (Rodriguez-Saona et al. 2009). Blueberry plants that had been exposed previously to herbivory by gypsy moth caterpillars (*Lymantria dispar*) demonstrated reduced larval damage to leaves on systemic unconnected branches upon subsequent challenge. But, if systemic branches were shielded from volatiles emitted by the original branch, no protective effect was observed (Rodriguez-Saona et al. 2009). Sagebrush bushes which are anatomically similar in structure were also observed to utilize volatile signals for regulation of systemic responses (Karban et al. 2006). When a sagebrush branch was mechanically damaged, proximal systemic branches that were downwind of the damage displayed significantly reduced herbivory. If airflow from the damaged branch was blocked, no reduction in herbivory on downwind systemic branches was observed. Trees face similar anatomical challenges, with relatively large distances separating one branch from another in terms of vasculature. As might be expected given this structure, systemic defenses in poplar tree leaves may be induced by exposure to volatiles emanating from leaves subjected to herbivory by gypsy moth larvae (Frost et al. 2007). However, shielding of receiver leaves from emitted volatiles did not prevent systemic defense responses from being activated, but did result in slower elicitation. This indicates not only that poplar utilizes both vasculature and volatile transmission of signaling to activate defenses, but also that volatile-mediated responses may occur on a more rapid timescale.

Together, experimental evidence indicates that while volatiles are important components of plant defense responses, they are also contributors to long-distance signaling for elicitation of systemic immunity to pathogens and herbivores. Although perception of volatiles and downstream signaling is not yet well-understood compared to message transmission through the vasculature, this does provide an important additional or alternative mechanism to activate systemic wound responses. The relative contributions of signals transmitted via vasculature versus aerial emission are likely to depend upon anatomy. Continuing research is needed to better understand volatile-regulated signal transduction.

7 Conclusions and Perspectives

While long-distance signaling in plant development (florigen) and during rapid seismonastic leaf movements in *Mimosa pudica* has been known for a long time, C. A. Ryan's seminal 1972 discovery of systemic wound signaling resulted in the identification of additional systemic signals that mediate defense responses against herbivores and inspired studies of systemic signaling in defense responses against pathogens (Chaps. 1–3).

We have discussed the many diverse signals that can function as systemic wound signals. The wide range of systemic signals is likely a function of the plant species, plant morphology, and ecological niche. Systemic wound signaling may also involve a coordinated action of various signals in a single plant, and systemic receiver tissues may integrate several input signals to activate a defense response.

Sometimes it is not possible to attribute a precise function to a systemic signal and it is not always known whether a candidate signal is actually translocated. Grafting experiments resulted in great progress in determining the role of JA in systemic wound signaling (Li et al. 2002, 2005; Schilmiller and Howe 2005). However, grafting experiments addressing hypotheses about systemic signals require a combination of different genotypes such as transgenic plants or mutants, and they are not available for a number of candidate signals. In addition, approaches have now been developed that allow for tracking systemic signals in real time, e.g., by use of luciferase imaging (Miller et al. 2009). In many studies, herbivory was mimicked by mechanical wounding, which is an inadequate substitute for herbivory in some plants. Future experiments should address this by employing herbivores, starting with eggs or early larval stages.

Plants have evolved amazing alternatives to the animal nervous system to coordinate diverse processes over long distances. The animal nervous system may have coevolved with the evolution of mobility. Furthermore, the development of the peripheral nervous system is determinate, reflecting the fixed body plans in animals. It is capable of reacting to stimuli within fractions of a second, which may decide between life and death of the animal. In contrast, since predator and stress avoidance or fight and flight are not options for sessile plants, different mechanisms of intraplant long-distance communication evolved that reflect the sessile lifestyle and phenotypic plasticity of plants.

Acknowledgements J.W.S. was supported by National Science Foundation grant IOS-0745545.

References

Adie B, Chico JM, Rubio-Somoza I, Solano R (2007) Modulation of plant defenses by ethylene. J Plant Growth Regul 26:160–177

Alarcon JJ, Malone M (1994) Substantial hydraulic signals are triggered by leaf-biting insects in tomato. J Exp Bot 45:953–957

Alborn HT, Turlings TCJ, Jones TH, Stenhagen G, Loughrin JH, Tumlinson JH (1997) An elicitor of plant volatiles from beet armyworm oral secretion. Science 276:945–949

Aloni B, Wyse RE, Griffith S (1986) Sucrose transport and phloem unloading in stem of *Vicia faba*: possible involvement of a sucrose carrier and osmotic regulation. Plant Physiol 81:482–486

Ament K, Krasikov V, Allmann S, Rep M, Takken FL, Schuurink RC (2006) Methyl salicylate production in tomato affects biotic interactions. Plant J 62:124–134

Amon P, Haas E, Sumper M (1998) The sex-inducing pheromone and wounding trigger the same set of genes in the multicellular green alga Volvox. Plant Cell 10:781–789

Anderson JP, Badruzsaufari E, Schenk PM, Manners JM, Desmond OJ, Ehlert C, Maclean DJ, Ebert PR, Kazan K (2004) Antagonistic interaction between abscisic acid and jasmonate-ethylene signaling pathways modulates defense gene expression and disease resistance in *Arabidopsis*. Plant Cell 16:3460–3479

Arimura G, Ozawa R, Shimoda T, Nishioka T, Boland W, Takabayashi J (2000) Herbivory-induced volatiles elicit defence genes in lima bean leaves. Nature 406(6795):512–515

Arimura G, Kost C, Boland W (2005) Herbivore-induced, indirect plant defences. Biochim Biophys Acta 1734:91–111

Arimura G, Shiojiri K, Karban R (2010) Acquired immunity to herbivory and allelopathy caused by airborne plant emissions. Phytochemistry 71:1642–1649

Avila CA, Arévalo-Soliz LM, Jia L, Navarre DA, Chen Z, Howe GA, Meng QW, Smith JE, Goggin FL (2012) Loss of function of *FATTY ACID DESATURASE7* in tomato enhances basal aphid resistance in a salicylate-dependent manner. Plant Physiol 158:2028–2041

Baldwin IT, Kessler A, Halitschke R (2002) Volatile signaling in plant–plant–herbivore interactions: what is real? Curr Opin Plant Biol 5:351–354

Baldwin IT, Halitschke R, Paschold A, von Dahl CC, Preston CA (2006) Volatile signaling in plant–plant interactions: "talking trees" in the genomics era. Science 311:812–815
Bari R, Jones JD (2009) Role of plant hormones in plant defence responses. Plant Mol Biol 69:473–488
Barroso JB, Corpas FJ, Carreras A et al (2006) Localization of S-nitrosoglutathione and expression of S-nitrosoglutathione reductase in pea plants under cadmium stress. J Exp Bot 57:1785–1793
Barton KE, Koricheva J (2010) The ontogeny of plant defense and herbivory: characterizing general patterns using meta-analysis. Am Nat 175:481–493
Baydoun EAH, Fry SC (1985) The immobility of pectic substances in injured tomato leaves and its bearing on the identity of the wound hormone. Planta 165:269–276
Beckers GJ, Spoel SH (2006) Fine-tuning plant defence signalling: salicylate versus jasmonate. Plant Biol (Stuttg) 8:1–10
Beckers GJM, Jaskiewicz M, Liu Y, Underwood WR, He SY, Zhang S, Conrath U (2009) Mitogen-activated protein kinases 3 and 6 are required for full priming of stress responses in *Arabidopsis thaliana*. Plant Cell 21:944–953
Berger B, Baldwin IT (2007) The hydroxyproline-rich glycopeptide systemin precursor *NapreproHypSys* does not play a central role in *Nicotiana attenuata's* anti-herbivore defense responses. Plant Cell Environ 30:1450–1464
Berger B, Baldwin IT (2009) Silencing the hydroxyproline-rich glycopeptide systemin precursor in two accessions of *Nicotiana attenuata* alters flower morphology and rates of self-pollination. Plant Physiol 149:1690–1700
Bergey D, Howe G, Ryan CA (1996) Polypeptide signaling for plant defensive genes exhibits analogies to defense signaling in animals. Proc Natl Acad Sci USA 93:12053–12058
Berner JM, van der Westhuizen AJ (2010) Inhibition of xanthine oxidase activity results in the inhibition of Russian wheat aphid-induced defense enzymes. J Chem Ecol 36:1375–1380
Birkenmeier GF, Ryan CA (1998) Wound signaling in tomato plants. Evidence that ABA is not a primary signal for defense gene activation. Plant Physiol 117:687–693
Bishop P, Makus DJ, Pearce G, Ryan CA (1981) Proteinase inhibitor inducing factor activity in tomato leaves resides in oligosaccharides enzymically released from cell walls. Proc Natl Acad Sci USA 78:3536–3640
Bishop P, Pearce G, Bryant JE, Ryan CA (1984) Isolation and characterization of the proteinase inhibitor inducing factor from tomato leaves: identity and activity of poly- and oligogalacturonide fragments. J Biol Chem 259:13172–13177
Boege K, Marquis RJ (2005) Facing herbivory as you grow up: the ontogeny of resistance in plants. Trends Ecol Evol 20:441–448
Breithaupt C, Kurzbauer R, Lilie H, Schaller A, Strassner J, Huber R, Macheroux P, Clausen T (2006) Crystal structure of 12-oxophytodienoate reductase 3 from tomato: self-inhibition by dimerization. Proc Natl Acad Sci USA 103:14337–14342
Broekgaarden C, Poelman EH, Steenhuis G, Voorrips RE, Dicke M, Vosman B (2008) Responses of *Brassica oleracea* cultivars to infestation by the aphid *Brevicoryne brassicae*: an ecological and molecular approach. Plant Cell Environ 31:1592–1605
Browse J (2009) The power of mutants for investigating jasmonate biosynthesis and signaling. Phytochemistry 70:1539–1546
Browse J, Howe GA (2008) New weapons and a rapid response against insect attack. Plant Physiol 146:832–838
Chaki M, Valderrama R, Fernández-Ocaña AM, Carreras A, Gómez-Rodríguez MV, Pedrajas JR, Begara-Morales JC, Sánchez-Calvo B, Luque F, Leterrier M, Corpas FJ, Barroso JB (2011) Mechanical wounding induces a nitrosative stress by down-regulation of GSNO reductase and an increase in S-nitrosothiols in sunflower (*Helianthus annuus*) seedlings. J Exp Bot 62:1803–1813
Chen F, D'Auria JC, Tholl D, Ross JR, Gershenzon J, Noel JP, Pichersky E (2003) An *Arabidopsis thaliana* gene for methylsalicylate biosynthesis, identified by a biochemical genomics approach, has a role in defense. Plant J 36:577–588

Chen H, Wilkerson CG, Kuchar JA, Phinney BS, Howe GA (2005) Jasmonate-inducible plant enzymes degrade essential amino acids in the herbivore midgut. Proc Natl Acad Sci USA 102:19237–19242

Chung HS, Niu Y, Browse J, Howe GA (2009) Top hits in contemporary JAZ: An update on jasmonate signaling. Phytochemistry 70:1547–1559

Constabel CP, Yip L, Ryan CA (1998) Prosystemin from potato, black nightshade, and bell pepper: primary structure and biological activity of predicted systemin polypeptides. Plant Mol Biol 36:55–62

Corrado G, Sasso R, Pasquariello M, Iodice L, Carretta A, Cascone P, Ariati L, Digilio MC, Guerrieri E, Rao R (2007) Systemin regulates both systemic and volatile signaling in tomato plants. J Chem Ecol 33:669–681

Couldridge C, Newbury HJ, Ford-Lloyd B, Bale J, Pritchard J (2007) Exploring plant responses to aphid feeding using a full *Arabidopsis* microarray reveals a small number of genes with significantly altered expression. Bull Entomol Res 97:523–532

de la Noval BM, Pérez E, Olalde V, Délano JP, Martínez N (2004) Inducción de β-1,3-glucanasa y quitinasas en plántulas de tomate por hongos micorrizógenos y sistemina. Cult Trop 2:5–12

de la Noval B, Pérez E, Martínez B, León O, Martínez-Gallardo N, Délano-Frier J (2007) Exogenous systemin has a contrasting effect on disease resistance in mycorrhizal tomato (*Solanum lycopersicum*) plants infected with necrotrophic or hemibiotrophic pathogens. Mycorrhiza 17:449–460

De Moraes CM, Lewis WJ, Pare PW, Alborn HT, Tumlinson JH (1998) Herbivore-infested plants selectively attract parasitoids. Nature 393:570–573

De Moraes CM, Mescher MC, Tumlinson JH (2001) Caterpillar-induced nocturnal plant volatiles repel conspecific females. Nature 410:577–580

de Vos M, Kim JH, Jander G (2007) Biochemistry and molecular biology of Arabidopsis–aphid interactions. Bioessays 29:871–883

Degenhardt DC, Refi-Hind S, Stratmann JW, Lincoln DE (2010) Systemin and jasmonic acid regulate constitutive and herbivore-induced systemic volatile emissions in tomato, *Solanum lycopersicum*. Phytochemistry 71:2024–2037

Díaz M, Achkor H, Titarenko E, Martínez MC (2003) The gene encoding glutathione-dependent formaldehyde dehydrogenase/GSNO reductase is responsive to wounding, jasmonic acid and salicylic acid. FEBS Lett 543:136–139

Dicke M, Agrawal AA, Bruin J (2003) Plants talk, but are they deaf? Trends Plant Sci 8:403–405

Diezel C, Kessler D, Baldwin IT (2011) Pithy protection: *Nicotiana attenuata*'s jasmonic acid-mediated defenses are required to resist stem-boring weevil larvae. Plant Physiol 155:1936–1946

Doares SH, Narvaez-Vasquez J, Conconi A, Ryan CA (1995) Salicylic acid inhibits synthesis of proteinase inhibitors in tomato leaves induced by systemin and jasmonic acid. Plant Physiol 108:1741–1746

Doherty HM, Selvendran RR, Bowles DJ (1988) The wound response of tomato plants can be inhibited by aspirin and related hydroxybenzoic acids. Physiol Mol Plant Pathol 33:377–384

Dombrecht B, Xue GP, Sprague SJ, Kirkegaard JA, Ross JJ, Reid JB, Fitt GP, Sewelam N, Schenk PM, Manners JM, Kazan K (2007) MYC2 differentially modulates diverse jasmonate-dependent functions in Arabidopsis. Plant Cell 19:2225–2245

Dombrowski JE (2003) Salt stress activation of wound-related genes in tomato plants. Plant Physiol 132:2098–3107

Dombrowski JE, Pearce G, Ryan CA (1999) Proteinase inhibitor-inducing activity of the prohormone prosystemin resides exclusively in the C-terminal systemin domain. Proc Natl Acad Sci USA 96:12947–12952

Dombrowski JE, Hind SR, Martin RC, Stratmann JW (2011) Wounding systemically activates a mitogen-activated protein kinase in forage and turf grasses. Plant Sci 180:686–693

Dong X (2004) NPR1, all things considered. Curr Opin Plant Biol 7:547–552

Doss RP, Oliver JE, Proebsting WM, Potter SW, Kuy S, Clement SL, Williamson RT, Carney JR, DeVilbiss ED (2000) Bruchins: insect-derived plant regulators that stimulate neoplasm formation. Proc Natl Acad Sci USA 97:6218–6223

Engelberth J, Alborn HT, Schmelz EA, Tumlinson JH (2004) Airborne signals prime plants against insect herbivore attack. Proc Natl Acad Sci USA 101:1781–1785

Erb M, Flors V, Karlen D, de Lange E, Planchamp C, D'Alessandro M, Turlings TC, Ton J (2009) Signal signature of aboveground-induced resistance upon belowground herbivory in maize. Plant J 59:292–302

Erb M, Köllner TG, Degenhardt J, Zwahlen C, Hibbard BE, Turlings TC (2011) The role of abscisic acid and water stress in root herbivore-induced leaf resistance. New Phytol 189:308–320

Espunya MC, De Michele R, Gómez-Cadenas A, Martínez MC (2012) S-Nitrosoglutathione is a component of wound- and salicylic acid-induced systemic responses in *Arabidopsis thaliana*. J Exp Bot 63:3219–3227

Farag MA, Paré PW (2002) C_6-Green leaf volatiles trigger local and systemic VOC emissions in tomato. Phytochemistry 61:545–554

Farmer EE (2001) Surface-to-air signals. Nature 411:854–856

Farmer EE, Ryan CA (1994) Interplant communication: airborne methyl jasmonate induces synthesis of proteinase inhibitors in plant leaves. Proc Natl Acad Sci USA 87:7713–7716

Feechan A, Kwon E, Yun BW, Wang Y, Pallas A, Loake GJ (2005) A central role for *S*-nitrosothiols in plant disease resistance. Proc Natl Acad Sci USA 102:8054–8059

Felix G, Boller T (1995) Systemin induces rapid ion fluxes and ethylene biosynthesis in *Lycopersicon peruvianum* cells. Plant Cell 7:381–389

Felton GW, Workman J, Duffey SS (1992) Avoidance of antinutritive plant defense: role of midgut pH in Colorado potato beetle. J Chem Ecol 18:571–583

Fisher DB (1990) Measurement of phloem transport rates by an indicator-dilution technique. Plant Physiol 94:455–462

Fraenkel GS (1959) The raison d'être of secondary plant substances. Science 129:466–1470

Frost CJ, Appel HM, Carlson JE, De Moraes CM, Mescher MC, Schultz JC (2007) Within-plant signalling via volatiles overcomes vascular constraints on systemic signalling and primes responses against herbivores. Ecol Lett 10:490–498

Gfeller A, Liechti R, Farmer EE (2010) Arabidopsis jasmonate signaling pathway. Sci Signal 3: cm4

Glauser G, Grata E, Dubugnon L, Rudaz S, Farmer EE, Wolfender J-L (2008) Spatial and temporal dynamics of jasmonate synthesis and accumulation in Arabidopsis in response to wounding. J Biol Chem 283:16400–16407

Glauser G, Dubugnon L, Mousavi SAR, Rudaz S, Wolfender J-L, Farmer EE (2009) Velocity estimates for signal propagation leading to systemic jasmonic acid accumulation in wounded Arabidopsis. J Biol Chem 284:34506–34513

Glendinning JI (2002) How do herbivorous insects cope with noxious secondary plant compounds in their diet? Entomol Exp Appl 104:15–25

Godard KA, White R, Bohlmann J (2008) Monoterpene-induced molecular responses in *Arabidopsis thaliana*. Phytochemistry 69:1838–1849

Green TR, Ryan CA (1972) Wound-induced proteinase inhibitor in plant leaves: a possible defense mechanism against insects. Science 175:776–777

Gupta JG (2011) Protein *S*-nitrosylation in plants: photorespiratory metabolism and NO signaling. Sci Signal 4:jc1

Gutsche AR, Heng-Moss TM, Higley LG, Sarath G, Mornhinweg DW (2009) Physiological responses of resistant and susceptible barley, *Horedum vulgare* to the Russian wheat aphid, *Diurpahis noxia* (Mordvilko). Arthropod Plant Interact 3:233–240

Harfouche AL, Shivaji R, Stocker R, Williams PW, Luthe DS (2006) Ethylene signaling mediates a maize defense response to insect herbivory. Mol Plant Microbe Interact 19:189–199

Heil M, Karban R (2009) Explaining evolution of plant communication by airborne signals. Trends Ecol Evol 25:137–144

Heil M, Silva Bueno JC (2007) Within-plant signaling by volatiles leads to induction and priming of an indirect plant defense in nature. Proc Natl Acad Sci USA 104:5467–5472
Heil M, Ton J (2008) Long-distance signalling in plant defence. Trends Plant Sci 13:264–272
Heil M, Lion U, Boland W (2008) Defense-inducing volatiles: in search of the active motif. J Chem Ecol 34:601–604
Heitz T, Bergey DA, Ryan CA (1997) A gene encoding a chloroplast-targeted lipoxygenase in tomato leaves is transiently induced by wounding, systemin, and methyl jasmonate. Plant Physiol 114:1085–1093
Heitz T, Widemann E, Lugan R, Miesch L, Ullmann P, Désaubry L, Holder E, Grausem B, Kandel S, Miesch M, Werck-Reichhart D, Pinot F (2012) Cytochromes P450 CYP94C1 and CYP94B3 catalyze two successive oxidation steps of plant hormone jasmonoyl-isoleucine for catabolic turnover. J Biol Chem 287:6296–6306
Hind SR, Malinowski R, Yalamanchili R, Stratmann JW (2010) Tissue-type specific systemin perception and the elusive systemin receptor. Plant Signal Behav 5:42–44
Hong JK, Yun BW, Kang JG, Raja MU, Kwon E, Sorhagen K, Chu C, Wang Y, Loake GJ (2008) Nitric oxide function and signalling in plant disease resistance. J Exp Bot 59:147–154
Howe GA (2004) Jasmonates as signals in the wound response. J Plant Growth Regul 23:223–237
Howe GA, Jander G (2008) Plant immunity to insect herbivores. Annu Rev Plant Biol 59:41–66
Howe GA, Lightner J, Browse J, Ryan CA (1996) An octadecanoid pathway mutant (JL5) of tomato is compromised in signaling for defense against insect attack. Plant Cell 8:2067–2077
Huang M, Sanchez-Moreiras AM, Abel C, Sohrabi R, Lee S, Gershenzon J, Tholl D (2012) The major volatile organic compound emitted from *Arabidopsis thaliana* flowers, the sesquiterpene (E)-β-caryophyllene, is a defense against a bacterial pathogen. New Phytol 193:997–1008
Huffaker A, Ryan CA (2007) Endogenous peptide defense signals in Arabidopsis differentially amplify signaling for the innate immune response. Proc Natl Acad Sci USA 104:10732–10736
Huffaker A, Pearce G, Ryan CA (2006) An endogenous peptide signal in Arabidopsis activates components of the innate immune response. Proc Natl Acad Sci USA 103:10098–10103
Huffaker A, Dafoe NJ, Schmelz EA (2011) ZmPep1, an ortholog of Arabidopsis elicitor peptide 1, regulates maize innate immunity and enhances disease resistance. Plant Physiol 155:1325–1338
Huffaker A, Pearce G, Veyrat N, Erb M, Turlings TC, Sartor R, Shen Z, Briggs S, Vaughan MM, Alborn HT, Teal PE, Schmelz EA (2013) Plant elicitor peptides are conserved signals regulating direct and indirect anti-herbivore defense. Proc Natl Acad Sci USA. Published online before print March 18, 2013, doi: 10.1073/pnas.1214668110
Jones CG, Hopper RF, Coleman JS, Krischik VA (1993) Control of systemically induced herbivore resistance by plant vascular architecture. Oecologia 93:452–456
Kallenbach M, Alagna F, Baldwin IT, Bonaventure G (2010) *Nicotiana attenuata* SIPK, WIPK, NPR1, and fatty acid-amino acid conjugates participate in the induction of jasmonic acid biosynthesis by affecting early enzymatic steps in the pathway. Plant Physiol 152:96–106
Kallenbach M, Bonaventure G, Gilardoni PA, Wissgott A, Baldwin IT (2012) *Empoasca* leafhoppers attack wild tobacco plants in a jasmonate-dependent manner and identify jasmonate mutants in natural populations. Proc Natl Acad Sci USA 109:E1548–E1557
Kandoth PK, Ranf S, Pancholi SS, Jayanty S, Walla MD, Miller W, Howe GA, Lincoln DE, Stratmann JW (2007) Tomato MAPKs LeMPK1, LeMPK2, and LeMPK3 function in the systemin-mediated defense response against herbivorous insects. Proc Natl Acad Sci USA 104:12205–12210
Karban R, Baldwin IT, Baxter KJ, Laue G, Felton GW (2000) Communication between plants: induced resistance in wild tobacco plants following clipping of neighboring sagebrush. Oecologia 125:66–71
Karban R, Shiojiri K, Huntzinger M, McCall AC (2006) Damage-induced resistance in sagebrush: volatiles are key to intra- and interplant communication. Ecology 87:922–930
Katsir L, Chung HS, Koo AJK, Howe GA (2008) Jasmonate signaling: a conserved mechanism of hormone sensing. Curr Opin Plant Biol 11:428–435
Kazan K, Manners JM (2008) Jasmonate signaling: toward an integrated view. Plant Physiol 146:1459–1468

Kempema LA, Cui X, Holzer FM, Walling LL (2007) Arabidopsis transcriptome changes in response to phloem-feeding silverleaf whitefly nymphs. Similarities and distinctions in responses to aphids. Plant Physiol 143:849–865

Kerchev PI, Fenton B, Foyer CH, Hancock RD (2012) Plant responses to insect herbivory: interactions between photosynthesis, reactive oxygen species and hormonal signalling pathways. Plant Cell Environ 35:441–453

Kessler A, Baldwin IT (2001) Defensive function of herbivore-induced plant volatile emissions in nature. Science 291:2141–2144

Kessler A, Baldwin IT (2002) Plant responses to insect herbivory: the emerging molecular analysis. Annu Rev Plant Biol 53:299–328

Kim CY, Liu Y, Thorne ET, Yang H, Fukushige H, Gassmann W, Hildebrand D, Sharp RE, Zhang S (2003) Activation of a stress-responsive mitogen-activated protein kinase cascade induces the biosynthesis of ethylene in plants. Plant Cell 15:2707–2718

Köllner TG, Held M, Lenk C, Hiltpold I, Turlings TC, Gershenzon J, Degenhardt J (2008) A maize (E)-beta-caryophyllene synthase implicated in indirect defense responses against herbivores is not expressed in most American maize varieties. Plant Cell 20:482–494

Koo AJK, Howe GA (2012) Catabolism and deactivation of the lipid-derived hormone jasmonoyl-isoleucine. Front Plant Sci 3:19

Koo AJ, Gao X, Jones AD, Howe GA (2009) A rapid wound signal activates the systemic synthesis of bioactive jasmonates in Arabidopsis. Plant J 59:974–986

Koo AJ, Cooke TF, Howe GA (2011) Cytochrome P450 CYP94B3 mediates catabolism and inactivation of the plant hormone jasmonoyl-L-isoleucine. Proc Natl Acad Sci USA 108:9298–9303

Koornneef A, Leon-Reyes A, Ritsema T, Verhage A, Den Otter FC, Van Loon LC, Pieterse CMJ (2008) Kinetics of salicylate-mediated suppression of jasmonate signaling reveal a role for redox modulation. Plant Physiol 147:1358–1368

Krol E, Mentzel T, Chinchilla D, Boller T, Felix G, Kemmerling B, Postel S, Arents M, Jeworutzki E, Al-Rasheid KA, Becker D, Hedrich R (2010) Perception of the Arabidopsis danger signal peptide 1 involves the pattern recognition receptor AtPEPR1 and its close homologue AtPEPR2. J Biol Chem 285:13471–13479

Kuĉśnierczyk A, Winge P, Jørstad TS, Troczyńska J, Rossiter JT, Bones AM (2008) Towards global understanding of plant defence against aphids – timing and dynamics of early *Arabidopsis* defence responses to cabbage aphid (*Brevicoryne brassicae*) attack. Plant Cell Environ 31:1097–1115

Lee GI, Howe GA (2003) The tomato mutant *spr1* is defective in systemin perception and the production of a systemic wound signal for defense gene expression. Plant J 33:567–576

Leitner M, Boland W, Mithöfer A (2005) Direct and indirect defences induced by piercing-sucking and chewing herbivores in *Medicago trunculata*. New Phytol 167:597–606

Leon J, Rojo E, Sanchez-Serrano JJ (2001) Wound signalling in plants. J Exp Bot 52:1–9

Leon-Reyes A, Spoel SH, De Lange ES, Abe H, Kobayashi M, Tsuda S, Millenaar FF, Welschen RA, Ritsema T, Pieterse CM (2009) Ethylene modulates the role of NONEXPRESSOR OF PATHOGENESIS-RELATED GENES1 in cross talk between salicylate and jasmonate signaling. Plant Physiol 149:1797–1809

Leon-Reyes A, Du Y, Koornneef A, Proietti S, Körbes AP, Memelink J, Pieterse CM, Ritsema T (2010) Ethylene signaling renders the jasmonate response of Arabidopsis insensitive to future suppression by salicylic Acid. Mol Plant Microbe Interact 23:187–197

Li C, Williams MM, Loh Y-T, Lee GI, Howe GA (2002a) Resistance of cultivated tomato to cell content-feeding herbivores is regulated by the octadecanoid-signaling pathway. Plant Physiol 130:494–503

Li L, Li C, Lee GI, Howe GA (2002b) Distinct roles for jasmonate synthesis and action in the systemic wound response of tomato. Proc Natl Acad Sci USA 99:6416–6421

Li C, Liu G, Xu C, Lee GI, Bauer P, Ling HQ, Ganal MW, Howe GA (2003) The tomato suppressor of prosystemin-mediated responses2 (Spr2) gene encodes a fatty acid desaturase required for the biosynthesis of jasmonic acid and the production of a systemic wound signal for defense gene expression. Plant Cell 15:1646–1661

Li L, Zhao Y, McCaig BC, Wingerd BA, Wang J, Whalon ME, Pichersky E, Howe GA (2004) The tomato homolog of CORONATINE-INSENSITIVE is required for the maternal control of seed maturation, jasmonate-signaled defense responses, and glandular trichome development. Plant Cell 16:126–143

Li CY, Schilmiller AL, Liu GH, Lee GI, Jayanty S, Sageman C (2005) Role of beta-oxidation in jasmonate biosynthesis and distal wound signaling in tomato. Plant Cell 17:971–986

Lindermayr C, Sell S, Müller B, Leister D, Durner J (2010) Redox regulation of the NPR1-TGA1 system of *Arabidopsis thaliana* by nitric oxide. Plant Cell 22:2894–2907

Lippert D, Chowrira S, Ralph SG, Zhuang J, Aeschliman D, Ritland C, Ritland K, Bohlmann J (2007) Conifer defense against insects: proteome analysis of Sitka spruce (*Picea sitchensis*) bark induced by mechanical wounding or feeding by white pine weevils (*Pissodes strodi*). Proteomics 7:248–270

Little D, Gouhier-Darimont C, Bruessow F, Reymond P (2007) Oviposition by pierid butterflies triggers defense responses in *Arabidopsis*. Plant Physiol 143:784–800

Liu Y, Zhang S (2004) Phosphorylation of 1-aminocyclopropane-1-carboxylic acid synthase by MPK6, a stress-responsive mitogen-activated protein kinase, induces ethylene biosynthesis in Arabidopsis. Plant Cell 16:3386–3399

Liu L, Hausladen A, Zeng M, Que L, Heitman J, Stamler JS (2001) A metabolic enzyme for S-nitrosothiol conserved from bacteria to humans. Nature 410:490–494

Lorenzo O, Solano R (2005) Molecular players regulating the jasmonate signalling network. Curr Opin Plant Biol 8:532–540

Lorenzo O, Chico JM, Sánchez-Serrano JJ, Solano R (2004) JASMONATE-INSENSITIVE1 encodes a MYC transcription factor essential to discriminate between different jasmonate-regulated defense responses in Arabidopsis. Plant Cell 16:1938–1950

Ma Y, Walker RK, Zhao Y, Berkowitz GA (2012) Linking ligand perception by PEPR pattern recognition receptors to cytosolic Ca^{2+} elevation and downstream immune signaling in plants. Proc Natl Acad Sci USA 109:19852–19857

Maffei M, Bossi S, Spiteller D, Mithofer A, Boland W (2004) Effects of feeding *Spodoptera littoralis* on lima bean leaves. I. Membrane potentials, intracellular calcium variations, oral secretions, and regurgitate components. Plant Physiol 134:1752–1762

Maffei ME, Mithöfer A, Arimura GI, Uchtenhagen H, Bossi S, Bertea CM, Cucuzza LS, Novero M, Volpe V, Quadro S, Boland W (2006) Effects of feeding *Spodoptera littoralis* on lima bean leaves. III. Membrane depolarization and involvement of hydrogen peroxide. Plant Physiol 140:1022–1035

Malone M (1993) Hydraulic signals. Philos Trans R Soc Lond, Ser B Biol Sci 341:33–39

Malone M (1996) Rapid, long-distance signal transmission in higher plants. In: Callow JA (ed) Advances in botanical research, vol 22. Academic, San Diego, CA, pp 163–228

Malone M, Alarcon JJ (1995) Only xylem-borne factors can account for systemic wound signaling in the tomato plant. Planta 196:740–746

Malone M, Palumbo L, Boari F, Monteleone M, Jones HG (1994) The relationship between wound-induced proteinase-inhibitors and hydraulic signals in tomato seedlings. Plant Cell Environ 17:81–87

Matsui K, Kurishita S, Hisamitsu A, Kajiwara T (2000) A lipid-hydrolysing activity involved in hexenal formation. Biochem Soc Trans 28:857–860

McConn M, Browse J (1996) The critical requirement for linolenic acid is pollen development, not photosynthesis, in an Arabidopsis mutant. Plant Cell 8:403–416

McGurl B, Pearce G, Orozco-Cárdenas M, Ryan CA (1992) Structure, expression, and antisense inhibition of the systemin precursor gene. Science 225:1570–1573

McGurl B, Orozco-Cárdenas M, Pearce G, Ryan CA (1994) Overexpression of the prosystemin gene in transgenic tomato plants generates a systemic signal that constitutively induces proteinase inhibitor synthesis. Proc Natl Acad Sci USA 91:9799–9902

Meindl T, Boller T, Felix G (1998) The plant wound hormone systemin binds with the N-terminal part to its receptor but needs the C-terminal part to activate it. Plant Cell 10:1561–1570

Mewis I, Tokuhisa JG, Schultz JC, Appel HM, Ulrichs C, Gershenzon J (2006) Gene expression and glucosinolate accumulation in *Arabidopsis thaliana* in response to generalist and specialist herbivores of different feeding guilds and the role of defense signaling pathways. Phytochemistry 67:2450–2462

Michereff MF, Laumann RA, Borges M, Michereff-Filho M, Diniz IR, Neto AL, Moraes MC (2011) Volatiles mediating a plant–herbivore-natural enemy interaction in resistant and susceptible soybean cultivars. J Chem Ecol 37:273–285

Miller G, Schlauch K, Tam R, Cortes D, Torres MA, Shulaev V, Dangl JL, Mittler R (2009) The plant NADPH oxidase RBOHD mediates rapid systemic signaling in response to diverse stimuli. Sci Signal 2:ra45

Mittler R, Vanderauwera S, Suzuki N, Miller G, Tognetti VB, Vandepoele K, Gollery M, Shulaev V, van Breusegem F (2011) ROS signaling: the new wave? Trends Plant Sci 16:300–309

Moloi MJ, van der Westhuizen AJ (2006) The reactive oxygen species are involved in resistance responses of wheat to the Russian wheat aphid. J Plant Physiol 163:1118–1125

Moloi MJ, van der Westhuizen AJ (2008) Antioxidative responses and the Russian wheat aphid (*Diuraphis noxia*) resistance responses in wheat (*Triticum aestivum*). Plant Biol 10:403–407

Moran PJ, Thompson GA (2001) Molecular responses to aphid feeding in Arabidopsis in relation to plant defense pathways. Plant Physiol 125:1074–1085

Mou Z, Fan W, Dong X (2003) Inducers of plant systemic acquired resistance regulate NPR1 function through redox changes. Cell 113:935–944

Narvaez-Vasquez J, Ryan CA (2002) The systemin precursor gene regulates both defensive and developmental genes in *Solanum tuberosum*. Proc Natl Acad Sci USA 99:15818–15821

Nárvaez-Vasquez J, Orozco-Cardenas ML, Ryan CA (1994) A sulfhydryl reagent modulates systemic signaling for wound-induced and systemin-induced proteinase inhibitor synthesis. Plant Physiol 105:725–730

Narvaez-Vasquez J, Pearce G, Ryan CA (2005) The plant cell wall matrix harbors a precursor of defense signaling peptides. Proc Natl Acad Sci USA 102:12974–12977

Narvaez-Vasquez J, Orozco-Cardenas ML, Ryan CA (2007) Systemic wound signaling in tomato plants is cooperatively regulated by multiple plant peptides. Plant Mol Biol 65:711–718

O'Donnell PJ, Calvert C, Atzorn R, Wasternack C, Leyser HMO, Bowles DJ (1996) Ethylene as a signal mediating the wound response of tomato plants. Science 274:1914–1917

Orians CM, Pomerleau J, Ricco R (2000) Vascular architecture generates fine scale variation in systemic induction of proteinase inhibitors in tomato. J Chem Ecol 26:471–485

Orozco-Cardenas M, Ryan CA (1999) Hydrogen peroxide is generated systemically in plant leaves by wounding and systemin via the octadecanoid pathway. Proc Natl Acad Sci USA 96:6553–6557

Orozco-Cárdenas ML, Ryan CA (2002) Nitric oxide negatively modulates wound signaling in tomato plants. Plant Physiol 130:487–493

Orozco-Cárdenas M, McGurl B, Ryan CA (1993) Expression of an antisense prosystemin gene in tomato plants reduces resistance toward *Manduca sexta* larvae. Proc Natl Acad Sci USA 90:8273–8276

Orozco-Cárdenas ML, Narváez-Vásquez J, Ryan CA (2001) Hydrogen peroxide acts as a second messenger for the induction of defense genes in tomato plants in response to wounding, systemin, and methyl jasmonate. Plant Cell 13:179–191

Park SJ, Huang Y, Ayoubi P (2006) Identification of expression profiles of sorghum genes in response to green-bug phloem-feeding using cDNA subtraction microarray analysis. Planta 223:932–947

Partida-Martinez LP, Heil M (2011) The microbe-free plant: fact or artifact? Front Plant Sci 2:100

Pauwels L, Barbero GF, Geerinck J, Tilleman S, Grunewald W, Perez AC, Chico JM, Bossche RV, Sewell J, Gil E, Garcia-Casado G, Witters E, Inze D, Long JA, De Jaeger G, Solano R,

Goossens A (2010) NINJA connects the co-repressor TOPLESS to jasmonate signalling. Nature 464:788–791
Pearce G (2011) Systemin, hydroxyproline-rich systemin and the induction of protease inhibitors. Curr Protein Pept Sci 12:399–408
Pearce G, Strydom D, Johnson S, Ryan CA (1991) A polypeptide from tomato leaves induced wound-inducible proteinase inhibitor proteins. Science 253:895–898
Pearce G, Johnson S, Ryan CA (1993) Purification and characterization from tobacco (*Nicotiana tabacum*) leaves of six small, wound-inducible, proteinase isoinhibitors of the potato inhibitor II family. Plant Physiol 102:639–644
Pearce G, Moura DS, Stratmann J, Ryan CA (2001) Production of multiple plant hormones from a single polyprotein precursor. Nature 411:817–820
Pearce G, Siems WF, Bhattacharya RC, Chen Y-C, Ryan CA (2007) Three hydroxyproline-rich glycopeptides from a single petunia polyproline precursor activate *defensin I*, a pathogen defense response gene. J Biol Chem 282:17777–17784
Pearce G, Yamaguchi Y, Barona G, Ryan CA (2010) A subtilisin-like protein from soybean contains an embedded cryptic signal that activates defense-related genes. Proc Natl Acad Sci USA 107:14921–14925
Peña-Cortés H, Fisahn J, Willmitzer L (1995) Signals involved in wound-induced proteinase inhibitor II gene expression in tomato and potato plants. Proc Natl Acad Sci USA 92:4106–4113
Peñuelas J, Llusià J (2004) Plant VOC emissions: making use of the unavoidable. Trends Ecol Evol 19:402–404
Peuke AD, Windt C, Van As H (2006) Effects of cold-girdling on flows in the transport phloem in *Ricinus communis*: is mass flow inhibited? Plant Cell Environ 29:15–25
Philippe RN, Ralph SG, Mansfield SD, Bohlmann J (2010) Transcriptome profiles of hybrid poplar (*Populus trichocarpa* × *deltoids*) reveal rapid changes in undamaged, systemic sink leaves after simulated feeding by forest tent caterpillar (*Malacosoma disstria*). New Phytol 188:787–802
Ponce De León I, Schmelz EA, Gaggero C, Castro A, ÁLvarez A, Montesano M (2012) *Physcomitrella patens* activates reinforcement of the cell wall, programmed cell death and accumulation of evolutionary conserved defence signals, such as salicylic acid and 12-oxo-phytodienoic acid, but not jasmonic acid, upon *Botrytis cinerea* infection. Mol Plant Pathol 13:960–974
Popescu SC, Popescu GV, Bachan S, Zhang Z, Gerstein M, Snyder M, Dinesh-Kumar SP (2009) MAPK target networks in *Arabidopsis thaliana* revealed using functional protein microarrays. Genes Dev 23:80–92
Qi Z et al (2010) Ca^{2+} signaling by plant *Arabidopsis thaliana* Pep peptides depends on AtPepR1, a receptor with guanylyl cyclase activity, and cGMP-activated Ca^{2+} channels. Proc Natl Acad Sci USA 107:21193–21198
Ralph SG, Yueh H, Friedmann M, Aeschliman D, Zeznik JA, Nelson CC, Butterfield YS, Kirkpatrick R, Liu J, Jones SJ, Marra MA, Douglas CJ, Ritland K, Bohlmann J (2006) Conifer defence against insects: microarray gene expression profiling of Sitka spruce (*Picea sitchensis*) induced by mechanical wounding or feeding by spruce budworms (*Choristoneura occidentalis*) or white pine weevils (*Pissodes strobi*) reveals large scale changes of the host transcriptome. Plant Cell Environ 29:1545–1570
Ralph SG, Chun HJE, Cooper D, Kirkpatrick R, Kolosova N, Gunter L, Tuskan GA, Douglas CJ, Holt RA, Jones SJM, Marra MA, Bohlmann J (2008) Analysis of 4664 high-quality sequence-finished poplar full-length cDNA clones and their utility for the discovery of genes responding to insect feeding. BMC Genomics 9:57
Ramadan A, Muroi A, Arimura G (2011) Herbivore-induced maize volatiles serve as priming cues for resistance against post-attack by the specialist armyworm *Mythimna separata*. J Plant Interact 6:SI155–SI158
Rasmann S, Köllner TG, Degenhardt J, Hiltpold I, Toepfer S, Kuhlmann U, Gershenzon J, Turlings TC (2005) Recruitment of entomopathogenic nematodes by insect-damaged maize roots. Nature 434:732–737

Rempt M, Pohnert G (2010) Novel acetylenic oxylipins from the moss *Dicranum scoparium* with antifeeding activity against herbivorous slugs. Angew Chem Int Ed 49:4755–4758

Ren F, Lu Y-T (2006) Overexpression of tobacco hydroxyproline-rich glycopeptide systemin precursor A in transgenic tobacco enhances resistance against *Heliocoverpa armigera* larvae. Plant Sci 171:286–292

Reymond P, Weber H, Damond M, Farmer EE (2000) Differential gene expression in response to mechanical wounding and insect feeding in Arabidopsis. Plant Cell 12:707–720

Rhodes JD, Thain JF, Wildon CD (1996) The pathway for systemic electrical signal conduction in the wounded tomato plant. Planta 200:50–57

Rhodes JD, Thain JF, Wildon CD (1999) Evidence for physically distinct systemic signalling pathways in the wounded tomato plant. Ann Bot 84:109–116

Rocco M, Corrado G, Arena S, D'Ambrosio C, Tortiglione C, Sellaroli S, Marra M, Rao R, Scaloni A (2008) The expression of tomato prosystemin gene in tobacco plants highly affects host proteomic repertoire. J Proteomics 71:176–185

Rocha-Granados MC (2004) Sobre-expresión de la prosistemina y sistemina de jitomate (*Lycopersicon esculentum*) en plantas de tabaco (*Nicotiana tabacum*) y su influencia sobre la resistencia a insectos. Ph.D. Thesis. Cinvestav-Unidad Irapuato, México

Rocha-Granados MDC, Sanchez-Hernandez C, Sanchez-Hernandez C, Martinez-Gallardo NA, Ochoa-Alejo N, Delano-Frier JP (2005) The expression of the hydroxyproline-rich glycopeptide systemin precursor A in response to (a) biotic stress and elicitors is indicative of its role in the regulation of the wound response in tobacco (*Nicotiana tubacum L.*). Planta 222:794–810

Rodriguez-Saona CR, Rodriguez-Saona LE, Frost CJ (2009) Herbivore-induced volatiles in the perennial shrub, *Vaccinium corymbosum*, and their role in inter-branch signaling. J Chem Ecol 35:163–175

Rustérucci C, Espunya MC, Díaz M, Chabannes M, Martínez MC (2007) S-Nitrosoglutathione reductase affords protection against pathogens in Arabidopsis, both locally and systemically. Plant Physiol 143:1282–1292

Ruther J, Kleier S (2005) Plant–plant signaling: ethylene synergizes volatile emission in *Zea mays* induced by exposure to (Z)-3-hexen-1-ol. J Chem Ecol 31:2217–2222

Ryan CA (2000) The systemin signaling pathway: differential activation of plant defensive genes. Biochim Biophys Acta 1477:112–121

Ryan CA, Moura DS (2002) Systemic signaling in plants: a new perception. Proc Natl Acad Sci USA 99:6519–6520

Ryan CA, Pearce G (2003) Systemins – A functionally defined family of peptide signals that regulate defensive genes in Solanaceae species. Proc Natl Acad Sci 100:14573–14577

Ryan CA, Huffaker A, Yamaguchi Y (2007) New insights into innate immunity in Arabidopsis. Cell Microbiol 9:1902–1908

Sato C, Seto Y, Nabeta K, Matsuura H (2009) Kinetics of the accumulation of jasmonic acid and its derivatives in distal leaves of tobacco (*Nicotiana tabacum* cv. Xanthi nc) and translocation of deuterium-labeled jasmonic acid from the wounding site to the distal site. Biosci Biotechnol Biochem 73:1962–1970

Sato C, Aikawa K, Sugiyama S, Nabeta K, Masuta C, Matsuura H (2011) Distal transport of exogenously applied jasmonoyl–isoleucine with wounding stress. Plant Cell Physiol 52:509–517

Schaller A, Oecking C (1999) Modulation of plasma membrane H+-ATPase activity differentially activates wound and pathogen defense responses in tomato plants. Plant Cell 11:263–272

Schaller A, Stintzi A (2009) Enzymes in jasmonate biosynthesis: structure, function, regulation. Phytochemistry 70:1532–1538

Schaller F, Schaller A, Stintzi A (2005) Biosynthesis and metabolism of jasmonates. J Plant Growth Regul 23:179–199

Schilmiller AL, Howe GA (2005) Systemic signaling in the wound response. Curr Opin Plant Biol 8:369–377

Schittko U, Baldwin IT (2003) Constraints to Herbivore-induced systemic responses: bidirectional signaling along orthostichies in *Nicotiana attenuata*. J Chem Ecol 29:763–770

Schittko U, Preston CA, Baldwin IT (2000) Eating the evidence? *Manduca sexta* larvae can not disrupt specific jasmonate induction in *Nicotiana attenuata* by rapid consumption. Planta 210:343–346

Schmeltz I (1971) Nicotine and other tobacco alkalioids. In: Jacobson MCD (ed) Naturally occurring insecticides. Mercel Dekker, New York, pp 99–136

Schmelz EA, Alborn HT, Tumlinson JH (2003) Synergistic interactions between volicitin, jasmonic acid and ethylene mediate insect-induced volatile emission in *Zea mays*. Physiol Plant 117:403–412

Schmelz EA, Carroll MJ, LeClere S, Phipps SM, Meredith J, Chourey PS, Alborn HT, Teal PE (2006) Fragments of ATP synthase mediate plant perception of insect attack. Proc Natl Acad Sci USA 103:8894–8899

Schmelz EA, Engelberth J, Alborn HT, Tumlinson JH 3rd, Teal PE (2009) Phytohormone-based activity mapping of insect herbivore-produced elicitors. Proc Natl Acad Sci USA 106:653–657

Schmidt S, Baldwin IT (2006) Systemin in black nightshade (*Solanum nigrum*). The tomato homologous polypeptide does not mediate direct defense responses. Plant Physiol 142:1751–1758

Schmidt S, Baldwin IT (2009) Down-regulation of systemin after herbivory is associated with increased root allocation and competitive ability in *Solanum nigrum*. Oecologia 159:473–482

Schommer C, Palatnik JF, Aggarwal P, Chételat A, Cubas P, Farmer EE, Nath U, Weigel D, Carrington JC (2008) Control of jasmonate biosynthesis and senescence by miR319 targets. PLoS Biol 6:e230

Schowalter TD, Hargrove WW, Crossley DA (1986) Herbivory in forested ecosystems. Annu Rev Entomol 31:177–196

Schwachtje J, Minchin PE, Jahnke S, van Dongen JT, Schittko U, Baldwin IT (2006) SNF1-related kinases allow plants to tolerate herbivory by allocating carbon to roots. Proc Natl Acad Sci USA 103:12935–12940

Seo S, Okamoto M, Seto H, Ishizuka K, Sano H, Ohashi Y (1995) Tobacco MAP kinase: a possible mediator in wound signal transduction pathways. Science 270:1988–1992

Seo S, Sano H, Ohashi Y (1999) Jasmonate-based wound signal transduction requires activation of WIPK, a tobacco mitogen-activated protein kinase. Plant Cell 11:289–298

Seo HS, Song JT, Cheong JJ, Lee YH, Lee YW, Hwang I, Lee JS, Choi YD (2001) Jasmonic acid carboxyl methyltransferase: a key enzyme for jasmonate-regulated plant responses. Proc Natl Acad Sci USA 98:4788–4793

Shiojiri K, Kishimoto K, Ozawa R, Kugimiya S, Urashimo S, Arimura G, Horiuchi J, Nishioka T, Matsui K, Takabayashi J (2006) Changing green leaf volatile biosynthesis in plants: an approach for improving plant resistance against both herbivores and pathogens. Proc Natl Acad Sci USA 103:16672–16676

Shulaev V, Silverman P, Raskin I (1997) Airborne signalling by methyl salicylate in plant pathogen resistance. Nature 385:718–721

Snoeren TA, Mumm R, Poelman EH, Yang Y, Pichersky E, Dicke M (2010) The herbivore-induced plant volatile methyl salicylate negatively affects attraction of the parasitoid *Diadegma semiclausum*. J Chem Ecol 36:479–489

Spoel SH, Loake GJ (2011) Redox-based protein modifications: the missing link in plant immune signalling. Curr Opin Plant Biol 14:358–364

Stahl E (1888) Pfianzen und Scbnecken. Biologische Studie ueber die Scbutzmittel der Pfianzen gegen Scbneckenfrass. Jenaische Zeitschrift fuer Naturwissenschaft 22:557–685

Stahlberg R, Cosgrove DJ (1997) The propagation of slow wave potentials in pea epicotyls. Plant Physiol 113:209–217

Stamler JS (1995) S-nitrosothiols and the bioregulatory actions of nitrogen oxides through reactions with thiol groups. Curr Top Microbiol Immunol 196:19–36

Stankovic B, Davies E (1996) Both action potentials and variation potentials induce proteinase inhibitor gene expression in tomato. FEBS Lett 390:275–279

Staswick PE, Tiryaki I (2004) The oxylipin signal jasmonic acid is activated by an enzyme that conjugates it to isoleucine in Arabidopsis. Plant Cell 16:2117–2127

Staswick PE, Serban B, Rowe M, Tiryaki I, Maldonado MT, Maldonado MC, Suza W (2005) Characterization of an Arabidopsis enzyme family that conjugates amino acids to indole-3-acetic acid. Plant Cell 17:616–627

Steppuhn A, Baldwin IT (2008) Induced defenses and the cost-benefit paradigm. In: Schaller A (ed) Induced plant resistance to herbivory. Springer, Germany, pp 61–83

Stitz M, Baldwin IT, Gaquerel E (2011) Diverting the flux of the JA pathway in *Nicotiana attenuata* compromises the plant's defense metabolism and fitness in nature and glasshouse. PLoS One 6:e25925

Stratmann JW (2003) Long distance run in the wound response—jasmonic acid is pulling ahead. Trends Plant Sci 8:247–250

Stratmann JW, Ryan CA (1997) Myelin basic protein kinase activity in tomato leaves is induced systemically by wounding and increases in response to systemin and oligosaccharide elicitors. Proc Natl Acad Sci USA 94:11085–11089

Stratmann J, Paputsoglu G, Oertel W (1996) Differentiation of *Ulva mutabilis* (Chlorophyta) gametangia and gamete release are controlled by extracellular inhibitors. J Phycol 32:1009–1021

Suza WP, Staswick PE (2008) The role of JAR1 in jasmonoyl-L-isoleucine production in Arabidopsis wound response. Planta 227:1221–1232

Suza WP, Rowe ML, Hamberg M, Staswick PE (2010) A tomato enzyme synthesizes (+)-7-iso-jasmonoyl-L-isoleucine in wounded leaves. Planta 231:717–728

Tada Y, Spoel SH, Pajerowska-Mukhtar K, Mou Z, Song J, Wang C, Zuo J, Dong X (2008) Plant immunity requires conformational changes of NPR1 via *S*-nitrosylation and thioredoxins. Science 321:952–956

Tejeda-Sartorius M, Martínez de la Vega O, Délano-Frier J (2008) Jasmonic acid influences mycorrhizal colonization in tomato plants by modifying the expression of genes involved in carbohydrate partitioning. Physiol Plant 133:339–353

Thaler JS, Bostock RM (2004) Interactions between abscisic-acid-mediated responses and plant resistance to pathogens and insects. Ecology 85:48–58

Thaler JS, Fidantsef AL, Bostock RM (2002) Antagonism between jasmonate- and salicylate-mediated induced plant resistance: effects of concentration and timing of elicitation on defense-related proteins, herbivore, and pathogen performance in tomato. J Chem Ecol 28:1131–1159

Thaler JS, Agrawal AA, Halitschke R (2010) Salicylate-mediated interactions between pathogens and herbivores. Ecology 91:1075–1082

Theodoulou FL, Job K, Slocombe SP, Footitt S, Holdsworth M, Baker A, Larson TR, Graham IA (2005) Jasmonoic acid levels are reduced in COMATOSE ATP-binding cassette transporter mutants. Implications for transport of jasmonate precursors into peroxisomes. Plant Physiol 137:835–840

Thines B, Katsir L, Melotto M, Niu Y, Mandaokar A, Liu G, Nomura K, He SY, Howe GA, Browse J (2007) JAZ repressor proteins are targets of the SCFCOI1 complex during jasmonate signalling. Nature 448:661–665

Thompson GA, Goggin FL (2006) Transcriptomics and functional genomics of plant defence induction by phloem-feeding insects. J Exp Bot 57:755–766

Thompson JN, Pellmyr O (1991) Evolution of oviposition behavior and host preference in Lepidoptera. Annu Rev Entomol 36:65–89

Tortiglione C, Fogliano V, Ferracane R, Fanti P, Pennacchio F, Monti LM, Rao R (2003) An insect peptide engineered into the tomato prosystemin gene is released in transgenic tobacco plants and exerts biological activity. Plant Mol Biol 53:891–902

Toth GB, Pavia H (2007) Induced herbivore resistance in seaweeds: a meta-analysis. J Ecol 95:425–434

Truitt CL, Pare PW (2004) In situ translocation of volicitin by beet armyworm larvae to maize and systemic immobility of the herbivore elicitor in planta. Planta 218:999–1007

Turlings TC, Tumlinson JH, Lewis WJ (1990) Exploitation of herbivore-induced plant odors by host-seeking parasitic wasps. Science 250:1251–1253

Van Dam NM, Horn M, Mareš M, Baldwin IT (2001) Ontogeny constrains systemic protease inhibitor response in *Nicotiana attenuata*. J Chem Ecol 27:547–568
van Eck L, Schultz T, Leach JE, Scofield SR, Peairs FB, Botha AM, Lapitan NHV (2010) Virus-induced gene silencing of *WRKY53* and an inducible *phenylalanine ammonia-lyase* in wheat reduces aphid resistance. Plant Biotechnol J 8:1023–1032
Verhage A, Vlaardingerbroek I, Raaymakers C, Van Dam NM, Dicke M, Van Wees SCM, Pieterse CMJ (2011) Rewiring of the jasmonate signaling pathway in Arabidopsis during insect herbivory. Front Plant Sci 2:47
von Dahl CC, Baldwin IT (2007) Deciphering the role of ethylene in plant-herbivore interactions. J Plant Growth Reg 26:201–209
Walling LL (2000) The myriad plant responses to herbivores. J Plant Growth Reg 19:195–216
Walling LL (2008) Avoiding effective defenses: strategies employed by phloem-feeding insects. Plant Physiol 146:859–866
Wasternack C (2007) Jasmonates: an update on biosynthesis, signal transduction and action in plant stress response, growth and development. Ann Bot 100:681–697
Weber H, Vick BA, Farmer EE (1997) Dinor-oxo-phytodienoic acid: a new hexadecanoid signal in the jasmonate family. Proc Natl Acad Sci USA 94:10473–10478
Wei Z, Hu W, Lin Q, Cheng X, Tong M, Zhu L, Chen R, He G (2009) Understanding rice plant resistance to the brown planthopper (*Nilaparvata lugens*): a proteomic approach. Proteomics 9:2798–2808
Wichard T, Göbel C, Feussner I, Pohnert G (2005) Unprecedented lipoxygenase/hydroperoxide lyase pathways in the moss *Physcomitrella patens*. Angew Chem Int Ed 44:158–161
Wildon DC, Thain JF, Minchin PEH, Gubb IR, Reilly AJ, Skipper YD, Doherty HM, O'Donnell PJ, Bowles DJ (1992) Electrical signalling and systemic proteinase inhibitor induction in the wounded plant. Nature 360:62–65
Will T, van Bel AJE (2008) Induction as well as suppression: How aphid saliva may exert opposite effects on plant defense. Plant Signal Behav 3:427–430
Wittstock U, Gershenzon J (2002) Constitutive plant toxins and their role in defense against herbivores and pathogens. Curr Opin Plant Biol 5:300–307
Wu J, Hettenhausen C, Meldau S, Baldwin IT (2007) Herbivory rapidly activates MAPK signaling in attacked and unattacked leaf regions but not between leaves of *Nicotiana attenuata*. Plant Cell 19:1096–1122
Wu J, Wang L, Baldwin I (2008) Methyl jasmonate-elicited herbivore resistance: does MeJA function as a signal without being hydrolyzed to JA? Planta 227:1161–1168
Wünsche H, Baldwin IT, Wu J (2011) S-Nitrosoglutathione reductase (GSNOR) mediates the biosynthesis of jasmonic acid and ethylene induced by feeding of the insect herbivore *Manduca sexta* and is important for jasmonate-elicited responses in *Nicotiana attenuata*. J Exp Bot 62:4605–4616
Yamaguchi Y, Huffaker A (2011) Endogenous peptide elicitors in higher plants. Curr Opin Plant Biol 14:351–357
Yamaguchi Y, Pearce G, Ryan CA (2006) The cell surface leucine-rich repeat receptor for AtPep1, an endogenous peptide elicitor in Arabidopsis, is functional in transgenic tobacco cells. Proc Natl Acad Sci USA 103:10104–10109
Yamaguchi Y, Huffaker A, Bryan AC, Tax FE, Ryan CA (2010) PEPR2 is a second receptor for the Pep1 and Pep2 peptides and contributes to defense responses in Arabidopsis. Plant Cell 22:508–522
Yamaguchi Y, Barona G, Ryan CA, Pearce G (2011) GmPep914, an eight-amino acid peptide isolated from soybean leaves, activates defense-related genes. Plant Physiol 156:932–942
Zangerl AR (1999) Locally-induced responses in plants: The ecology and evolution of restrained defense. In: Agrawal AA, Tuzun S, Bent E (eds) Induced plant defenses against pathogens and herbivores. APS Press, St. Paul, Minnesota, pp 231–249
Zangerl AR, Rutledge CE (1996) The probability of attack and patterns of constitutive and induced defense: a test of optimal defense theory. Am Nat 147:599–608
Zimmermann MR, Maischak H, Mithofer A, Boland W, Felle HH (2009) System potentials, a novel electrical long-distance apoplastic signal in plants, induced by wounding. Plant Physiol 149:1593–1600

Calcium as a Trigger and Regulator of Systemic Alarms and Signals along the Phloem Pathway

Aart J.E. van Bel and Jens B. Hafke

Abstract This chapter explores how Ca^{2+} transmembrane movements of individual cells along the phloem pathway could collectively act as a basis for systemic signalling. Sieve elements play a pivotal role in this unifying, speculative concept of Ca^{2+}-triggered signalling which comprises three principal assumptions: (1) Arrays of sieve elements provide a self-amplifying, least-resistance route for electrical potential waves as the inductors of various Ca^{2+} signatures in cells along the phloem pathway. (2) Ca^{2+}-based systemic signalling occurs in three successive, partly overlapping waves that are distinct in timescale, nature, and site of origin. (3) The second and third waves of signalling require Ca^{2+}-induced occlusion of the symplasmic connections in sieve tubes and adjoining tissues.

Keywords Action potentials • Ca^{2+} compartmentation • Ca^{2+} hotspots • Ca^{2+}-permeable channels • Electrical potential waves • Phloem • Sieve tubes • Systemic signalling • Variation potentials

1 Introduction

Under natural conditions, plants have to cope with a broad range of abiotic and biotic stimuli that must be sensed to adapt to environmental changes. Stimuli in the cellular environment are monitored by a vast battery of chemical sensors complemented by the action of ligand-activated, voltage-dependent, or mechano-sensitive channels

A.J.E. van Bel (✉)
Plant Cell Biology Research Group, Institute of General Botany, Justus-Liebig University Giessen, Gießen, Germany
e-mail: aart.v.bel@bot1.bio.uni-giessen.de

J.B. Hafke
Plant Cell Physiology Group, Institute of Plant Physiology, Justus-Liebig University Giessen, Gießen, Germany

F. Baluška (ed.), *Long-Distance Systemic Signaling and Communication in Plants*, Signaling and Communication in Plants 19, DOI 10.1007/978-3-642-36470-9_18,

(Hedrich 2012). These sensors mediate Ca^{2+} influx directly or indirectly as the first event in a chain of intracellular responses. The cytosolic Ca^{2+} is elevated according to stimulus-specific and spatio-temporal patterns named Ca^{2+} signatures (Webb et al. 1996; Ng and McAinsh 2003; McAinsh and Pittman 2009; Kudla et al. 2010) that are decoded and relayed by signalling molecules into a multitude of downstream events. Influx of Ca^{2+} also initiates so-called receptor potentials (e.g. Williams and Pickard 1972b) that can trigger self-propagating potential waves moving long distances along excitable cells.

In short, Ca^{2+} influx being one of the prime events in the transfer of information from the cellular environment to the cell interior is performed by Ca^{2+}-permeable channels. Thus far, no Ca^{2+}-selective channels in the plasma membrane of plant cells have been identified with certainty (Kudla et al. 2010); hence the more general term "Ca^{2+}-permeable" is used here (e.g. Sanders et al. 2002). In general, Ca^{2+} influx triggers a multidimensional web of intracellular cascades leading to changes in gene expression and/or metabolism, thereby conferring adaptive reactions (Sanders et al. 2002; White and Broadley 2003; McAinsh and Pittman 2009; Kudla et al. 2010). We will explore how trans-membrane Ca^{2+} movements of individual cells could collectively act as a basis for systemic signalling on the whole-plant level.

In the concept advanced here, sieve elements play a pivotal role for several reasons. Sieve element–companion cell complexes—the sieve-tube modules—are the conductive elements of the phloem and responsible for systemic transmission of electrical potential waves (EPWs; Samejima and Sibaoka 1983; Fromm 1991; Wildon et al. 1992; Fromm and Spanswick 1993; Rhodes et al. 1996; Fromm and Lautner 2006; Furch et al. 2007). Sieve elements have the capacity of a domino-like influx of Ca^{2+} ions along the pathway in response to a wide range of stimuli. The Ca^{2+} influx results in a transiently symplasmic isolation of sieve elements and probably of the adjoining parenchymatous cells, conferring a temporary autonomy of vascular cells. It is speculated that Ca^{2+} influx into vascular parenchyma cells triggers cell-specific signal cascades operating during this autonomous period to produce a diversity of chemical signals that are released into and translocated by the phloem stream.

2 Evolutionary Aspects of Calcium-Based Systemic Signalling

2.1 Evolution of Sieve Element–Companion Cell Complexes

Sieve elements are axially oriented cells that have become specialised in long-distance transport during evolution. The evolutionary development is derived from a phylogenetic comparison of mosses, seedless vascular plants, and seed plants, and it appears that sieve-element evolution is largely reflected by the ontogeny of sieve elements in angiosperms (van Bel and Knoblauch 2000; van Bel 2003).

To serve long-distance transport, the cellular equipment is reduced to a plasma membrane envelope lined with a thin margin of gelous cytoplasm (mictoplasm) with a limited number of organelles. This layer which is in open connection with the luminal sap has been redefined as mictoplasm (van Bel 2003) for practical reasons (it is the layer in which Ca^{2+} concentrations are strongly enhanced during passage of EPWs). Originally, the mictoplasm was defined as the mixture of cytoplasmic contents and the luminal sieve-element fluid arising from disintegration of the vacuole (Engleman 1965).

That the mictoplasm is in direct contact with the liquid in the lumen implies that the vacuole dissolves during sieve-element ontogeny in angiosperms (van Bel 2003). Several other cellular devices are "sacrificed" during development in support of the transport function; nucleus, ribosomes, and Golgi apparatus are degraded (Behnke and Sjolund 1990). The remnants are ER stacks, parietal structural proteins, and a special form of plastids, the sieve-element plastids, being appreciably smaller than chloroplasts (Behnke and Sjolund 1990). The ER is organised in aggregates of regularly stacked ER that is often oriented perpendicular to the plasma membrane (e.g. Sjolund and Shih 1983; Ehlers et al. 2000). It is still unclear if a complete cytoskeleton is present in sieve elements (Chaffey and Barlow 2002), but cytoskeleton proteins do occur in sieve-tube saps (Barnes et al. 2004; Walz et al. 2004; Giavalisco et al. 2006; Aki et al. 2008). As a result of their limited cellular outfit, sieve elements must almost fully rely on the activities of the associate companion cell for their survival (van Bel 2003).

Each sieve element is flanked by one to four companion cells (Esau 1969) that originate from the same cambial precursor as the sieve element. As inferred from dye coupling experiments, temporary symplasmic isolation of the cambial sieve element and companion cell precedes developmental specialisation (van Bel and van Rijen 1994). Towards the end of the transient seclusion, specialised plasmodesmata arise at the interface between sieve element and companion cell. These plasmodesmata—named pore-plasmodesma units (PPUs; van Bel and Kempers 1997)—are branched at the companion-cell side. They possess the ability to traffic macromolecules between companion cell and sieve elements (Imlau et al. 1999; Lucas et al. 2001, 2009).

2.2 *Evolution of Propagation of Action Potentials*

It appears that most electrical signals in plants are initiated by Ca^{2+} influx. Hence, it is crucial for a better understanding of Ca^{2+}-mediated systemic signalling to explore which sieve-element devices enable processing of local and systemic electrical signals. Since the sieve-element plasma membrane is still intact, the basics of Ca^{2+} traffic via the sieve-element plasma membrane and the coincident electrical signalling may not strongly differ from that in parenchyma cells.

Parenchyma cells have the ability to electrotonically spread polarisation waves which can move symplasmically to a restricted number of non-excitable neighbour cells (e.g. Brinckmann and Lüttge 1974; Overall and Gunning 1982; van Bel and van Rijen 1994; Holdaway-Clarke et al. 1996). The intercellular current is strongly dampened by the plasmodesmal bottlenecks so that the signals dissipate within a few cells distance (Overall and Gunning 1982). During evolution, however, diverse arrays of parenchyma cells in various plant species must have developed the capability to amplify electric signals so that they can be propagated via plasmodesmata over considerable distances to affect remote processes.

As a first example, the respiration of the ovary increases following a pollen-triggered action potential (AP) that is transmitted down the style (Sinyukin and Britikov 1967). Similarly, EPWs were observed after stimulation of *Hibiscus* stigmas by heat or cold shocks which affect the ovary metabolism (Fromm et al. 1995). Furthermore, touching the trigger hairs of the trap leaves of *Dionaea* causes an EPW (Sibaoka 1966), inducing rapid closure of the leaf halves (Hodick and Sievers 1989). Bending the trigger hairs provokes change in mechanical pressure inducing Ca^{2+} release into the cytosol of the sensory cells (Hodick and Sievers 1988), possibly released from whorls of ER (Buchen et al. 1983). The Ca^{2+} elevation coincides with a membrane depolarisation from -160 to -50 mV (Hodick and Sievers 1988). The resultant EPW is probably propagated via the numerous plasmodesmata in the basal walls of the sensory cells (Buchen et al. 1983). As a last example, the plasma membrane of the epidermal cells is depolarised when the glandular heads of *Drosera* leaf tentacles are touched by insects (Williams and Pickard 1972a). If the depolarisation exceeds a certain voltage threshold, a train of APs is initiated to propagate along the tentacle stalk (Williams and Pickard 1972a). Abundant plasmodesmal connectivity in the transverse walls of the stalk cells allows rapid AP propagation towards the stalk base (Williams and Pickard 1974), which, upon AP arrival, triggers bending of the tentacles (Williams and Pickard 1972b).

During evolution, sieve elements appear to have developed similar plasma membrane devices for acceleration of the propagation velocity and extension of the distance range of electrical signals to achieve remote actions. Moreover, the electrical resistance in the walls between sieve elements has been strongly reduced by transforming plasmodesmata into sieve pores (van Bel 2003), which facilitates longitudinal electrical propagation.

3 Electrical Potential Waves

3.1 Definitions

Plants possess diverse types of EPWs mainly propagated along the vascular pathways (Pickard 1973; Rhodes et al. 1996; Stahlberg and Cosgrove 1997; Davies 2006; Davies and Stankovic 2006; Stahlberg et al. 2006; Furch et al. 2007; Hafke et al. 2009; Zimmermann et al. 2009). Fast transmission of electrical signals is achieved by action potentials (APs) and slower transmission by variation potentials (VPs), while a third type (system potentials, SPs) has recently been discovered (Zimmermann et al. 2009). Quite often mixed EPWs are recorded, e.g. as result of overlapping APs and VPs, which impedes proper signal analysis (Stahlberg et al. 2006; Davies and Stankovic 2006; Furch et al. 2007; Hafke et al. 2009).

The EPW types are distinguished based on kinetic parameters such as amplitude, duration, and profile, altogether designated as (signal) signatures (e.g. Stahlberg et al. 2006). Distinction between APs and VPs based solely upon their kinetic characteristics (Stankovic et al. 1997; Dziubinska et al. 2001) might be doubtful since VPs often mimic AP kinetics (Stahlberg et al. 2006; Furch et al. 2008). Therefore, additional criteria such as stimulus specificity and strength required to signal triggering are invoked (Stahlberg et al. 2006).

However, a strict distinction between APs and VPs is sometimes hardly feasible. Strong injuries, for instance, often trigger supplementary APs, resulting in composite signals of APs and VPs (Davies and Stankovic 2006; Furch et al. 2007, 2009; Hafke et al. 2009). For instance, burning triggers composite signals (Houwink 1935; van Sambeek and Pickard 1976; Roblin 1985; Roblin and Bonnemain 1985; Stankovic et al. 1997, 1998) when recorded at the close proximity of the burning site (Davies and Stankovic 2006; Furch et al. 2007, 2009, 2010; Hafke et al. 2009). With increasing distance from the burning site, the signals drift apart as result of the different propagation velocities (Davies et al. 1991; Stankovic et al. 1998; Davies and Stankovic 2006; Furch et al. 2010).

3.2 Action Potentials

APs are triggered by nonwounding and natural stimuli (Trebacz et al. 2006; Stahlberg et al. 2006) such as the fast thigmonastic movements in touch-sensitive plants (Sibaoka 1969, 1991), the movements of the carnivorous Venus flytrap (Hodick and Sievers 1988, 1989) and waterwheel plant (Iijima and Sibaoka 1981), and the complex tentacle movements in sundews (Williams and Pickard 1972a, b; Williams and Spanswick 1976). Distant treatment with ice water also induces APs (Fromm 1991; Fromm and Bauer 1994) as do drought-mimicking treatments of roots (Fromm and Fei 1998).

APs are characterised by spike-like changes of the resting membrane potential (Stahlberg and Cosgrove 1997). A period of less than one minute needed for the onset of depolarisation and the completion of repolarisation identifies an EPW as an AP most of the times (Pickard 1973; Fromm 1991). Like in animals, APs seem to be all-or-nothing events (Fromm and Spanswick 1993; Pyatygin et al. 2008). Below a critical membrane potential threshold, an AP or self-amplifying electrical signal with a defined amplitude and velocity being independent of the stimulus strength starts propagating through the plant (Zawadzki et al. 1991; Fromm 1991; Wildon et al. 1992; Fromm and Spanswick 1993; Stankovic et al. 1998; Davies and Stankovic 2006; Fromm and Lautner 2006).

The fast AP kinetics depends on the orchestrated activity of voltage-dependent ion channels (Lunevsky et al. 1983; Okihara et al. 1991; Zawadzki et al. 1991; Wayne 1994; Fromm and Bauer 1994; Davies 2004). The ionic mechanisms appear to bear a strong resemblance (Trebacz et al. 2006) throughout the entire plant kingdom (Iijima and Sibaoka 1981, 1982, 1985; Hodick and Sievers 1988; Fromm and Spanswick 1993; Trebacz et al. 1994; Fromm and Bauer 1994; Opritov et al. 2002; Krol et al. 2003, 2004; Fisahn et al. 2004; Felle and Zimmermann 2007; Furch et al. 2007, 2009) and in some algal taxa (Lunevsky et al. 1983; Okihara et al. 1991; Homann and Thiel 1994; Thiel et al. 1997).

An initial membrane depolarisation is triggered by a transient increase in cytoplasmic Ca^{2+} concentration brought about by gating of Ca^{2+}-permeable channels in the plasma membrane. The enhanced Ca^{2+} concentration leads to a Cl^- efflux via Ca^{2+}-dependent anion channels also located in the plasma membrane (Lunevsky et al. 1983; Tsutsui et al. 1986; Okihara et al. 1991; Homann and Thiel 1994). There are various indications that additional Ca^{2+} is released from intracellular compartments during the transient Ca^{2+} increase. Cellular second messengers such as inositol 1,4,5-triphosphate (IP_3, Gilroy et al. 1990), inositol hexakisphosphate (IP_6, Lemtiri-Chlieh et al. 2003), nicotinic acid adenine dinucleotide phosphate (NAADP; Navazio et al. 2000), or cyclic ADP-ribose (cADPR, Leckie et al. 1998) trigger Ca^{2+} mobilisation from endomembrane systems such as ER and vacuoles.

Voltage-dependent K^+ channels open up just before the steady-state potential for Cl^- ions is reached and mediate K^+ efflux in order to repolarise the membrane potential (Homann and Thiel 1994; Thiel et al. 1997). Since the full repolarisation is limited by the Nernst potential for K^+, re-establishment of the membrane resting level is boosted by the activity of the electrogenic proton pumps (Kishimoto et al. 1985).

3.3 *Variation Potentials*

VPs are mostly triggered by damaging treatments including burning (Houwink 1935; van Sambeek and Pickard 1976; Roblin 1985; Wildon et al. 1992; Stankovic et al. 1998; Mancuso 1999; Furch et al. 2007, 2010; Hafke et al. 2009), hot water (van Sambeek and Pickard 1976), vigorous cutting (Mancuso 1999; Furch et al. 2008; Hafke et al. 2009), strong crushing (van Sambeek and Pickard 1976), electroshocks (Pickard and Minchin 1990), and injury by chewing insects (Alarcon and Malone 1994).

VPs differ from APs in various ways. First of all, VPs do not obey the all-or-nothing law: the signals are commensurate with the stimulus strength and last for periods of 10 s up to 30 min (Stahlberg and Cosgrove 1997; Stahlberg et al. 2006). The propagation velocities are 5–10 times slower than with APs (Stahlberg and Cosgrove 1997), and the amplitude drops rapidly along the transmission path

(Davies 2004; Stahlberg et al. 2005, 2006). Thus, VPs attenuate with decreasing distance from the site of stimulus and finally fade away (van Sambeek and Pickard 1976).

While the fast repolarisation during APs is mediated by ion channels, slow repolarisation of VPs might originate from the shutdown of electrogenic proton pumps at the plasma membrane as indicated by pH-dependent fluorochromes and the ineffectiveness of ion channel blockers (Stahlberg and Cosgrove 1992, 1996). The mechanism of proton-pump inhibition during VP generation is not understood in detail but might be due to Ca^{2+} inhibition. It is known that the plasma-membrane proton pump is inhibited by an increase in cytosolic Ca^{2+} in the physiological range (Kinoshita et al. 1995). Perhaps, proton-pump activity is equally suppressed by Ca^{2+} in APs, but hardly detectable due to a lower rate of Ca^{2+} influx.

Generation and propagation of VPs have been observed only in intact plants, whereas APs can propagate in isolated plant organs (Stahlberg et al. 2006). Therefore, VP generation in sieve elements seems to be linked to physical processes in xylem vessels. Relaxation of the negative pressure propagating as a wave through the xylem vessels provides the likely basis for VP generation (Stahlberg and Cosgrove 1997). That the basis of VPs originates from hydraulic changes in the dead xylem vessels was demonstrated by the fact that VPs can pass dead or poisoned areas in contrast to APs (Stahlberg et al. 2006).

4 Consequences of Symplasmic Organisation of the Sieve-Tube Tracks for EPW Propagation

4.1 Action Potentials

Selective and rapid propagation of action potentials in neurons of mammals is achieved by insulation through glia cells. Neurons and glia cells are electrically coupled via numerous connexons and these neuron–glia complexes are electrically insulated from surrounding cells by virtue of a myelin sheath and a lack of gap junctions (Fields and Stevens-Graham 2002). Selective propagation of APs via arrays of sieve elements hints at the existence of electrical barriers towards other cells. In analogy to the "symplasmic isolation" of neurons, sieve element–companion cell complexes may be insulated due to structural barriers.

The electrical conductivity of the sieve-element plasma membrane, the longevity of the sieve elements and the low electrical conductivity of the large sieve pores make sieve tubes ideal conduits for long-distance electrical signalling (van Bel and Ehlers 2005). Moreover, strong electrical coupling between sieve elements and companion cells and a poor electrical coupling with other phloem cells argue in favour of a neuron-like insulation (van Bel and van Rijen 1994). Symplasmic connections between sieve elements and companion cells are indeed abundant, whereas those between sieve elements and phloem parenchyma cells are absent along most of the phloem pathway (Kempers et al. 1998). Plasmodesmata between companion cells and phloem parenchyma cells are sparse and closed under source-limiting conditions (Patrick and Offler 1996; Hafke et al. 2005). These structural conditions would fit electrical insulation.

Yet, excess photoassimilates are unloaded along the pathway under sink-limiting conditions (Patrick and Offler 1996). The necessity to quickly fill axial storage compartments implies a lack of permanent insulation (Patrick and Offler 1996). In addition, electrical currents are expected to pass plasmodesmata with exclusion limits less than small molecules which would make the companion cells electrically leaky through the few plasmodesmata towards the phloem parenchyma cells (Kempers et al. 1998) when they are not tightly closed. Thus, there is a good chance that electrical isolation of sieve element–companion cell complexes is incomplete.

Lateral transmission of electrical currents diverted from action potentials likely occurs via plasmodesmata (Rhodes et al. 1996; Trebacz et al. 2006). Loss of electrical current was inferred from depolarisations of adjacent parenchyma cells coincident with the passage of burning-triggered EPWs along sieve tubes (Rhodes et al. 1996) and from the decreasing reactivity of aphids with increasing distance from the site of burning (Furch et al. 2010). The latter conclusion should be drawn with care, since burning-triggered EPWs comprise VP components (Dziubinska et al. 2001; Hlavackova et al. 2006; Furch et al. 2010) which dampen with the distance (Davies 2004; Stahlberg et al. 2005).

That longitudinal transmission can be limited to organs or regions within the plants (Trebacz et al. 2006) indicates the existence of electrical barriers probably of anatomical origin. For unknown reasons, APs in lupine propagate both acropetally and basipetally along the stem, but do not reach roots or leaves (Zawadzki 1980). In specialised plants such as *Mimosa pudica*, long distances are covered by APs owing to an insulating sclerenchyma sheath around the sieve element–companion cell complexes (Fleurat-Lessard and Roblin 1982) which may be functionally analogous to the myelin sheath. This structural shield is interrupted in pulvini, where many plasmodesmata provide ample symplasmic access to flexor parenchyma cells (Fleurat-Lessard et al. 1978). The inherent abundant electrical coupling facilitates current leakage from the sieve elements towards the flexor cells, which readily react by loss here of osmotic substances (Fleurat-Lessard et al. 1978). The massive current loss here may exemplify hardly observable current leakage from sieve elements during AP passage along the phloem pathway in non-specialised plants. As a consequence, action potentials diverted to cells symplasmically connected with companion cells probably cause reduced rates of Ca^{2+} influx there (e.g. Rhodes et al. 1996).

4.2 Variation Potentials

As for VPs, the situation is more complicated from the structural, functional, and organisational viewpoint. Like others (e.g. Pyatygin et al. 2008), we tend to believe that VPs are not propagated along the sieve tubes, but represent serial depolarisations originating from the vascular parenchyma cells mimicking electrical propagation via the sieve tubes (Malone 1996; van Bel et al. 2011a).

Serial disturbance of the hydraulic equilibrium in the xylem leads to water uptake by the parenchyma cells lining the xylem vessels. The resultant increase in turgor causes membrane depolarisation of these cells (Malone and Stankovic 1991; Stahlberg and Cosgrove 1992, 1997; Mancuso 1999; Davies 2006). How this information is passed on to the sieve tubes depends on the available tools and the structural conditions in the lateral pathways between vascular cells. It is likely that the receptor potentials are triggered via mechano-sensitive Ca^{2+} permeable channels in the xylem parenchyma cells, but we can only speculate on the subsequent scenario of lateral electrical transmission.

Given its hydraulic and chemical components, mechano-sensitive and ligand-activated Ca^{2+} permeable channels have been postulated to be engaged in VP generation (Davies 1993, 2006; Stankovic et al. 1997; Davies and Stankovic 2006; Stahlberg et al. 2006). The first possibility for lateral propagation is that the receptor potential activates voltage-dependent channels and is communicated by waves of voltage changes via plasmodesmata to adjacent cells and ultimately to the sieve tubes. The second option is that the turgor of all vascular cells towards and including the sieve tubes may also be increased by intake of water after cutting as argued for cucurbit exudation mechanisms (Zimmermann et al. 2013). Successive longitudinal turgor changes then may produce trains of receptor potentials in lateral direction. According to this scenario, VP propagation may result from the concerted action of mechano-sensitive channels that perceive local turgor changes, and the resultant receptor potentials mimic electrical propagation in lateral direction. As a third alternative, Ca^{2+} influx may induce signalling cascades conferring production of chemical signals (Ricca 1916: van Sambeek and Pickard 1976; van Sambeek et al. 1976; Boari and Malone 1993; Malone 1996; Stahlberg and Cosgrove 1997; Mancuso 1999; Pyatygin et al. 2008). Oligosaccharides as well as the peptide systemin in solanacean plant species (Narvaez-Vazquez and Ryan 2004) are among potential second messengers triggering VP-like depolarisations (Thain et al. 1995; Moyen and Johannes 1996) inducing Ca^{2+} influx via ligand-activated Ca^{2+}-permeable channels (CNGCs; reviewed by Kudla et al 2010). In view of the concurrence of diverse Ca^{2+}-permeable channels in plasma membranes (Kudla et al. 2010), a combination of the above scenarios is likely.

Lateral transmission of electrical or chemical information would require an open state of the plasmodesmata (van Bel et al. 2011a), in particular, of the few at the interface between phloem parenchyma cells and sieve element–companion cell complexes (Kempers et al. 1998). If, however, lateral information is mediated by serial, lateral waves of increased turgor, plasmodesmal opening may be unnecessary and even contra-productive. As a last speculative remark, Ca^{2+} influx into the sieve elements is amplified by a corrobative effect of both activation modes during mixed EPWs.

5 Deployment of Carriers and Pumps Involved in EPW Propagation Via Sieve Tubes

Little information is available on the location and nature of ion channels and pumps involved in AP transmission via the phloem. Voltage-dependent dihydropyridine-sensitive Ca^{2+}-permeable channels were localised to the phloem tissues of tobacco and water lettuce (Volk and Franceschi 2000). Similarly, voltage-dependent Ca^{2+}-permeable channels were localised in more detail to the plasma membrane and ER membranes of sieve elements of broad bean (Furch et al. 2009).

Whilst information on Cl^- channels is entirely lacking, limited knowledge on voltage-dependent K^+ channels in sieve elements is available. Phloem-localised K^+ channels of the AKT2/3 type were electrophysiologically characterised and linked to AP depolarisation (Marten et al. 1999; Bauer et al. 2000; Lacombe et al. 2000; Deeken et al. 2002). Weak inward rectifying currents matching the characteristics of AKT2/3 channels were recorded in sieve-element protoplasts of broad bean (Hafke et al. 2007). The increasing conductance of AKT2/3 channels at more alkaline pH values (Marten et al. 1999) as well as the extracellular alkalisation during transmission of SPs (Zimmermann et al. 2009) point to the engagement of AKT2/3 channels in membrane repolarisation. Moreover, passive K^+ fluxes contribute to a stabilisation of the membrane potential of sieve elements by charge compensation (Marten et al. 1999; Deeken et al. 2002; van Bel and Hafke 2005). Hence, several authors have attributed a role to AKT2/3 channels in phloem loading and unloading of sucrose (Marten et al. 1999; Lacombe et al. 2000; Deeken et al. 2002; Gajdanowicz et al. 2011).

As for proton pumps, immunological methods localised H^+-ATPases to sieve elements and companion cells in transport phloem of castor bean (Langhans et al. 2001). VPs could be triggered by mechano-sensitive channels (Davies 2006; Davies and Stankovic 2006) as found in the plasma membrane of various cell types (e.g. Cosgrove and Hedrich 1991). To the best of our knowledge, the presence of mechano-sensitive channels has not been documented yet for vascular parenchyma cells. Mechano-sensitive Ca^{2+}-permeable channels were postulated to reside on the plasma membrane of intact sieve elements (Knoblauch et al. 2001; Furch et al. 2009) and sieve-element protoplasts (Hafke et al. 2007). A transient increase in Ca^{2+} ions in sieve elements through Ca^{2+}-permeable channels was interpreted to occur during the VP phase of mixed EPWs (Furch et al. 2009).

6 Putative Nature of Ca^{2+}-permeable Channels in Sieve Elements

In animals, highly selective Ca^{2+} channels mediate Ca^{2+} fluxes at the plasma membrane (Tsien et al. 1987; Tsien and Tsien 1990), whereas non-selective cation channels (NSCCs) are responsible for Ca^{2+} fluxes in plants (Demidchik and

Maathuis 2007; McAinsh and Pittman 2009). At least three principal classes of NSCCs are presumably involved in the generation of stimulus-specific Ca^{2+} elevations: hyperpolarisation-activated (HACCs; Gelli and Blumwald 1997; Hamilton et al. 2000; Kiegle et al. 2000), depolarisation-activated (DACCs; Thuleau et al. 1994; Thion et al. 1998; Carpaneto et al. 2007; White 2009), and mechano-sensitive (MSCs; Cosgrove and Hedrich 1991; Ding and Pickard 1993; Dutta and Robinson 2004) Ca^{2+}-permeable channels. To date, only MSCs have been identified in the sieve-element plasma membrane (Knoblauch et al. 2001; Hafke et al. 2007).

HACCs may catalyse a long-lasting Ca^{2+} influx into sieve elements as in root cells (Demidchik et al. 2002; Demidchik and Maathuis 2007) during the second EPW phase after remote burning (Furch et al. 2009). In many cells, the resting value of the membrane potential is more positive than the activation voltage of HACCs, but the activation voltage of HACCs can shift to more positive membrane potentials, increasing the intracellular Ca^{2+} level (Demidchik et al. 2002; Demidchik and Maathuis 2007; McAinsh and Pittman 2009). Such changes may be due to Ca^{2+} influx via voltage-independent nonspecific cation channels (VI-NSCCs; White and Broadley 2003; Demidchik and Maathuis 2007). Similarly, mechano-sensitive Ca^{2+}-permeable channels in sieve elements that are briefly activated during VPs brought about by distant burning could shift the activation voltage of HACCs to a more positive membrane potential.

MSCs may act as primary temperature sensors (Minorsky and Spanswick 1989; Monroy and Dhindsa 1995; Plieth et al. 1999; White 2009) as demonstrated by the gradually increasing activity of MSCs below 20 °C (Ding and Pickard 1993). MSCs activated by tensile forces (Demidchik and Maathuis 2007) could have mediated Ca^{2+} influx into sieve elements in reaction to temperature shocks (Thorpe et al. 2010).

DACCs might also be engaged in cold-induced Ca^{2+} influx (Plieth 1999; Plieth et al. 1999; White 2009). A DACC type called the maxi cation channel was proposed to be responsible for the generation of complex temperature-dependent Ca^{2+} signatures (White and Ridout 1999; White 2004, 2009) and is therefore a potential player in temperature perception by sieve elements.

A fourth group of Ca^{2+}-permeable channels includes the ligand-activated channels (CNCGs: Navazio et al. 2000; Lemtiri-Chlieh et al. 2003) which are suspected to occur on tonoplast and ER membranes in plants. The wealth of potential ligands associated with VP generation (see Sect. 18.4) renders a plasma-membrane location of CNCGs highly plausible. In conclusion, circumstantial evidence (Knoblauch et al. 2001; Hafke et al. 2007; Furch et al. 2009; Thorpe et al. 2010) seems to indicate the presence of the above-mentioned classes of Ca^{2+}-permeable channels in the plasma membrane of sieve element–companion cell complexes.

7 Ca^{2+} Hotspots in Sieve Elements

In legumes, Ca^{2+} concentration changes in sieve elements were monitored in vivo by native bio-indicators. Their sieve elements contain the so-called forisomes (Knoblauch et al. 2001), giant module-structured protein bodies up to 100 μm in length (Schwan et al. 2009; Peters et al 2010; Tuteja et al. 2010), the tips of which are residing in the mictoplasm. Upon Ca^{2+} elevation in sieve elements, forisomes disperse and reach up to six times their original volume (Knoblauch et al. 2003, 2012; Peters et al. 2006) so that the sieve element, and hence the sieve-tube flow, is blocked (Knoblauch et al. 2012). After removal of Ca^{2+}, probably by Ca^{2+} efflux facilitators (Kudla et al. 2010), forisomes re-contract to their original size (Knoblauch et al. 2003; Furch et al. 2007, 2009). During passage of an EPW, forisomes disperse instantaneously and re-contract after 10–20 min, indicative of a sudden Ca^{2+} influx and gradual Ca^{2+} removal (Furch et al. 2007, 2009).

The reversible forisome reaction has been shown in vivo and in vitro (Knoblauch et al. 2001, 2003; Furch et al. 2007, 2009). Their function may lie in the ability to sieve-tube occlusion in response to tissue damage to save the precious contents of the sieve-tube sap and to prevent the invasion of phytopathogens (van Bel 2003). Exemplary for the impact of Ca^{2+} influx provoked by sieve-element injury, it was demonstrated by in vitro forisomes that Ca^{2+} binding proteins in aqueous aphid saliva prevent blocking of the sieve tubes in response to injury by the penetrating stylet (Will et al. 2007).

In broad bean, Ca^{2+} levels in sieve-tube sap collected via cut aphid stylets (Fisher and Frame 1984) and in the sieve-element mictoplasm were in the range between 50 and 100 nmol (Furch et al. 2009). These concentrations were in the same range as measured in the cytoplasm of other cells (Malho et al. 1998; Trewavas 1999), but sharply contrasted the values (in the micromole to millimole range) obtained in previous studies on sieve-tube sap composition (Fromm and Bauer 1994; Brauer et al. 1998). Concentrations in the low nanomolar range indicate a classical role for Ca^{2+} as signalling ion (White and Broadley 2003; Kudla et al. 2010) in sieve elements (Furch et al. 2009). Remote burning stimuli triggered prolonged EPWs along sieve tubes (Furch et al. 2007, 2009; Hafke et al. 2009) that coincided with elevated levels of 200–500 nM in the mictoplasm and in the proximity of the sieve plates and are accompanied by forisome dispersion (Furch et al. 2009). However, the measured Ca^{2+} levels do not meet the 50 μmol threshold needed for forisome dispersion (Knoblauch et al. 2001, 2005; Furch et al. 2009). This discrepancy suggests that special conditions are required for forisome dispersion in vivo.

Ca^{2+} concentrations exceeding 50 μmol might only exist temporarily at the cytoplasmic mouth of the Ca^{2+}-permeable channel (Trewavas 1999). Such high local Ca^{2+} concentrations have been found in animal cells (Llinàs et al. 1992, 1995), but have not been documented yet for plant cells including sieve elements. The fluorescent reporters (i.e. Oregon Green-488-BAPTA-1; Furch et al. 2009) used thus far for determination of the sieve-element Ca^{2+} concentration act as buffering mobile dyes and, hence, dissipate strong Ca^{2+} microgradients (Bolsover and Silver 1991; Malho et al. 1998; Demuro and Parker 2006). Thus, the absence of forisome reaction with Ca^{2+} buffers such as 1,2-bis(2-aminophenpxy)ethane-N,N,N′,N′-tetra acetic acid (BAPTA) can be interpreted as indirect evidence in favour of the existence of Ca^{2+} concentration microdomains (Ca^{2+} hotspots) in sieve elements

(Furch et al. 2009; Hafke et al. 2009). Such hotspots could also be meaningful for callose deposition, since Ca^{2+} concentrations required for callose synthase activity are much higher than those in the sieve-tube sap (Colombani et al. 2004).

A prerequisite for establishing Ca^{2+} hotspots is a local aggregation of Ca^{2+} channels (White and Broadley 2003). The highest Ca^{2+} channel densities were found in the vicinity of the sieve plates and around the pore-plasmodesma units (e.g. van Bel and Kempers 1997). Their aggregation is probably necessary to satisfy the high demand of Ca^{2+} needed for occlusion reactions (Furch et al. 2009; Hafke et al. 2009). Voltage-dependent Ca^{2+} channels were localised to the sieve-element plasma membrane and ER membranes in sieve elements (Furch et al. 2009) so that voltage changes in the plasma membrane (Furch et al. 2007, 2009) could directly activate voltage-dependent channels at the ER (Klüsener et al. 1995; Klüsener and Weiler 1999; McAinsh and Pittman 2009).

This implies that Ca^{2+} ions can be recruited from the ER cisternae being intercellular Ca^{2+} storage compartments (Sjolund and Shih 1983; Buchen et al. 1983). Second messengers such as the inositol phosphates IP_3 (inositol triphospate; Gilroy et al. 1990) and IP_6 (myo inositol hexakisphosphate; Lemtiri-Chlieh et al. 2003) or cADPR (cyclic ADP-ribose; Leckie et al. 1998) may play a pivotal role in the signalling chain to release Ca^{2+} ions from endomembrane systems (Lemtiri-Chlieh et al. 2003; Kudla et al. 2010). Inhibition of APs by the phospholipase C inhibitor neomycin—phospholipase C is involved in the formation of IP_3—has been reported for a variety of plant species (Krol et al. 2003, 2004). In sieve elements, Ca^{2+}-dependent forisome dispersion (Knoblauch et al. 2001) may depend on Ca^{2+} liberation from the ER, which provides indirect evidence for an involvement of intracellular Ca^{2+} stores in sieve-element occlusion (Furch et al. 2009; Hafke et al. 2009; Thorpe et al. 2010).

Moreover, an elevated Ca^{2+} concentration in the mictoplasmic matrix may contribute to an increased Ca^{2+} efflux from the ER stacks. In analogy to the Ca^{2+}-induced Ca^{2+} release at the tonoplast (Bewell et al. 1999; Sanders et al. 2002), Ca^{2+} influx via the plasma membrane may trigger additional Ca^{2+} release into the mictoplasm via presumptive Ca^{2+}-dependent Ca^{2+} channels on the ER membranes (Furch et al. 2009; Hafke et al. 2009). Ca^{2+} recruitment from internal Ca^{2+} stores is an established event during temperature shocks (Knight et al. 1996; Gong et al. 1998; White and Broadley 2003). Additional circumstantial evidence (Furch et al. 2009; Thorpe et al. 2010; van Bel et al. 2011a) also points to the ER as an important Ca^{2+} store in sieve elements, which may be a major reason why the ER stacks have been retained during sieve-element evolution (van Bel 2003). All in all, it seems that Ca^{2+} hotspots are created where high densities of Ca^{2+} channels in the plasma membrane are accompanied by an abundance of ER stacks (Hafke et al. 2009). The frequently perpendicular orientation of the ER stacks (Ehlers et al. 2000) may be meaningful for hotspot development as proposed in the next paragraph.

The relationship between the forisome position inside sieve elements and their reactivity to stimuli also points the existence of Ca^{2+} hotspots (Furch et al. 2009). In keeping with the Ca^{2+} channel density, dispersion reactivity increases with the position inside the sieve element and the degree of contact with the plasma membrane (Furch et al. 2009). Forisomes must be positioned in such a way to optimally exploit the Ca^{2+} hotspots (Hafke et al. 2009). The forked forisome ends (Hafke et al. 2007) which disperse first during weaker forisome response might be positioned between the ER stacks inside the local microdomains where Ca^{2+} concentrations are in the range of the threshold value for forisome dispersion. Supportive of this idea is that the forisome ends are inserted into the interstices of the ER stacks where an unstirred micro-environment provides the conditions for the existence of Ca^{2+} hotspots (Furch et al. 2009; Hafke et al. 2009). A narrow mechanical or electrical coupling between plasma membrane and ER membranes (Hepler et al. 1990) in sieve elements, facilitated by macromolecular anchors of unknown nature (Ehlers et al. 2000), would be indispensable for forisome response.

Furthermore, it may be crucial that forisomes are tethered to the cytoskeleton to ensure their position. In this frame, the connection between cytoskeletal elements and Ca^{2+} channels as found in other plant cells and the inherent cytoskeleton involvement in Ca^{2+}-dependent signal cascades may be of major significance (Trewavas and Malho 1997; Mazars et al. 1997; Drøbak et al. 2004; Davies and Stankovic 2006). The activity of DACCs (Mazars et al. 1997; Thion et al. 1998) and MSCs (Wang et al. 2004; Zhang et al. 2007) is modulated by microtubules and microfilaments respectively.

However, a cytoskeleton may not exist in sieve elements due to developmental degradation (Parthasaraty and Pesacreta 1980; Thorsch and Esau 1981; Evert 1990). This claim contrasts with more recent observations that favour the existence of at least an incomplete version of the cytoskeleton in sieve elements (e.g. Chaffey and Barlow 2002). That cytoskeleton disruptors such as latrunculin and oryzalin suppress depolarisation, Ca^{2+} influx, and forisome dispersion in response to cold shocks corroborates the view that a cytoskeleton mediates Ca^{2+}-induced influx and forisome responses (Hafke et al. unpublished results).

8 Ca^{2+} Processing in Sieve Elements

Irrespective of the different origin and regardless of the engagement of voltage-dependent, mechano-sensitive, or ligand-activated ion channels (Stahlberg and Cosgrove 1997; Davies 2006; Zimmermann et al. 2009), the initial influx of Ca^{2+} ions (Trebacz et al. 2006; Davies and Stankovic 2006; Demidchik and Maathuis 2007; McAinsh and Pittman 2009) into sieve elements (Furch et al. 2009; Hafke et al. 2009; Thorpe et al. 2010) is an essential common feature of EPWs. Ca^{2+} ions activate numerous intracellular signalling cascades with the implication that Ca^{2+}-permeable channels function as relay stations, transforming long-distance electrical messages into local intracellular chemical signals (Demidchik and Maathuis 2007; Furch et al. 2009; Hafke et al. 2009). The question arises as to what extent sieve elements have preserved intracellular signalling cascades during evolution. Typical Ca^{2+} compartments in plants cells such nuclei, mitochondria, dictyosomes, and chloroplasts that often play a role in signalling (Pittman and Hirschi 2003) are absent or highly reduced in sieve elements (Sjölund 1997; van Bel 2003).

Ca^{2+} changes in the mictoplasm may resemble those induced by various local stimuli such as cold shocks in other cell types (Malho et al. 1998; Plieth 2001). Elevated Ca^{2+} levels brought about by abrupt chilling are of special interest, because most studies on stimulus-induced Ca^{2+} influx used cooling regimes (Knight et al. 1996; Plieth 1999, 2001; Plieth et al. 1999; White 2009). Both the magnitude of the electrical response (Minorsky and Spanswick 1989; White 2009) and the elevation of Ca^{2+} concentration (Plieth 1999; Plieth et al. 1999) were related to the chilling rate, extent, and duration (White 2009). Forisome responses were related to the cooling rate in sieve elements of broad bean (Thorpe et al. 2010), which implicates Ca^{2+} responses similar to those in root cells (Plieth 1999; Plieth et al. 1999; White 2009).

In parenchymatous cells, nature, intensity, and strength of local stimuli are transformed in specific Ca^{2+} signatures that could originate from the interplay between Ca^{2+}-permeable channels in the plasma membrane and endomembrane systems with intervention of the cytoskeleton (Webb et al. 1996; Malho et al. 1998; Trewavas 1999; Ng and McAinsh 2003; White and Broadley 2003; Demidchik and Maathuis 2007; McAinsh and Pittman 2009). Besides Ca^{2+} influx channels, Ca^{2+} efflux facilitators such as Ca^{2+}-ATPases at the plasma membrane and endomembranes or Ca^{2+} exchangers (McAinsh and Pittman 2009; Kudla et al. 2010) could modulate Ca^{2+} signatures during signalling (Pittman and Hirschi 2003; McAinsh and Pittman 2009). Due to the absence of a vacuolar compartment, Ca^{2+}-permeable channels and pumps at the tonoplast cannot be involved in Ca^{2+} processing in sieve elements (Sjölund 1997; van Bel et al. 2011a).

It has been speculated that soluble Ca^{2+} buffering proteins fine-tune and shape Ca^{2+} transients during signalling (McAinsh and Pittman 2009). Given the wealth of soluble proteins in sieve-tube sap (e.g. Nakamura et al. 1993; Lin et al. 2009; Gaupels et al. 2012; Dinant and Lucas 2013), this type of Ca^{2+} sequestration may be of paramount importance for Ca^{2+} signalling in sieve elements. Together with Ca^{2+} buffering in the cytosol, Ca^{2+} binding proteins associated with the cytoskeleton could act as modulators of Ca^{2+} signatures (Malho et al. 1998). This mode of action would be a viable option for Ca^{2+} regulation in sieve elements, provided that a cytoskeleton is present.

9 Consequences of Ca^{2+} Influx into Sieve Elements

9.1 Sieve-Element Occlusion Mechanisms

Coincident with the passage of EPWs triggered by remote burning, sieve-plate occlusion is induced when the Ca^{2+} concentration in sieve elements rises above a critical threshold as exemplified by forisome dispersion (Furch et al. 2009; Hafke et al. 2009). Sieve-element occlusion turned out to occur quicker than the onset of callose production (Furch et al. 2007, 2010). Apparently, the time-lagged callose deposition under the control of *AtCal7* (Xie et al. 2011; Barratt et al. 2011) is preceded by a more rapid occlusion mechanism through constitutive proteins. In intact *Vicia faba* plants, forisomes disperse within seconds and re-contract within 10–20 min after distant burning (Furch et al. 2007, 2009). By the time that a forisome had re-contracted, probably due to active removal of Ca^{2+} ions by membrane pumps, callose deposition reached its maximum followed by a slower degradation over the following hours (Furch et al. 2007, 2008, 2010). Both modes of occlusion are under the control of Ca^{2+}, the difference being that protein-mediated occlusion may have a lower Ca^{2+} threshold (Furch et al. 2009; Hafke et al. 2009).

A dual sieve-plate occlusion mechanism was also found in intact *Cucurbita maxima* plants (Furch et al. 2010). Apparent coagulation of phloem protein 1 (PP1) and phloem protein 2 (PP2) several centimetres away from the site of burning preceded callose deposition (Furch et al. 2010). As demonstrated for a few species, callose is gradually degraded over several hours so that PPUs first return to their open state and then the sieve pores follow (Furch et al. 2007, 2008, 2010). Its occurrence in systematically distant families suggests that dual occlusion may be widespread. In this safety design, protein occlusion guarantees instantaneous sieve-plate sealing that bridges the time until callose deposition is completed (van Bel et al. 2011a).

Although the contours of the dual occlusion are getting shape, a fair number of questions are still to be addressed: (1) Most likely, not every sieve-element protein-clogging reaction is Ca^{2+} dependent. The forisomes that comprise SEO proteins (Pélissier et al. 2008), a widespread family among cotyledons (Rüping et al. 2010; Anstead et al. 2012; Ernst et al. 2012; Jekat et al. 2012), are Ca^{2+} responsive. However, the sieve-plate sealing PP1 and PP2 in cucurbits do not belong to the SEO (sieve element occlusion) family (Ernst et al. 2012) and may react to reactive oxygen species (Alosi et al. 1988). (2) Since structural phloem-specific proteins are virtually absent in grasses, protein occlusion may be of minor importance there (van Bel 2003), unless water-soluble proteins are able to coagulate in response to injury (Will et al. 2009). (3) There might be a vast spectrum of Ca^{2+} thresholds needed for occlusion reactions in diverse plant species (Furch et al. 2007, 2008, 2009, 2010), which is particularly relevant for VP effects because they are positively related to the stimulus strength (Stahlberg and Cosgrove 1997; Stahlberg et al. 2006). (4) The capacity to remove Ca^{2+} from the sieve element may be the decisive determinant of reversible/definitive sieve-tube occlusion which demands research on the existence of Ca^{2+} efflux facilitators (reviewed by Kudla et al. 2010) and their regulation at the sieve-element plasma membrane.

9.2 *Impact on the Symplasmic Organisation of Phloem*

After passage of an EPW of sufficient strength, the symplasmic organisation of the phloem is probably changed transiently (van Bel et al. 2011a). EPW-induced Ca^{2+}

influx into the vascular parenchyma cells may cause the production of callose collars around the plasmodesmata (e.g. Tucker 1990; Kauss and Jeblick 1991; Radford et al. 1998; Holdaway-Clarke et al. 2000; Sivaguru et al. 2000, 2005; Michard et al. 2011). Occlusion of the sieve pores and PPUs complemented by the likely closure of plasmodesmata between vascular cells would make phloem cells autonomous units, temporarily less sensitive of input from neighbour cells. Without the usual interactive events, cells may be able to switch to more innate cascades implementing more cell-specific genetic or metabolic programmes. Once the temporary segregation of phloem cells is lifted, the stimulus-induced products are released into sieve elements for systemic transport as soon as the lifeline between companion cells and sieve elements has been restored (van Bel et al. 2011a). Analogous effects of plasmodesmal closure on cellular autonomy have been demonstrated for the development of stomatal guard cells (Palevitz and Hepler 1985), the divergent development of sieve element and companion cell (van Bel and van Rijen 1994), the formation of symplasmic domains (Ehlers et al. 1999), the synchronisation of mitotic activity (Ehlers and Kollmann 2000), and the explosive elongation of cotton hairs cells (Ruan et al. 2001).

9.3 Impact on Cellular Metabolism of the Respective Phloem Cells

Effects of sudden Ca^{2+} influx and Ca^{2+} signatures depend on the equipment of recipient cells, which promises an immense variety of responses (Kudla et al. 2010). To give a few established examples, sieve elements respond to Ca^{2+} influx by sieve-plate occlusion (Furch et al. 2007, 2009), pulvinar flexor cells in *Mimosa pudica* become capable of massive water release (Fleurat-Lessard and Bonnemain 1978), and companion cells start producing abundant amounts of NO in *Vicia faba* (Gaupels et al. 2008), or contribute to an increased cold resistance in *Cucurbita* (Retivin et al. 1997). How phloem parenchyma cells can act as intermediates between electrical and chemical long-distance signalling has been demonstrated by the EPW-induced production of systemin (Narvaez-Vazquez and Ryan 2004). Translocation of systemin gives rise to expression of proteinase-inhibitor genes in target tissues (Wildon et al. 1992; Stankovic and Davis 1997).

10 Ca^{2+}-Triggered Systemic Signalling May Comprise Three Consecutive Steps

One might speculate on of EPW-induced waves of Ca^{2+} influx into sieve elements and their adjoining parenchyma cells have an impact on the whole-plant level. The current loss in *Mimosa pudica* (Fleurat-Lessard and Bonnemain 1978) may reflect a general function of APs in plants different from that in animals. APs in plants do not seem primarily engaged in propagation of electro-chemical signals generated by minor ion displacements. Instead, gating of ion channels in plants causes massive displacement of ions (Pyatygin et al. 2008) not only along the sieve tubes, but also in the adjoining cells as speculated before. Besides Ca^{2+} influx, ion movements may play a part in regulating ion homeostasis (e.g. Mummert and Gradmann 1991; Trebacz et al. 1994; Zimmermann and Felle 2009), because K^+ and Cl^- fluxes exceed by far the theoretical ion charges required for loading the membrane capacitance (Fromm and Spanswick 1993; Trebacz et al. 1994).

The signals transmit information over long distances (Trebacz et al. 2006; Stahlberg et al. 2006) to exert remote control on gene expression (Wildon et al. 1992; Pena Cortes et al. 1995; Herde et al. 1995; Stankovic and Davies 1997; Stankovic et al. 1998) and a multitude of physiological processes such as growth, respiration, water supply, photosynthesis, and synthesis of a cohort of signalling substances (e.g. jasmonic acid, salicylic acid, and free radicals) in distant target tissues (reviewed by Fromm and Lautner 2006; Trebacz et al. 2006; Pyatygin et al. 2008). This broad spectrum of distant responses to a variety of stimuli has been reported without detailed knowledge of the underlying mechanisms. Here, a modest attempt is made to conceive a common mechanistic platform for the multiple forms of systemic signalling (e.g. Dempsey and Klessig 2012). Central to this unifying concept is that Ca^{2+} fluxes elicit signalling cascades in all vascular cells along the pathway with the restriction that the responses fade away when the Ca^{2+} thresholds to elicit the respective cascades are not met.

The view has been advanced that phloem-borne signalling events occur in waves (van Bel and Ehlers 2005). We explore the possibility here that Ca^{2+}-based signalling goes through—roughly spoken—three partly overlapping phases being distinct in timescale, nature, and site of origin.

The first stage (timescale: seconds to minutes) includes the delivery of Ca^{2+} ions into the cells along the phloem pathway as the direct result of longitudinally propagating EPWs and their associate lateral short-distance diversions. The Ca^{2+} influx confers immediate occlusion of the symplasmic connections. Cell-specific Ca^{2+} signatures induce immediate pro-active responses, in particular in the sink ends, to an imminent change and are dependent on the stimulus, i.e. dissimilar signatures may be obtained with the respective Ca^{2+} channels having their activities linked with different cytoskeleton components (Mazars et al. 1997; Thion et al. 1998; Wang et al. 2004; Zhang et al. 2007).

The second stage (timescale: minutes to hours) includes the period of symplasmic isolation. In sieve elements, Ca^{2+} ions are partly sequestered by constitutive Ca^{2+} binding proteins such as Ca^{2+}-dependent protein kinases (Nakamura et al. 1993; Yoo et al. 2002; Gaupels et al. 2012) in the stagnant sieve-element lumina. These complexes may be translocated as long-distance signals as soon as the sieve pores have been re-opened. Simultaneously, various signalling cascades are initiated in the parenchyma cells. For instance, specific Ca^{2+} binding proteins (White and Broadley 2003; Kudla et al. 2010) are attached to the cytoskeleton (Malho et al. 1998). In this way, information provided by Ca^{2+}

signatures is decoded and transformed into protein–protein interactions, resulting in Ca^{2+}-dependent phosphorylation cascades and transcriptional responses that lead to downstream reactions (Luan et al. 2002; Sanders et al. 2002; Kudla et al. 2010).

The third stage (timescale: hours to days) includes implementation of Ca^{2+} effects following the transient autonomy of the vascular cells. The multitude of potential combinations of Ca^{2+} influx and the variety of the cell types involved potentiate the complexity of the processes and provide a wealth of possibilities (Kudla et al. 2010; Dempsey and Klessig 2012). Production of numerous systemic signal types discussed in other chapters may well depend on concerted Ca^{2+} influxes and corresponding Ca^{2+} signatures in diverse vascular cell types. Among others, expression of several resistance-associated genes (Kudla et al. 2010), production of several RNA types (Kehr and Buhtz 2013), and synthesis of proteins (Dinant and Lucas 2013) and lipidic substances (Guelette et al. 2012) may be essentially related to Ca^{2+} influx. Whether Ca^{2+} influx is directly related to the synthesis of jasmonic acid (Fisahn et al. 2004) and/or salicylic acid is uncertain to date, but there is no doubt that Ca^{2+} is engaged in the actions of jasmonic acid (Munemasa et al. 2011) and salicylic acid (Du et al. 2009; Boursiac et al. 2010).

For the impact of Ca^{2+} signals on the production of macromolecules, the reader is referred to an excellent review (Kudla et al. 2010), but a few examples are given here. Ca^{2+} signals are converted into transcriptional responses for a fair number of genes (Kaplan et al. 2006), which may comprise about 3 % of the protein-coding genes in *Arabidopsis* (Kudla et al. 2010). Many of these expression responses depend on Ca^{2+} regulation of their transcription factors (e.g. Finkler et al. 2007). As an interesting note in the context of this chapter, one of the transcription factors interacts with the promoter of *AtEDS1*, a regulator of the synthesis of salicylic acid (Du et al. 2009).

Macromolecules produced in the vascular parenchyma and released into the sieve-tube sap via PPUs (Lucas et al. 2001; Chen and Kim 2006; Lough and Lucas 2006; Ding and Itaya 2007) might find their way to target cells by molecular tagging so that compounds required for local and remote use can be distinguished (van Bel et al. 2011b). In this way (van Bel et al. 2011b), macromolecules may be recognised to remain in the sieve element into which they had been released, or move to either companion cells along the pathway (Fisher et al. 1992; Golecki et al. 1999), or to sink cells. Presumably, macromolecules are trafficked back into companion cells by the aid of non-cell autonomous agents (Schulz 1999; Itaya et al 2000; Lucas et al. 2009), whereas they enter sink cells via permanently widened plasmodesmata (Fisher and Cash-Clarke 2000) each of which may demand differential entrance codes (Foster et al. 2002).

References

Aki T, Shigyo M, Nakano R, Yoneyama T, Yanagisawa S (2008) Nano scale proteomics revealed the presence of regulatory proteins including three FT-like proteins in phloem and xylem saps from rice. Plant Cell Physiol 49:767–790

Alarcon JJ, Malone M (1994) Substantial hydraulic signals are triggered by leaf-biting insects in tomato. J Exp Bot 45:953–957

Alosi MC, Melroy DL, Park RB (1988) The regulation of gelation of phloem exudate from *Cucurbita* fruit by dilution, glutathione, and glutathione reductase. Plant Physiol 86:1089–1094

Anstead JA, Froehlich DR, Knoblauch M, Thompson GA (2012) Arabidopsis P-protein filament formation requires both AtSEO1 and AtSEO2. Plant Cell Physiol 53:1033–1042

Barnes A, Bale J, Constantinidou C, Ashton P, Jones A, Pritchard J (2004) Determining protein identity from sieve element sap in *Ricinus communis* L. by quadrupole time of flight (Q-TOF) mass spectrometry. J Exp Bot 55:1473–1481

Barratt DHP, Kolling K, Graf A, Pike M, Calder G, Findlay K, Zeeman SC, Smith AM (2011) Callose synthase GSL7 is necessary for normal phloem transport and inflorescence growth in *Arabidopsis*. Plant Physiol 155:328–341

Bauer CS, Hoth S, Haga K, Phillipar K, Aoki N, Hedrich R (2000) Differential expression and regulation of K^+ channels in the maize coleoptile: molecular and biophysical analysis of cells isolated from cortex and vasculature. Plant J 24:139–145

Behnke HD, Sjolund RD (1990) Sieve elements Comparative structure, induction and development. Springer, Berlin

Bewell MA, Maathuis FJM, Allen GJ, Sanders D (1999) Calcium-induced calcium release mediated by voltage activated cation channel in vacuolar vesicles from red beet. FEBS Lett 458:41–44

Boari F, Malone M (1993) Rapid and systemic hydraulic signals are induced by localized wounding in a wide range of species. J Exp Bot 44:741–746

Bolsover S, Silver RA (1991) Artifacts in calcium measurements: recognition and remedies. Trends Cell Biol 1:71–74

Boursiac Y, Lee SM, Romanowsky S, Blank R, Sladek C, Chung WS, Harper JF (2010) Disruption of the vacuolar calcium-ATPases in Arabidopsis results in the activation of a salicylic acid-dependent programmed cell death pathway. Plant Physiol 154:1158–1171

Brauer M, Zhong W-J, Schobert C, Sanders D, Komor E (1998) Free calcium ion concentration in the sieve-tube sap of *Ricinus communis* L. seedling. Planta 206:103–107

Brinckmann E, Lüttge U (1974) Lichtabhängige Membranpotentialschwankungen und deren interzelluläre Weiterleitung bei panaschierten Photosynthesemutanten von *Oenothera*. Planta 119:45–57

Buchen B, Hensel D, Sievers A (1983) Polarity in mechanoreceptor cells of trigger hairs of *Dionaea muscipula* Ellis. Planta 158:458–468

Carpaneto A, Ivashikina N, Levchenko V, Krol E, Jeworutzki E, Zhu J-K, Hedrich R (2007) Cold transiently activates calcium-permeable channels in Arabidopsis mesophyll cells. Plant Physiol 143:487–494

Chaffey N, Barlow P (2002) Myosin, microtubules, and microfilaments: co-operation between cytoskeletal components during cambial cell division and vascular differentiation in trees. Planta 214:526–536

Chen X-Y, Kim J-Y (2006) Transport of macromolecules through plasmodesmata and the phloem. Physiol Plant 126:560–571

Colombani A, Djerbi S, Bessueille L, Blomqvist K, Ohlsson A, Berglund T, Teeri TT, Bulone V (2004) In vitro synthesis of (1→3)-β-D-glucan (callose) and cellulose by detergent extracts of membranes from cell suspension cultures of hybrid aspen. Cellulose 11:313–327

Cosgrove DJ, Hedrich R (1991) Stretch-activated chloride, potassium, and calcium channels coexisting in plasma membranes of guard cells of *Vicia faba* L. Planta 186:143–153

Davies E (1993) Intercellular and intracellular signals in plants and their transduction via the membrane-cytoskeleton interface. Semin Cell Biol 4:139–147

Davies E (2004) New functions for electrical signals in plants. New Phytol 161:607–610

Davies E (2006) Electrical signals in plants: facts and hypotheses. In: Volkov AG (ed) Plant electrophysiology. Springer, Berlin, pp 407–422

Davies E, Stankovic B (2006) Electrical signals, the cytoskeleton and gene expression: a hypothesis on the coherence of the cellular responses to environmental insult. In: Baluska F, Mancuso S, Volkmann D (eds) Communication in plants. Springer, Berlin, pp 309–320

Davies E, Zawadzki T, Witters D (1991) Electrical activity and signal transmission in plants: how do plants know? In: Penel C, Grepin H (eds) Plant signalling, plasma membrane and change of state. University of Geneva, Geneva, pp 119–137

Deeken R, Geiger D, Fromm J, Koroleva O, Ache P, Langenfeld-Heyser R, Sauer N, May ST, Hedrich R (2002) Loss of AKT2/3 potassium channel affects sugar loading into the phloem of Arabidopsis. Planta 216:334–344

Demidchik V, Maathuis FJM (2007) Physiological roles of non-selective cation channels in plants: from salt stress to signaling and development. New Phytol 175:387–404

Demidchik V, Bowen HC, Maathuis FJM, Shabala SN, Tester MA, White PJ, Davies JM (2002) *Arabidopsis thaliana* root nonselective cation channels mediate calcium uptake and are involved in growth. Plant J 32:799–808

Dempsey DA, Klessig D (2012) SOS – too many signals for systemic acquired resistance. Trends Plant Sci 17:538–545

Demuro A, Parker I (2006) Imaging single-channel calcium microdomains. Cell Calcium 40:413–422

Dinant S, Lucas WJ (2013) Sieve elements: puzzling activities deciphered through proteomics studies. In: Thompson GA, van Bel AJE (eds) Phloem. Molecular cell biology, systemic communication, biotic interactions. Wiley, Cichester, pp 157–185

Ding B, Itaya A (2007) Control of directional macromolecular trafficking across specific cellular boundaries: a key to integrative plant biology. J Integr Plant Biol 49:1227–1234

Ding JP, Pickard BG (1993) Modulation of mechanosensitive calcium-selective cation channels by temperature. Plant J 3:713–720

Drøbak BK, Franklin-Tong VE, Staiger CJ (2004) The role of the actin cytoskeleton in plant cell signaling. New Phytol 163:13–30

Du L, Ali GS, Simons KA, Hou J, Yang T, Reddy AS, Poovaiah BW (2009) Ca^{2+}/calmodulin regulates salicylic acid-mediated plant immunity. Nature 437:741–745

Dutta R, Robinson KR (2004) Identification and characterization of stretch-activated ion channels in pollen protoplasts. Plant Physiol 135:1398–1406

Dziubinska H, Trebasz K, Zawadzki T (2001) Transmission route for action potentials and variation potentials in *Helianthus annuus* L. J Plant Physiol 158:1167–1172

Ehlers K, Kollmann R (2000) Synchronization of mitotic activity in protoplast-derived *Solanum nigrum* L. microcalluses is correlated with plasmodesmal connectivity. Planta 210:269–278

Ehlers K, Binding H, Kollmann R (1999) The formation of symplasmic domains by plugging of plasmodesmata: a general event in plant morphogenesis. Protoplasma 209:181–192

Ehlers K, Knoblauch M, van Bel AJE (2000) Ultrastructural features of well-preserved and injured sieve elements: minute clamps keep the phloem transport conduits free for mass flow. Protoplasma 214:80–92

Engleman E (1965) Sieve elements of *Impatiens sultanii*. II Developmental aspects. Ann Bot 29:103–104

Ernst AM, Jekat SB, Zielonka S, Müller B, Neumann U, Rüping B, Twyman RM, Krzyzanek V, Prüfer D, Noll G (2012) Sieve element occlusion (SEO) genes encode structural phloem proteins involved in wound sealing of phloem. Proc Natl Acad Sci USA 109:E1980–E1989

Esau K (1969) The phloem. Encyclopedia of plant anatomy, vol 5. Bornträger, Berlin

Evert RF (1990) Dicotyledons. In: Behnke HD, Sjolund RD (eds) Sieve elements. Comparative structure, induction and development. Springer, Berlin, pp 103–137

Felle HH, Zimmermann MR (2007) Systemic signalling in barley through action potentials. Planta 226:203–214

Fields RD, Stevens-Graham B (2002) New insights into neuron-glia communication. Science 298:556–562

Finkler A, Ashery-Padan R, Fromm H (2007) CAMTAs; calmodulin-binding transcription factors from plants to human. FEBS Lett 581:3893–3898

Fisahn J, Herde O, Willmitzer L, Pena-Cortes H (2004) Analysis of the transient increase in cytosolic Ca^{2+} during the action potential of higher plants with high temporal resolution: requirement of Ca^{2+} transients for induction of jasmonic acid biosynthesis and PINII gene expression. Plant Cell Physiol 45:456–459

Fisher DB, Cash-Clarke CE (2000) Sieve tube unloading and post-phloem transport of fluorescent tracers and proteins injected into sieve tubes via severed aphid stylets. Plant Physiol 123:125–137

Fisher DB, Frame JM (1984) A guide to the use of exuding stylet technique in phloem physiology. Planta 161:385–393

Fisher DB, Wu K, Ku MSB (1992) Turnover of soluble proteins in wheat sieve tube. Plant Physiol 100:587–592

Fleurat-Lessard P, Bonnemain JL (1978) Structural and ultrastructural characteristics of the vascular apparatus of the sensitive plant *Mimosa pudica* L. Protoplasma 94:127–143

Fleurat-Lessard P, Roblin G (1982) Comparative histocytology of the petiole and the main pulvinus in *Mimosa pudica* L. Ann Bot 50:83–92

Foster TM, Lough TJ, Emerson SJ, Lee RH, Bowman JL, Foster RLS, Lucas WJ (2002) A surveillance system regulates selective entry of RNA into the shoot apices. Plant Cell 14:1497–1508

Fromm J (1991) Control of phloem unloading by action potentials in *Mimosa*. Plant Physiol 83:529–553

Fromm J, Bauer T (1994) Action potentials in maize sieve tubes change phloem translocation. J Exp Bot 45:463–469

Fromm J, Fei H (1998) Electrical signaling and gas exchange in maize plants of drying soil. Plant Sci 132:203–213

Fromm J, Lautner S (2006) Characteristics and functions of phloem-transmitted electrical signals in higher plants. In: Baluska F, Mancuso S, Volkmann D (eds) Communication in plants. Springer, Berlin, pp 321–332

Fromm J, Spanswick R (1993) Characteristics of action potential in willow (*Salix viminalis* L). J Exp Bot 44:1119–1125

Fromm J, Hajirezaei M, Wilke I (1995) The biochemical response of electrical signaling in the reproductive system of *Hibiscus* plants. Plant Physiol 109:375–384

Furch ACU, Hafke JB, Schulz A, van Bel AJE (2007) Ca^{2+}-mediated remote control of reversible sieve tube occlusion in *Vicia faba*. J Exp Bot 58:2827–2838

Furch ACU, Hafke JB, van Bel AJE (2008) Plant- and stimulus-specific variations in remote-controlled sieve-tube occlusion. Plant Signal Behav 3:858–861

Furch ACU, van Bel AJE, Fricker MD, Felle HH, Fuchs M, Hafke JB (2009) Sieve element Ca^{2+} channels as relay stations between remote stimulus and sieve tube occlusion. Plant Cell 21:2118–2132

Furch ACU, Zimmermann MR, Will T, Hafke JB, van Bel AJE (2010) Remote-controlled stop of mass flow by biphasic occlusion in *Cucurbita maxima*. J Exp Bot 61:3697–3708

Gajdanowicz P, Michard E, Sandmann M, Rocha M, Correa LGG, Ramirez-Aguilar SJ, Gomez-Porras JL, Gonzales W, Thibaud JB, van Dongen JT, Dreyer I (2011) Potassium gradients serve as a mobile energy source in plant vascular tissues. Proc Natl Acad Sci USA 108:864–869

Gaupels F, Furch ACU, Will T, Muir K, Kogel K-H, van Bel AJE (2008) Nitric oxide generation in *Vicia faba* phloem cells reveals them to be sensitive detectors as well as possible systemic transducers of stress signals. New Phytol 178:634–646

Gaupels F, Sarioglu H, Beckman M, Hause B, Spannagl M, Draper J, Lindermayr C, Durner J (2012) Deciphering systemic wound response of the pumpkin extrafascicular phloem by metabolomics and stable-isotope protein labeling. Plant Physiol 160:2285–2299

Gelli A, Blumwald E (1997) Hyperpolarization-activated Ca^{2+}-permeable channels in the plasma membrane of tomato cells. J Membr Biol 155:35–45

Giavalisco P, Kapitza K, Kolasa A, Buhtz A, Kehr J (2006) Towards the proteome of *Brassica napus* phloem sap. Proteomics 6:896–909

Gilroy S, Read ND, Trewavas AJ (1990) Elevation of cytoplasmic calcium by caged calcium or caged inositol triphosphate initiates stomatal closure. Nature 346:769–771

Golecki B, Schulz A, Thompson GA (1999) Translocation of structural P-proteins in the phloem. Plant Cell 11:127–140

Gong M, van der Luit AH, Knight MR, Trewavas AJ (1998) Heat-shock induced changes in intracellular Ca^{2+} level in tobacco seedlings in relation to thermotolerance. Plant Physiol 116:429–437

Guelette BS, Benning UF, Hoffmann-Benning S (2012) Identification of lipids and lipid-binding proteins in phloem exudates from *Arabidopsis*. J Exp Bot 63:3603–3616

Hafke JB, van Amerongen JK, Kelling F, Furch ACU, Gaupels F, van Bel AJE (2005) Thermodynamic battle for photosynthate acquisition between sieve tubes and adjoining parenchyma in transport phloem. Plant Physiol 138:1527–1537

Hafke JB, Furch ACU, Reitz MU, van Bel AJE (2007) Functional sieve element protoplasts. Plant Physiol 145:703–711

Hafke JB, Furch ACU, Fricker MD, van Bel AJE (2009) Forisome dispersion in *Vicia faba* is triggered by Ca^{2+} hotspots created by concerted action of diverse Ca^{2+} channels in sieve elements. Plant Signal Behav 4:968–972

Hamilton DWA, Hills A, Kohler B, Blatt MR (2000) Ca^{2+} channels at the plasma membrane of stomatal guard cells are activated by hyperpolarization and abscisic acid. Proc Natl Acad Sci USA 97:4967–4972

Hedrich R (2012) Ion channels in plants. Physiol Rev 92:1777–1811

Hepler PK, Palevitz BA, Lancelle SA, McCauley MM, Lichtscheidl L (1990) Cortical endoplasmic reticulum in plants. J Cell Sci 96:355–373

Herde O, Fuss H, Pena-Cortes H, Fisahn J (1995) Proteinase inhibitor II gene expression induced by electrical stimulation and control of photosynthetic activity in tomato plants. Plant Cell Physiol 36:737–742

Hlavackova V, Krchnak P, Naus J, Novak O, Spundova M, Strnad M (2006) Electrical and chemical signals involved in short-term systemic photosynthetic responses of tobacco plants to local burning. Planta 225:235–244

Hodick D, Sievers A (1988) The action potential of *Dionaea muscipula* Ellis. Planta 174:8–18

Hodick D, Sievers A (1989) On the mechanism of trap closure of Venus flytrap (*Dionaea muscipula* Ellis). Planta 179:32–42

Holdaway-Clarke TL, Walker NA, Overall R (1996) Measurement of the electrical resistance of plasmodesmata and membranes of corn suspension cells. Planta 199:537–544

Holdaway-Clarke TL, Walker NA, Hepler PK, Overall RL (2000) Physiological elevations in cytoplasmic free calcium by cold or microinjection result in transient closure of higher plant plasmodesmata. Planta 210:329–335

Homann U, Thiel G (1994) Cl^- and K^+ channel currents during the action potential in *Chara*: simultaneous recording of membrane voltage and patch currents. J Membr Biol 141:297–309

Houwink AL (1935) The conduction of excitation in *Mimosa pudica*. Recueil des Travaux Botaniques Neerlandais 32:51–91

Iijima T, Sibaoka T (1981) Action potentials in the trap-lobes of *Aldrovanda vesiculosa*. Plant Cell Physiol 22:1595–1601

Iijima T, Sibaoka T (1982) Propagation of action potentials over the trap-lobes of *Aldrovanda vesiculosa*. Plant Cell Physiol 23:679–688

Iijima T, Sibaoka T (1985) Membrane potentials in excitable cells of *Aldrovanda vesiculosa* trap-lobes. Plant Cell Physiol 26:1–13

Imlau A, Truernit E, Sauer N (1999) Cell-to-cell and long-distance trafficking of the green fluorescent protein in the phloem and symplastic unloading of the protein into the sinks. Plant Cell 11:309–322

Itaya A, Liang G, Woo Y-M, Nelson RS, Ding B (2000) Nonspecific intercellular protein trafficking probed by green-fluorescent proteins in plants. Protoplasma 213:165–175

Jekat SB, Ernst A, Zielonka S, Noll G, Prüfer D (2012) Interactions among tobacco sieve element occlusion (SEO) proteins. Plant Signal Behav 7:1918–1920

Kaplan B, Davydov O, Knight H, Galon Y, Knight MR, Fluhr R, Fromm H (2006) Rapid transcriptome changes induced by cytosolic Ca^{2+} transients reveal ABRE-related sequences as Ca^{2+} responsive cis elements in Arabidopsis. Plant Cell 18:2733–2748

Kauss H, Jeblick W (1991) Induced Ca^{2+} uptake and callose synthesis in suspension-cultured cells of *Cataranthus roseus* are decreased by the protein phosphatase inhibitor okadaic acid. Physiol Plant 81:309–312

Kehr J, Buhtz A (2013) Endogenous RNA constituents of the phloem and their possible roles in long-distance signalling. In: Thompson GA, van Bel AJE (eds) Phloem molecular cell biology, systemic communication, biotic interactions. Wiley, Cichester, pp 186–208

Kempers R, Ammerlaan A, van Bel AJE (1998) Symplasmic constriction and ultrastructural features of the sieve element/companion cell complex in the transport phloem of apoplasmically and symplasmically loading species. Plant Physiol 116:271–278

Kiegle E, Gilliham M, Haseloff J, Tester M (2000) Hyperpolarisation activated calcium currents found only in cells from the elongation zone of *Arabidopsis thaliana* roots. Plant J 21:225–229

Kinoshita T, Nishimura M, Shimazaki K (1995) Cytosolic concentration of Ca^{2+} regulates the plasma membrane H^+-ATPase in guard cells of Fava bean. Plant Cell 7:1333–1342

Kishimoto U, Takeuchi Y, Ohkawa T, Kami-ike N (1985) A kinetic analysis of the electrogenic pump of *Chara corallina*. III. Pump acivity during action potential. J Membr Biol 86:27–36

Klüsener B, Weiler EW (1999) A calcium-selective channel from root-tip endomembranes of garden cress. Plant Physiol 119:1399–1405

Klüsener B, Boheim G, Liß H, Engelberth J, Weiler EW (1995) Gadolinium-sensitive, voltage-dependent calcium release channels in the endoplasmic reticulum of a higher plant mechanoreceptor organ. EMBO J 14:2708–2714

Knight H, Trewavas AJ, Knight MR (1996) Cold calcium signaling in Arabidopsis involves two cellular pools and a change in calcium signature after acclimation. Plant Cell 8:489–503

Knoblauch M, Peters WS, Ehlers K, van Bel AJE (2001) Reversible calcium-regulated stopcocks in legume sieve tubes. Plant Cell 13:1221–1230

Knoblauch M, Noll G, Müller T, Prüfer D, Schneider-Hüther I, Scharner D, van Bel AJE, Peters WS (2003) ATP-independent contractile proteins from plants. Nat Mater 2:600–603

Knoblauch M, Noll GA, Müller T, Prüfer D, Schneider-Hüther I, Scharner D, van Bel AJE, Peters WS (2005) Corrigendum. Nat Mater 4:353

Knoblauch M, Stubenrauch M, van Bel AJE, Peters WS (2012) Forisome performance in artificial sieve tubes. Plant Cell Environ 15:1419–1427

Krol E, Dziubinska H, Trebacz K (2003) Low-temperature-induced transmembrane potential changes in the liverwort *Conocephalum conicum*. Plant Cell Physiol 44:527–533

Krol E, Dziubinska H, Trebacz K (2004) Low-temperature-induced transmembrane potential changes in mesophyll cells of *Arabidopsis thaliana*, *Helianthus annuus* and *Vicia faba*. Physiol Plant 120:265–270

Kudla J, Batistic O, Hashimoto K (2010) Calcium signals: the lead currency of plant information processing. Plant Cell 22:541–563

Lacombe B, Pilot G, Michard E, Gaymard F, Sentenac H, Thibaud J-B (2000) A shaker-like K^+ channel with weak rectification is expressed in both source and sink phloem tissues of Arabidopsis. Plant Cell 12:837–851

Langhans M, Ratajczak R, Lützelschwab M, Michalke W, Wächter R, Fischer-Schliebs E, Ullrich CE (2001) Immunolocalization of plasma membrane H^+-ATPase and tonoplast-type pyrophosphatase of the sieve element-companion cell complex in the stem of *Ricinus communis* L. Planta 213:11–19

Leckie CP, McAinsh MR, Allen GJ, Sanders D, Hetherington AM (1998) Abscisic acid-induced stomatal closure mediated by cyclic ADP-ribose. Proc Natl Acad Sci USA 95:15837–15842

Lemtiri-Chlieh F, MacRobbie EA, Webb AA, Manison NF, Brownlee C, Skepper JN, Chen J, Prestwich GD, Brearley CA (2003) Inositol hexakisphosphate mobilizes an endomembrane store of calcium in guard cells. Proc Natl Acad Sci USA 100:10091–10095
Lin MK, Lee YJ, Lough BS, Phinney WJ, Lucas WJ (2009) Analysis of the pumpkin phloem proteome provides insights into angiosperm sieve tube function. Mol Cell Proteomics 8:343–356
Llinàs R, Sugimori M, Silver RB (1992) Microdomains of high calcium concentration in a presynaptic terminal. Science 256:677–679
Llinàs R, Sugimori M, Silver RB (1995) The concept of calcium concentration microdomains in synaptic transmission. Neuropharmacology 34:1443–1451
Lough T, Lucas WJ (2006) Integrative plant biology. Role of phloem long-distance macromolecular trafficking. Annu Rev Plant Biol 57:203–232
Luan S, Kudla J, Rodriguez-Concepcion M, Yalovsky S, Gruissem W (2002) Calmodulins and calcineurin B-like proteins: calcium sensors for specific signal response coupling in plants. Plant Cell 14:389–400
Lucas WJ, Yoo B-C, Kragler F (2001) RNA as a long-distance information macromolecule in plants. Nat Rev Mol Cell Biol 2:849–857
Lucas WJ, Ham B-K, Kim J-Y (2009) Plasmodesmata – bridging the gap between neighbouring plant cells. Trends Cell Biol 19:495–503
Lunevsky VZ, Zherelova OM, Vostrikov IY, Berestovsky GN (1983) Excitation of *Characeae* cell membranes as a result of activation of calcium and chloride channels. J Membr Biol 72:43–58
Malho R, Moutinho A, van der Luit A, Trewavas AJ (1998) Spatial characteristics of calcium signalling: the calcium wave as a basic unit in plant cell calcium signalling. Philos Trans R Soc Lond B Biol Sci 353:1463–1473
Malone M (1996) Rapid, long-distance signal transmission in higher plants. Adv Bot Res 22:163–228
Malone M, Stankovic B (1991) Surface potentials and hydraulic signals in wheat leaves following localized wounding by heat. Plant Cell Environ 14:431–436
Mancuso S (1999) Hydraulic and electrical transmission of wound-induced signals in *Vitis vinifera*. Aust J Plant Physiol 26:55–61
Marten I, Hoth S, Deeken R, Ketchum KA, Hoshi T, Hedrich R (1999) *AKT3*, a phloem-localised K^+ channel is blocked by protons. Proc Natl Acad Sci USA 96:7581–7586
Mazars C, Thion L, Thuleau P, Graziana A, Knight MR, Moreau M, Ranjeva R (1997) Organization of cytoskeleton controls the changes in cytosolic calcium of cold-shocked *Nicotiana plumbaginifolia* protoplasts. Cell Calcium 22:413–420
McAinsh MR, Pittman JK (2009) Shaping the calcium signature. New Phytol 181:275–294
Michard E, Lima PT, Borges F, Silva A, Portes MT, Carvalho JE, Gilliham M, Liu L-H, Obermeyer G, Feijo JA (2011) Glutamate receptor-like genes form Ca^{2+} channels in pollen tubes and are regulated by pistil D-serin. Science 332:434–437
Minorsky PV, Spanswick RM (1989) Electrophysiological evidence for a role for calcium in temperature sensing by roots of cucumber seedlings. Plant Cell Environ 12:137–143
Monroy AF, Dhindsa RS (1995) Low temperature signal transduction: induction of cold acclimation-specific genes of alfalfa by calcium at 25°C. Plant Cell 7:321–331
Moyen C, Johannes E (1996) Systemin transiently depolarizes the tomato mesophyll cell membrane and antagonizes fusicoccin induced extracellular acidification of mesophyll tissue. Plant Cell Environ 19:464–470
Mummert H, Gradmann D (1991) Action potentials in Acetabularia: measurement and simulation of voltage-gated fluxes. J Membr Biol 124:265–273
Munemasa S, Hossain MA, NakamuraY MIC, Nurata Y (2011) The Arabidopsis calcium-dependent protein kinase, CPK6, functions as a positive regulator of methyl jasmonate signalling in guard cells. Plant Physiol 155:553–561
Nakamura S, Hayashi H, Mori S, Chino M (1993) Protein phosphorylation in the sieve tubes of rice plants. Plant Cell Physiol 34:927–933

Narvaez-Vazquez J, Ryan CA (2004) The cellular localization of prosystemin: a functional role for phloem parenchyma in systemic wound signalling. Planta 218:360–369

Navazio L, Bewell MA, Siddiqua A, Dickinson GD, Galione A, Sanders D (2000) Calcium release from the endoplasmic reticulum of higher plants is elicited by the NADP metabolite nicotinic acid adenine dinucleotide phosphate. Proc Natl Acad Sci USA 97:8693–8698

Ng CKY, McAinsh MR (2003) Encoding specificity in plant calcium signalling: hot-spotting the ups and downs and waves. Ann Bot 92:477–485

Okihara K, Ohkawa T, Tsutsui I, Kasai M (1991) A Ca^{2+}- and voltage-dependent Cl–sensitive anion channel in the Chara plasmalemma: a patch clamp study. Plant Cell Physiol 32:593–601

Opritov VA, Pyatygin SS, Vodeneev VA (2002) Direct coupling of action potential generation in cells of a higher plant (*Cucurbita pepo*) with the operation of an electrogenic pump. Russ J Plant Physiol 49:142–147

Overall R, Gunning BES (1982) Intercellular communication in *Azolla* roots. I Electrical coupling. Protoplasma 111:151–160

Palevitz BA, Hepler PK (1985) Changes in dye coupling of stomatal cells of *Allium* and *Commelina* demonstrated by microinjection of Lucifer yellow. Planta 164:473–479

Parthasaraty MV, Pesacreta TC (1980) Microfilaments in plant vascular cells. Can J Bot 58:807–815

Patrick JW, Offler CE (1996) Post-sieve element transport of photoassimilates in sink regions. J Exp Bot 47:1165–1177

Pélissier HC, Peters WS, Collier R, van Bel AJE, Knoblauch M (2008) GFP tagging of sieve element occlusion (SEO) proteins in green fluorescent forisomes. Plant Cell Physiol 49:1699–1710

Pena Cortes H, Fishan J, Willmitzer L (1995) Signals involved in wound-induced proteinase inhibitor II gene expression in tomato and potato plants. Proc Natl Acad Sci USA 92:4106–4113

Peters WS, van Bel AJE, Knoblauch M (2006) The geometry of the forisome-sieve element-sieve plate complex in the phloem of *Vicia faba* L. leaflets. J Exp Bot 57:3091–3098

Peters WS, Haffer D, Hanakam CB, van Bel AJE, Knoblauch M (2010) Legume phylogeny and the evolution of a unique contractile apparatus that regulates phloem transport. Am J Bot 97:797–808

Pickard B (1973) Action potentials in plants. Bot Rev 39:172–201

Pickard WF, Minchin PEH (1990) The transient inhibition of phloem translocation in *Phaseolus vulgaris* by abrupt temperature drops, vibration, and electric shock. J Exp Bot 41:1361–1369

Pittman JK, Hirschi KD (2003) Don't shoot the (second) messenger: endomembrane transporters and binding proteins modulate cytosolic Ca^{2+} levels. Curr Opin Plant Biol 6:257–262

Plieth C (1999) Temperature sensing by plants: calcium-permeable channels as primary sensors – a model. J Membr Biol 172:121–127

Plieth C (2001) Plant calcium signaling and monitoring: pros and cons and recent experimental approaches. Protoplasma 218:1–23

Plieth C, Hansen U-P, Knight H, Knight MR (1999) Temperature sensing by plants: the primary characteristics of signal perception and calcium response. Plant J 18:491–497

Pyatygin SS, Opritov VA, Vodeneev VA (2008) Signaling role of action potential in higher plants. Russ J Plant Physiol 55:285–291

Radford JE, Vesk M, Overall RL (1998) Callose deposition at plasmodesmata. Protoplasma 201:30–37

Retivin VG, Opritov VA, Fedulina SB (1997) Generation of action potentials induces preadaptation of *Cucurbita pepo* L. stem tissues to freezing injury. Russ J Plant Physiol 44:432–442

Rhodes JD, Thain JF, Wildon DC (1996) The pathway for systemic electrical signal conduction in the wounded tomato plant. Planta 200:50–57

Ricca U (1916) Soluzione d'un problema di fisiologia: la propagazione di stimulo nella *Mimosa*. Nuovo G Bot Ital 23:51–170

Roblin G (1985) Analysis of the variation potential induced by wounding in plants. Plant Cell Physiol 26:455–461
Roblin G, Bonnemain J-L (1985) Propagation in *Vicia faba* stem of a potential variation by wounding. Plant Cell Physiol 26:1273–1283
Ruan Y-L, Llewellyn DJ, Furbank RT (2001) The control of single-celled cotton fiber elongation by developmentally reversible gating of plasmodesmata and coordinate expression of sucrose and K^+ transport genes and expansin. Plant Cell 13:47–60
Rüping B, Ernst AM, Jekat SB, Nordzieke S, Reineke AR, Müller B, Bornberg-Bauer E, Prüfer D, Noll GA (2010) Molecular and phylogenetic characterization of the sieve element occlusion gene family in Fabaceae and non-Fabaceae plants. BMC Plant Biol 10:219
Samejima M, Sibaoka T (1983) Identification of the excitable cells in the petiole of *Mimosa pudica* by intracellular injection of procion yellow. Plant Cell Physiol 24:33–39
Sanders D, Pelloux J, Brownlee C, Harper JF (2002) Calcium at the crossroads of signaling. Plant Cell 14:401–417
Schulz A (1999) Physiological control of plasmodesmatal gating. In: van Bel AJE, van Kesteren WJP (eds) Plasmodesmata. Structure, function, role in cell communication. Springer, Berlin, pp 173–204
Schwan S, Menzel M, Fritzsche M, Heilmann A, Spohn U (2009) Micromechanical measurements of P-protein aggregates (forisomes) from *Vicia faba* plants. Biophys Biochem 139:99–105
Sibaoka T (1966) Action potentials in plant organs. Symp Soc Exp Biol 20:49–74
Sibaoka T (1969) Physiology of rapid movements in plants. Annu Rev Plant Physiol 20:165–184
Sibaoka T (1991) Rapid plant movements triggered by action potentials. J Plant Res 104:73–95
Sinyukin AM, Britikov EA (1967) Action potentials in the reproductive system of plants. Nature 215:1278–1280
Sivaguru M, Fujiwara T, Samaj J, Baluska F, Yang Z, Osawa H, Maeda T, Mori T, Volkmann D, Matsumoto M (2000) Aluminium-induced 1→3-β-D-glucan inhibits cell-to-cell trafficking of molecules through plasmodesmata. A new mechanism of aluminium toxicity in plants. Plant Physiol 124:991–1005
Sivaguru M, Yamamoto Y, Rengel Z, Ahn S-J, Matsumoto M (2005) Early events responsible for aluminium toxicity symptoms in suspension-cultured tobacco cells. New Phytol 165:99–109
Sjölund RD (1997) The phloem sieve element: a river runs through it. Plant Cell 9:1137–1146
Sjolund RD, Shih CY (1983) Freeze-fracture analysis of phloem structure in plant tissue cultures. I The sieve element reticulum. J Ultrastr Res 82:111–121
Stahlberg R, Cosgrove DJ (1992) Rapid alterations in growth rate and electrical potentials upon stem excision in pea seedlings. Planta 187:523–531
Stahlberg R, Cosgrove DJ (1996) Induction and ionic basis of slow wave potentials in seedlings of *Pisum sativum L.* Planta 200:416–425
Stahlberg R, Cosgrove DJ (1997) The propagation of slow wave potentials in pea epicotyls. Plant Physiol 113:209–217
Stahlberg R, Cleland RE, Van Volkenburgh E (2005) Decrement and amplification of slow wave potentials during their propagation in *Helianthus annuus* L. shoots. Planta 220:550–558
Stahlberg R, Cleland RE, Van Volkenburgh E (2006) Slow wave potentials – a propagating electrical signal unique to higher plants. In: Baluska F, Mancuso S, Volkmann D (eds) Communication in plants. Springer, Berlin, pp 291–308
Stankovic B, Davies E (1997) Intercellular communication in plants: electrical stimulation of proteinase inhibitor gene expression in tomato. Planta 202:402–406
Stankovic B, Zawadzki T, Davies E (1997) Characterization of the variation potential in sunflower. Plant Physiol 115:1083–1088
Stankovic B, Witters DL, Zawadzki T, Davies E (1998) Action and variation potentials in sunflower: An analysis of their relationship and distinguishing characteristics. Physiol Plant 103:51–58
Thain JF, Gubb IR, Wildon DC (1995) Depolarization of tomato leaf cells by oligogalacturonide elicitors. Plant Cell Environ 18:211–214

Thiel G, Homann U, Plieth C (1997) Ion channel activity during the action potential in *Chara*: a new insight with new techniques. J Exp Bot 48:609–622

Thion L, Mazars C, Nacry P, Bouchez D, Moreau M, Ranjeva R, Thuleau P (1998) Plasma membrane depolarization-activated calcium channels stimulated by microtubule-depolymerizing drugs in wild-type *Arabidopsis thaliana* protoplasts, display constitutively large activities and a longer half-life in *ton 2* mutant cells affected in the organization of the cortical microtubules. Plant J 13:603–610

Thorpe MR, Furch ACU, Minchin PEH, Föller J, van Bel AJE, Hafke JB (2010) Rapid cooling triggers forisome dispersion just before phloem transport stops. Plant Cell Environ 33:259–271

Thorsch J, Esau K (1981) Nuclear degeneration and the association of endoplasmic reticulum with the nuclear envelope and microtubules in maturing sieve elements of *Gossypium hirsutum*. J Ultrastr Res 74:195–204

Thuleau P, Moreau M, Schroeder JI, Ranjeva R (1994) Recent advances in the regulation of plant calcium channels: evidence for regulation by G-proteins, the cytoskeleton and second messengers. Curr Opin Plant Biol 1:424–427

Trebacz K, Simonis W, Schönknecht G (1994) Cytoplasmic Ca^{2+}, K^+, Cl^- and NO_3^- activities in the liverwort *Conocephalum conicum* L. at rest and during action potentials. Plant Physiol 106:1073–1084

Trebacz K, Dziubinska H, Krol E (2006) Electrical signals in long-distance communication in plants. In: Baluska F, Mancuso S, Volkmann D (eds) Communication in plants. Springer, Berlin, pp 277–290

Trewavas A (1999) Le calcium, c'est la vie: Calcium makes waves. Plant Physiol 120:1–6

Trewavas AJ, Malho R (1997) Signal perception and transduction: the origin of the phenotype. Plant Cell 9:1181–1195

Tsien RW, Tsien RY (1990) Calcium channels, stores and oscillations. Annu Rev Cell Biol 6:715–760

Tsien RW, Hess P, Mc Cleskey EW, Rosenberg RL (1987) Calcium channels: mechanisms of selectivity, permeation, and block. Annu Rev Biophys Biophys Chem 16:265–790

Tsutsui I, Ohkawa T, Nagai R, Kishimoto U (1986) Inhibition of Cl^- channel activation in *Chara corallina* membrane by lanthanum ion. Plant Cell Physiol 27:1197–1200

Tucker EB (1990) Calcium-loaded 1,2-bis(2-aminophenoxy)ethane-N, N, N′, N′-tetraacetic acid blocks cell-to-cell diffusion of carboxyfluorescein in staminal hairs of Setcreasea purpurea. Planta 182:34–38

Tuteja N, Umate P, van Bel AJE (2010) Forisomes: calcium-powered protein complexes with potential as 'smart' biomaterials. Trends Biotechnol 28:102–110

van Bel AJE (2003) The phloem, a miracle of ingenuity. Plant Cell Environ 26:125–149

van Bel AJE, Ehlers K (2005) Electrical signalling via plasmodesmata. In: Oparka KJ (ed) Plasmodesmata. Blackwell, Oxford, pp 263–278

van Bel AJE, Hafke JB (2005) Physiochemical determinants of phloem transport. In: Holbrook NM, Zwieniecki M (eds) Vascular transport in plants. Elsevier, Burlington, pp 19–44

van Bel AJE, Kempers R (1997) The pore/plasmodesm unit; key element in the interplay between sieve element and companion cell. Progr Bot 58:278–291

van Bel AJE, Knoblauch M (2000) Sieve element and companion cell: the story of the comatose patient and the hyperactive nurse. Aust J Plant Physiol 27:477–487

van Bel AJE, van Rijen HVM (1994) Microelectrode-recorded development of the symplasmic autonomy of the sieve element/companion cell complex in the stem phloem of *Lupinus luteus* L. Planta 192:165–175

van Bel AJE, Knoblauch M, Furch ACU, Hafke JB (2011a) $(\text{Questions})^n$ on phloem biology. 1. Electropotential waves, Ca^{2+} fluxes and cellular cascades along the propagation pathway. Plant Sci 181:210–218

van Bel AJE, Knoblauch M, Furch ACU, Hafke JB, Patrick JW (2011b) $(\text{Questions})^n$ on phloem biology. 2. Mass flow, molecular hopping, distribution patterns and macromolecular signalling. Plant Sci 181:325–330 van Sambeek JW, Pickard BG (1976) Mediation of rapid electrical, metabolic, transpirational, and photosynthetic changes by factors released from wounds. I. Variation potentials and putative action potentials. Can J Bot 54:2642–2650

van Sambeek JW, Pickard BG, Ulbright CE (1976) Mediation of rapid electrical, metabolic, transpirational, and photosynthetic changes by factors released from wounds. II. Mediation of the variation potential by Ricca's factor. Can J Bot 54:2651–2661

Volk G, Franceschi VR (2000) Localization of a calcium-channel-like protein in the sieve element plasma membrane. Aust J Plant Physiol 27:779–786

Walz C, Giavalisco P, Schad M, Juenger M, Klose J, Kehr J (2004) Proteomics of cucurbit phloem exudates reveals a network of defence proteins. Phytochemistry 65:1795–1804

Wang YF, Fan L-M, Zhang W-Z, Zhang W, Wu W-H (2004) Ca^{2+}- permeable channels in the plasma membrane of Arabidopsis pollen are regulated by actin microfilaments. Plant Physiol 136:3892–3904

Wayne R (1994) The excitability of plant cells: with special emphasis on Characean internodal cells. Bot Rev 60:265–367

Webb AAR, McAinsh MR, Taylor JE, Hetherington AM (1996) Calcium ions as intracellular second messengers in higher plants. Adv Bot Res 22:45–96

White PJ (2004) Calcium signals in root cells: the roles of plasma membrane calcium channels. Biologia 59:77–83

White PJ (2009) Depolarization-activated calcium channels shape the calcium signatures induced by low-temperature stress. New Phytol 183:6–8

White PJ, Broadley MR (2003) Calcium in plants. Ann Bot 92:487–511

White PJ, Ridout MS (1999) An energy-barrier model for the permeation of monovalent and divalent cations through the maxi cation channel in the plasma membrane of rye roots. J Membr Biol 168:63–75

Wildon DC, Thain JF, Minchin PEH, Gubb IR, Reilly AJ, Skipper YD, Doherty HM, O'Donnell PJ, Bowles DJ (1992) Electrical signalling and systemic proteinase inhibitor induction in the wounded plant. Nature 360:62–65

Will T, Tjallingii WF, Thönnessen A, van Bel AJE (2007) Molecular sabotage of plant defense by aphids. Proc Natl Acad Sci USA 104:10536–10541

Will T, Kornemann SR, Furch ACU, Tjallingii WF, van Bel AJE (2009) Aphid saliva counteracts sieve-tube occlusion: a universal phenomenon? J Exp Biol 22:3305–3312

Williams SE, Pickard BG (1972a) Properties of action potentials in *Drosera* tentacles. Planta 103:222–240

Williams SE, Pickard BG (1972b) Receptor potentials and action potentials in *Drosera* tentacles. Planta 103:193–221

Williams SE, Pickard BG (1974) Connections and barriers between cells of *Drosera* tentacles in relation to their electrophysiology. Planta 116:1–6

Williams SE, Spanswick RM (1976) Propagation of the neuroid action potential of the carnivorous plant *Drosera*. J Comp Physiol 108:211–223

Xie B, Wang X, Zhu M, Zhang Z, Hong Z (2011) *CalS7* encodes a callose synthase responsible for callose deposition in the phloem. Plant J 65:1–14

Yoo BC, Lee JY, Lucas WJ (2002) Analysis of the complexity of protein kinases within the phloem sieve tube system. J Biol Chem 277:15325–15332

Zawadzki T (1980) Action potentials in Lupinus angustifolius L. shoots. V. Spread of excitation in the stem, leaves, and root. J Exp Bot 31:1371–1377

Zawadzki T, Davies E, Dziubinska H, Trebacz K (1991) Characteristics of action potentials in *Helianthus annuus*. Physiol Plant 83:601–604

Zhang W, Fan L, Wu W (2007) Osmo-sensitive and stretch-activated calcium-permeable channels in *Vicia faba* guard cells are regulated by actin dynamics. Plant Physiol 143:1140–1151

Zimmermann MR, Felle HH (2009) Dissection of heat-induced systemic signals: superiority of ion fluxes to voltage changes in substomatal cavities. Planta 229:539–547

Zimmermann MR, Maischak H, Mithöfer A, Boland W, Felle HH (2009) System potentials, a novel electrical long-distance apoplastic signal in plants, induced by wounding. Plant Physiol 149:593–600

Zimmermann MR, Hafke JB, van Bel AJE, Furch ACU (2013) The interaction of xylem and phloem during exudation and wound occlusion in *Cucurbita maxima*. Plant Cell Environ 36:337–347

The Role of Volatiles in Plant–Plant Interactions

Cesar R. Rodriguez-Saona, Mark C. Mescher, and Consuelo M. De Moraes

Abstract Plants respond to herbivory by emitting volatile organic compounds (VOCs), which mediate diverse ecological interactions between plants and other organisms. Almost three decades after it was first proposed that plants respond to VOCs from injured neighbors, this phenomenon is now well established and has been documented across multiple levels of biological organization (i.e., molecular, biochemical, and ecological). Recent studies have also shown that herbivore-induced VOCs can play a role in within-plant communication. In general, VOCs appear frequently to prime defenses in plants, enhancing plant responses to subsequent herbivore attack. The mechanisms underlying such effects remain largely unknown, though we have recently begun to learn more about the genes involved in plant–plant signaling. This chapter summarizes our current knowledge about the role of VOCs in plant-to-plant interactions. By synthesizing these findings, our chapter intends to point out gaps in existing research, in particular the need for further studies under natural conditions.

Keywords Volatile organic compounds • Within-plant communication • Herbivore-induced plant volatiles • Priming • Plant signaling • Trophic interactions

---

C.R. Rodriguez-Saona (✉)
Department of Entomology, Rutgers University, 125A Lake Oswego Road, Chatsworth, NJ 08019, USA
e-mail: crodriguez@aesop.rutgers.edu

M.C. Mescher • C.M. De Moraes
Department of Entomology, Penn State University, 535 ASI Building, University Park, PA 16802, USA
e-mail: mcmescher@psu.edu; czd10@psu.edu

F. Baluška (ed.), *Long-Distance Systemic Signaling and Communication in Plants*, Signaling and Communication in Plants 19, DOI 10.1007/978-3-642-36470-9_19,

1 Introduction

Plants are known to emit volatile organic compounds (VOCs) from both aboveground vegetative and reproductive tissues (e.g., leaves, flowers, stems) and belowground roots (Knudsen et al. 1993; Steeghs et al. 2004; Dudareva et al. 2004). Constitutive VOC emissions are often limited, but emissions can increase dramatically in response to environmental stresses, including herbivore feeding (e.g., Takabayashi and Dicke 1996; Dicke and Vet 1999). Volatiles induced by herbivory are commonly referred to as herbivore-induced plant volatiles and often differ qualitatively from those induced by mechanical damage (Boland et al. 1999; Dicke 1999), reflecting in large part the recognition by plants of biochemical elicitors in the oral secretions of feeding herbivores (Alborn et al. 1997).

Herbivore-induced VOCs are emitted at the site of feeding (locally) but also from parts of damaged plants that are connected via the vascular system to the damaged site (Dicke et al. 1990; Turlings and Tumlinson 1992; Röse et al. 1996). This suggests that internal signals move through the plant to activate systemic VOC emissions (Takabayashi et al. 1991; Dicke et al. 1993). However, these internal signals are often restricted by the strength of vascular connectivity within plants (Orians et al. 2000; Viswanathan and Thaler 2004; Orians 2005; Rodriguez-Saona and Thaler 2005), suggesting a potential role of VOCs as external signals in systemic-induced responses (Frost et al. 2007; Heil 2010; Holopainen and Blande 2012).

Once emitted, however, plant VOCs convey "public" information that can be perceived not only by the emitting plant but also by other organisms. Indeed, constitutive and induced plant volatiles play various ecological roles in interactions between plants and other organisms including pollinators, pathogens, predators, parasitoids, and herbivores (Fig. 1) (Arimura et al. 2009; Hare 2011). For example, herbivore-induced VOCs can attract natural enemies of herbivores (Dicke and Sabelis 1988; Turlings et al. 1990; Takabayashi and Dicke 1996; De Moraes et al. 1998; Dicke and Vet 1999; Sabelis et al. 1999), provide protection against pathogens (Shulaev et al. 1997; Nakamura and Hatanaka 2002; Shiojiri et al. 2006), and repel herbivores (Bernasconi et al. 1998; De Moraes et al. 2001; Heil 2004), potentially protecting plants against their antagonists (Kessler and Baldwin 2001).

As discussed below, the role of VOCs in mediating interactions between plants has been explored and debated for nearly three decades (e.g., Baldwin and Schultz 1983; Rhoades 1983; Fowler and Lawton 1985), and plant responses to volatile cues are now well documented (e.g., Karban et al. 2000; Karban 2001; Heil and Karban 2010). Volatile-mediated interactions among plants are frequently referred to as plant–plant communication (also dubbed "talking trees" by the popular press). Karban (2008) proposed three criteria for recognizing that plants "communicate": First, the cue should cause a rapid response in the receiving plant; second, the cue needs to be plastic and the response to be conditional on receiving the cue; and third, the emission of a cue (signal) needs to be "intentional" and beneficial. These criteria depart somewhat from the distinction frequently drawn in the (animal) communication literature (e.g., Maynard-Smith and Harper 2003) between "cues"—which can be any information-bearing feature of the environment

Fig. 1 Volatile organic compounds (VOCs) released from plants play several important roles in interactions between plants and other organisms. Herbivore-induced VOCs protect plants against their enemies (1) and attracted natural enemies of herbivores (2). These interactions occur both aboveground as well as belowground (3) and can be mediated through internal systemic signals moving within the plant (*red broken arrow*). Herbivore-induced VOCs also serve as cues in within-plant signaling (A) and inter- and intraspecific plant–plant signaling (B). Although most studies so far investigate aboveground VOC signals, it is likely that VOC signals from roots are involved in root–root communication (C)

exploited by an organism as a guide to future action—and "signals," which are adaptive for the transmitting organism via their effects on the behavior of the receiver. Under the latter paradigm, "communication" is generally construed as successful signaling (i.e., adaptive transmission of information). Thus, whether particular VOC-mediated interactions constitute plant–plant communication depends on the precise terminology employed. However, this chapter will leave such semantic distinctions largely to one side while summarizing and discussing available evidence bearing on the role of plant volatiles in conveying information within and between plants.

2 Aboveground Signaling

The first evidence that plants respond to cues emitted by their neighboring damaged plants came from early papers by Rhoades (1983), who reported increased resistance in Sitka willow (*Salix sitchensis* Sanson ex. Bong) trees in the field when near trees were attacked by tent caterpillars, and Baldwin and Schultz (1983), who reported that undamaged poplar (*Populus x euroamericana* (Dole) Guinier) ramets and sugar maple (*Acer saccharum* Marshall) seedlings exposed to mechanically damaged plants increase their level of defenses (i.e., phenolic compounds) under controlled laboratory conditions. Although the specific cues involved were not identified in either study, the authors argued that airborne cues might be involved. These early studies on plant–plant "communication" were criticized for methodological flaws and failure to rule out alternative hypotheses, and consequently the phenomenon of plant response to olfactory cues remained controversial (Fowler and Lawton 1985; Bruin et al. 1995; Karban and Baldwin 1997; Dicke and Bruin 2001). Haukioja et al. (1985) later reported that *Epirrita autumnata* L. (autumnal moth) performance correlated positively with the distance of its birch-tree host from the closest tree defoliated in the previous year; however, the mechanism again remained undocumented. And other studies found no evidence for plant–plant communication (e.g., Myers and Williams 1984).

In the 1990s, a few studies conducted under carefully controlled laboratory conditions provided somewhat more conclusive evidence for plant–plant interactions and ignited new research on this topic. Farmer and Ryan (1990) reported that tomato plant exposure (in airtight chambers) to methyl jasmonate (MeJA)—a VOC known to activate plant defenses—induced production of proteinase inhibitor proteins (which can function as anti-herbivore defenses [Broadway and Duffey 1986]). Similarly, when tomato plants were placed near sagebrush, *Artemisia tridentata* Nutt.—a plant shown to emit methyl jasmonate—in chambers, proteinase inhibitor levels were increased in the tomato leaves. Later, Bruin et al. (1992) found that oviposition by herbivorous mites was reduced on cotton seedlings exposed to volatiles from mite-infested seedlings. And Shulaev et al. (1997) showed that exposure to the VOC methyl salicylate (MeSA) activates resistance in neighboring tobacco against tobacco mosaic virus as well as expression of pathogenesis-related defensive genes.

Further evidence for volatile-mediated plant–plant effects was provided by a number of field studies conducted in the early 2000s. Dolch and Tscharntke (2000) found that subsequent defoliation of alder (*Alnus glutinosa* L.) trees by the alder leaf beetle increased with distance from initially defoliated trees, indicating that defoliation increases resistance in neighboring undamaged trees. In a 5-year study, Karban et al. (2000) reported reduced herbivory by grasshoppers and increased activity of polyphenol oxidase in wild tobacco plants (*Nicotiana attenuata* Torr. ex. S. Watson) grown within 15 cm of sagebrush that had been manually clipped or damaged by herbivores—experiments blocking air or soil contact between sagebrush and tobacco indicated that airborne rather than soilborne cues were involved. Furthermore, these effects were not explained by herbivores avoiding clipped sagebrush (Karban and Baxter 2001). Although data were variable across 5 years, tobacco plants neighboring clipped sagebrush plants produced more flowers and seed-bearing capsules than plants near unclipped neighbors (Karban and Maron 2002), indicating potential fitness benefits to plants exposed to herbivore-induced VOCs. Similar results were found in naturally rooted plants when grown 10 cm from clipped or herbivore-damaged sagebrush (Karban et al. 2003). Sagebrush plants near clipped conspecific neighbors also experienced increased resistance compared with those near unclipped neighbors (Karban et al. 2004), and such interactions can occur at distances up to at least 60 cm (Karban et al. 2006). But three forb species (lomatium, lupine, and valerian) were unaffected by exposure to neighboring clipped sagebrush (Karban et al. 2004), suggesting that plant–plant communication is not a universal phenomenon and might be restricted to particular inducible plant species or combinations of emitting and receiving plants.

Plant–plant interactions mediated by VOCs can also influence multi-trophic interactions involving other organisms. For instance, Bruin et al. (1992) found increased predatory mite attraction to cotton seedlings previously exposed to volatiles from mite-infested plants. Under laboratory conditions, Choh et al. (2006) found that Lima bean (*Phaseolus lunatus* L.) plants had increased extrafloral nectar secretions after exposure to herbivorous spider mite-induced volatiles, which in turn reduced dispersal of predatory mites on exposed plants. Under natural conditions, Kost and Heil (2006) also showed that extrafloral nectar secretion is induced by VOCs released from herbivore-damaged Lima bean plants or by synthetic blends of these volatiles—leading to greater attraction of natural enemies, reduction of herbivore damage, and increased production of inflorescences and leaves (Kost and Heil 2006). Similarly, exposure to VOCs from herbivore-infested hybrid aspen (*Populus tremula* [L.] × *Populus tremuloides* [Michx.]) saplings induced secretion of sugars from extrafloral nectars but not emission of volatile terpenes (Li et al. 2012). However, when subsequently challenged by herbivores, plants previously exposed to VOCs from infested plants released more terpenes than unexposed plants (Li et al. 2012), indicating that VOCs from infested plants primed volatile emissions from neighboring plants (see Sect. 7).

In recent years, significant advances in molecular techniques have allowed a better understanding on the mechanisms underlying plant–plant communication (Baldwin et al. 2006). For example, Arimura et al. (2000), using cDNA microarray technology, demonstrated that exposure to herbivore-induced VOCs activates 227 genes in Lima bean leaves. Exposure of Lima bean leaves to *Tetranychus urticae* Koch-induced volatiles, *T. urticae* infestation, and artificial wounding increased the transcription of the genes involved in ethylene biosynthesis, i.e., S-adenosylmethionine (SAM) synthetase and 1-aminocyclopropane-1-carboxylic acid oxidase (Arimura et al. 2002). Peng et al. (2005), using RT-PCR analysis, found that the expression of three plant defensive genes [pathogenesis-related protein (*PR-1*), β-1,3-glucanase (*BGL2*), and phenylalanine ammonia lyase (*PAL*)] was upregulated in undamaged tomato and tobacco plants in response to exposure to either tomato or tobacco damaged by the cotton bollworm, *Helicoverpa armigera* (Hübner). Although transcriptional changes in plant defenses might not necessarily lead to phenotypic changes, they demonstrate that plants are influenced by volatile signals.

There is now evidence of plant olfactory responses to VOCs in more than ten plant species (Heil and Karban 2010; Helms et al. 2012). It seems likely that such responses are frequently adaptive for the receiving plants, although only four studies to date have addressed the fitness benefits of plant response to volatile cues (Karban and Maron 2002; Kost and Heil 2006; Pearse et al. 2012; Karban et al. 2012). Karban et al. (2012) examined the long-term consequences (over 12 years) of plant exposure to VOCs in sagebrush; although survival was not consistent, plants near clipped neighbors had more branches and these branches produced more inflorescences than those near unclipped neighbors. As further discussed below, the fitness effects of plant-to-plant VOC emission for emitting plants are less certain, and it is possible that most cases of VOC-mediated effects among plants reflect eavesdropping by the receiving plant on volatile cues that evolved from a within-plant wound signal (e.g., Karban et al. 2006; Frost et al. 2007) or other ecological functions.

3 Belowground Signaling

Plant–plant interactions are known to be mediated through the belowground transmission of secondary metabolites (Fig. 1). For instance, plants commonly produce root exudates that can have allelopathic effects on neighboring plants, i.e., inhibitory effects on germination or development (e.g., Bais et al. 2004; Jarchow and Cook 2009). However, relative to aboveground signaling, our knowledge of the potential role of VOCs in belowground interactions (e.g., between damaged and undamaged roots) remains quite limited. A few studies have shown that signals from roots of herbivore-infested plants can affect tri-trophic interactions in neighboring plants. For example, Chamberlain et al. (2001) found that uninfested broad bean (*Vicia faba* L.) plants grown in soil alongside pea aphid-infested plants became relatively more attractive to the parasitoid *Aphidius ervi* Haliday than plants grown with uninfested controls. Similarly, Dicke and Dijkman (2001) found that uninfested Lima bean plants became more attractive to predatory mites when roots were placed in distilled water with roots from spider mite-infested plants than when placed with roots from uninfested plants. And, intact broad bean plants became attractive to *A. ervi* when placed in a hydroponic solution previously used to grow pea aphid-infested plants, whereas no attraction was observed when intact plants were placed in a solution previously used to grow intact plants (Guerrieri et al. 2002). Although the mechanism remains unknown, these studies suggest that compounds (possibly VOCs) emitted from roots of herbivore-infested plants may affect aboveground VOC emissions from neighboring plants.

4 Volatile Signaling Within Plants and Among Kin

Although it is relatively easy to understand why plants should eavesdrop on their neighbors, it is less clear why a plant should send messages to neighboring plants that might benefit potential competitors. Heil and Karban (2010) proposed four not

Table 1 Evidence for airborne intraplant signaling

Common name	Scientific name	Inducing agent	Response induced in receiver	Type of response	Experimental conditions	References
Sagebrush	*Artemisia tridentata*	Mechanical	Increase resistance to herbivores	Activation	Natural	Karban et al. (2006)
Lima bean	*Phaseolus lunatus*	Herbivory	Increase extrafloral nectar secretion	Priming	Natural	Heil and Silva Bueno (2007)
Hybrid polar	*Populus deltoides* × *nigra*	Herbivory	Increase volatile emissions	Priming	Laboratory	Frost et al. (2007)
Blueberry	*Vaccinium corymbosum*	Herbivory	Increase resistance to herbivores	Activation	Laboratory	Rodriguez-Saona et al. (2009)
			Increase volatile emissions	Priming	Laboratory	
Tea	*Camellia sinensis*	Jasmonic acid	Increase unidentified metabolites	Activation	Laboratory	Dong et al. (2011)

mutually exclusive explanations for the role of VOCs: (1) that VOCs defend plants against biotic and abiotic stresses, (2) that VOCs have synergistic effects on other types of defenses, (3) that VOCs play a role in within-plant signaling, and (4) that VOCs play a role as between plant signals that evolve via kin or group selection. Under hypothesis (1), the primary function of VOCs from herbivore-damaged plants is to directly protect plants against their enemies by negatively affecting herbivores and pathogens, or indirectly provide protection to plants by attracting the natural enemies of herbivores, as indicated above; perception of these volatiles by neighbors becomes secondary. Under hypothesis (2), VOCs induce various direct and indirect defenses in plants that can act synergistically in plant protection. For example, VOCs can induce direct defenses that reduce the growth of caterpillars on plants while also attracting natural enemies of these herbivores (Dicke and Van Loon 2000). Numerous studies provide evidence supporting hypotheses (1) and (2) (see reviews by Unsicker et al. 2009; Karban 2011), and these will not be discussed in depth here. In the following, we will discuss evidence for hypotheses (3) and (4).

Recent studies document the role of VOCs in intraplant signaling in at least five systems (Table 1). In these cases, the recipient of the information conveyed by volatiles is the emitting plant itself (Fig. 1). First proposed by Farmer (2001) and Orians (2005), evidence for intraplant communication was first empirically tested by Karban et al. (2006), who reported increased induced resistance when branches of sagebrush, *A. tridentata*, were exposed to volatiles from neighboring clipped branches within a plant. The authors found no evidence for systemic-induced resistance, indicating limited vascular connections between branches, and thus concluded that sagebrush relies on external volatile cues rather than internal signals for activation of systemic-induced resistance. Plant ontogeny affected the strength of intraplant signaling in sagebrush: systemic-induced resistance between branches was observed on young plants but not on old plants (Shiojiri and Karban 2006). And young plants were both more effective emitters and more responsive receivers of volatile cues (Shiojiri and Karban 2006). Interestingly, two closely related species, *Artemisia cana* Pursh and *A. douglasiana* Besser, exhibited systemic-induced resistance among neighboring branches that did not require volatile cues, although communication between stems required a volatile cue in *A. douglasiana* (Shiojiri and Karban 2008). Further studies on this system showed that cues released from clipped branches remained active for up to 3 days (Shiojiri et al. 2009) and that cues must be received for an extended time (> 1h) before responding (Shiojiri et al. 2012a).

In another system, Heil and Silva Bueno (2007) reported increased extrafloral nectar secretion from undamaged leaves when exposed to VOCs from leaves (of the same shoot) damaged by chrysomelid beetles in Lima bean, *P. lunatus*. VOCs also primed extrafloral nectar secretion such that the response was enhanced following damage (Heil and Silva Bueno 2007). Distance from the emitter affected extrafloral nectar secretion in Lima bean such that no benefit was observed at distances greater than 50 cm (Heil and Adame-Álvarez 2010). Recently, Dong et al. (2011) used metabolomics to show that exposure of tea (*Camellia sinensis* (L.)) leaves to VOCs from adjacent leaves (induced by jasmonic acid) triggers dramatic changes in metabolite composition.

In laboratory studies, Frost et al. (2007) and Rodriguez-Saona et al. (2009) showed that VOCs induced by gypsy moth larvae prime volatile emissions in adjacent undamaged leaves within hybrid poplar saplings and blueberry plants, respectively, potentially overcoming vascular constraints on the internal transmission of wound signals. Indeed it may be hypothesized that plants with limited vascular connectivity should be more responsive to VOCs. This is the case in most studies documented thus far. For example, shrubs like sagebrush and blueberries that grow in arid and resource-poor environments and as a result have limited (sectored) vascular connections might use volatile cues to coordinate systemic responses.

While the potential adaptive benefits of within-plant signaling are clear, inter-plant communication might also be adaptive if the receiving plants are clones of, or have positive genetic relatedness to, the emitter. Conversely, receiving individuals might plausibly benefit from discriminating between signals arising from self/kin and other individuals. Karban and Shiojiri (2009) reported that sagebrush plants that received volatile cues from genetically identical cuttings later accumulated less natural herbivore damage than plants that received cues from non-self cuttings. However, Pearse et al. (2012) reported contradictory effects from field studies with *Achyrachaena mollis* Schauer (blow wives), in which plants with damaged neighbors subsequently suffered *increased* herbivory and reduced fitness, but only if the damaged plant was a close relative. These findings are currently difficult to interpret but clearly suggest that plant responses to volatile cues are context dependent and may be influenced by variation in the genotype of the emitting individual.

5 Volatile Cues from Undamaged Plants

As compared to VOCs from damaged plants, the effects of VOCs from undamaged plants in plant–plant interactions have received much less attention. Glinwood et al. (2004) found that settling by the bird cherry-oat aphid, *Rhopalosiphum padi* (L.), was reduced on barley (*Hordeum vulgare* L.) plants after exposure to volatiles from Canada thistle (*Cirsium arvense* (L.) Scop.) or bull thistle (*C. vulgare* (Savi) Ten.). Attraction of *R. padi* to barley species was also reduced after exposure to *Cirsium*, indicating changes in volatiles from barley after exposure. Moreover, even within the same plant species, Glinwood et al. (2009) found that aphid, *R. padi*, preference for a certain barley cultivar was reduced when exposed to a different barley cultivar. These interactions can also affect higher trophic levels. The ladybird beetle *Coccinella septempunctata* L. and the parasitoid *Aphidius colemani* Viereck were more attracted to barley exposed to a different cultivar than unexposed barley (Glinwood et al. 2009). In field studies, Ninkovic et al. (2011) found that *C. septempunctata* preferred specific combinations of different barley genotypes as compared with pure stands. Although the volatiles involved remain uncharacterized, these studies show intra- and interspecies communication between undamaged plants and possible consequences in multi-trophic interactions.

Another interesting case of plant exploitation of olfactory cues from neighboring plants involves host location by obligately parasitic plants in the genus *Cuscuta* (dodder). Runyon et al. (2006) demonstrated that *C. pentagona* Engelm seedlings exhibited directed growth toward the odors of tomato plants (*Solanum lycopersicum* L.) and several individual host-derived compounds. Moreover, the parasite seedlings could discriminate between the odors of host and nonhost species.

6 Chemistry of VOC Emissions Mediating Plant–Plant Interactions

If plants can communicate, what "language" do they use? VOC emissions are frequently complex blends of many different compounds (Dudareva et al. 2006), whose composition can vary not only across plant species but also among genotypes (Degen et al. 2004; Snoeren et al. 2010). However, it is likely that blends are more similar within species than between species. Thus, it is quite plausible that receiving plants can distinguish the VOC emissions of conspecifics from those of other plant species and even discriminate among the emissions of different genotypes (as discussed above). To date, however, we know very little about exactly which features of VOC blends are perceived by receiving plants or about how this information is processed to activate a response. While the responses of insects and other animals to olfactory cues often rely on complex qualitative features of the volatile blend encountered (Cunningham et al. 2006; de Bruyne and Baker 2008), research to date suggests that plants frequently respond to individual VOC compounds. These compounds come from a variety of biosynthetic pathways (Paré and Tumlinson 1999) and have diverse physiological and ecological functions (Kesselmeier and Staudt 1999; Hare 2011).

Several possible VOC elicitors of plant defenses have been suggested to be involved in plant–plant signaling. These include the phytohormones ethylene (O'Donnell et al. 1996; Tscharntke et al. 2001; Pierik et al. 2004), MeJA (Farmer and Ryan 1990; Tscharntke et al. 2001; Glinwood et al. 2007), and MeSA (Shulaev et al. 1997; Glinwood et al. 2007; Yi et al. 2009). For example, Pierik et al. (2004) identified ethylene as a VOC responsible for shade avoidance in a plant's response to the presence of neighbors. Ethylene is also emitted from Lima bean leaves when exposed to volatiles induced by spider mite feeding (Arimura et al. 2002). MeJA and MeSA were mentioned previously to be implicated in signaling between plants.

Probably the VOCs most commonly associated as signals in between- and within-plant communication are green leaf volatiles, such as (*Z*)-3-hexenol, (*E*)-2-hexenal, and (*Z*)-3-hexenyl acetate. These six carbon VOCs are derived from linolenic acid via the lipoxygenase pathway (Paré and Tumlinson 1999). Under laboratory conditions, Arimura et al. (2001) found that these three green leaf volatiles induce the expression of defense genes in exposed Lima bean plants. Similar studies have shown that (*Z*)-3-hexenol triggers expression of defense genes in *Arabidopsis thaliana* (L.) Heynh. (Wei and Kang 2011). Ethylene synergizes with (*Z*)-3-hexenol for enhanced VOC emissions in corn plants (Ruther and Kleier 2005). Exposure to (*E*)-2-hexenal also activates genes involved in plant defense in Arabidopsis (Bate and Rothstein 1998) and activates local and systemic VOC emissions in tomatoes (Farag and Paré 2002). In Lima bean, exposure to (*Z*)-3-hexenyl acetate induced increased extrafloral nectar secretion (Kost and Heil 2006); however, exposure to other structurally related compounds such as (*E*)-3-hexenyl acetate, 5-hexenyl acetate, and (*Z*)-3-hexenylisovalerate (not normally released from attacked plants) also elicited a similar response (Heil et al. 2008), suggesting a lack of chemical specificity. In contrast, Ruther and Furstenau (2005) found that the strength of induced volatile emissions after exposure to green leaf volatiles is influenced by the position or configuration of the double bond but not by the functional group. Emission of green leaf volatiles from plants is not specific to herbivore feeding damage since they are also released following mechanical tissue damage. Thus, it is somewhat unlikely that they provide highly reliable information to receiving plants about herbivore presence and identity.

Volatile terpenes (e.g., monoterpenes, sesquiterpene, and homoterpenes) are often induced de novo after herbivore damage (Paré and Tumlinson 1997; Paré and Tumlinson 1999); compared with green leaf volatiles, they may thus provide more reliable signals to receiving plants. For example, *A. thaliana* exposure to monoterpenes (e.g., myrcene and ocimene volatiles) resulted in changes in plant transcriptome and induction of MeJA accumulation (Godard et al. 2008). Belowground, two terpene VOCs induced by root herbivory, (*E*)-β-caryophyllene (Rasmann et al. 2005) and pregeijerene (1,5-dimethylcyclodeca-1,5,7-triene) (Ali et al. 2012), and shown to attract entomopathogenic nematodes, have recently been identified from corn and citrus roots, respectively. However, whether these or other VOCs emitted from roots are involved in belowground plant–plant communication is still unknown.

To date, the evidence suggests that plant responses to damage-associated VOCs are not necessarily species specific—notwithstanding ecological studies (discussed above) suggesting that plants may respond differently to volatile signals from emitting plants with different genotypes and possibly even discriminate between signals deriving from self and other (Karban and Shiojiri 2009; Pearse et al. 2012). To further advance our knowledge on the chemical signals involved in plant–plant communication, more studies are clearly needed to document the specific VOC cues and signals responsible for eliciting plant responses, and the specificity of these responses under natural conditions.

7 Induction Versus Priming

The responses from receiving plants to VOC emissions are likely influenced by the strength of the volatile signal or cue. Plants are sessile, and whether or not they will respond to emissions (e.g., from damaged neighbors) under field situations will depend on factors such as distance from the emitting plant (Karban et al. 2000, 2003, 2006; Frost et al. 2008a; Heil and Adame-Álvarez 2010) as well as the duration (Shiojiri et al. 2009; Shiojiri et al. 2012a; Shiojiri et al. 2012b; Girón-Calva et al. 2012) and dose (Girón-Calva et al. 2012) of exposure to the active signals—with shorter distances from the emitting plant and prolonged exposure times at high concentrations of the active signals most likely to elicit a stronger response from the receiving plant. Environmental conditions will also influence the strength of the chemical signals. For example, air pollutants such as ozone can breakdown volatile chemicals and reduce the distance over which the signals are effective, thus interfering with plant–plant communication (Blande et al. 2010).

The strength of the signal may provide information to the receiving plants on the risk of herbivory. Thus, the recipient of the information can adjust their defenses accordingly. For instance, the strength of the signal can be one determinant of whether the receiving plant will activate defenses or merely be "primed." Priming is a "cost-effective" strategy where plants initiate a state of readiness in response to a signal that allows for enhanced induced resistance only after attack has occurred (Frost et al. 2008b; Choudhary et al. 2008). Preston et al. (2004) showed that exposure of MeJA at dosages equivalent to those released by sagebrush does not elicit resistance in wild tobacco, *N. attenuata*, plants. Later field and laboratory studies by Kessler et al. (2006) demonstrated that exposure to volatiles from clipped sagebrush primes (rather than activates) wild tobacco plants for induced chemical defenses. There are now several reports that VOCs prime not only direct defenses but also indirect defenses such as the emission of herbivore-induced VOCs and extrafloral nectar secretion in plants (Choh et al. 2004; Choh and Takabayashi 2006; Heil and Kost 2006; Choh and Takabayashi 2007; Peng et al. 2011). Priming is also a key defense strategy in within-plant signaling (Table 1). Heil and Ton (2008) proposed that VOC signals serve as a fast way to prime defenses in plants (priming phase), which is then followed by the activation of defenses from arrival of delayed vascular signals or the plant's enemy (defense phase).

Several VOCs have been shown to prime plants. Among them, the green leaf volatiles ((*Z*)-3-hexenal, (*Z*)-3-hexenol, and (*Z*)-3-hexenyl acetate) prime neighboring corn seedling for enhanced inducible defenses (e.g., jasmonic acid and volatile sesquiterpene production) (Engelberth et al. 2004). Frost et al. (2008c) found that hybrid poplar saplings exposed to (*Z*)-3-hexenyl acetate have increased jasmonic acid and terpene emissions following gypsy moth larval feeding than unexposed saplings. Nonanal primed the expression of pathogenesis-related genes in Lima bean (Yi et al. 2009).

Studies that expose plants to synthetic VOCs often have the problem of using unrealistically high concentrations. An alternative approach employs transgenic plants that are modified in their ability to emit or receive the signals. For example, Paschold et al. (2006) found that transformed "emitter" wild tobacco plants genetically silenced in the production of green leaf volatiles or terpene volatiles were unable to communicate with untransformed "receiver" plant neighbors. Using transgenic tobacco plants that constitutively overexpress (*E*)-β-ocimene synthase, Muroi et al. (2011) demonstrated that exposure to these transgenic plants primes induced defenses in Lima bean and corn plants. Interestingly, uninfested plants exposed to VOCs from infested plants previously exposed to transgenic tobacco plants also had enhanced indirect defenses in response to herbivory (Muroi et al. 2011; Arimura et al. 2012).

8 Conclusions

While the phenomenon of plant olfaction long remained controversial, there is now compelling evidence that plants can perceive and respond to volatiles and that plant VOC emissions mediate ecologically relevant information within individual plants and between close neighbors. Nevertheless, many questions remain unanswered. To date the mechanisms by which plants perceive and respond to olfactory cues remain largely unknown—although a recent study by Zebelo et al. (2012) showed that tomato plants exposed to herbivore-induced VOCs respond with a strong depolarization of the plasma membrane potential, which increased with increasing green leaf volatile concentrations; these VOCs also induced a strong increase in cytosolic calcium flux. Moreover, we know relatively little about the specific qualitative and quantitative features of VOC blends that plants perceive or the specificity of the information conveyed in particular interactions. Once the mechanisms of plant olfaction and volatile-mediated defense priming and induction are better understood, it may be possible to exploit this knowledge to enhance the management of agricultural crops, by conserving volatile-signaling traits in cultivated species and potentially even by engineering plants with inducible (or permeable) defenses keyed to olfactory cues associated with the presence of particular pest species.

The broader ecological and evolutionary significance of plant olfaction and volatile-mediated plant–plant interactions remains poorly characterized. More studies are needed (particularly under field conditions) to determine the prevalence and ecological importance of plant-to-plant VOC transmission. While plants are clearly responsive to volatile cues, it is not clear how frequently or reliably volatiles transmit information between plants under field conditions. Such information transfer likely depends on the distance between the receiver and emitter, the duration of the exposure, and the dose of exposure. In addition, volatile signals in nature are likely to be variable (due to abiotic factors such as temperature, wind, radiation), unstable (due to the action of air pollutants), and potentially unreliable (due to the fact that various environmental stresses induce the release of common VOC emissions such as green leaf volatiles). We also currently have limited information about the range of olfactory cues to which plants can be responsive.

As discussed here, most research to date has focused on the role of damage-induced plant volatiles as cues or signals indicating the likelihood of subsequent herbivory. But, as noted above, plants can also perceive and respond to constitutive plant VOC emissions, including the case of some parasitic plants that exploit the odors of potential host plants as host-location cues (Runyon et al. 2006). A recent study by Helms et al. (2013) furthermore suggests that plant may sometimes directly perceive VOCs emitted by arthropod herbivores and respond by priming their defense responses.

Finally, further elucidation of the fitness consequences, under natural conditions, of volatile-mediated plant-to-plant interactions for both emitting and receiving plants will help to elucidate the adaptive functions of volatile emissions and evaluate hypotheses regarding the likely trajectories by which plant olfactory and volatile-signaling systems have evolved.

References

Alborn HT, Turlings TCJ, Jones TH, Stenhagen G, Loughrin JH, Tumlinson JH (1997) An elicitor of plant volatiles from beet armyworm oral secretion. Science 276:945–949

Ali JG, Alborn HT, Campos-Herrera R, Kaplan F, Duncan LW, Rodriguez-Saona C, Koppenhöfer AM, Stelinski LL (2012) Subterranean, herbivore-induced plant volatile increases biological control activity of multiple beneficial nematode species in distinct habitats. PLoS One 7: e38146

Arimura G-I, Tashiro K, Kuhara S, Nishioka T, Ozawa R, Takabayashi J (2000) Gene responses in bean leaves induced by herbivory and by herbivore-induced volatiles. Biochem Biophys Res Commun 277:305–310

Arimura G-I, Ozawa R, Horiuchi J, Nishioka T, Takabayashi J (2001) Plant–plant interactions mediated by volatiles emitted from plants infested by spider mites. Biochem Syst Ecol 29:1049–1061

Arimura G-I, Ozawa R, Nishioka T, Boland W, Koch T, Kühnemann F, Takabayashi J (2002) Herbivore-induced volatiles induce the emission of ethylene in neighboring lima bean plants. Plant J 29:87–98

Arimura G-I, Matsui K, Takabayashi J (2009) Chemical and molecular ecology of herbivore-induced plant volatiles: proximate factors and their ultimate functions. Plant Cell Physiol 50:911–923

Arimura G-i, Muroi A, Nishihara M (2012) Plant–plant–plant communications, mediated by (*E*)-β-ocimene emitted from transgenic tobacco plants, prime indirect defense responses of lima beans. J Plant Interact 7:193–196

Bais HP, Park SW, Weir TL, Callaway RM, Vivanco JM (2004) How plants communicate using the underground information superhighway. Trends Plant Sci 9:26–32

Baldwin IT, Schultz JC (1983) Rapid changes in tree leaf chemistry induced by damage: evidence for communication between plants. Science 221:277–279

Baldwin IT, Halitschke R, Paschold A, von Dahl CC, Preston CA (2006) Volatile signaling in plant–plant interactions: talking trees in the genomics era. Science 311:812–815

Bate NJ, Rothstein SJ (1998) C6-volatiles derived from the lipoxygenase pathway induce a subset of defense-related genes. Plant J 16:561–569

Bernasconi ML, Turlings TCJ, Ambrosetti L, Bassetti P, Dorn S (1998) Herbivore-induced emissions of maize volatiles repel the corn leaf aphid, *Rhopalosiphum maidis*. Entomol Exp Appl 87:133–142

Blande JD, Holopainen JK, Li T (2010) Air pollution impedes plant-to-plant communication by volatiles. Ecol Lett 13:1172–1181

Boland W, Koch T, Krumm T, Piel J, Jux A (1999) Induced biosynthesis of insect semiochemicals in plants. In: Chadwick DJ, Goode J (eds) Insect–plant interactions and induced plant defence, Novartis Foundation Symposium 223. Wiley, Chicester, pp 110–126

Broadway RM, Duffey SS (1986) Plant proteinase inhibitors: mechanism of action and effect on the growth and digestive physiology of larval *Heliothis zea* and *Spodoptera exigua*. J Insect Physiol 32:827–833

Bruin J, Dicke M, Sabelis MW (1992) Plants are better protected against spider-mites after exposure to volatiles from infested conspecifics. Experientia 48:525–529

Bruin J, Sabelis MW, Dicke M (1995) Do plants tap SOS signals from their infested neighbors? TREE 10:167–170

Chamberlain K, Guerrieri E, Pennacchio F, Pettersson J, Pickett JA, Poppy GM, Powell W, Wadhams LJ, Woodcock CW (2001) Can aphid-induced plant signals be transmitted aerially and through the rhizosphere? Biochem Syst Ecol 29:1063–1074

Choh Y, Takabayashi J (2006) Herbivore-induced extra-floral nectar production in Lima bean plants enhanced by previous exposure to volatiles from infested conspecifics. J Chem Ecol 32:2073–2077

Choh Y, Takabayashi J (2007) Duration of priming of two indirect plant defenses. Plant Signal Behav 2:13–14

Choh Y, Shimoda T, Ozawa R, Dicke M, Takabayashi J (2004) Lima bean leaves exposed to volatiles from herbivore-induced conspecific plants emit carnivore attractants: active or passive response? J Chem Ecol 30:1305–1317

Choh Y, Kugimiya S, Takabayashi J (2006) Induced production of extrafloral nectar in intact lima bean plants in response to volatiles from spider mite-infested conspecific plants as a possible indirect defense against spider mites. Oecologia 147:455–460

Choudhary DK, Johri BN, Prakash A (2008) Volatiles as priming agents that initiate plant growth and defense responses. Curr Sci 94:595–604

Cunningham JP, Moore CJ, Zalucki MP, Cribb BW (2006) Insect odour perception: recognition of odour components by flower foraging moths. Proc R Soc B 273:2035–2040

de Bruyne M, Baker TC (2008) Odor detection in insects: volatile codes. J Chem Ecol 34:882–897

De Moraes CM, Lewis WJ, Pare PW, Alborn HT, Tumlinson JH (1998) Herbivore-infested plants selectively attract parasitoids. Nature 393:570–573

De Moraes CM, Mescher MC, Tumlinson JH (2001) Caterpillar-induced nocturnal plant volatiles repel conspecific females. Nature 410:577–580

Degen T, Dillmann C, Marion-Poll F, Turlings TCJ (2004) High genetic variability of herbivore-induced volatile emission within a broad range of maize inbred lines. Plant Physiol 135:1928–1938

Dicke M (1999) Are herbivore-induced plant volatiles reliable indicators of herbivore identity to foraging carnivorous arthropods? Entomol Exp Appl 91:131–142

Dicke M, Bruin J (2001) Chemical information transfer between plants: back to the future. Biochem Syst Ecol 29:981–994

Dicke M, Dijkman H (2001) Within-plant circulation of systemic elicitor of induced defence and release from roots of elicitor that affects neighbouring plants. Biochem Syst Ecol 29:1075–1087

Dicke M, Sabelis MW (1988) How plants obtain predatory mites as bodyguards. Neth J Zool 38:148–165

Dicke M, Van Loon JJA (2000) Multitrophic effects of herbivore-induced plant volatiles in an evolutionary context. Entomol Exp Appl 97:237–249

Dicke M, Vet LEM (1999) Plant–carnivore interactions: evolutionary and ecological consequences for plant, herbivore and carnivore. In: Olff H, Brown VK, Drent RH (eds) Herbivores: between plants and predators. Blackwell, Oxford, pp 483–520

Dicke M, van Beek TA, Posthumus MA, Ben Dom N, van Bokhoven H, de Groot AE (1990) Isolation and identification of volatile kairomone that affects acarine predator-prey interactions: involvement of host plant in its production. J Chem Ecol 16:381–396
Dicke M, van Baarlen P, Wessels R, Dijkman H (1993) Herbivory induces systemic production of plant volatiles that attract predators of the herbivore: extraction of endogenous elicitor. J Chem Ecol 19:581–599
Dolch R, Tscharntke T (2000) Defoliation of alders (*Alnus glutinosa*) affects herbivory by leaf beetles on undamaged neighbours. Oecologia 125:504–511
Dong F, Yang Z, Baldermann S, Sato Y, Asai T, Watanabe N (2011) Herbivore-induced volatiles from tea (*Camellia sinensis*) plants and their involvement in intraplant communication and changes in endogenous nonvolatile metabolites. J Agric Food Chem 59:13131–13135
Dudareva N, Pichersky E, Gershenzon J (2004) Biochemistry of plant volatiles. Plant Physiol 135:1893–1902
Dudareva N, Negre F, Nagegowda DA, Orlova I (2006) Plant volatiles: recent advances and future perspectives. Crit Rev Plant Sci 25:417–440
Engelberth J, Alborn HT, Schmelz EA, Tumlinson JH (2004) Airborne signals prime plants against insect herbivore attack. Proc Natl Acad Sci USA 101:1781–1785
Farag MA, Paré PW (2002) C-6-green leaf volatiles trigger local and systemic VOC emissions in tomato. Phytochemistry 61:545–554
Farmer EE (2001) Surface-to-air signals. Nature 411:854–856
Farmer EE, Ryan CA (1990) Interplant communication: airborne methyl jasmonate induces synthesis of proteinase inhibitors in plant leaves. Proc Natl Acad Sci USA 87:7713–7716
Fowler SV, Lawton JH (1985) Rapidly induced defenses and talking trees: the devil's advocate position. Am Nat 126:181–195
Frost CJ, Appel HM, Carlson JE, De Moraes CM, Mescher MC, Schultz JC (2007) Within-plant signalling via volatiles overcomes vascular constraints on systemic signalling and primes responses against herbivores. Ecol Lett 10:490–498
Frost CJ, Mescher MC, Carlson JE, De Moraes CM (2008a) Why do distance limitations exist on plant–plant signaling via airborne volatiles? Plant Signal Behav 3:466–468
Frost CJ, Mescher MC, Carlson JE, De Moraes CM (2008b) Plant defense priming against herbivores: getting ready for a different battle. Plant Physiol 146:818–824
Frost CJ, Mescher MC, Dervinis C, Davis JM, Carlson JE, De Moraes CM (2008c) Priming defense genes and metabolites in hybrid poplar by the green leaf volatile cis-3-hexenyl acetate. New Phytol 180:722–734
Girón-Calva PS, Molina-Torres J, Heil M (2012) Volatile dose and exposure time impact perception in neighboring plants. J Chem Ecol 38:226–228
Glinwood R, Ninkovic V, Pettersson J, Ahmed E (2004) Barley exposed to aerial allelopathy from thistles (*Cirsium* spp) becomes less acceptable to aphids. Ecol Entomol 29:188–195
Glinwood R, Gradin T, Karpinska B, Ahmed E, Johsson L, Ninkovic V (2007) Aphid acceptance of barley exposed to volatile phytochemicals differs between plants exposed in daylight and darkness. Plant Signal Behav 2:321–326
Glinwood R, Ahmed E, Qvarfordt E, Ninkovic V, Pettersson J (2009) Airborne interactions between undamaged plants of different cultivars affect insect herbivores and natural enemies. Arthropod–Plant Interact 3:215–224
Godard KA, White R, Bohlmann J (2008) Monoterpene-induced molecular responses in *Arabidopsis thaliana*. Phytochemistry 69:1838–1849
Guerrieri E, Poppy GM, Powell W, Rao R, Pennacchio F (2002) Plant-to-plant communication mediating in-flight orientation of *Aphidius ervi*. J Chem Ecol 28:1703–1715
Hare JD (2011) Ecological role of volatiles produced by plants in response to damage by herbivorous insects. Annu Rev Entomol 56:161–180
Haukioja E, Suomela J, Neuvonen S (1985) Long-term inducible resistance in a birch foliage: triggering cues and efficacy on a defoliator. Oecologia 65:363–369

Heil M (2004) Direct defense or ecological costs? Responses of herbivorous beetles to volatiles released by wild Lima bean (*Phaseolus lunatus*). J Chem Ecol 30:1289–1295

Heil M (2010) Within-plant signalling by volatiles triggers systemic defences. In: Baluška F, Ninkovic V (eds) Plant communication from an ecological perspective. Springer, Berlin, pp 99–112

Heil M, Adame-Álvarez RM (2010) Short signalling distances make plant communication a soliloquy. Biol Lett 6:843–845

Heil M, Karban R (2010) Explaining the evolution of plant communication by airborne signals. Trends Ecol Evol 25:137–144

Heil M, Kost C (2006) Priming of indirect defences. Ecol Lett 9:813–817

Heil M, Silva Bueno JC (2007) Within-plant signaling by volatiles leads to induction and priming of an indirect plant defense in nature. Proc Natl Acad Sci 104:5467–5472

Heil M, Ton J (2008) Long-distance signalling in plant defence. Trends Plant Sci 13:264–272

Heil M, Lion U, Boland W (2008) Defense-inducing volatiles: In search of the active motif. J Chem Ecol 34:601–604

Helms AM, De Moraes CM, Tooker JF, Mescher MC (2013) Exposure of *Solidago altissima* plants to volatile emissions of an insect antagonist (*Eurosta solidaginis*) deters subsequent herbivory. Proc Natl Acad Sci USA 110:199–204

Holopainen J, Blande J (2012) Molecular plant volatile communication. In: López-Larrea C (ed) Sensing in nature, vol 739. Springer, New York, pp 17–31

Jarchow ME, Cook BJ (2009) Allelopathy as a mechanism for the invasion of *Typha angustifolia*. Plant Ecol 204:113–124

Karban R (2001) Communication between sagebrush and wild tobacco in the field. Biochem Syst Ecol 29:995–1005

Karban R (2008) Plant behaviour and communication. Ecol Lett 11:727–739

Karban R (2011) The ecology and evolution of induced resistance against herbivores. Funct Ecol 25:339–347

Karban R, Baldwin IT (1997) Induced responses to herbivory. University of Chicago Press, Chicago

Karban R, Baxter K (2001) Induced resistance in wild tobacco with clipped sagebrush neighbors: The role of herbivore behavior. J Insect Behav 14:147–156

Karban R, Maron J (2002) The fitness consequences of interspecific eavesdropping between plants. Ecology 83:1209–1213

Karban R, Shiojiri K (2009) Self-recognition affects plant communication and defense. Ecol Lett 12:502–506

Karban R, Baldwin I, Baxter K, Laue G, Felton G (2000) Communication between plants: induced resistance in wild tobacco plants following clipping of neighboring sagebrush. Oecologia 125:66–71

Karban R, Maron J, Felton GW, Ervin G, Eichenseer H (2003) Herbivore damage to sagebrush induces resistance in wild tobacco: evidence for eavesdropping between plants. Oikos 100:325–332

Karban R, Huntzinger M, McCall AC (2004) The specificity of eavesdropping on sagebrush by other plants. Ecology 85:1846–1852

Karban R, Shiojiri K, Huntzinger M, McCall AC (2006) Damage-induced resistance in sagebrush: volatiles are key to intra- and interplant communication. Ecology 87:922–930

Karban R, Ishizaki S, Shiojiri K (2012) Long-term demographic consequences of eavesdropping for sagebrush. J Ecol 100:932–938

Kesselmeier J, Staudt M (1999) Biogenic volatile organic compounds (VOC): an overview on emission, physiology and ecology. J Atmos Chem 33:23–88

Kessler A, Baldwin IT (2001) Defensive function of herbivore-induced plant volatile emissions in nature. Science 291:2141–2144

Kessler A, Halitschke R, Diezel C, Baldwin IT (2006) Priming of plant defense responses in nature by airborne signaling between *Artemisia tridentata* and *Nicotiana attenuata*. Oecologia 148:280–292

Knudsen JT, Tollsten L, Bergstrom G (1993) Floral scents – a checklist of volatile compounds isolated by head-space techniques. Phytochemistry 33:253–280

Kost C, Heil M (2006) Herbivore-induced plant volatiles induce an indirect defence in neighbouring plants. J Ecol 94:619–628

Li T, Holopainen JK, Kokko H, Tervahauta AI, Blande JD (2012) Herbivore-induced aspen volatiles temporally regulate two different indirect defences in neighbouring plants. Funct Ecol 26:1176–1185

Maynard-Smith J, Harper D (2003) Animal signals. Oxford University Press, Oxford and New York

Muroi A, Ramadan A, Nishihara M, Yamamoto M, Ozawa R, Takabayashi J, Arimura G-I (2011) The composite effect of transgenic plant volatiles for acquired immunity to herbivory caused by inter-plant communications. PLoS One 6:e24594

Myers JH, Williams KS (1984) Does tent caterpillar attack reduce the food quality of red alder foliage? Oecologia 62:74–79

Nakamura S, Hatanaka A (2002) Green-leaf-derived C6-aroma compounds with potent antibacterial action that act on both gram-negative and gram-positive bacteria. J Agric Food Chem 50:7639–7644

Ninkovic V, Al Abassi S, Ahmed E, Glinwood R, Pettersson J (2011) Effect of within-species plant genotype mixing on habitat preference of a polyphagous insect predator. Oecologia 166:391–400

O'Donnell PJ, Calvert C, Atzorn R, Wasternack C, Leyser HMO, Bowles DJ (1996) Ethylene as a signal mediating the wound response of tomato plants. Science 274:1914–1917

Orians C (2005) Herbivores, vascular pathways, and systemic induction: facts and artifacts. J Chem Ecol 31:2231–2242

Orians CM, Pomerleau J, Ricco R (2000) Vascular architecture generates fine scale variation in systemic induction of proteinase inhibitors in tomato. J Chem Ecol 26:471–485

Paré PW, Tumlinson JH (1997) De novo biosynthesis of volatiles induced by insect herbivory in cotton plants. Plant Physiol 114:1161–1167

Paré PW, Tumlinson JH (1999) Plant volatiles as a defense against insect herbivores. Plant Physiol 121:325–331

Paschold A, Halitschke R, Baldwin IT (2006) Using 'mute' plants to translate volatile signals. Plant J 45:275–291

Pearse IS, Porensky LM, Yang LH, Stanton ML, Karban R, Bhattacharyya L, Cox R, Dove K, Higgins A, Kamoroff C, Kirk T, Knight C, Koch R, Parker C, Rollins H, Tanner K (2012) Complex consequences of herbivory and interplant cues in three annual plants. PLoS One 7: e38105

Peng JY, Li ZH, Xiang H, Huang JH, Jia SH, Miao XX, Huang YP (2005) Preliminary studies on differential defense responses induced during plant communication. Cell Res 15:187–192

Peng J, van Loon JJA, Zheng S, Dicke M (2011) Herbivore-induced volatiles of cabbage (*Brassica oleracea*) prime defence responses in neighbouring intact plants. Plant Biol 13:276–284

Pierik R, Whitelam GC, Voesenek LACJ, de Kroon H, Visser EJW (2004) Canopy studies on ethylene-insensitive tobacco identify ethylene as a novel element in blue light and plant–plant signaling. Plant J 38:310–319

Preston C, Laue G, Baldwin IT (2004) Plant–plant signaling: Application of trans- or cis-methyl jasmonate equivalent to sagebrush releases does not elicit direct defenses in native tobacco. J Chem Ecol 30:2193–2214

Rasmann S, Köllner TG, Degenhardt J, Hiltpold I, Toepfer S, Kuhlmann U, Gershenzon J, Turlings TCJ (2005) Recruitment of entomopathogenic nematodes by insect-damaged maize roots. Nature 434:732–737

Rhoades DF (1983) Responses of alder and willow to attack by tent caterpillars and webworms: evidence for pheromonal sensitivity of willows. In: Hedin PA (ed) Plant resistance to insects. American Chemical Society, Washington, pp 55–68

Rodriguez-Saona C, Thaler JS (2005) Herbivore-induced responses and patch heterogeneity affect abundance of arthropods on plants. Ecol Entomol 30:156–163

Rodriguez-Saona CR, Rodriguez-Saona LE, Frost CJ (2009) Herbivore-induced volatiles in the perennial shrub, *Vaccinium corymbosum*, and their role in inter-branch signaling. J Chem Ecol 35:163–175

Röse USR, Manukian A, Heath RR, Tumlinson JH (1996) Volatile semiochemicals released from undamaged cotton leaves: a systemic response of living plants to caterpillar damage. Plant Physiol 111:487–495

Runyon JB, Mescher MC, De Moraes CM (2006) Volatile chemical cues guide host location and host selection by parasitic plants. Science 313:1964–1967

Ruther J, Fürstenau B (2005) Emission of herbivore-induced volatiles in absence of a herbivore-response of *Zea mays* to green leaf volatiles and terpenoids. Z Naturforschung C 60:743–756

Ruther J, Kleier S (2005) Plant–plant signaling: ethylene synergizes volatile emission in *Zea mays* induced by exposure to (Z)-3-hexen-1-ol. J Chem Ecol 31:2217–2222

Sabelis M, Janssen A, Pallini A, Venzon M, Bruin J, Drukker B, Scutareanu P (1999) Behavioral responses of predatory and herbivorous arthropods to induced plant volatiles: from evolutionary ecology to agricultural applications. In: Agrawal AA, Tuzun S, Bent E (eds) Induced plant defenses against pathogens and herbivores. APS Press, Saint Paul, MN, pp 269–296

Shiojiri K, Karban R (2006) Plant age, communication, and resistance to herbivores: young sagebrush plants are better emitters and receivers. Oecologia 149:214–220

Shiojiri K, Karban R (2008) Vascular systemic induced resistance for *Artemisia cana* and volatile communication for *Artemisia douglasiana*. Am Midl Nat 159:468–477

Shiojiri K, Kishimoto K, Ozawa R, Kugimiya S, Urashimo S, Arimura G, Horiuchi J, Nishioka T, Matsui K, Takabayashi J (2006) Changing green leaf volatile biosynthesis in plants: an approach for improving plant resistance against both herbivores and pathogens. Proc Natl Acad Sci USA 103:16672–16676

Shiojiri K, Karban R, Ishizaki S (2009) Volatile communication among sagebrush branches affects herbivory: timing of active cues. Arthropod–Plant Interact 3:99–104

Shiojiri K, Karban R, Ishizaki S (2012a) Prolonged exposure is required for communication in sagebrush. Arthropod–Plant Interact 6:197–202

Shiojiri K, Ozawa R, Matsui K, Sabelis MW, Takabayashi J (2012b) Intermittent exposure to traces of green leaf volatiles triggers a plant response. Sci Rep 2:378

Shulaev V, Silverman P, Raskin I (1997) Airborne signalling by methyl salicylate in plant pathogen resistance. Nature 385:718–721

Snoeren TAL, Kappers IF, Broekgaarden C, Mumm R, Dicke M, Bouwmeester HJ (2010) Natural variation in herbivore-induced volatiles in *Arabidopsis thaliana*. J Exp Bot 61:3041–3056

Steeghs M, Bais HP, de Gouw J, Goldan P, Kuster W, Northway M, Fall R, Vivanco JM (2004) Proton-transfer-reaction mass spectrometry (PTR-MS) as a new tool for real time analysis of root-secreted volatile organic compounds (VOCs) in *Arabidopsis thaliana*. Plant Physiol 135:47–58

Takabayashi J, Dicke M (1996) Plant–carnivore mutualism through herbivore-induced carnivore attractants. Trends Plant Sci 1:109–113

Takabayashi J, Dicke M, Posthumus M (1991) Variation in composition of predator-attracting allelochemicals emitted by herbivore-infested plants: relative influence of plant and herbivore. Chemoecology 2:1–6

Tscharntke T, Thiessen S, Dolch R, Boland W (2001) Herbivory, induced resistance and interplant signal transfer in *Alnus glutinosa*. Biochem Syst Ecol 29:1025–1047

Turlings TCJ, Tumlinson JH (1992) Systemic release of chemical signals by herbivore-injured corn. Proc Natl Acad Sci USA 89:8399–8402

Turlings TCJ, Tumlinson JH, Lewis WJ (1990) Exploitation of herbivore-induced plant odors by host-seeking parasitic wasps. Science 250:1251–1253

Unsicker SB, Kunert G, Gershenzon J (2009) Protective perfumes: the role of vegetative volatiles in plant defense against herbivores. Curr Opin Plant Biol 12:479–485

Viswanathan DV, Thaler JS (2004) Plant vascular architecture and within-plant spatial patterns in resource quality following herbivory. J Chem Ecol 30:531–543

Wei J, Kang L (2011) Roles of (Z)-3-hexenol in plant–insect interactions. Plant Signal Behav 6:369–371

Yi H-S, Heil M, Adame-Alvarez RM, Ballhorn DJ, Ryu CM (2009) Airborne induction and priming of plant defenses against a bacterial pathogen. Plant Physiol 151:2152–2161

Zebelo SA, Matsui K, Ozawa R, Maffei ME (2012) Plasma membrane potential depolarization and cytosolic calcium flux are early events involved in tomato (*Solanum lycopersicon*) plant-to-plant communication. Plant Sci 196:93–100

Index

F. Baluška (ed.), *Long-Distance Systemic Signaling and Communication in Plants*, Signaling and Communication in Plants 19, DOI 10.1007/978-3-642-36470-9,

MIX
Papier aus verantwortungsvollen Quellen
Paper from responsible sources
FSC® C105338

If you have any concerns about our products,
you can contact us on
ProductSafety@springernature.com

In case Publisher is established outside the EU,
the EU authorized representative is:
Springer Nature Customer Service Center GmbH
Europaplatz 3, 69115 Heidelberg, Germany

Printed by Libri Plureos GmbH
in Hamburg, Germany